Time Series

Recently Published Titles

Statistical Analysis of Financial Data
With Examples in R
James Gentle

Statistical Rethinking
A Bayesian Course with Examples in R and STAN, Second Edition
Richard McElreath

Statistical Machine Learning
A Model-Based Approach
Richard Golden

Randomization, Bootstrap and Monte Carlo Methods in Biology
Fourth Edition
Bryan F. J. Manly, Jorje A. Navarro Alberto

Principles of Uncertainty, Second Edition
Joseph B. Kadane

Beyond Multiple Linear Regression
Applied Generalized Linear Models and Multilevel Models in R
Paul Roback, Julie Legler

Bayesian Thinking in Biostatistics
Gary L. Rosner, Purushottam W. Laud, and Wesley O. Johnson

Linear Models with Python
Julian J. Faraway

Modern Data Science with R, Second Edition
Benjamin S. Baumer, Daniel T. Kaplan, and Nicholas J. Horton

Probability and Statistical Inference
From Basic Principles to Advanced Models
Miltiadis Mavrakakis and Jeremy Penzer

Bayesian Networks
With Examples in R, Second Edition
Marco Scutari and Jean-Baptiste Denis

Times Series
Modeling, Computation, and Inference, Second Edition
Raquel Prado, Marco A. R. Ferreira, and Mike West

For more information about this series, please visit: https://www.crcpress.com/Chapman--Hall/CRC-Texts-in-Statistical-Science/book-series/CHTEXSTASCI

Time Series

Modeling, Computation, and Inference

Second Edition

Raquel Prado

Marco A. R. Ferreira

Mike West

CRC Press is an imprint of the
Taylor & Francis Group, an **informa** business
A CHAPMAN & HALL BOOK

First edition published 2021
by CRC Press
6000 Broken Sound Parkway NW, Suite 300, Boca Raton, FL 33487-2742

and by CRC Press
2 Park Square, Milton Park, Abingdon, Oxon, OX14 4RN

CRC Press is an imprint of Taylor & Francis Group, LLC

Library of Congress Cataloging-in-Publication Data

ISBN: 978-1-498-74702-8 (hbk)
ISBN: 978-1-032-04004-2 (pbk)
ISBN: 978-1-351-25942-2 (ebk)

Typeset in CMR10
by KnowledgeWorks Global Ltd.

Contents

Preface

This book aims to integrate mainstream modeling approaches in time series with a range of significant recent developments in methodology and applications of time series analysis. We present overviews of several classes of models and related methodology for inference, statistical computation for model fitting and assessment, and forecasting. The book focuses mainly on time domain approaches while covering core topics and theory in the frequency domain, and connections between the two are often explored. Statistical analysis and inference involves likelihood and Bayesian methodologies, with a strong emphasis on using modern, simulation-based approaches for statistical parameter estimation, model fitting, and prediction; ranges of models and analyses are developed using Bayesian approaches and tools including Markov chain Monte Carlo and sequential Monte Carlo methods that define nowadays standard methodology.

Time series model theory and methods are illustrated with examples and case studies involving problems and data arising from a variety of applied fields, including signal processing, biomedical studies, finance, econometrics, and the environmental sciences. The book has three major aims: (1) to serve as a graduate textbook on Bayesian time series modeling and analysis; (2) to provide a broad range of references on state-of-the-art approaches to univariate and multivariate time series analysis, serving as an informed guide to the recent literature and a handbook for researchers and practitioners in applied areas that require sophisticated tools for analyzing challenging time series problems; and (3) to contact ranges of traditional as well as new and emerging topics that lie at research frontiers. Most of the material presented in Chapters 1 to 5, as well as selected topics from Chapters 6 to 11, are suitable as the core material for a one-term/semester or a one-quarter graduate course in time series analysis. Alternatively, a course might be structured to cover material on models and methods for univariate time series analysis based on Chapters 1 to 7 at greater depth in one course, with material and supplements related to the multivariate time series models and methods of Chapters 8 to 11 as a second course. Then, most chapters also contact more advanced topics and link to research areas with open questions.

Contents

The book presents a selective coverage of core and more advanced and recent topics in the very broad field of time series analysis. As one of the oldest and richest areas of statistical science, and a field that contacts applied interests across a huge spectrum of science, social science, and engineering applications, "time series" simply cannot be comprehensively covered in any single text. Our aim, to the contrary, is to present, summarize, and overview core models and methods, complementing the pedagogical development with a selective range of recent research developments and applications that exemplify the growth of time series analysis into new areas based on these core foundations. The flavor of examples and case studies reflects our own interests and experiences in time series research and applications in collaborations with researchers from other fields, and we aim to convey some of the interest in, and utility of, the modeling approaches through these examples. Readers and students with backgrounds in statistical inference and some exposure to applied statistics and computation should find the book accessible.

Chapter 1 offers an introduction and a brief review of Bayesian inference, including Markov chain Monte Carlo (MCMC) methods. Chapter 2 presents autoregressive moving average models (ARMA) from a Bayesian perspective and illustrates these models with several examples. Chapter 3 discusses some theory and methods of frequency domain approaches, including harmonic regression models and their relationships with the periodogram and Bayesian spectral analysis. Some multivariate extensions are explored later in Chapters 8 and 9 in contexts of analyzing multiple and multivariate time series. Chapter 4 reviews dynamic models and methods for inference and forecasting for this broad and flexible class of models. More specifically, this chapter includes a review of the dynamic linear models (DLMs) of West and Harrison (1997), discusses extensions to nonlinear and non-Gaussian dynamic models, and reviews key developments of MCMC for filtering, parameter learning, and smoothing. Chapter 5 concerns issues of model specification and posterior inference in a particular class of DLMs: the broadly useful and widely applied class of time-varying autoregressive models. Theory and methods related to time series decompositions into interpretable latent processes, and examples in which real data sets are analyzed, are included. Chapter 6 covers recent developments of sequential Monte Carlo methods for general state-space models. Chapter 7 reviews a selection of topics involving statistical mixture models in time series analysis, focusing on multiprocess models and univariate stochastic volatility models. Chapter 8 illustrates the analysis of multiple time series with common underlying structure and motivates some of the multivariate models that are developed later in Chapters 9 and 10. Chapter 9 discusses multivariate ARMA models, focusing on vector autoregressive (VAR) models, time series decompositions within this class of models, and mixtures of VAR models. Chapter 10 discusses a range of multivariate dynamic linear models, models and methods for time-varying, stochastic covariance matrices related to stochastic volatility, and contacts research frontiers in discussion of multivariate dynamic graphical models and other recent developments. The latter include contact with models and perspectives on problems of modeling and forecasting for increasingly large, complex, and hierarchically structured time series in commercial and other areas. Chapter 11 details developments of dynamic modeling with latent factor structures, a central area of time series methodology that has been heavily driven by advances in Bayesian methodology for dynamic models.

A collection of problems is included at the end of each chapter. Some of the chapters also include appendices that provide relevant supplements on statistical distribution theory and other mathematical aspects.

Acknowledgments

We recognize a number of colleagues for their impact on our thinking and eventual contributions to the broad field of time series modeling and forecasting, and directly or indirectly on the evolution of this text. Gabriel Huerta and Giovanni Petris provided material inputs that led to revisions of the core text material, and suggested some of the problems listed at the end of Chapters 1, 2, and 3. We thank Carlos Carvalho, Hedibert Lopes, Abel Rodríguez, and several other anonymous reviewers, as well as many colleagues at the University of California Santa Cruz (UCSC), Virginia Tech, and Duke University, and students from courses at UCSC, Virginia Tech, and Duke over many years, for their continued input as well as just day-to-day interactions that have had impact on the evolution of the core material presented.

Several of the data sets analyzed in the book come from collaborations with researchers in other fields, and such collaborations have been (and, we hope and expect) will continue to be critical to developments in modeling and methodology. Among many others, we are most appreciative of past contributions of collaborators including Dr. Andrew D. Krystal, Dr. Jose M. Quintana, and Dr. Leonard Trejo. We acknowledge the support and facilities at the Department of Statistics and the Baskin School of Engineering at

UCSC, the Department of Statistics at Virginia Tech, and the Department of Statistical Science at Duke University. We would also like to acknowledge the support of the Statistical and Applied Mathematical Science Institute (SAMSI) in North Carolina. In particular, some of the sections in Chapter 6 were written while Raquel Prado was visiting SAMSI as a participant of the 2008–2009 program on sequential Monte Carlo methods. We also acknowledge grants from the National Science Foundation, the National Institutes of Health, and a number of nongovernmental organizations and companies that have, over many years, provided support for our research that has contributed, directly and indirectly, to the development of models and methods presented in this book.

Raquel Prado, Marco A. R. Ferreira, and Mike West
December 2020

Authors

Raquel Prado is professor in the Department of Statistics at the Baskin School of Engineering at the University of California Santa Cruz (UCSC), USA. Her main research areas are time series analysis and Bayesian modeling, with a focus on analysis of large-dimensional nonstationary time series data and applications to biomedical signal processing and brain imaging.

Dr. Prado leads NSF- and NIH-funded projects, including multi-institutional and multi-disciplinary collaborative projects. She has supervised over 20 graduate students at UCSC and other academic institutions. Her former students work in academia, high tech companies, national laboratories, and local government agencies.

Dr. Prado is past president of the International Society for Bayesian Analysis (ISBA). She is an ISBA fellow and a fellow of the American Statistical Association (ASA). She has served on several committees at ASA and ISBA and is currently a member of the Committee on Applied and Theoretical Statistics (CATS) of the National Academies of Sciences, Engineering and Medicine.

Marco A. R. Ferreira is an associate professor in the Department of Statistics at Virginia Tech, where he served from 2016-2020 as the Director of Graduate Programs. Dr. Ferreira has served the statistics profession in editorial boards of multiple scientific journals including the journal *Bayesian Analysis*, in several committees of ISBA and ASA, as well as in scientific committees of numerous domestic and international conferences.

Dr. Ferreira's current research areas include dynamic models for time series and spatiotemporal data, multiscale models, objective Bayesian methods, stochastic search algorithms, and statistical computation. Major areas of application include bioinformatics, finance, and environmental science. His research is, and has been, funded by grants from the National Science Foundation. Marco has advised over 10 PhD students and postdocs and has published over 50 scientific papers. His former students and postdocs work in academic, industrial, and government positions.

Mike West holds a Duke University Distinguished Chair as the Arts & Sciences Professor of Statistics & Decision Sciences in the Department of Statistical Science, where he led the development of statistics from 1990-2002. A past president of the International Society for Bayesian Analysis (ISBA), Mike has served the international statistics profession in founding roles for ISBA and in other professional organizations and institutions. Dr. West's research and teaching activities are in Bayesian analysis in ranges of interlinked areas: theory and methods of dynamic models in time series analysis, multivariate analysis, latent structure, high-dimensional inference and computation, quantitative and computational decision analysis, stochastic computational methods, and statistical computing, among other topics. Interdisciplinary R&D has ranged across applications in signal processing, finance, econometrics, climatology, systems biology, genomics and neuroscience, among other areas. His main current interests are in macroeconomic forecasting and policy decisions, financial econometric forecasting and decisions, dynamic network studies in IT/commerce, and large-scale forecasting and decision problems in business and industry.

Dr. West has received a number of international awards for research and professional service, and multiple distinguished speaking awards. He has been, and continues to be, a statistical consultant for various companies, banks, government agencies, and academic centers, co-founder of a biotech company, and past, current advisor, or board member for several financial and IT companies. Dr. West teaches in academia and through short courses, works with and advises many undergraduates and master's students, and has mentored over 60 primary PhD students and postdoctoral associates, most of whom are now in academic, industrial, or government positions involving advanced statistical research.

Chapter 1

Notation, definitions, and basic inference

This chapter discusses key goals of time series analysis with motivating examples from different applied areas. Notation and key concepts related to time series processes are introduced, including the characterization of stationary processes. This is followed by a brief review on likelihood and Bayesian modeling and inference tools, which includes a primer on simulation-based methods for posterior inference within the Bayesian framework. The modeling and inference tools are illustrated for the class of first-order autoregressive processes.

1.1 Problem Areas and Objectives

The expression *time series data*, or *time series*, usually refers to a set of observations collected sequentially in time. These observations could have been collected at equally spaced time points. In this case we use the notation y_t with $(t = \ldots, -1, 0, 1, 2, \ldots)$; i.e., the set of observations is indexed by t, the time at which each observation was taken. If the observations were not taken at equally spaced points, then we use the notation y_{t_i}, with $i = 1, 2, \ldots$.

A *time series process* is a stochastic process or a collection of random variables y_t indexed in time. Note that y_t will be used throughout the book to denote a random variable or an actual realization of the time series process at time t. We use the notation $\{y_t, t \in \mathcal{T}\}$, or simply $\{y_t\}$, to refer to the time series process. If $\mathcal{T}$ is of the form $\{t_i, i \in \mathbb{N}\}$, with $\mathbb{N}$ the natural numbers, then the process is a discrete-time random process, and if $\mathcal{T}$ is an interval in the real line, or a collection of intervals in the real line, then the process is a continuous-time random process. In this framework, a time

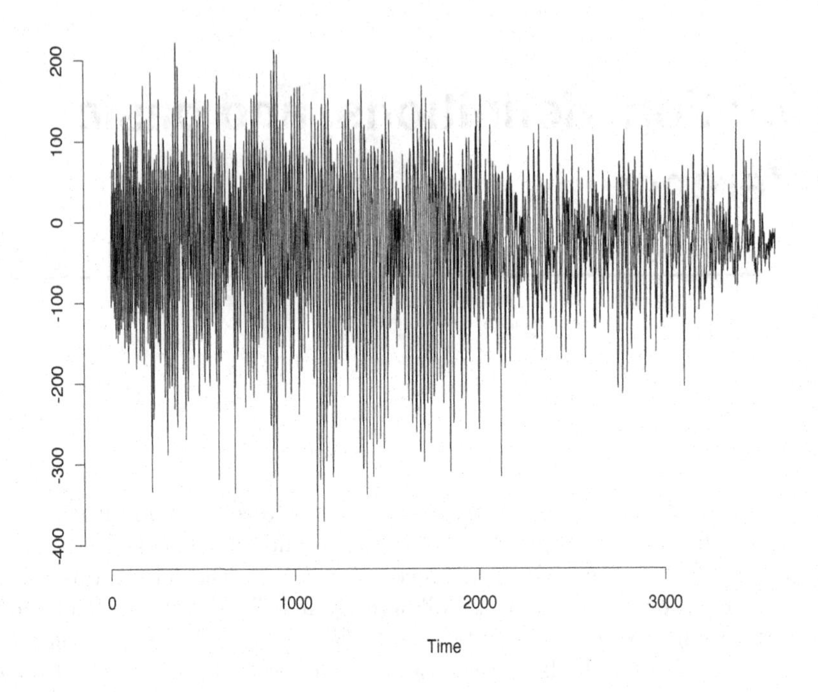

Figure 1.1 *EEG series (units in millivolts). The EEG was recorded at channel F_3 from a subject who received ECT.*

series data set $y_t, (t = 1, \dots, T)$, also denoted by $y_{1:T}$, is just a collection of T equally spaced realizations of some time series process.

In many statistical models the assumption that the observations are realizations of independent random variables is key. In contrast, time series analysis is concerned with describing the dependence among the elements of a sequence of random variables.

At each time t, y_t can be a scalar quantity, such as the total amount of rainfall collected at a certain location in a given day t, or it can be a k-dimensional vector containing k scalar quantities that were recorded simultaneously. For instance, if the total amount of rainfall and the average temperature at a given location are measured in day t, we have $k = 2$ scalar quantities and a two-dimensional vector of observations $\mathbf{y}_t = (y_{1,t}, y_{2,t})'$. In general, for k scalar quantities recorded at time t, we have a realization $\mathbf{y}_t$ of a vector process $\{\mathbf{y}_t, t \in \mathcal{T}\}$, with $\mathbf{y}_t = (y_{1,t}, \dots, y_{k,t})'$.

Figure 1.1 displays a portion of an electroencephalogram (EEG) recorded on a patient's scalp under certain electroconvulsive therapy (ECT) conditions. ECT is a treatment for patients under major clinical depression

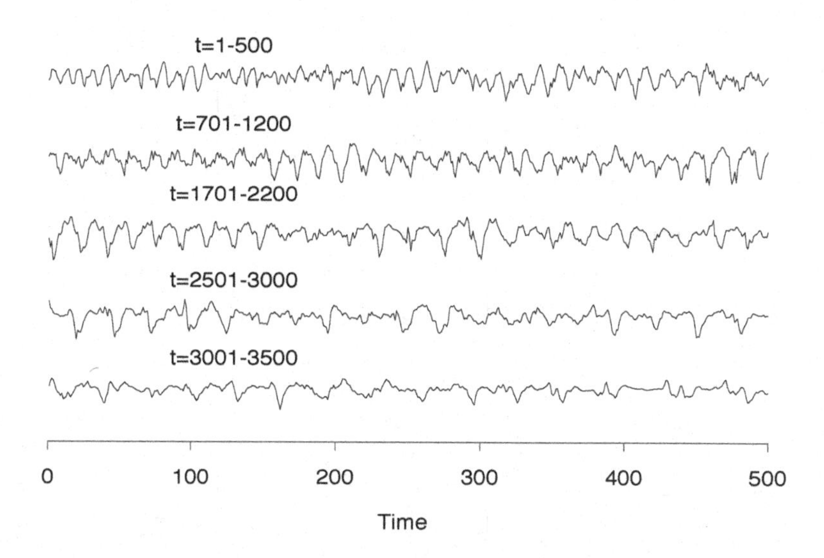

Figure 1.2 *Sections of the EEG trace displayed in Figure 1.1.*

(Krystal, Prado, and West 1999). When ECT is applied to a patient, seizure activity appears and can be recorded via electroencephalograms. The data correspond to one of 19 EEG series recorded simultaneously at different locations over the scalp. The main objective in analyzing these signals is the characterization of the clinical efficacy of ECT in terms of particular features that can be inferred from the recorded EEG traces. The data are fluctuations in electrical potential taken at a sampling rate of 256 Hz (i.e., 256 observations per second). For a more detailed description of these data and a full statistical analysis, see West, Prado, and Krystal (1999), Krystal, Prado, and West (1999), and Prado, West, and Krystal (2001).

From the time series analysis viewpoint, the objective here is modeling the data to provide useful insight about the underlying processes driving the multiple series during a seizure episode. Studying the differences and commonalities among the 19 EEG channels is also key. Univariate time series models for each individual EEG series could be explored and used to investigate relationships across the 19 channels (Chapters 2, 5, and 8). Multivariate time series analyses (Chapters 9 and 10)—in which the observed series, $\mathbf{y}_t$, is a 19-dimensional vector whose elements are the observed voltage levels measured at the 19 scalp locations at each time t—can also be considered. Uncovering the common latent structure that may underlie the 19 EEG time series over time can be achieved by decomposing these observed EEGs into simpler latent non observable components. Such latent

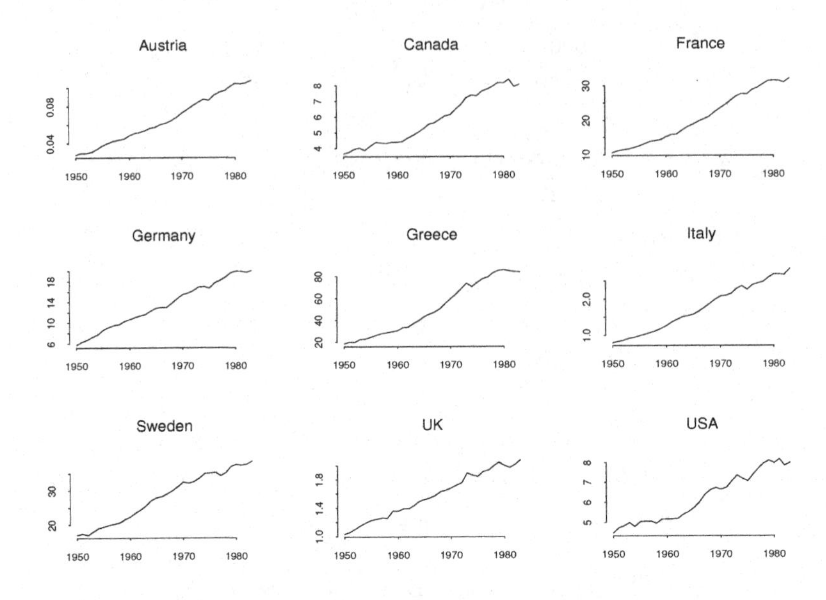

Figure 1.3 *International annual GDP time series.*

components can be obtained via time series decompositions derived from a specific state-space modeling framework (Chapters 5 and 8), or by explicitly modeling them as latent factors in a dynamic factor model (Chapter 11).

These EEG series display a quasiperiodic behavior that changes dynamically in time, as shown in Figure 1.2, where different portions of the EEG trace shown in Figure 1.1 are displayed. In particular, it is clear that the relatively high-frequency components that appear initially are slowly decreasing toward the end of the series. Any time series model used to describe these data should take into account their nonstationary and quasiperiodic structure. We discuss various modeling alternatives for analyzing these data in the subsequent chapters, including the class of time-varying autoregressions and some multichannel models.

Figure 1.3 shows the annual per capita GDP (gross domestic product) time series for Austria, Canada, France, Germany, Greece, Italy, Sweden, UK, and USA from 1950 to 1983. Goals of the analysis include forecasting turning points and comparing characteristics of the series across the national economies. Univariate and multivariate analyses of the GDP data can be considered.

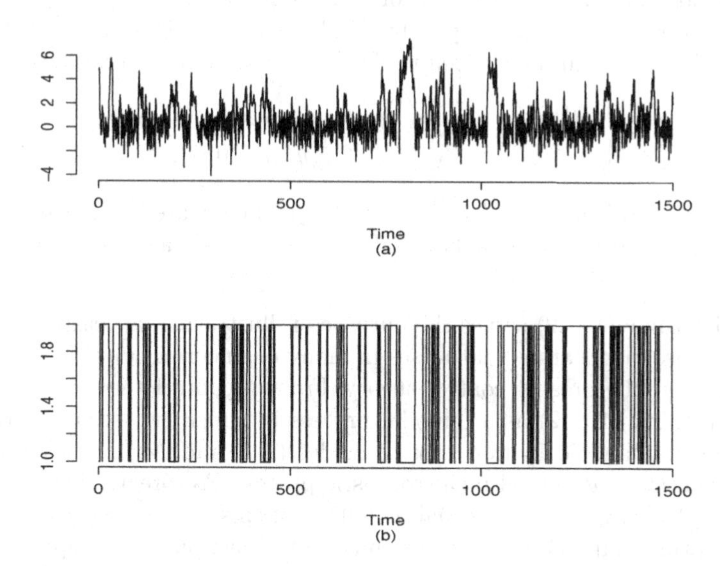

Figure 1.4 *(a): Simulated time series y_t; (b) Indicator variable δ_t with $\delta_t = 1$ if y_t was sampled from $\mathcal{M}_1$ and $\delta_t = 2$ if y_t was sampled from $\mathcal{M}_2$.*

One of the main differences between any time series analysis of the GDP series and any time series analysis of the EEG series, regardless of the type of models used in such analyses, lies in the objectives. As mentioned above, one of the goals in analyzing the GDP data is forecasting future outcomes of the series for the several countries given the observed values. In the EEG study previously described, there is no interest in forecasting future values of the series given the observed traces; instead, the objective is finding an appropriate model that describes the structure of the series and its latent components.

Other objectives of time series analysis include monitoring a time series in order to detect possible "on-line" (real time) changes. This is important for control purposes in engineering, industrial, and medical applications. For instance, consider a time series generated from the process $\{y_t\}$ with

$$y_t = \begin{cases} 0.9y_{t-1} + \epsilon_t^{(1)}, & y_{t-1} > 1.5 \quad (\mathcal{M}_1) \\ -0.3y_{t-1} + \epsilon_t^{(2)}, & y_{t-1} \leq 1.5 \quad (\mathcal{M}_2), \end{cases} \tag{1.1}$$

where $\epsilon_t^{(1)} \sim N(0, v_1)$, $\epsilon_t^{(2)} \sim N(0, v_2)$, and $v_1 = v_2 = 1$. Figure 1.4 (a) shows a time series plot of 1,500 observations simulated according to (1.1). Figure 1.4 (b) displays the values of an indicator variable, δ_t, with $\delta_t = 1$ if y_t was generated from $\mathcal{M}_1$, and $\delta_t = 2$ if y_t was generated from $\mathcal{M}_2$. Model (1.1) is a *threshold autoregressive (TAR) model* with two regimes that belongs to the broader class of mixture models (see Chapter 7). TAR models were initially developed by H. Tong (Tong 1983; Tong 1990). In particular, (1.1) can be written in the following, more general, form

$$y_t = \begin{cases} \phi^{(1)} y_{t-1} + \epsilon_t^{(1)}, & \theta + y_{t-d} > 0 \quad (\mathcal{M}_1) \\ \phi^{(2)} y_{t-1} + \epsilon_t^{(2)}, & \theta + y_{t-d} \leq 0 \quad (\mathcal{M}_2), \end{cases} \tag{1.2}$$

with $\epsilon_t^{(1)} \sim N(0, v_1)$ and $\epsilon_t^{(2)} \sim N(0, v_2)$. These are nonlinear models and the interest lies in making inferences on d, θ, and the parameters $\phi^{(1)}, \phi^{(2)}, v_1$, and v_2.

The TAR model (1.2) serves the purpose of illustrating, at least for a very simple case, a situation that arises in many engineering applications, particularly in the area of control theory. From a control theory viewpoint, we can think of (1.2) as a bimodal process in which two scenarios of operation are handled by two control modes ($\mathcal{M}_1$ and $\mathcal{M}_2$). In each mode the evolution is governed by a stochastic process. Autoregressions of order one, or AR(1) models (a formal definition of this type of process is given later in this chapter), were chosen in this example, but more sophisticated structures can be considered. The transitions between the modes occur when the series crosses a specific threshold and so, we can talk about an internally triggered mode switch. In an externally triggered mode switch, the moves are defined by external variables. In terms of the goals of time series analysis in this case we can consider two possible scenarios. In many control settings where the transitions between modes occur in response to a controller's actions, the current state is always known, and so, the learning process can be split into two: learning the stochastic models that control each mode conditional on the fact that we know in which mode we are—i.e., inferring $\phi^{(1)}, \phi^{(2)}, v_1$, and v_2—and learning the transition rule, that is, making inferences about d and θ assuming we know the values $\delta_{1:T}$. In other control settings for which the mode transitions do not occur in response to a controller's actions, it is necessary to simultaneously infer the parameters associated to the stochastic models that describe each mode and the transition rule. In this case we want to estimate $\phi^{(1)}, \phi^{(2)}, v_1, v_2, \theta$, and d conditioning only on the observed data $y_{1:T}$. Depending on the application, it may also be necessary to achieve parameter learning from the time series sequentially in time. Methods for sequential state and parameter learning in time series models are discussed throughout this book.

Clustering also arises as the primary goal in many applications. For example, a common scenario is one in which a collection of N time series generated from a relatively small number of processes, say K, with $K << N$, are available. It is not known a priori which time series are generated from which processes, and so the main objective of the analysis consists on grouping the time series into K clusters according to their spectral characteristics. Some references in this area include Kakizawa, Shumway, and Taniguchi (1998), Huan, Ombao, and Stoffer (2004), Gao, Ombao, and Ho (2009), Pamminger and Frühwirth-Schnatter (2010), and Nieto-Barajas and Contreras-Cristán (2014).

Finally, we may use time series techniques to describe serial dependencies between parameters of a given model with additional structure. For example, we could have a linear regression model of the form $y_t = \beta_0 + \beta_1 x_t + \epsilon_t$, for which ϵ_t does not exhibit the usual independent structure $\epsilon_t \sim N(0, v)$ for all t, but instead, the probability distribution of ϵ_t depends on $\epsilon_{t-1}, \ldots, \epsilon_{t-k}$ for some integer $k > 0$.

1.2 Stochastic Processes and Stationarity

Many time series models are based on the assumption of stationarity. Intuitively, a stationary time series process is a process whose behavior does not depend on when we start to observe it. In other words, different sections of the series will look roughly the same at intervals of the same length. Here we provide two widely used definitions of stationarity.

A time series process $\{y_t, t \in \mathcal{T}\}$ is *completely* or *strongly stationary* if, for any sequence of times $t_1, t_2, \ldots, t_n$, and any lag h with $h = 0, \pm 1, \pm 2, \ldots$, the probability distribution of the vector $(y_{t_1}, \ldots, y_{t_n})'$ is identical to the probability distribution of the vector $(y_{t_1+h}, \ldots, y_{t_n+h})'$.

In practice it is very difficult to verify that a process is strongly stationary and so, the notion of *weak* or *second-order stationarity* arises. A process is said to be weakly stationary, or second-order stationary if, for any sequence of times $t_1, \ldots, t_n$, and any integer lag h, all the first and second joint moments of $(y_{t_1}, \ldots, y_{t_n})'$ exist and are equal to the first and second joint moments of $(y_{t_1+h}, \ldots, y_{t_n+h})'$. If $\{y_t\}$ is second-order stationary, we have that

$$E(y_t) = \mu, \quad V(y_t) = v, \quad Cov(y_t, y_s) = \gamma(t - s), \tag{1.3}$$

where μ, v are constant, independent of t and $\gamma(t-s)$ is also independent of t and s, depending only on the length of the interval between time points. It is also possible to define stationarity up to order m in terms of the m joint moments (see for example Priestley 1994).

If the first two moments exist, complete stationarity implies second-order stationarity, but the converse is not necessarily true. If $\{y_t\}$ is a Gaussian process, i.e., if for any sequence of time points $t_1, \ldots, t_n$ the vector $(y_{t_1}, \ldots, y_{t_n})'$ follows a multivariate normal distribution, strong and weak stationarity are equivalent (see Shumway and Stoffer 2017 for a proof).

1.3 Autocorrelation and Cross-correlation

The first step in a statistical analysis often consists on performing a descriptive study of the data in order to summarize their main features. One of the most widely used descriptive techniques in time series data analysis is that of exploring the correlation patterns displayed by a series, or a couple of series, at different time points. This is done by plotting the sample autocorrelation and cross-correlation values, which are estimates of the autocorrelation and cross-correlation functions.

We begin by defining the concepts of autocovariance, autocorrelation, and cross-correlation functions. We then show how to estimate these functions from data. Let $\{y_t, t \in \mathcal{T}\}$ be a time series process. The autocovariance function of $\{y_t\}$ is defined as follows:

$$\gamma(s,t) = Cov\{y_t, y_s\} = E\{(y_t - \mu_t)(y_s - \mu_s)\}, \tag{1.4}$$

for all s, t, with $\mu_t = E(y_t)$. For stationary processes $\mu_t = \mu$ for all t and the covariance function depends on $|t - s|$ only. In this case we can write the autocovariance as a function of a particular time lag h, i.e.,

$$\gamma(h) = Cov\{y_t, y_{t-h}\}. \tag{1.5}$$

The autocorrelation function (ACF) is then given by

$$\rho(s,t) = \frac{\gamma(s,t)}{\sqrt{\gamma(t,t)\gamma(s,s)}}. \tag{1.6}$$

For stationary processes, the ACF can be written in terms of a lag h:

$$\rho(h) = \frac{\gamma(h)}{\gamma(0)}. \tag{1.7}$$

The ACF measures the linear dependence between a value of the time series process at time t and past or future values of such process. It inherits the properties of any correlation function–$\rho(h)$ always takes values in the interval $[-1, 1]$. In addition, $\rho(h) = \rho(-h)$ and, if y_t and y_{t-h} are independent, then $\rho(h) = 0$.

It is also possible to define the cross-covariance and cross-correlation functions of two univariate time series. If $\{y_t\}$ and $\{z_t\}$ are two time series processes, the cross-covariance is defined as

$$\gamma_{y,z}(s,t) = E\{(y_t - \mu_{y_t})(z_s - \mu_{z_s})\}, \tag{1.8}$$

for all s, t, and the cross-correlation is then given by

$$\rho_{y,z}(s,t) = \frac{\gamma_{y,z}(s,t)}{\sqrt{\gamma_{y,y}(t,t)\gamma_{z,z}(s,s)}}. \tag{1.9}$$

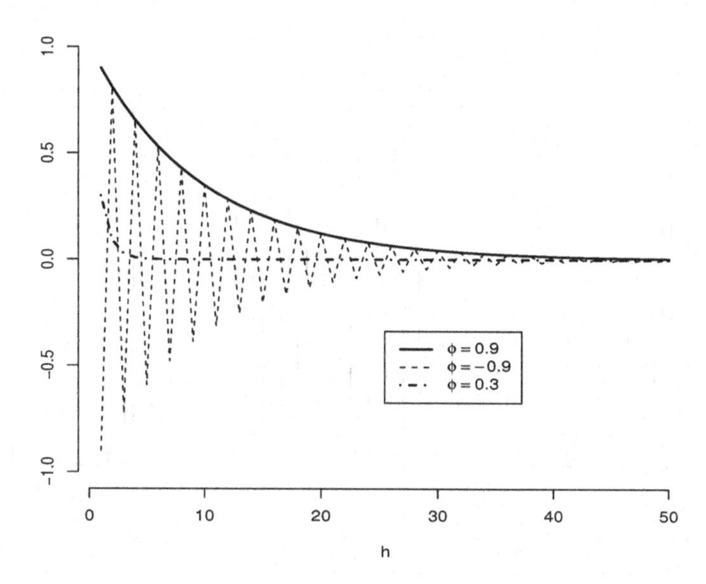

Figure 1.5 *Autocorrelation functions for AR*(1) *processes with parameters* 0.9, −0.9, *and* 0.3.

If both processes are stationary, we can write the cross-covariance and cross-correlation functions in terms of a lag value h. This is

$$\gamma_{y,z}(h) = E\{(y_t - \mu_y)(z_{t-h} - \mu_z)\} \tag{1.10}$$

and

$$\rho_{y,z}(h) = \frac{\gamma_{y,z}(h)}{\sqrt{\gamma_y(0)\gamma_z(0)}}. \tag{1.11}$$

Example 1.1 *White noise.* Consider a process such that $y_t \sim N(0, v)$ for all t, with $Cov(y_t, y_s) = 0$ if $t \neq s$. In this case $\gamma(0) = v$, $\gamma(h) = 0$ for all $h \neq 0$, and so, $\rho(0) = 1$ and $\rho(h) = 0$ for all $h \neq 0$.

Example 1.2 *First-order autoregression or AR*(1). In Chapter 2 we formally define and study the properties of general autoregressions of order p, or AR(p) processes. Here, we illustrate some properties of the simplest AR process, the AR(1). Consider a process such that $y_t = \phi y_{t-1} + \epsilon_t$ with $\epsilon_t \sim N(0, v)$ for all t. It is possible to show (see Problem 1 in this chapter) that, if $|\phi| < 1$, $\gamma(h) = \phi^{|h|}\gamma(0)$ for $h = 0, \pm 1, \pm 2, \ldots$, with $\gamma(0) = \frac{v}{(1-\phi^2)}$, and $\rho(h) = \phi^{|h|}$. Figure 1.5 displays the ACFs of AR(1) processes with parameters $\phi = 0.9, \phi = -0.9$ and $\phi = 0.3$, for lag values $h = 1 : 50$. For negative values of ϕ the ACF has an oscillatory behavior. In addition,

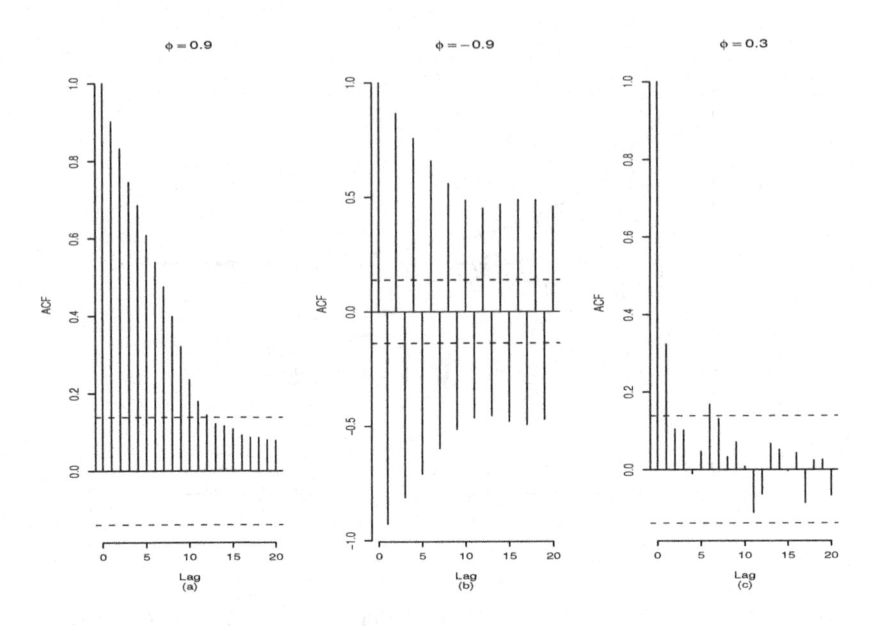

Figure 1.6 *Sample autocorrelations for AR processes with parameters 0.9, −0.9, and 0.3 (graphs (a), (b), and (c), respectively).*

the rate of decay of the ACF is a function of ϕ. The closer $|\phi|$ gets to the unity the lower the rate of decay is (e.g., compare the ACFs for $\phi = 0.9$ and $\phi = 0.3$). This is related to the characterization of stationary AR(1) processes as discussed in Chapter 2. An AR(1) process is stationary if and only if $|\phi| < 1$. This condition can also be written as a function of the characteristic root of the process. An AR(1) is stationary if and only if the root of the characteristic polynomial, $\Phi(u)$ with $\Phi(u) = 1 - \phi u$, lies outside the unit circle. This happens if and only if $|\phi| < 1$.

We now show how to estimate the autocovariance, autocorrelation, cross-covariance, and cross-correlation functions from data. Assume we have data $y_{1:T}$. The usual estimate of the autocovariance function is the sample autocovariance, which, for $h > 0$, is given by

$$\hat{\gamma}(h) = \frac{1}{T} \sum_{t=1}^{T-h} (y_{t+h} - \bar{y})(y_t - \bar{y}), \tag{1.12}$$

where $\bar{y} = \sum_{t=1}^{T} y_t/T$ is the sample mean. We can then obtain estimates of the autocorrelation function as $\hat{\rho}(h) = \frac{\hat{\gamma}(h)}{\hat{\gamma}(0)}$, for $h = 0, 1, \ldots$. Similarly, estimates of the cross-covariance and cross-correlation functions can be obtained. The sample cross-covariance is given by

$$\hat{\gamma}_{y,z}(h) = \frac{1}{T}\sum_{t=1}^{T-h}(y_{t+h} - \bar{y})(z_t - \bar{z}), \tag{1.13}$$

and so, the sample cross-correlation is given by

$$\hat{\rho}_{y,z}(h) = \hat{\gamma}_{y,z}(h) \Big/ \sqrt{\hat{\gamma}_y(0)\hat{\gamma}_z(0)}\,.$$

Example 1.3 *Sample ACFs of AR*(1) *processes.* Figure 1.6 displays the sample autocorrelation functions of simulated AR(1) processes with parameters $\phi = 0.9$, $\phi = -0.9$, and $\phi = 0.3$. The sample ACFs were computed based on a sample of $T = 200$ data points. For $\phi = 0.9$ and $\phi = 0.3$, the corresponding sample ACFs decay with the lag. The oscillatory form of the ACF for the process with $\phi = -0.9$ is captured by the corresponding sample ACF.

The estimates given in (1.12) and (1.13) are not unbiased estimates of the autocovariance and cross-covariance functions. Results related to the distributions of the sample autocorrelation and the sample cross-correlation functions appear, for example, in Shumway and Stoffer (2017).

1.4 Smoothing and Differencing

As mentioned before, many time series models are built under the stationarity assumption. Several descriptive techniques have been developed to study the stationary properties of a time series so that an appropriate model can then be applied to the data. For instance, looking at the sample autocorrelation function may be helpful in identifying some features of the data. However, in many practical scenarios the data are realizations from one or several nonstationary processes. In this case, methods that aim to eliminate the nonstationary components are often used. The idea is to separate the nonstationary components from the stationary ones so that the latter can be carefully studied via traditional time series models such as, for example, the ARMA (autoregressive moving average) models that will be discussed in subsequent chapters.

We review some commonly used methods for extracting nonstationary components from a time series. We do not attempt to provide a comprehensive list of such methods. Instead, we just list and summarize a few of them. We view these techniques as purely descriptive.

Many descriptive time series methods are based on the notion of *smoothing* the data, that is, decomposing the series as a sum of two components: a so called "smooth" component, plus another component that includes all the features of the data that are left unexplained by the smooth component. This is similar to the "signal plus noise" concept used in signal processing. The main difficulty with this approach lies in deciding which features of the data are part of the signal or the smooth component, and which ones are part of the noise.

One way of smoothing a time series is by moving averages (see Kendall, Stuart, and Ord 1983; Kendall and Ord 1990; Chatfield 1996; and Diggle 1990 for detailed discussions and examples). If we have data $y_{1:T}$, we can smooth them by applying an operation of the form

$$z_t = \sum_{j=-q}^{p} a_j y_{t+j}, \quad t = (q+1) : (T-p), \tag{1.14}$$

with p and q nonnegative integers, and where the a_js are weights such that $\sum_{j=-q}^{p} a_j = 1$. It is generally assumed that $p = q$, $a_j \geq 0$ for all j and $a_j = a_{-j}$. The order of the moving average in this case is $2p + 1$. The first question that arises when applying a moving average to a series is how to choose p and the weights. The simplest alternative is choosing a low value of p and equal weights. The higher the value of p, the smoother z_t is going to be. Other alternatives include successively applying a simple moving average with equal weights, or choosing the weights in such a way that a particular feature of the data is highlighted. For example, if a given time series recorded monthly displays a trend plus a yearly cycle, choosing a moving average with $p = 6$, $a_6 = a_{-6} = 1/24$, and $a_j = 1/12$ for $j = 0, \pm 1, \ldots, \pm 5$ would diminish the impact of the periodic component, emphasizing the trend (see Diggle 1990 for an example).

Figure 1.7 (a) shows monthly values of a Southern Oscillation Index (SOI) time series during 1950–1995. This series consists of 540 observations of the SOI computed as the difference of the departure from the long term monthly mean sea level pressures at Tahiti in the South Pacific and Darwin in Northern Australia. The index is one measure of the so called "El Niño-Southern Oscillation"—an event of critical importance and interest in climatological studies in recent decades. The fact that most of the observations in the last part of the series take negative values is related to a recent warming in the tropical Pacific. Figures 1.7 (b) and (c) show two smoothed series obtained via moving averages of orders 3 and 9, respectively, with equal weights. As explained before, we can see that the higher the order of the moving average the smoother the resulting series is.

Other ways to smooth a time series include fitting a linear regression to remove a trend or, more generally, fitting a polynomial regression; fitting a

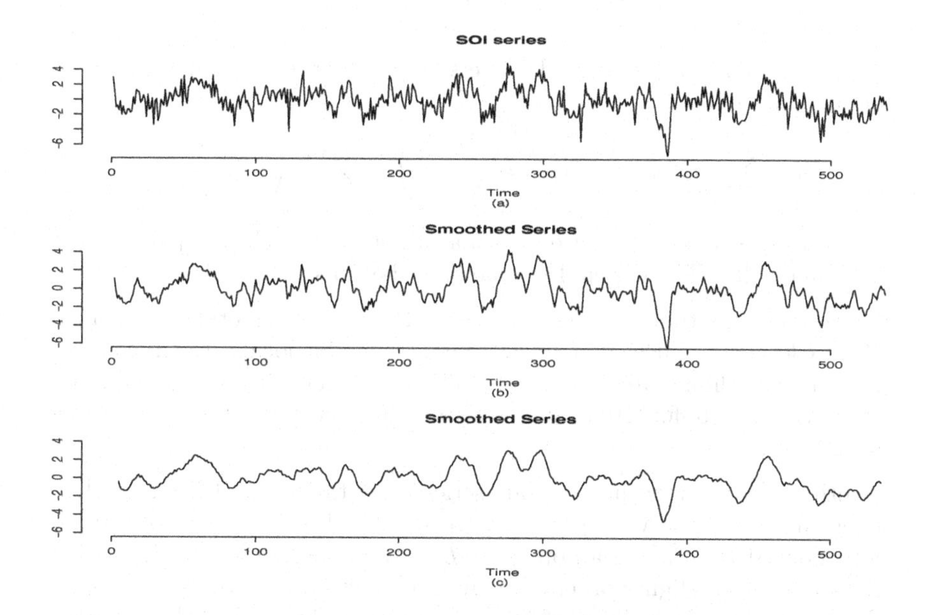

Figure 1.7 *(a): Southern oscillation index (SOI) time series; (b): Smoothed series obtained using a moving average of order 3 with equal weights; (c): Smoothed series obtained using a moving average of order 9 with equal weights.*

harmonic regression to remove periodic components; and performing kernel or spline smoothing.

Smoothing by polynomial regression consists on fitting a polynomial to the series. In other words, we want to estimate the parameters of the model

$$y_t = \beta_0 + \beta_1 t + \cdots + \beta_p t^p + \epsilon_t,$$

where ϵ_t is usually assumed as a sequence of zero mean, independent Gaussian random variables. Similarly, fitting harmonic regressions provides a way to remove cycles from a time series. So, if we want to remove periodic components with frequencies $w_1, \ldots, w_p$, we need to estimate $a_1, b_1, \ldots, a_p, b_p$ in the model

$$\begin{aligned} y_t &= a_1 \cos(2\pi w_1 t) + b_1 \sin(2\pi w_1 t) + \cdots \\ &\quad + a_p \cos(2\pi w_p t) + b_p \sin(2\pi w_p t) + \epsilon_t. \end{aligned}$$

In both cases the smoothed series would then be obtained as $\hat{y}_t$, with $\hat{y}_t = \hat{\beta}_0 + \hat{\beta}_1 t + \cdots + \hat{\beta}_p t^p$, and $\hat{y}_t = \hat{a}_1 \cos(2\pi w_1 t) + \hat{b}_1 \sin(2\pi w_1 t) + \cdots + \hat{a}_p \cos(2\pi w_p t) + \hat{b}_p \sin(2\pi w_p t)$, respectively, where $\hat{\beta}_i$, $\hat{a}_i$, and $\hat{b}_i$ are point estimates of the parameters. Usually $\hat{\beta}_i$ and $\hat{a}_i, \hat{b}_i$ are obtained by least squares estimation.

In kernel smoothing a smoothed version, z_t, of the original series y_t is obtained as follows:

$$z_t = \sum_{i=1}^{T} w_t(i) y_t, \quad w_i(t) = K\left(\frac{t-i}{b}\right) \Big/ \sum_{j=1}^{T} K\left(\frac{t-j}{b}\right),$$

where $K(\cdot)$ is a kernel function, such as a normal kernel. The parameter b is a bandwidth. The larger the value of b, the smoother z_t is.

Cubic and smoothing splines, as well as the *lowess* smoother (Cleveland 1979; Cleveland and Devlin 1988; lowess stands for locally weighted scatterplot smoothing) are also commonly used smoothing techniques. See Shumway and Stoffer (2017) for details and illustrations on these smoothing techniques.

Another way of smoothing a time series is by taking its differences. Differencing provides a way to remove trends. The first difference of a series y_t is defined in terms of an operator D that produces the transformation $Dy_t = y_t - y_{t-1}$. Higher-order differences are defined by successively applying the operator D. Differences can also be defined in terms of the backshift operator B, with $By_t = y_{t-1}$, and so $Dy_t = (1-B)y_t$. Higher-order differences can be written as $D^d y_t = (1-B)^d y_t$.

In connection with the methods presented here, it is worth mentioning that wavelet decompositions have been widely used in recent years for smoothing time series. Vidakovic (1999) and Percival and Walden (2006) present statistical approaches to modeling by wavelets. Wavelets are basis functions that are used to represent other functions. They are analogous to the sines and cosines in the Fourier transformation. One of the advantages of using wavelets bases, as opposed to Fourier representations, is that they are localized in frequency and time, and so, they are suitable for dealing with nonstationary signals that display jumps and other abrupt changes.

1.5 A Primer on Likelihood and Bayesian Inference

Assume that we have collected T observations, $y_{1:T}$, of a scalar time series process $\{y_t\}$. Suppose that for each y_t we have a probability distribution that can be written as a function of some parameter, or collection of parameters, namely $\boldsymbol{\theta}$, in such a way that the dependence of y_t on $\boldsymbol{\theta}$ is described in terms of a probability density function $p(y_t|\boldsymbol{\theta})$. If we think of $p(y_t|\boldsymbol{\theta})$ as a function of $\boldsymbol{\theta}$, rather than a function of y_t, we refer to it as the likelihood function. Using Bayes' theorem it is possible to obtain the posterior density function of $\boldsymbol{\theta}$ given y_t, $p(\boldsymbol{\theta}|y_t)$, as the product of the likelihood and the prior density $p(\boldsymbol{\theta})$, i.e.,

$$p(\boldsymbol{\theta}|y_t) = \frac{p(\boldsymbol{\theta})p(y_t|\boldsymbol{\theta})}{p(y_t)}, \tag{1.15}$$

with $p(y_t) = \int p(\boldsymbol{\theta})p(y_t|\boldsymbol{\theta})d\boldsymbol{\theta}$. $p(y_t)$ defines the so-called predictive density function. The prior distribution offers a way to incorporate our prior beliefs about $\boldsymbol{\theta}$ and Bayes' theorem allows us to update such beliefs after observing the data.

Bayes' theorem can also be used in a sequential way as follows: Before collecting any data, prior beliefs about $\boldsymbol{\theta}$ are expressed in a probabilistic form via $p(\boldsymbol{\theta})$. Assume that we then collect our first observation at time $t = 1$, y_1, and we obtain $p(\boldsymbol{\theta}|y_1)$ using Bayes' theorem. Once y_2 is observed we can obtain $p(\boldsymbol{\theta}|y_{1:2})$ via Bayes' theorem as $p(\boldsymbol{\theta}|y_{1:2}) \propto p(\boldsymbol{\theta})p(y_{1:2}|\boldsymbol{\theta})$. Now, if y_1 and y_2 are conditionally independent on $\boldsymbol{\theta}$, we can write $p(\boldsymbol{\theta}|y_{1:2}) \propto p(\boldsymbol{\theta}|y_1)p(y_2|\boldsymbol{\theta})$, i.e., the posterior of $\boldsymbol{\theta}$ given y_1 becomes a prior distribution before observing y_2. Similarly, $p(\boldsymbol{\theta}|y_{1:T})$ can be obtained in a sequential way, if all the observations are independent. However, in time series analysis the observations are not independent. For example, a common assumption is that each observation at time t depends only on $\boldsymbol{\theta}$ and the observation taken at time $t-1$. In this case we have

$$p(\boldsymbol{\theta}|y_{1:T}) \propto p(\boldsymbol{\theta})p(y_1|\boldsymbol{\theta}) \prod_{t=2}^{T} p(y_t|y_{t-1}, \boldsymbol{\theta}). \tag{1.16}$$

General models in which y_t depends on an arbitrary number of past observations will be studied in subsequent chapters. We now consider an example in which the posterior distribution has the form (1.16).

Example 1.4 *The AR(1) model.* We consider again the AR(1) process. The model parameters in this case are given by $\boldsymbol{\theta} = (\phi, v)'$. Now, for each time $t > 1$, the conditional likelihood is $p(y_t|y_{t-1}, \boldsymbol{\theta}) = N(y_t|\phi y_{t-1}, v)$. In addition, it can be shown that $y_1 \sim N(0, v/(1-\phi^2))$ if the process is stationary (see Problem 1 in Chapter 2) and so, the likelihood is given by

$$p(y_{1:T}|\boldsymbol{\theta}) = \frac{(1-\phi^2)^{1/2}}{(2\pi v)^{T/2}} \exp\left\{-\frac{Q^*(\phi)}{2v}\right\}, \tag{1.17}$$

with

$$Q^*(\phi) = y_1^2(1-\phi^2) + \sum_{t=2}^{T}(y_t - \phi y_{t-1})^2. \tag{1.18}$$

The posterior density is obtained via Bayes' rule and so

$$p(\boldsymbol{\theta}|y_{1:T}) \quad \propto \quad p(\boldsymbol{\theta})\frac{(1-\phi^2)^{1/2}}{(2\pi v)^{T/2}} \exp\left\{\frac{-Q^*(\phi)}{2v}\right\}.$$

We can also use the conditional likelihood $p(y_{2:T}|\boldsymbol{\theta}, y_1)$ as an approximation to the likelihood (see Box, Jenkins, Reinsel, and Ljung 2015 A7.4 for a justification), which leads to the following posterior density,

$$p(\boldsymbol{\theta}|y_{1:T}) \propto p(\boldsymbol{\theta})v^{-(T-1)/2} \exp\left\{\frac{-Q(\phi)}{2v}\right\}, \tag{1.19}$$

with $Q(\phi) = \sum_{t=2}^{T}(y_t - \phi y_{t-1})^2$. Several choices of $p(\boldsymbol{\theta})$ can be considered and will be discussed later. In particular, it is common to assume a prior structure such that $p(\boldsymbol{\theta}) = p(v)p(\phi|v)$, or $p(\boldsymbol{\theta}) = p(v)p(\phi)$.

Another important class of time series models is that in which parameters are indexed in time. In this case each observation is related to a parameter, or a set of parameters, say $\boldsymbol{\theta}_t$, that evolve over time. The so-called class of Dynamic Linear Models (DLMs) considered in Chapter 4 deals with models of this type. In such framework it is necessary to define a process that describes the evolution of $\boldsymbol{\theta}_t$ over time. As an example, consider the time-varying AR model of order one, or TVAR(1), given by

$$\begin{aligned} y_t &= \phi_t y_{t-1} + \epsilon_t, \\ \phi_t &= \phi_{t-1} + \nu_t, \end{aligned}$$

where ϵ_t and ν_t are independent in time and mutually independent, with $\epsilon_t \sim N(0, v)$ and $\nu_t \sim N(0, w)$. Some distributions of interest are the posterior distributions at time t, $p(\phi_t|y_{1:t})$ and $p(v|y_{1:t})$, the backward filtering or smoothing distributions $p(\phi_t|y_{1:T})$, and the h-steps ahead forecast distribution $p(y_{t+h}|y_{1:t})$. Details on how to find these distributions for rather general DLMs are given in Chapter 4.

1.5.1 ML, MAP, and LS Estimation

It is possible to obtain point estimates of the model parameters by maximizing the likelihood function or the full posterior distribution. A variety of methods and algorithms have been developed to achieve this goal. We briefly discuss some of these methods. In addition, we illustrate how these methods work in the simple AR(1) case.

A point estimate of $\boldsymbol{\theta}$, $\hat{\boldsymbol{\theta}}$ can be obtained by maximizing the likelihood function $p(y_{1:T}|\boldsymbol{\theta})$ with respect to $\boldsymbol{\theta}$. In this case we use the notation $\hat{\boldsymbol{\theta}} = \boldsymbol{\theta}_{\text{ML}}$. Similarly, if instead of maximizing the likelihood function we maximize the posterior distribution $p(\boldsymbol{\theta}|y_{1:T})$, we obtain the maximum a posteriori estimate for $\boldsymbol{\theta}$, $\hat{\boldsymbol{\theta}} = \boldsymbol{\theta}_{\text{MAP}}$.

Often, the likelihood function and the posterior distribution are complicated nonlinear functions of $\boldsymbol{\theta}$ and so it is necessary to use methods such as the Newton–Raphson algorithm or the scoring method to obtain the maximum likelihood estimator (MLE) or the maximum a posteriori (MAP) estimator. In general, the Newton–Raphson algorithm can be summarized as follows. Let $g(\boldsymbol{\theta})$ be the function of $\boldsymbol{\theta} = (\theta_1, \ldots, \theta_k)'$ that we want to maximize, and $\hat{\boldsymbol{\theta}}$ be the maximum. At iteration m of the Newton–Raphson algorithm we obtain $\boldsymbol{\theta}^{(m)}$, an approximation to $\hat{\boldsymbol{\theta}}$, as follows:

$$\boldsymbol{\theta}^{(m)} = \boldsymbol{\theta}^{(m-1)} - \left[g''(\boldsymbol{\theta}^{(m-1)})\right]^{-1} \times \left[g'(\boldsymbol{\theta}^{(m-1)})\right], \tag{1.20}$$

where $g'(\boldsymbol{\theta})$ and $g''(\boldsymbol{\theta})$ denote the first- and second-order partial derivatives of the function g, i.e., $g'(\boldsymbol{\theta})$ is a k-dimensional vector given by $g'(\boldsymbol{\theta}) = \left(\frac{\partial g(\boldsymbol{\theta})}{\partial \theta_1}, \ldots, \frac{\partial g(\boldsymbol{\theta})}{\partial \theta_k}\right)'$, and $g''(\boldsymbol{\theta})$ is a $k \times k$ matrix of second-order partial derivatives whose ij-th element is given by $\left[\frac{\partial g^2(\boldsymbol{\theta})}{\partial \theta_i \partial \theta_j}\right]$, for $i, j = 1:k$. Under certain conditions this algorithm produces a sequence $\boldsymbol{\theta}^{(1)}, \boldsymbol{\theta}^{(2)}, \ldots,$ that will converge to $\hat{\boldsymbol{\theta}}$. In particular, it is important to begin with a good starting value $\boldsymbol{\theta}^{(0)}$, since the algorithm does not necessarily converge for values in regions where $-g''(\cdot)$ is not positive definite. An alternative method is the scoring method, which involves replacing $g''(\boldsymbol{\theta})$ in (1.20) by the matrix of expected values $E(g''(\boldsymbol{\theta}))$.

In many practical scenarios, especially when dealing with models that have very many parameters, it is not useful to summarize the inferences in terms of the joint posterior mode. Instead, summaries are made in terms of marginal posterior modes, that is, the posterior modes for subsets of model parameters. Let us say that we can partition our model parameters in two sets, $\boldsymbol{\theta}_1$ and $\boldsymbol{\theta}_2$, so that $\boldsymbol{\theta} = (\boldsymbol{\theta}_1', \boldsymbol{\theta}_2')'$, and assume we are interested in $p(\boldsymbol{\theta}_2|y_{1:T})$. The EM (Expectation-Maximization) algorithm proposed in Dempster, Laird, and Rubin (1977) is useful when dealing with models for which $p(\boldsymbol{\theta}_2|y_{1:T})$ is hard to maximize directly, but it is relatively easy to work with $p(\boldsymbol{\theta}_1|\boldsymbol{\theta}_2, y_{1:T})$ and $p(\boldsymbol{\theta}_2|\boldsymbol{\theta}_1, y_{1:T})$. The EM algorithm can be described as follows:

1. Start with some initial value $\boldsymbol{\theta}_2^{(0)}$.
2. For $m = 1, 2, \ldots$
   - Compute $E^{(m-1)}[\log p(\boldsymbol{\theta}_1, \boldsymbol{\theta}_2|y_{1:T})]$ given by the expression

$$\int \log p(\boldsymbol{\theta}_1, \boldsymbol{\theta}_2|y_{1:T}) p(\boldsymbol{\theta}_1|\boldsymbol{\theta}_2^{(m-1)}, y_{1:T}) d\boldsymbol{\theta}_1. \tag{1.21}$$

   This is the E-step.
   - Set $\boldsymbol{\theta}_2^{(m)}$ to the value that maximizes (1.21). This is the M-step.

At each iteration the algorithm satisfies that $p(\boldsymbol{\theta}_2^{(m)}|y_{1:T}) \geq p(\boldsymbol{\theta}_2^{(m-1)}|y_{1:T})$. There is no guarantee that the EM algorithm converges to the mode; in the case of multimodal distributions the algorithm may converge to a local mode. Various alternatives have been considered to avoid getting stuck in a local mode, such as running the algorithm with several different random initial points, or using simulated annealing methods. Some extensions of the EM algorithm include the ECM (expectation-conditional-maximization) algorithm, the ECME (expectation-conditional-maximization-either, a variant of the ECM in which either the log-posterior density or the expected log-posterior density is maximized) and the SEM (supplemented EM) algorithms (see Gelman, Carlin, Stern, Dunson, Vehtari, and Rubin 2014 and references therein) and stochastic versions of the EM algorithm such as the MCEM (Monte Carlo EM, see Wei and Tanner 1990).

Example 1.5 *ML, MAP, and LS estimators for the AR(1) model.* Consider the AR(1) model $y_t = \phi y_{t-1} + \epsilon_t$, with $\epsilon_t \sim N(0, 1)$. In this case $v = 1$ and $\theta = \phi$. The conditional MLE is found by maximizing $\exp\{-Q(\phi)/2\}$ or, equivalently, by minimizing $Q(\phi)$. Therefore, we obtain $\hat{\phi} = \phi_{\text{ML}} = \sum_{t=2}^{T} y_t y_{t-1} / \sum_{t=2}^{T} y_{t-1}^2$. Similarly, the MLE for the unconditional likelihood function is obtained by maximizing $p(y_{1:T}|\phi)$ or, equivalently, by minimizing the expression

$$-0.5[\log(1 - \phi^2) - Q^*(\phi)].$$

Newton–Raphson or scoring methods can be used to find $\hat{\phi}$. As an illustration, the conditional and unconditional ML estimators were found for

100 samples from an AR(1) with $\phi = 0.9$. Figure 1.8 shows a graph with the conditional and unconditional log-likelihood functions (solid and dotted lines respectively). The points correspond to the maximum likelihood estimators with $\hat{\phi} = 0.9069$ and $\hat{\phi} = 0.8979$ being the MLEs for the conditional and unconditional likelihoods, respectively. For the unconditional case, a Newton–Raphson algorithm was used to find the maximum. The algorithm converged after five iterations with a starting value of 0.1.

Figure 1.9 shows the log-posterior densities of ϕ under Gaussian priors of the form $\phi \sim N(\mu, c)$, for $\mu = 0$, $c = 1.0$ (left panel) and $c = 0.01$ (right panel). Note that this prior does not impose any restriction on ϕ and so it gives nonnegative probability to values of ϕ that lie in the nonstationary region. It is possible to choose priors on ϕ whose support is the stationary region. This will be considered in Chapter 2. Figure 1.9 illustrates the effect of the prior on the MAP estimators. For a prior $\phi \sim N(0, 1)$, the MAP estimators are $\hat{\phi}_{\text{MAP}} = 0.9051$ and $\hat{\phi}_{\text{MAP}} = 0.8963$ for the conditional and unconditional likelihoods, respectively. When a smaller value of c is considered, or in other words, when the prior distribution is more concentrated around zero, then the MAP estimates shift toward the prior mean. For a prior $\phi \sim N(0, 0.01)$, the MAP estimators are $\hat{\phi}_{\text{MAP}} = 0.7588$ and $\hat{\phi}_{\text{MAP}} = 0.7550$ for the conditional and unconditional likelihoods, respectively.

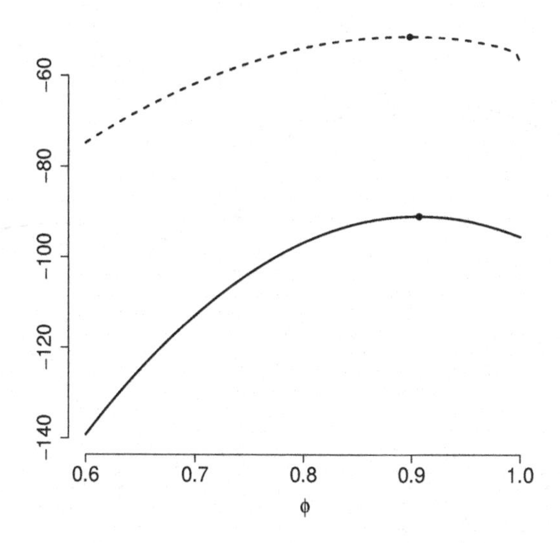

Figure 1.8 *Conditional and unconditional log-likelihoods (solid and dashed lines, respectively) based on 100 observations simulated from an* $AR(1)$ *with* $\phi = 0.9$.

It is also possible to obtain the least squares estimators for the conditional and unconditional likelihoods. For the conditional case, the least squares (LS) estimator is obtained by minimizing the conditional sum of squares $Q(\phi)$, and so in this case $\phi_{\text{ML}} = \phi_{\text{LS}}$. In the unconditional case, the LS estimator is found by minimizing the unconditional sum of squares $Q^*(\phi)$, and so the LS and the ML estimators do not coincide.

1.5.2 *Traditional Least Squares*

Likelihood and Bayesian approaches for fitting linear autoregressions rely on very standard methods of linear regression analysis. Therefore, some review of the central ideas and results in regression is in order and given here. This introduces notation and terminology that will be used throughout the book.

A linear model with a univariate response variable and $p > 0$ regression variables (otherwise predictors or covariates) has the form

$$y_i = \mathbf{f}_i'\boldsymbol{\beta} + \epsilon_i,$$

for $i = 1, 2, \ldots$, where y_i is the i-th observation on the response variable, and has corresponding values of the regressors in the design vector $\mathbf{f}_i' = (f_{i1}, \ldots, f_{ip})$. The design vectors are assumed known and fixed prior to

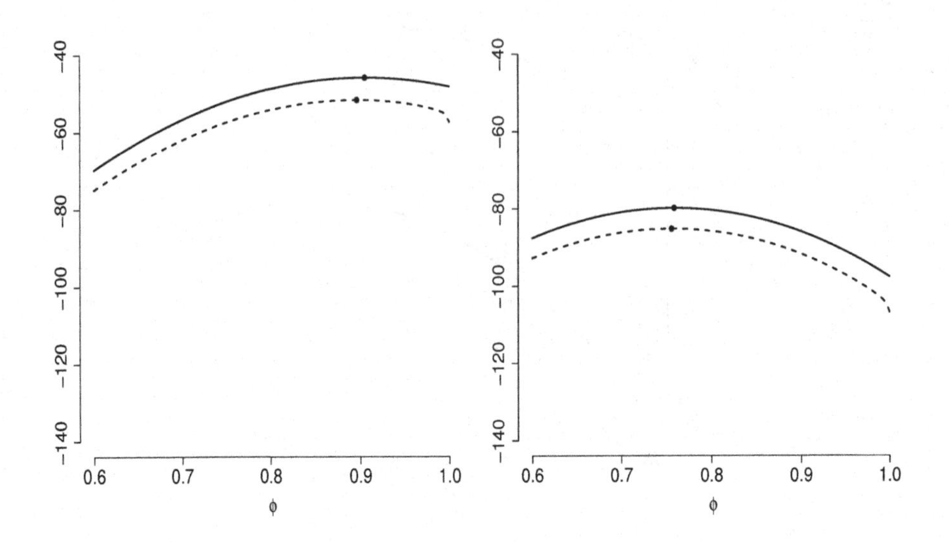

Figure 1.9 *Conditional and unconditional log-posterior densities (solid and dashed lines, respectively) based on 100 observations simulated from an* $AR(1)$ *with* $\phi = 0.9$*. The posterior densities were obtained with priors of the form* $\phi \sim N(0, c)$*, for* $c = 1$ *(left panel) and* $c = 0.01$ *(right panel).*

observing the corresponding responses. The error terms ϵ_i are assumed independent and normal, distributed as $N(\epsilon_i|0, v)$ with some variance v. The regression parameter vector $\boldsymbol{\beta} = (\beta_1, \dots, \beta_p)'$ is to be estimated, along with the error variance. Now assume we have a set of n responses denoted as $\mathbf{y} = (y_1, \dots, y_n)'$. We note that this notation is general and so, the responses are not necessarily temporally indexed and n is not necessarily equal to T. The model for $\mathbf{y}$ is

$$\mathbf{y} = \mathbf{F}'\boldsymbol{\beta} + \boldsymbol{\epsilon}, \tag{1.22}$$

where $\mathbf{F}$ is the known $p \times n$ design matrix with i-th column $\mathbf{f}_i$. In addition, $\boldsymbol{\epsilon} = (\epsilon_1, \dots, \epsilon_n)'$, with $\boldsymbol{\epsilon} \sim N(\boldsymbol{\epsilon}|0, v\mathbf{I}_n)$, and $\mathbf{I}_n$ the $n \times n$ identity matrix. The sampling distribution is defined as

$$p(\mathbf{y}|\mathbf{F}, \boldsymbol{\beta}, v) = \prod_{i=1}^{n} N(y_i|\mathbf{f}_i'\boldsymbol{\beta}, v) = (2\pi v)^{-n/2} \exp(-Q(\mathbf{y}, \boldsymbol{\beta})/2v),$$

where $Q(\mathbf{y}, \boldsymbol{\beta}) = (\mathbf{y} - \mathbf{F}'\boldsymbol{\beta})'(\mathbf{y} - \mathbf{F}'\boldsymbol{\beta}) = \sum_{i=1}^{n}(y_i - \mathbf{f}_i'\boldsymbol{\beta})^2$. This gives a likelihood function for $(\boldsymbol{\beta}, v)$. We can also write $Q(\mathbf{y}, \boldsymbol{\beta})$ as

$$Q(\mathbf{y}, \boldsymbol{\beta}) = (\boldsymbol{\beta} - \hat{\boldsymbol{\beta}})'\mathbf{F}\mathbf{F}'(\boldsymbol{\beta} - \hat{\boldsymbol{\beta}}) + R,$$

where $\hat{\boldsymbol{\beta}} = (\mathbf{F}\mathbf{F}')^{-1}\mathbf{F}\mathbf{y}$ and $R = (\mathbf{y} - \mathbf{F}'\hat{\boldsymbol{\beta}})'(\mathbf{y} - \mathbf{F}'\hat{\boldsymbol{\beta}})$. This assumes that $\mathbf{F}$ is of full rank p, otherwise an appropriate linear transformation of the design vectors can be used to reduce $\mathbf{F}$ to a full rank matrix and the model decreases in dimension. Here $\hat{\boldsymbol{\beta}}$ is the MLE of $\boldsymbol{\beta}$ and the residual sum of squares R gives the MLE of v as R/n; a more usual estimate of v is $s^2 = R/(n-p)$, with $n-p$ being the associated degrees of freedom.

1.5.3 Full Bayesian Analysis

We summarize some aspects of various Bayesian approaches for fitting linear models, including reference and conjugate analyses. Nonconjugate analyses may lead to posterior distributions that are not available in closed form. Therefore, nonconjugate inferential approaches often rely on obtaining random draws from the posterior distribution using Markov chain Monte Carlo methods, which will be used a good deal later in this book. Some key references are the books of Box and Tiao (1973) and Zellner (1996). The book of Greenberg (2008) provides an excellent introduction to Bayesian statistics and econometrics using a simulation-based approach.

1.5.3.1 Reference Bayesian Analysis

Reference Bayesian analysis is based on the traditional reference (improper) prior $p(\boldsymbol{\beta}, v) \propto 1/v$. The corresponding posterior density is $p(\boldsymbol{\beta}, v|\mathbf{y}, \mathbf{F}) \propto p(\mathbf{y}|\mathbf{F}, \boldsymbol{\beta}, v)/v$ and has the following features:

- The marginal posterior for $\boldsymbol{\beta}$ is a multivariate Student-t with $n-p$ degrees of freedom. It has mode $\hat{\boldsymbol{\beta}}$, scale matrix $s^2(\mathbf{F}\mathbf{F}')^{-1}$, and density
$$p(\boldsymbol{\beta}|\mathbf{y}, \mathbf{F}) = c(n,p)|\mathbf{F}\mathbf{F}'|^{1/2}\{1 + (\boldsymbol{\beta} - \hat{\boldsymbol{\beta}})'\mathbf{F}\mathbf{F}'(\boldsymbol{\beta} - \hat{\boldsymbol{\beta}})/(n-p)s^2\}^{-n/2}$$
with $c(n,p) = \Gamma(n/2)/[\Gamma((n-p)/2)(s^2\pi(n-p))^{p/2}]$, where $\Gamma(\cdot)$ is the gamma function. When n is large, the posterior is approximately normal, $N(\boldsymbol{\beta}|\hat{\boldsymbol{\beta}}, s^2(\mathbf{F}\mathbf{F}')^{-1})$. Note also that, given v, the conditional posterior for $\boldsymbol{\beta}$ is exactly normal, namely $N(\boldsymbol{\beta}|\hat{\boldsymbol{\beta}}, v(\mathbf{F}\mathbf{F}')^{-1})$.
- The marginal posterior for v is inverse gamma with parameters $(n-p)/2$ and $(n-p)s^2/2$, or $(v|\mathbf{y}, \mathbf{F}) \sim IG((n-p)/2, (n-p)s^2/2)$.
- The total sum of squares of the responses $\mathbf{y}'\mathbf{y} = \sum_{i=1}^{n} y_i^2$ factorizes as $\mathbf{y}'\mathbf{y} = R + \hat{\boldsymbol{\beta}}'\mathbf{F}\mathbf{F}'\hat{\boldsymbol{\beta}}$. The sum of squares explained by the regression is $\mathbf{y}'\mathbf{y} - R = \hat{\boldsymbol{\beta}}'\mathbf{F}\mathbf{F}'\hat{\boldsymbol{\beta}}$; this is also called the fitted sum of squares, and a larger value implies a smaller residual sum of squares and, in this sense, a closer fit to the data.
- Under a proper prior distribution for $(\boldsymbol{\beta}, v)$ the marginal density of $(\mathbf{y}|\mathbf{F})$

can be obtained as

$$p(\mathbf{y}|\mathbf{F}) = \int p(\mathbf{y}|\mathbf{F}, \boldsymbol{\beta}, v)p(\boldsymbol{\beta}, v)d\boldsymbol{\beta}dv.$$

Note that the reference prior used here is improper, invalidating the calculation of a proper marginal density for $(\mathbf{y}|\mathbf{F})$. However, one can still obtain an expression for $p(\mathbf{y}|\mathbf{F})$ up to a proportionality constant as

$$p(\mathbf{y}|\mathbf{F}) = \int \frac{p(\mathbf{y}|\mathbf{F}, \boldsymbol{\beta}, v)}{v}\, d\boldsymbol{\beta}dv \propto \frac{\Gamma((n-p)/2)}{\pi^{(n-p)/2}}|\mathbf{F}\mathbf{F}'|^{-1/2}R^{-(n-p)/2}.$$

This can also be written as

$$p(\mathbf{y}|\mathbf{F}) \propto \frac{\Gamma((n-p)/2)}{\pi^{(n-p)/2}}|\mathbf{F}\mathbf{F}'|^{-1/2}(\mathbf{y}'\mathbf{y})^{(p-n)/2}\{1 - \hat{\boldsymbol{\beta}}'\mathbf{F}\mathbf{F}'\hat{\boldsymbol{\beta}}/(\mathbf{y}'\mathbf{y})\}^{(p-n)/2}.$$

For large n, the term $\{1-\hat{\boldsymbol{\beta}}'\mathbf{F}\mathbf{F}'\hat{\boldsymbol{\beta}}/(\mathbf{y}'\mathbf{y})\}^{(p-n)/2}$ in the above expression is approximately $\exp(\hat{\boldsymbol{\beta}}'\mathbf{F}\mathbf{F}'\hat{\boldsymbol{\beta}}/2r)$ where $r = \mathbf{y}'\mathbf{y}/(n-p)$.

Some additional comments:

- For models with the same number of parameters that differ only through $\mathbf{F}$, the corresponding observed data densities will tend to be larger for those models with larger values of the explained sum of squares $\hat{\boldsymbol{\beta}}'\mathbf{F}\mathbf{F}'\hat{\boldsymbol{\beta}}$ (though the determinant term plays a role too). Otherwise, $p(\mathbf{y}|\mathbf{F})$ also depends on the parameter dimension p.
- *Orthogonal regression.* If $\mathbf{F}\mathbf{F}' = k\mathbf{I}_p$ for some k, then everything simplifies. Write $\mathbf{f}_j^*$ for the j-th column of $\mathbf{F}'$, and β_j for the corresponding component of the parameter vector $\boldsymbol{\beta}$. Then $\hat{\boldsymbol{\beta}} = (\hat{\beta}_1, \ldots, \hat{\beta}_p)'$ where each $\hat{\beta}_j$ is the individual MLE from a model on $\mathbf{f}_j^*$ alone, i.e., $\mathbf{y} = \mathbf{f}_j^*\beta_j + \boldsymbol{\epsilon}$, and the elements of $\boldsymbol{\beta}$ are uncorrelated under the posterior T distribution. The explained sum of squares partitions into a sum of individual pieces too, namely $\hat{\boldsymbol{\beta}}'\mathbf{F}\mathbf{F}'\hat{\boldsymbol{\beta}} = \sum_{j=1}^{p} \mathbf{f}_j^{*\prime}\mathbf{f}_j^*\hat{\beta}_j^2$, and so calculations and interpretations are easy.

Example 1.6 *Reference analysis in the AR(1) model.* For the conditional likelihood using the notation above we have $\mathbf{y} = (y_2, \ldots, y_T)'$, $\mathbf{F} = (y_1, \ldots, y_{T-1})$ and the reference prior $p(\phi, v) \propto 1/v$. The MLE for ϕ is $\phi_{\text{ML}} = \sum_{t=2}^{T} y_{t-1}y_t / \sum_{t=1}^{T-1} y_t^2$. Under the reference prior $\phi_{\text{MAP}} = \phi_{\text{ML}}$. The residual sum of squares is given by

$$R = \sum_{t=2}^{T} y_t^2 - \frac{(\sum_{t=2}^{T} y_t y_{t-1})^2}{\sum_{t=1}^{T-1} y_t^2},$$

and so $s^2 = R/(T-2)$ estimates v. The marginal posterior distribution

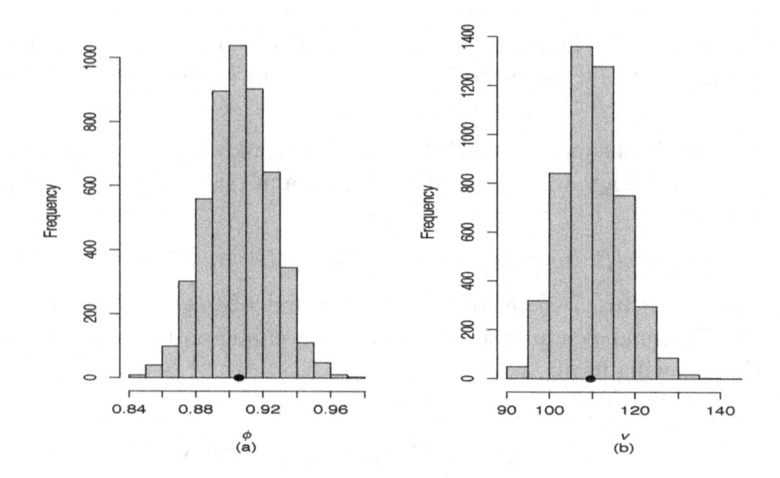

Figure 1.10 *(a)* $p(\phi|\mathbf{y}, \mathbf{F})$; *(b)* $p(v|\mathbf{y}, \mathbf{F})$.

of ϕ is a univariate Student-t distribution with $T - 2$ degrees of freedom, centered at ϕ_{ML} with scale $s^2(\mathbf{F}\mathbf{F}')^{-1}$, i.e.,

$$(\phi|\mathbf{y}, \mathbf{F}) \sim t_{(T-2)}\left(m, \frac{C}{T-2}\right),$$

where

$$m = \frac{\sum_{t=2}^{T} y_{t-1} y_t}{\sum_{t=1}^{T-1} y_t^2}$$

and

$$C = \frac{\sum_{t=2}^{T} y_t^2 \sum_{t=2}^{T} y_{t-1}^2 - \left(\sum_{t=2}^{T} y_t y_{t-1}\right)^2}{\left(\sum_{t=1}^{T-1} y_t^2\right)^2}.$$

Finally, the posterior for v is a scaled inverse chi-squared with $T-2$ degrees of freedom and scale s^2, i.e., $Inv{-}\chi^2(v|T-2, s^2)$ or, equivalently, an inverse gamma with parameters $(T-2)/2$ and $(T-2)s^2/2$, $IG(v|(T-2)/2, (T-2)s^2/2)$.

As an illustration, a reference analysis was performed for a time series of 500 points simulated from an AR(1) model with $\phi = 0.9$ and $v = 100$. Figures 1.10 (a) and (b) display the marginal posterior densities of $(\phi|\mathbf{y}, \mathbf{F})$ and $(v|\mathbf{y}, \mathbf{F})$ based on 5,000 samples from the joint posterior of ϕ and v. The circles in the histogram indicate ϕ_{ML} and s^2, respectively.

1.5.3.2 Conjugate Bayesian Analysis

Let $p(y_t|\boldsymbol{\theta})$ be a likelihood function. A class Π of prior distributions forms a *conjugate family* if the posterior $p(\boldsymbol{\theta}|y_t)$ belongs to the class Π for every prior $p(\boldsymbol{\theta})$ in Π.

Consider again the model $\mathbf{y} = \mathbf{F}'\boldsymbol{\beta} + \boldsymbol{\epsilon}$, with $\mathbf{F}$ a known $p \times n$ design matrix and $\boldsymbol{\epsilon} \sim N(\boldsymbol{\epsilon}|0, v\mathbf{I}_n)$. In a conjugate Bayesian analysis for this model priors of the form

$$p(\boldsymbol{\beta}, v) = p(\boldsymbol{\beta}|v)p(v) = N(\boldsymbol{\beta}|\mathbf{m}_0, v\mathbf{C}_0) \times IG(v|n_0/2, d_0/2) \qquad (1.23)$$

are taken with $\mathbf{m}_0$ a vector of dimension p and $\mathbf{C}_0$ a $p \times p$ matrix. Both $\mathbf{m}_0$ and $\mathbf{C}_0$ are known quantities. The corresponding posterior distribution has the following form:

$$\begin{aligned} p(\boldsymbol{\beta}, v|\mathbf{y}, \mathbf{F}) \quad \propto \quad & v^{-[(p+n+n_0)/2+1]} \times \\ & e^{-[(\boldsymbol{\beta}-\mathbf{m}_0)'\mathbf{C}_0^{-1}(\boldsymbol{\beta}-\mathbf{m}_0)+(\mathbf{y}-\mathbf{F}'\boldsymbol{\beta})'(\mathbf{y}-\mathbf{F}'\boldsymbol{\beta})+d_0]/2v}. \end{aligned}$$

This analysis has the following features:

- $(\mathbf{y}|\mathbf{F}, v) \sim N(\mathbf{F}'\mathbf{m}_0, v(\mathbf{F}'\mathbf{C}_0\mathbf{F} + \mathbf{I}_n))$ and $(\mathbf{y}|\mathbf{F})$ follows a multivariate Student-t distribution, i.e., $(\mathbf{y}|\mathbf{F}) \sim T_{n_0}[\mathbf{F}'\mathbf{m}_0, d_0(\mathbf{F}'\mathbf{C}_0\mathbf{F} + \mathbf{I}_n)/n_0]$.
- The posterior distribution of $\boldsymbol{\beta}$ given v is Gaussian, $(\boldsymbol{\beta}|\mathbf{y}, \mathbf{F}, v) \sim N(\mathbf{m}, v\mathbf{C})$, with

  $$\begin{aligned} \mathbf{m} &= \mathbf{m}_0 + \mathbf{C}_0\mathbf{F}[\mathbf{F}'\mathbf{C}_0\mathbf{F} + \mathbf{I}_n]^{-1}(\mathbf{y} - \mathbf{F}'\mathbf{m}_0) \\ \mathbf{C} &= \mathbf{C}_0 - \mathbf{C}_0\mathbf{F}[\mathbf{F}'\mathbf{C}_0\mathbf{F} + \mathbf{I}_n]^{-1}\mathbf{F}'\mathbf{C}_0, \end{aligned}$$

  or, defining $\mathbf{e} = \mathbf{y} - \mathbf{F}'\mathbf{m}_0, \mathbf{Q} = \mathbf{F}'\mathbf{C}_0\mathbf{F} + \mathbf{I}_n$, and $\mathbf{A} = \mathbf{C}_0\mathbf{F}\mathbf{Q}^{-1}$ we can also write $\mathbf{m} = \mathbf{m}_0 + \mathbf{A}\mathbf{e}$ and $\mathbf{C} = \mathbf{C}_0 - \mathbf{A}\mathbf{Q}\mathbf{A}'$.
- $(v|\mathbf{y}, \mathbf{F}) \sim IG(n^*/2, d^*/2)$ with $n^* = n + n_0$ and

  $$d^* = (\mathbf{y} - \mathbf{F}'\mathbf{m}_0)'\mathbf{Q}^{-1}(\mathbf{y} - \mathbf{F}'\mathbf{m}_0) + d_0.$$
- $(\boldsymbol{\beta}|\mathbf{y}, \mathbf{F}) \sim T_{n^*}[\mathbf{m}, d^*\mathbf{C}/n^*]$.

Example 1.7 *Conjugate analysis in the AR(1) model using the conditional likelihood.* Assume we choose a prior of the form $\phi|v \sim N(0, v)$ and $v \sim IG(n_0/2, d_0/2)$, with n_0 and d_0 known. Then, $p(\phi|\mathbf{y}, \mathbf{F}, v) \sim N(m, vC)$ with

$$m = \frac{\sum_{t=1}^{T-1} y_t y_{t+1}}{\sum_{t=1}^{T-1} y_t^2 + 1}, \quad C = \frac{1}{1 + \sum_{t=1}^{T-1} y_t^2},$$

$(v|\mathbf{y}, \mathbf{F}) \sim IG(n^*/2, d^*/2)$ with $n^* = T + n_0 - 1$ and

$$d^* = \sum_{t=2}^{T} y_t^2 - \frac{\left(\sum_{t=1}^{T-1} y_t y_{t+1}\right)^2}{\sum_{t=1}^{T-1} y_t^2 + 1} + d_0.$$

1.5.4 Nonconjugate Bayesian Analysis

For the general regression model, the reference and conjugate priors produce joint posterior distributions that have closed analytical forms. However, in many scenarios it is either not possible or not desirable to work with a conjugate prior or with a prior that leads to a posterior distribution that can be written in analytical form. In these cases it might be possible to use analytical or numerical approximations to the posterior. Another alternative consists on summarizing the inference by obtaining random draws from the posterior distribution. Sometimes it is possible to obtain such draws by direct simulation, but often this is not the case, and so methods such as Markov chain Monte Carlo (MCMC) are used.

Consider again the AR(1) model under the full likelihood (1.17). No conjugate prior is available in this case. Furthermore, a prior of the form $p(\phi, v) \propto 1/v$ does not produce a posterior distribution in closed form. In fact, the joint posterior distribution is such that

$$p(\phi, v|y_{1:T}) \propto v^{-(T/2+1)}(1-\phi^2)^{1/2} \exp\left\{\frac{-Q^*(\phi)}{2v}\right\}. \tag{1.24}$$

Several approaches could be considered to summarize this posterior distribution. For example, we could use a normal approximation to the distribution $p(\phi, v|y_{1:T})$ centered at the ML or MAP estimates of (ϕ, v). In general, the normal approximation to a posterior distribution $p(\boldsymbol{\theta}|y_{1:T})$ is given by

$$p(\boldsymbol{\theta}|y_{1:T}) \approx N(\hat{\boldsymbol{\theta}}, v(\hat{\boldsymbol{\theta}})), \tag{1.25}$$

with $\hat{\boldsymbol{\theta}} = \boldsymbol{\theta}_{\text{MAP}}$ and $v(\boldsymbol{\theta})^{-1} = -\frac{\partial^2}{\partial\boldsymbol{\theta}\partial\boldsymbol{\theta}'} \log p(\boldsymbol{\theta}|y_{1:T})$.

Alternatively, it is possible to use iterative MCMC methods to obtain samples from $p(\phi, v|y_{1:T})$. We summarize two of the most widely used MCMC methods below: the Metropolis algorithm and the Gibbs sampler. For full consideration of MCMC methods see, for example, Gamerman and Lopes (2006) and Robert and Casella (2005).

1.5.5 Posterior Sampling

1.5.5.1 The Metropolis-Hastings Algorithm

Assume that our target posterior distribution, $p(\boldsymbol{\theta}|y_{1:T})$, can be computed up to a normalizing constant. The Metropolis-Hastings algorithm (Metropolis et al. 1953, Hastings 1970) creates a sequence of random draws $\boldsymbol{\theta}^{(1)}$, $\boldsymbol{\theta}^{(2)}, \ldots,$ whose distributions converge to the target distribution. Each sequence can be considered as a Markov chain whose stationary distribution is $p(\boldsymbol{\theta}|y_{1:T})$. The sampling algorithm can be summarized as follows:

- Draw a starting point $\boldsymbol{\theta}^{(0)}$ with $p(\boldsymbol{\theta}^{(0)}|y_{1:T}) > 0$ from a starting distribution $p_0(\boldsymbol{\theta})$.
- For $m = 1, 2, \ldots$
  1. Sample a candidate $\boldsymbol{\theta}^*$ from a jumping distribution $J(\boldsymbol{\theta}^*|\boldsymbol{\theta}^{(m-1)})$. If the distribution J is symmetric, i.e., if $J(\boldsymbol{\theta}_a|\boldsymbol{\theta}_b) = J(\boldsymbol{\theta}_b|\boldsymbol{\theta}_a)$ for all $\boldsymbol{\theta}_a, \boldsymbol{\theta}_b$, and m, then we refer to the algorithm as the Metropolis algorithm. If J_m is not symmetric, we refer to the algorithm as the Metropolis-Hastings algorithm.
  2. Compute the importance ratio
  $$r = \frac{p(\boldsymbol{\theta}^*|y_{1:T})/J(\boldsymbol{\theta}^*|\boldsymbol{\theta}^{(m-1)})}{p(\boldsymbol{\theta}^{(m-1)}|y_{1:n})/J(\boldsymbol{\theta}^{(m-1)}|\boldsymbol{\theta}^*)}.$$
  3. Set
  $$\boldsymbol{\theta}^{(m)} = \begin{cases} \boldsymbol{\theta}^* & \text{with probability} = \min(r, 1) \\ \boldsymbol{\theta}^{(m-1)} & \text{otherwise.} \end{cases}$$

An ideal jumping distribution is one that is easy to sample from and makes the evaluation of the importance ratio easy. In addition, the jumping distributions $J(\cdot|\cdot)$ should be such that each jump moves a reasonable distance in the parameter space so that the random walk is not too slow, and also, the jumps should not be rejected too often.

1.5.5.2 Gibbs Sampling

Assume $\boldsymbol{\theta}$ has k components, i.e., $\boldsymbol{\theta} = (\boldsymbol{\theta}_1, \ldots, \boldsymbol{\theta}_k)$. The Gibbs sampler (Geman and Geman 1984) can be viewed as a special case of the Metropolis-Hastings algorithm for which the jumping distribution at each iteration m is a function $p(\boldsymbol{\theta}_j^*|\boldsymbol{\theta}_{-j}^{(m-1)}, y_{1:T})$, where $\boldsymbol{\theta}_{-j}$ denotes a vector with all the components of $\boldsymbol{\theta}$ except for component $\boldsymbol{\theta}_j$. In other words, for each component of $\boldsymbol{\theta}$ we do a Metropolis-Hastings step for which the jumping distribution is given by

$$J_j(\boldsymbol{\theta}^*|\boldsymbol{\theta}^{(m-1)}) = \begin{cases} p(\boldsymbol{\theta}_j^*|\boldsymbol{\theta}_{-j}^{(m-1)}, y_{1:T}) & \text{if } \boldsymbol{\theta}_{-j}^* = \boldsymbol{\theta}_{-j}^{(m-1)} \\ 0 & \text{otherwise,} \end{cases}$$

and so $r = 1$ and every jump is accepted.

If it is not possible to sample from $p(\boldsymbol{\theta}_j^*|\boldsymbol{\theta}_{-j}^{(m)}, y_{1:T})$ an approximation, say $g(\boldsymbol{\theta}_j^*|\boldsymbol{\theta}_{-j}^{(m-1)})$, can be considered. However, in this case it is necessary to compute the Metropolis acceptance ratio r.

1.5.5.3 Convergence

In theory, a value from the posterior distribution of $(\boldsymbol{\theta}|y_{1:T})$ is obtained by MCMC when the number of iterations of the chain approaches infinity. In practice, a value obtained after a sufficiently large number of iterations is taken as a draw from the target posterior distribution of $(\boldsymbol{\theta}|y_{1:T})$. How can we determine how many MCMC iterations are enough to obtain convergence? As pointed out in Gamerman and Lopes (2006), there are two general approaches to the study of convergence. One is probabilistic and it consists on measuring distances and bounds on distribution functions generated from a chain. So, for example, it is possible to measure the total variation distance between the distribution of the chain at iteration i and the target distribution of $(\boldsymbol{\theta}|y_{1:T})$. An alternative approach consists on studying the convergence of the chain from a statistical perspective. This approach is easier and more practical than the probabilistic one; however, it cannot guarantee convergence.

There are several ways of monitoring convergence from a statistical viewpoint, ranging from graphical displays of the MCMC traces for all or some of the model parameters or functions of such parameters, to sophisticated statistical tests. As mentioned before, one of the two main problems with simulation-based iterative methods is deciding whether the chain has reached convergence, i.e., if the number of iterations is large enough to guarantee that the available samples are drawn from the target posterior distribution. In addition, large within-sequence correlation may lead to inferences that are not precise enough. In other words, if M draws from a chain with very large within-sequence correlation are used to represent the posterior distribution, the "effective" number of draws used in such representation is far smaller than M. Some well-known tests to assess convergence are implemented as R packages (R Core Team 2018), such as Bayesian Output Analysis (BOA, Smith 2007) and Convergence Diagnosis and Output Analysis for MCMC (CODA, Plummer, Best, Cowles, and Vines 2006). Specifically, these packages include convergence diagnostics such as the Brooks, Gelman, and Rubin diagnostic for a list of sequences (Brooks and Gelman 1998; Gelman and Rubin 1992), which monitors the mixing of the simulated sequences by comparing the within and between variance of the sequences; the Geweke diagnostic (1992) and Heidelberger and Welch diagnostic (1983), which are based on sequential testing of portions of the simulated chains to determine if they correspond to samples from the same distribution; and the Raftery and Lewis method (Raftery and Lewis 1992), which considers the problem of how many iterations are needed to estimate a particular posterior quantile from a single MCMC chain. BOA and CODA also provide the user with some descriptive plots of the chains—e.g., autocorrelations, density, means, and trace plots—as well as plots of some of the convergence diagnostics.

Example 1.8 *A Metropolis-Hastings for an AR(1) model.* Consider again the AR(1) model with the unconditional likelihood (1.17) and a prior of the form $p(\phi, v) \propto 1/v$. A MCMC algorithm to obtain samples from the posterior distribution is described below. For each iteration $m = 1, 2, \ldots$

- Sample $v^{(m)}$ from $(v|\phi, y_{1:T}) \sim IG(T/2, Q^*(\phi)/2)$. Note that this is a Gibbs step and so every draw will be accepted.
- Sample $\phi^{(m)}$ using a Metropolis step with a Gaussian jumping distribution, i.e., at iteration m we draw a candidate sample ϕ^* from a Gaussian distribution centered at $\phi^{(m-1)}$, that is,

$$\phi^* \sim N\left(\phi^{(m-1)}, c\right),$$

  with c a constant. The value of c controls the acceptance rate of the algorithm. In practice, target acceptance rates usually go from 25% to 40%. See for instance Gelman, Carlin, Stern, Dunson, Vehtari, and Rubin (2014), Chapter 11, for a discussion on how to set the value of c.

In order to illustrate the MCMC methodology, we considered 500 observations generated from an AR(1) model with coefficient $\phi = 0.9$ and variance $v = 1.0$. The MCMC scheme above was implemented in order to achieve posterior estimation of the model parameters based on the 500 synthetic observations. Figures 1.11 (a) and (b) display the traces of the model parameters for two chains of 1,000 MCMC samples. Several values of c were considered and the value $c = 0.0005$ was chosen because it led to a Metropolis acceptance rate of approximately 35%. The starting values for the chains were set at $v^0 = 0.1$, $\phi^0 = 0.5$, and $v^0 = 3$, $\phi^0 = 0.0$. No convergence problems are apparent from these pictures. Figures 1.11 (c) and (d) show the posterior distributions for ϕ and v based on 450 samples of one of the MCMC chains taken every other iteration after a burn-in period of 100 iterations. The early iterations of a MCMC output are usually discarded in order to eliminate, or diminish as much as possible, the effect of the starting distribution. These are referred to as burn-in iterations. The length of the burn-in period varies greatly depending on the context and complexity of the MCMC sampler.

1.6 Appendix

1.6.1 The Uniform Distribution

A random variable x follows a uniform distribution in the interval (a, b), with $a < b$, and so $x \sim U(a, b)$, or $p(x) = U(x|a, b)$, if its density function

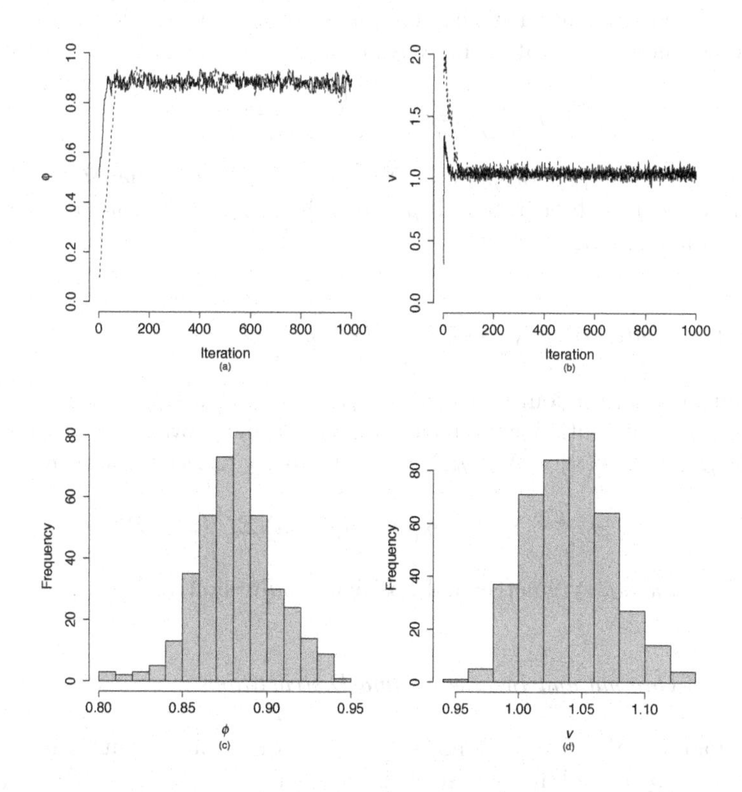

Figure 1.11 *Panels (a) and (b) show traces of 1,000 MCMC samples of the parameters ϕ and v, respectively. The draws from two chains are displayed. Solid lines correspond to samples from a chain with starting values of $(\phi^{(0)}, v^0) = (0.5, 0.1)$ and the dashed lines correspond to samples from a chain with starting values of $(\phi^{(0)}, v^{(0)}) = (0, 3)$. Panels (c) and (d) show histograms of 450 samples from the marginal posterior distributions of ϕ and v. The samples were taken every other MCMC iteration after a burn-in period of 100 iterations.*

is given by

$$p(x) = \frac{1}{(b-a)}, \quad x \in [a, b].$$

In addition, $E(x) = (a + b)/2$ and $V(x) = (b - a)^2/12$.

1.6.2 The Univariate Normal Distribution

A real-valued random variable x follows a normal distribution with mean μ and variance $v > 0$ if its density is given by

$$p(x) = \frac{1}{\sqrt{2\pi v}} \exp\left[-\frac{(x-\mu)^2}{2v}\right].$$

We use $x \sim N(\mu, v)$, or $p(x) = N(x|\mu, v)$, to denote that x follows a univariate normal distribution. If $\mu = 0$ and $v = 1$, we say that x follows a standard normal distribution.

1.6.3 The Multivariate Normal Distribution

A k-dimensional random vector $\mathbf{x} = (x_1, \ldots, x_k)'$ that follows a multivariate normal distribution with mean $\boldsymbol{\mu}$ and variance-covariance matrix $\boldsymbol{\Sigma}$, $\mathbf{x} \sim N(\boldsymbol{\mu}, \boldsymbol{\Sigma})$, or $p(\mathbf{x}) = N(\mathbf{x}|\boldsymbol{\mu}, \boldsymbol{\Sigma})$, has a density function given by

$$p(\mathbf{x}) = (2\pi)^{-k/2}|\boldsymbol{\Sigma}|^{-1/2} \exp\left[-\frac{1}{2}(\mathbf{x}-\boldsymbol{\mu})'\boldsymbol{\Sigma}^{-1}(\mathbf{x}-\boldsymbol{\mu})\right].$$

Here $\boldsymbol{\Sigma}$ is a $k \times k$ symmetric and positive definite matrix.

1.6.4 The Gamma and Inverse-gamma Distributions

A random variable x that follows a gamma distribution with shape parameter $\alpha > 0$ and inverse scale parameter $\beta > 0$, $x \sim G(\alpha, \beta)$, or $p(x) = G(x|\alpha, \beta)$, has a density of the form

$$p(x) = \frac{\beta^\alpha}{\Gamma(\alpha)} x^{\alpha-1} e^{-\beta x}, \quad x > 0,$$

where $\Gamma(\cdot)$ is the gamma function. In addition, $E(x) = \alpha/\beta$ and $V(x) = \alpha/\beta^2$.

If $\frac{1}{x} \sim G(\alpha, \beta)$, then x follows an inverse-gamma distribution, $x \sim IG(\alpha, \beta)$, or $p(x) = IG(x|\alpha, \beta)$ with

$$p(x) = \frac{\beta^\alpha}{\Gamma(\alpha)} x^{-(\alpha+1)} e^{-\beta/x}, \quad x > 0.$$

In this case $E(x) = \beta/(\alpha - 1)$ for $\alpha > 1$ and $V(x) = \beta^2/[(\alpha-1)^2(\alpha-2)]$ for $\alpha > 2$.

1.6.5 The Exponential Distribution

A random variable x with an exponential distribution with parameter $\beta > 0$, $x \sim Exp(\beta)$, or $p(x) = Exp(x|\beta)$, has density

$$p(x) = \beta e^{-\beta x}, \quad x > 0.$$

This distribution is the same as $G(x|1, \beta)$.

1.6.6 The Chi-square Distribution

x follows a chi-square distribution with $\nu > 0$ degrees of freedom, $x \sim \chi^2_\nu$, if its density is given by

$$p(x) = \frac{2^{-\nu/2}}{\Gamma(\nu/2)} x^{\nu/2-1} e^{-x/2}, \quad x > 0.$$

This distribution is the same as $G(x|\nu/2, 1/2)$.

1.6.7 The Inverse Chi-square Distributions

x is said to follow an inverse chi-squared distribution with $\nu > 0$ degrees of freedom, $x \sim Inv - \chi^2_\nu$, if $x \sim IG(\nu/2, 1/2)$. Also, x is said to follow a scaled inverse chi-squared distribution with ν degrees of freedom and scale $s > 0$, i.e., $x \sim Inv - \chi^2(\nu, s^2)$, if $x \sim IG(\nu/2, \nu s^2/2)$.

1.6.8 The Univariate Student-t Distribution

A real-valued random variable x follows a Student-t distribution with $\nu > 0$ degrees of freedom, location μ, and scale $\sigma > 0$, $x \sim t_\nu(\mu, \sigma^2)$, if its density is

$$p(x) = \frac{\Gamma((\nu+1)/2)}{\Gamma(\nu/2)\sqrt{\nu\pi}\sigma} \left[1 + \frac{1}{\nu}\left(\frac{x-\mu}{\sigma}\right)^2\right]^{-(\nu+1)/2}.$$

In addition, $E(x) = \mu$ for $\nu > 1$ and $V(x) = \nu\sigma^2/(\nu - 2)$ for $\nu > 2$.

1.6.9 The Multivariate Student-t Distribution

A random vector $\mathbf{x}$ of dimension k follows a multivariate Student-t distribution with $\nu > 0$ degrees of freedom, location $\boldsymbol{\mu}$, and scale matrix $\boldsymbol{\Sigma}$, $\mathbf{x} \sim T_\nu(\boldsymbol{\mu}, \boldsymbol{\Sigma})$, if its density is given by

$$p(\mathbf{x}) = \frac{\Gamma((\nu+k)/2)}{\Gamma(\nu/2)(\nu\pi)^{k/2}} |\boldsymbol{\Sigma}|^{-1/2} \left[1 + \frac{1}{\nu}(\mathbf{x} - \boldsymbol{\mu})'\boldsymbol{\Sigma}^{-1}(\mathbf{x} - \boldsymbol{\mu})\right]^{-(\nu+k)/2}.$$

Here $\boldsymbol{\Sigma}$ is a $k \times k$ symmetric and positive definite matrix. In addition, $E(\mathbf{x}) = \boldsymbol{\mu}$ for $\nu > 1$ and $V(\mathbf{x}) = \nu\boldsymbol{\Sigma}/(\nu - 2)$, for $\nu > 2$.

1.7 Problems

1. Show that the autocorrelation function of an autoregression of order one with AR parameter ϕ, such that $|\phi| < 1$, and zero mean uncorrelated Gaussian innovations with variance v is $\rho(h) = \phi^{|h|}$ for $h = 0, \pm 1, \pm 2, \ldots$
2. Consider the AR(1) model $y_t = \phi y_{t-1} + \epsilon_t$, with $\epsilon_t \sim N(0, v)$.
   (a) Find the MLE of (ϕ, v) for the conditional likelihood.
   (b) Find the MLE of (ϕ, v) for the unconditional likelihood (1.17).
   (c) Assume that v is known. Find the MAP estimator of ϕ under a uniform prior $p(\phi) = U(\phi|-1, 1)$ for the conditional and unconditional likelihoods.
3. Show that the expressions for $p(\boldsymbol{\beta}|\mathbf{y}, \mathbf{F})$ and $p(\mathbf{y}|\mathbf{F})$ under the reference analysis in Section 1.5.3 are those given on page 21.
4. Show that the distributions of $(\phi|\mathbf{y}, \mathbf{F})$ and $(v|\mathbf{y}, \mathbf{F})$ obtained for the AR(1) reference analysis using the conditional likelihood are those given in Example 1.6.
5. Show that the expressions for the distributions of $(\mathbf{y}|\mathbf{F}, v)$, $(v|\mathbf{y}, \mathbf{F})$, and $(\boldsymbol{\beta}|\mathbf{y}, \mathbf{F})$ under the conjugate analysis in Section 1.5.3 are those given on page 24.
6. Show that the distributions of $(\phi|\mathbf{y}, \mathbf{F})$ and $(v|\mathbf{y}, \mathbf{F})$ obtained for the AR(1) conjugate analysis using the conditional likelihood are those given in Example 1.7.
7. Consider the following models:

$$
\begin{aligned}
y_t &= \phi_1 y_{t-1} + \phi_2 y_{t-2} + \epsilon_t, &(1.26)\\
y_t &= a\cos(2\pi\omega_0 t) + b\sin(2\pi\omega_0 t) + \epsilon_t, &(1.27)
\end{aligned}
$$

   with $\epsilon_t \sim N(0, v)$ and $\omega_0 > 0$ fixed.
   (a) Sample $T = 200$ observations from each model using your favorite choice of the parameters. Make sure your choice of (ϕ_1, ϕ_2) in model (1.26) lies in the stationary region. That is, choose ϕ_1 and ϕ_2 such that $-1 < \phi_2 < 1$, $\phi_1 < 1 - \phi_2$, and $\phi_1 > \phi_2 - 1$.
   (b) Find the MLEs of the parameters in models (1.26) and (1.27). Use the conditional likelihood for model (1.26).
   (c) Find the MAP estimators of the model parameters under the reference prior. Again, use the conditional likelihood for model (1.26).
   (d) Sketch $p(v|\mathbf{y}, \mathbf{F})$ and $p(\phi_1, \phi_2|\mathbf{y}, \mathbf{F})$ for model (1.26).
   (e) Sketch $p(a, b|\mathbf{y}, \mathbf{F})$ and $p(v|\mathbf{y}, \mathbf{F})$ for model (1.27).

(f) Perform a conjugate Bayesian analysis, i.e., repeat (c) to (e) assuming conjugate prior distributions in both models. Study the sensitivity of the posterior distributions to the choice of the hyperparameters in the prior.

8. Refer to the conjugate analysis of the AR(1) model in Example 1.7. Using the fact that $(\phi|\mathbf{y}, \mathbf{F}, v) \sim N(m, vC)$, find the posterior mode of v via the EM algorithm.

9. Sample $T = 1{,}000$ observations from model (1.1) with $d = 1$ using your preferred choice of values for θ, $\phi^{(i)}$, and v_i for $i = 1, 2$, with $|\phi^{(i)}| < 1$ and $\phi^{(1)} \neq \phi^{(2)}$. Assuming that d is known and using prior distributions of the form $p(\phi^{(i)}) = N(0, c)$, for $i = 1, 2$ and some $c > 0$, $p(\theta) = U(\theta| - a, a)$ and $p(v) = IG(\alpha_0, \beta_0)$, with $a > 0$, $\alpha_0 > 0$, and $\beta_0 > 0$, obtain samples from the joint posterior distribution by implementing a Metropolis-Hastings algorithm.

Chapter 2

Traditional time domain models

Autoregressive time series models are central to stationary time series data analysis and, as components of larger models or in suitably modified and generalized forms, underlie nonstationary time-varying models. The concepts and structure of linear autoregressive models also provide important background material for appreciation of nonlinear models. This chapter discusses model forms and inference for autoregressions and related topics. This is followed by discussion of the class of stationary autoregressive, moving average models.

2.1 Structure of Autoregressions

Consider the time series of equally spaced quantities y_t, for $t = 1, 2, \ldots,$ arising from the model

$$y_t = \sum_{j=1}^{p} \phi_j y_{t-j} + \epsilon_t, \tag{2.1}$$

where ϵ_t is a sequence of uncorrelated error terms and the ϕ_js are constant parameters. This is a sequentially defined model; y_t is generated as a function of past values, parameters, and errors. The ϵ_ts are termed innovations, and are assumed to be conditionally independent of past values of the series. They are also often assumed normally distributed, $N(\epsilon_t|0, v)$, and so they are independent. This is a standard autoregressive model framework, or AR(p), where p is the order of the autoregression.

AR models may be viewed from a purely empirical standpoint; the data are assumed related over time and the AR form is about the simplest class of empirical models for exploring dependencies. A more formal motivation is, of course, based on the genesis in stationary stochastic process theory. Here we proceed to inference in this model class.

The sequential definition of the model and its Markovian nature imply a sequential structuring of the data density

$$p(y_{1:T}) = p(y_{1:p}) \prod_{t=p+1}^{T} p(y_t | y_{(t-p):(t-1)}) \tag{2.2}$$

for any $T > p$. The leading term is the joint density of the p initial values of the series, as yet undefined. Here the densities are conditional on $(\phi_1, \ldots, \phi_p, v)$, though this is not made explicit in the notation. If the first p values of the series are known and viewed as fixed constants, the conditional density of $\mathbf{y} = (y_T, y_{T-1}, \ldots, y_{p+1})'$ given the first p values is

$$\begin{aligned} p(y_{(p+1):T} | y_{1:p}) &= \prod_{t=p+1}^{T} p(y_t | y_{(t-p):(t-1)}) = \prod_{t=p+1}^{T} N(y_t | \mathbf{f}_t' \boldsymbol{\phi}, v) \\ &= N(\mathbf{y} | \mathbf{F}' \boldsymbol{\phi}, v \mathbf{I}_{T-p}) = p(\mathbf{y} | \mathbf{F}, \boldsymbol{\phi}, v), \end{aligned} \tag{2.3}$$

where $\boldsymbol{\phi} = (\phi_1, \ldots, \phi_p)'$, $\mathbf{f}_t = (y_{t-1}, \ldots, y_{t-p})'$, and $\mathbf{F}$ is a $p \times (T-p)$ matrix given by $\mathbf{F} = [\mathbf{f}_T, \ldots, \mathbf{f}_{p+1}]$. This has a linear model form and so the standard estimation methods discussed in Chapter 1 apply.

Extending the model to include a nonzero mean μ for each y_t gives $y_t = \mu + (\mathbf{f}_t - \mu \mathbf{l})' \boldsymbol{\phi} + \epsilon_t$ where $\mathbf{l} = (1, \ldots, 1)'$, or $y_t = \beta + \mathbf{f}_t' \boldsymbol{\phi} + \epsilon_t$ where $\beta = (1 - \mathbf{l}' \boldsymbol{\phi}) \mu$. Other practically useful extensions of (2.3) include models with additional regression terms for the effects of independent regressor variables on the series, differing variances for the ϵ_ts over time, and non-normal error distributions.

2.1.1 Stationarity in AR Processes

An AR(p) process y_t is *stable* (see, e.g., Lütkepohl 2005) if all the roots of the autoregressive characteristic polynomial, defined as

$$\Phi(u) = 1 - \sum_{j=1}^{p} \phi_j u^j,$$

have moduli greater than unity. That is, y_t is stable if $\Phi(u) = 0$ only when $|u| > 1$. A stable AR process can be written as a one-sided linear process dependent only on present and past ϵ_ts, i.e., if y_t is stable, we can write

$$y_t = \Psi(B) \epsilon_t = \sum_{j=0}^{\infty} \psi_j \epsilon_{t-j}, \tag{2.4}$$

for some values of ψ_js with $\psi_0 = 1$, and $\sum_{j=0}^{\infty} |\psi_j| < \infty$. Here B denotes the backshift operator such that $B^j \epsilon_t = \epsilon_{t-j}$ and $\Psi(B) = 1+\psi_1 B+\psi_2 B^2+\cdots+\psi_j B^j+\cdots$. Processes that can be written as linear combinations of current and past ϵ_ts are referred to as causal processes by some authors (e.g., Shumway and Stoffer 2017). In addition, stability implies stationarity and so, the stability condition is often referred to as the stationarity condition in the time series literature (e.g., see Priestley 1994, Kendall and Ord 1990, Hamilton 1994, and Tiao 2001a, among others). For example, Priestley (1994) refers to the stationarity condition in the context of AR processes as the condition in which stationary AR processes can be expressed in the form (2.4). In this book we also refer to stationary AR processes as those that are stable and can be written as linear filters of current and past ϵ_ts.

The autoregressive characteristic polynomial can also be written as $\Phi(u) = \prod_{j=1}^{p}(1 - \alpha_j u)$, so that its roots are the reciprocals of the α_js. The α_js may be real-valued or may appear as pairs of complex conjugates. Either way, if $|\alpha_j| < 1$ for all j, the process is stationary.

As mentioned in Example 1.2, when $p = 1$ the condition $-1 < \phi_1 < 1$ implies stationarity, and so in this case the stationary distribution of y_t is $N(y_t|0, v/(1 - \phi_1^2))$. At the boundary $\phi_1 = 1$ the model becomes a nonstationary random walk. The bivariate stationary distribution of $(y_t, y_{t-1})'$ is normal with correlation $\rho(1) = \phi_1$; that of $(y_t, y_{t-h})'$ for any h is $\rho(h) = \phi_1^{|h|}$. A positive autoregressive parameter ϕ_1 leads to a process that wanders away from the stationary mean of the series, with such excursions being more extensive when ϕ_1 is closer to unity; $\phi_1 < 0$ leads to more oscillatory behavior about the mean. When $p = 2$, $y_t = \phi_1 y_{t-1} + \phi_2 y_{t-2} + \epsilon_t$, the stationarity condition implies that parameter values must lie in the region $-1 < \phi_2 < 1$, $\phi_1 < 1 - \phi_2$, and $\phi_1 > \phi_2 - 1$. Further discussion appears in Section 2.4.

2.1.2 *State-Space Representation of an AR(p)*

The state-space or dynamic linear model (DLM) representation of an AR(p) model has utility in both exploring mathematical structure and, as we shall see later, in inference and data analysis. One version of this representation of (2.1) is simply

$$y_t = \mathbf{F}'\mathbf{x}_t \quad (2.5)$$

$$\mathbf{x}_t = \mathbf{G}\mathbf{x}_{t-1} + \boldsymbol{\omega}_t, \quad (2.6)$$

where $\mathbf{x}_t = (y_t, y_{t-1}, \ldots, y_{t-p+1})'$ is the state vector at time t. The innovation at time t appears in the error vector $\boldsymbol{\omega}_t = (\epsilon_t, 0, \ldots, 0)'$. In addition,

$\mathbf{F} = (1, 0, \ldots, 0)'$ and

$$\mathbf{G} = \begin{pmatrix} \phi_1 & \phi_2 & \phi_3 & \cdots & \phi_{p-1} & \phi_p \\ 1 & 0 & 0 & \cdots & 0 & 0 \\ 0 & 1 & 0 & \cdots & 0 & 0 \\ \vdots & & & \ddots & 0 & \vdots \\ 0 & 0 & \cdots & \cdots & 1 & 0 \end{pmatrix}. \tag{2.7}$$

The expected behavior of the future of the process may be exhibited through the forecast function $f_t(h) = E(y_{t+h}|y_{1:t})$ as a function of integers $h > 0$ for any fixed "origin" $t \geq p$, conditional on the most recent p values of the series in the current state vector $\mathbf{x}_t = (y_t, y_{t-1}, \ldots, y_{t-p+1})'$. In this case, we have $f_t(h) = \mathbf{F}'\mathbf{G}^h\mathbf{x}_t$. The form is most easily appreciated in cases when the matrix $\mathbf{G}$ has distinct eigenvalues, real and/or complex. It easily follows (see Problem 4 in this chapter) that these eigenvalues are precisely the reciprocal roots of the autoregressive polynomial equation $\Phi(u) = 0$, namely the α_j above. Then, we can write

$$f_t(h) = \sum_{j=1}^{p} c_{tj}\alpha_j^h, \tag{2.8}$$

where the c_{tj}s are (possibly complex-valued) constants depending on $\boldsymbol{\phi}$ and the current state $\mathbf{x}_t$, and the α_js are the p distinct eigenvalues/reciprocal roots. Each c_{tj} coefficient is given by $c_{tj} = d_j e_{tj}$. The d_j and e_{tj} values are the elements of the p-dimensional vectors $\mathbf{d} = \mathbf{E}'\mathbf{F}$ and $\mathbf{e}_t = \mathbf{E}^{-1}\mathbf{x}_t$, where $\mathbf{E}$ is the eigenmatrix of $\mathbf{G}$, i.e., $\mathbf{E}$ is the $p \times p$ matrix whose columns are the eigenvectors in order corresponding to the eigenvalues.

The form of the forecast function depends on the combination of real and complex eigenvalues of $\mathbf{G}$. Suppose, for example, that α_j is real and positive; the contribution to the forecast function is then $c_{tj}\alpha_j^h$. If the process is such that $|\alpha_i| < 1$ for all i, this function of h decays exponentially to zero, monotonically if $\alpha_j > 0$, otherwise oscillating between consecutive positive and negative values. If $|\alpha_j| > 1$, the forecast function is explosive. The relative contribution to the overall forecast function is measured by the decay rate and the initial amplitude c_{tj}, the latter depending explicitly on the current state, and therefore having different impact at different times as the state varies in response to the innovations sequence.

In the case of complex eigenvalues, the fact that $\mathbf{G}$ is real-valued implies that any complex eigenvalues appear in pairs of complex conjugates. Suppose, for example, that α_1 and α_2 are complex conjugates $\alpha_1 = r\exp(i\omega)$ and $\alpha_2 = r\exp(-i\omega)$ with modulus r and argument ω. Then, the corresponding complex factors c_{t1} and c_{t2} are conjugate, $a_t\exp(\pm i b_t)$, and the resulting contribution to $f_t(h)$, which must be real-valued, is

$$c_{t1}\alpha_1^h + c_{t2}\alpha_2^h = 2a_t r^h \cos(\omega h + b_t).$$

Hence, ω determines the constant frequency of a sinusoidal oscillation in the forecast function, the corresponding wavelength or period being $\lambda = 2\pi/\omega$. In a model where the stationary condition holds—i.e., when $r < 1$—the sinusoidal oscillations over times $t+h$, with $h > 0$, are subject to exponential decay through the damping factor r^h. In cases with $r > 1$, the sinusoidal variation explodes in amplitude as r^h increases. The factors a_t and b_t determine the relative amplitude and phase of the component. The amplitude factor $2a_t$ measures the initial magnitude of the contribution of this term to the forecast function, quite separately from the decay factor r. At a future time epoch $t^* > t$, the new state vector $\mathbf{x}_{t^*}$ will define an updated forecast function $f_{t^*}(h)$ with the same form as (2.8), but with updated coefficients depending on $\mathbf{x}_{t^*}$, and so affecting the factors a_{t^*} and b_{t^*}. Therefore, as time evolves, the relative amplitudes and phases of the individual components vary according to the changes in state induced by the sequence of innovations.

Generally, the forecast function (2.8) is a linear combination of exponentially decaying or exploding terms, and decaying or exploding factors multiplying sinusoids of differing periods. Returning to the model (2.1), this basic expected behavior translates into a process that has the same form but in which, at each time point, the innovation ϵ_t provides a random shock to the current state of the process. This describes a process that exhibits such exponentially damped or exploding behavior, possibly with quasiperiodic components, but in which the amplitudes and phases of the components are randomly varying over time in response to the innovations.

2.1.3 *Characterization of AR(2) Processes*

The special case of $p = 2$ is illuminating and of practical importance. The process is stationary if $-1 < \phi_2 < 1$, $\phi_1 < 1 - \phi_2$, and $\phi_1 > \phi_2 - 1$. In such cases, the quadratic characteristic polynomial $\Phi(u) = 0$ has reciprocal roots α_1 and α_2 lying within the unit circle. These define the following:

- Two real roots when $\phi_1^2 + 4\phi_2 \geq 0$, in which case the forecast function decays exponentially.
- A pair of complex conjugate roots $r\exp(\pm i\omega)$ when $\phi_1^2 + 4\phi_2 < 0$. The roots have modulus $r = \sqrt{-\phi_2}$ and argument given by $\cos(\omega) = \phi_1/2r$. The forecast function behaves as an exponentially damped cosine.

We already know that $-2 < \phi_1 < 2$ for stationarity; for complex roots, we have the additional restriction to $-1 < \phi_2 < -\phi_1^2/4$. So, in these cases, the model $y_t = \phi_1 y_{t-1} + \phi_2 y_{t-2} + \epsilon_t$ represents a quasicyclical process, behaving as a damped sine wave of fixed period $2\pi/\omega$, but with amplitude and phase randomly varying over time in response to the innovations. A large variance v induces greater degrees of variation in this dynamic, quasicyclical process. If v is very small, or were to become zero at some point, the process would decay to zero in amplitude due to the damping factor. On the boundary of this region at $\phi_2 = -1$, the modulus is $r = 1$ and the forecast function is sinusoidal with no damping; in this case, $\phi_1 = 2\cos(\omega)$. So, for $|\phi_1| < 2$, the model $y_t = \phi_1 y_{t-1} - y_{t-2} + \epsilon_t$ is that of a sinusoid with randomly varying amplitude and phase; with a small or zero innovation variance v the sinusoidal form sustains, representing essentially a fixed sine wave of constant amplitude and phase. It is easily seen that the difference equation $y_t = 2\cos(\omega)y_{t-1} - y_{t-2}$ defines, for given initial values, a sine wave of period $2\pi/\omega$.

2.1.4 Autocorrelation Structure of an AR(p)

The autocorrelation structure of an AR(p) is given in terms of the solution of the homogeneous difference equation

$$\rho(h) - \phi_1\rho(h-1) - \cdots - \phi_p\rho(h-p) = 0, \quad h > 0. \tag{2.9}$$

In general, if $\alpha_1, \cdots, \alpha_r$ denote the reciprocal roots of the characteristic polynomial $\Phi(u)$, where each root has multiplicity $m_1, \ldots, m_r$ and $\sum_{i=1}^r m_i = p$, then, the general solution to (2.9) is

$$\rho(h) = \alpha_1^h p_1(h) + \alpha_2^h p_2(h) + \cdots + \alpha_r^h p_r(h), \quad h > 0, \tag{2.10}$$

where $p_j(h)$ is a polynomial of degree $m_j - 1$.

For example, in the AR(2) case we have the following scenarios:

- The characteristic polynomial has two different real roots, each one with multiplicity $m_1 = m_2 = 1$. Then, the autocorrelation function has the form

  $$\rho(h) = a\alpha_1^h + b\alpha_2^h, \quad h > 0,$$

  where a and b are constants and α_1, α_2 are the reciprocal roots. Under the stationarity condition, this autocorrelation function decays exponentially as h goes to infinity and, as we saw before, this behavior is shared by the forecast function. The constants a and b are determined by specifying two initial conditions such as $\rho(0) = 1$ and $\rho(-1) = \phi_1/(1-\phi_2)$.
- The characteristic polynomial has one real root with multiplicity $m_1 = 2$, and so the autocorrelation function is given by

  $$\rho(h) = (a + bh)\alpha_1^h, \quad h > 0,$$

  where a and b are constants and α_1 is the reciprocal root. Under the stationarity condition, this autocorrelation function also decays exponentially as h goes to infinity.

- The characteristic polynomial has two complex conjugate roots. In this case the reciprocal roots can be written as $\alpha_1 = r\exp(i\omega)$ and $\alpha_2 = r\exp(-i\omega)$, and so the autocorrelation function is

$$\rho(h) = ar^h \cos(h\omega + b), \quad h > 0,$$

where a and b are constants. Under the stationarity condition, the autocorrelation and forecast functions behave as an exponentially damped cosine.

Example 2.1 *ACFs of AR(2) processes.*

We simulated 500 data points from three different AR(2) processes with the following parameter choices: (a) $\phi_1 = 0.1$, $\phi_2 = 0.8$; (b) $\phi_1 = 1.8$, $\phi_2 = -0.81$; (c) $\phi_1 = 1.2$, $\phi_2 = -0.9$. Figure 2.1 displays the simulated data from these 3 AR(2) processes (left plots), along with the corresponding autocorrelation functions, $\rho(h)$s for $h = 0 : 30$ (center plots), and the sample autocorrelation functions (right plots) obtained from the observed simulated data.

The AR(2) process in (a) has two real reciprocal roots $r_1 = 0.946$ and $r_2 = -0.846$ and so, the ACF has the form $\rho(h) = a(0.946)^h + b(-0.846)^h$ for $h \geq 0$ and some constants a and b. The root $r_1 = 0.946$ dominates the ACF, however, root r_2 is negative and still relatively large in magnitude, significantly contributing to the oscillatory behavior observed in the data and the ACF. The process in (b) has one real reciprocal root with value $r = 0.9$ and multiplicity $m = 2$; therefore, the ACF has a form $(a + bh)0.9^h$. Finally, the process in (c) has a pair of complex reciprocal roots each with modulus $r = 0.949$ and period $\lambda = 2\pi/\omega = 7.09$. The ACF is given by $a(0.949)^h \cos(2\pi h/7.09 + b)$, which has a damped sinusoidal form. The forecast functions for these processes will have a similar structure (not shown).

2.1.5 The Partial Autocorrelation Function

The autocorrelation and forecast functions summarize important features of autoregressive processes. We introduce another function that will provide additional information about autoregressions: the partial autocorrelation function, or PACF. We start by defining the general form of the PACF and we then see that the partial autocorrelation coefficients of a stationary AR(p) process are zero after lag p. This fact has important consequences in estimating the order of an autoregression, at least informally. In practice, it

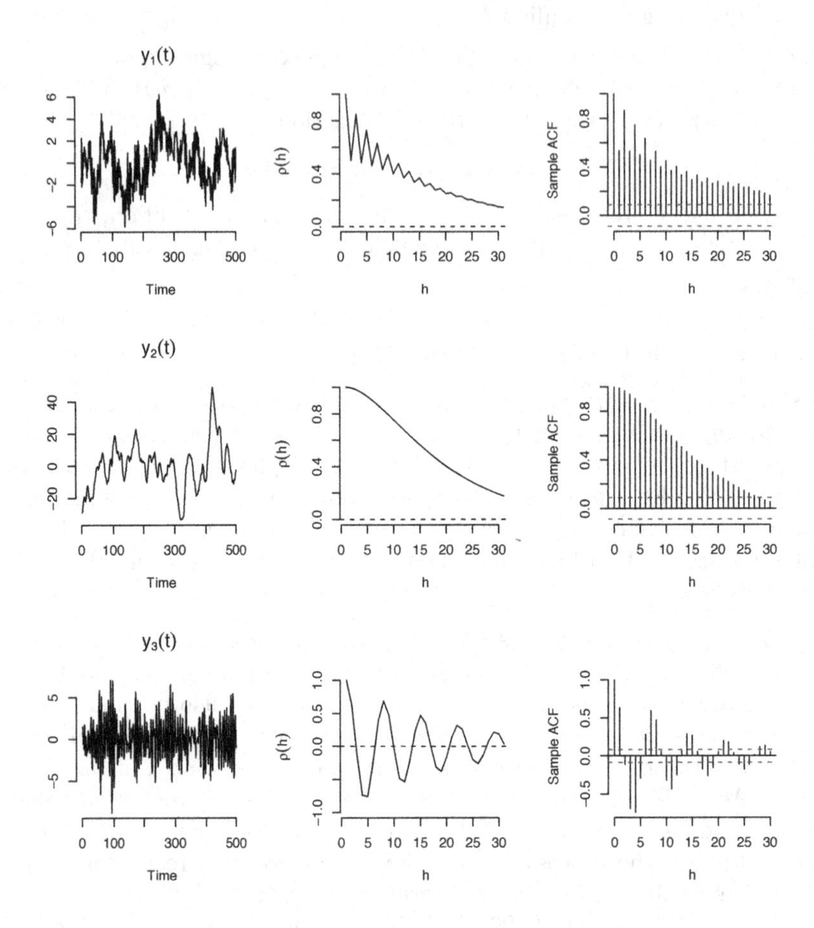

Figure 2.1 *Simulated data from $AR(2)$ processes and corresponding true and sample autocorrelation functions. Top row: $AR(2)$ with two distinct real roots. Center row: $AR(2)$ with one real root with multiplicity 2. Bottom row: $AR(2)$ with a pair of complex roots.*

is possible to decide if an autoregression may be a suitable model for a given time series by looking at the estimated PACF plot. If the series was originally generated by an AR(p) model, its estimated partial autocorrelation coefficients should not be significant after the p-th lag.

The PACF of a process is defined in terms of the partial autocorrelation coefficients at lag h, denoted by $\phi(h, h)$. The PACF coefficient at lag h is a function of the so-called best linear predictor of y_h given $y_{h-1}, \ldots, y_1$.

Specifically, this best linear predictor, denoted by y_h^{h-1}, has the form $y_h^{h-1} = \beta_1 y_{h-1} + \cdots + \beta_{h-1} y_1$, where $\boldsymbol{\beta} = (\beta_1, \ldots, \beta_{h-1})'$ is chosen to minimize the mean square linear prediction error, $E(y_h - y_h^{h-1})^2$. If y_0^{h-1} is the minimum mean square linear predictor of y_0 based on $y_1, \cdots, y_{h-1}$ and the process is stationary, it can be shown that y_0^{h-1} is given by $y_0^{h-1} = \beta_1 y_1 + \cdots + \beta_{h-1} y_{h-1}$, with the same β_js in the expression of y_h^{h-1}. The PACF is then written in terms of $\phi(h, h)$, for $h = 1, 2, \cdots$, which are given by

$$\phi(h,h) = \begin{cases} \rho(y_1, y_0) = \rho(1) & h = 1 \\ \rho(y_h - y_h^{h-1}, y_0 - y_0^{h-1}) & h > 1, \end{cases} \tag{2.11}$$

where $\rho(y_i, y_j)$ denotes the correlation between y_i and y_j. If $\{y_t\}$ follows an AR(p), it is possible to show that $\phi(h, h) = 0$ for $h > p$ (for a proof see for example Shumway and Stoffer 2017, Chapter 3).

Using some properties of the best linear predictors it is also possible to show that the autocorrelation coefficients satisfy the following equation,

$$\Gamma_n \boldsymbol{\phi}_n = \boldsymbol{\gamma}_n, \tag{2.12}$$

where Γ_n is an $n \times n$ matrix whose elements are the autocovariances $\{\gamma(j-h)\}_{j,h=1}^n$. $\boldsymbol{\phi}_n$ and $\boldsymbol{\gamma}_n$ are n-dimensional vectors given by $\boldsymbol{\phi}_n = (\phi(n,1), \ldots, \phi(n,n))'$ and $\boldsymbol{\gamma}_n = (\gamma(1), \ldots, \gamma(n))'$. If Γ_n is nonsingular, then we can write $\boldsymbol{\phi}_n = \Gamma_n^{-1} \boldsymbol{\gamma}_n$. As discussed later in Section 2.3, these equations provide a way to get estimates of $\boldsymbol{\phi}_p$ for AR(p) processes based on the sample autocovariance function. Alternatively, when dealing with stationary autoregressive processes, it is possible to find estimates of $\boldsymbol{\phi}_p$ more efficiently using the Durbin-Levinson recursion (Levinson 1947; Durbin 1960). as follows. For $n = 0$, begin with $\phi(0, 0) = 0$. Then, for any $n \geq 1$,

$$\phi(n,n) = \frac{\rho(n) - \sum_{h=1}^{n-1} \phi(n-1,h)\rho(n-h)}{1 - \sum_{h=1}^{n-1} \phi(n-1,h)\rho(h)},$$

with

$$\phi(n,h) = \phi(n-1,h) - \phi(n,n)\phi(n-1,n-h),$$

for $n \geq 2$ and $h = 1 : (n-1)$. Note that in the case of stationary AR(p) processes $\phi(p, 1), \ldots, \phi(p, p)$ correspond, respectively, to the AR coefficients $\phi_1, \ldots, \phi_p$ and $\phi(h, h) = 0$ for all $h > p$.

The sample PACF can be obtained using the Durbin-Levison recursion by substituting the autocovariances and autocorrelations by the sample autocovariances and the sample autocorrelations $\hat{\gamma}(\cdot)$ and $\hat{\rho}(\cdot)$, respectively. The sample PACF coefficients are denoted by $\hat{\phi}(h, h)$.

2.2 Forecasting

In traditional time series analysis, the one-step-ahead prediction of y_{t+1}, i.e., the forecast of y_{t+1} given $y_{1:t}$, is given by

$$y_{t+1}^t = \phi(t,1)y_t + \phi(t,2)y_{t-1} + \cdots + \phi(t,t)y_1, \tag{2.13}$$

with $\boldsymbol{\phi}_t = (\phi(t,1), \cdots, \phi(t,t))'$ the solution of (2.12) at $n = t$. The mean square error (MSE) of the one-step-ahead prediction is given by

$$MSE_{t+1}^t = E(y_{t+1} - y_{t+1}^t)^2 = \gamma(0) - \boldsymbol{\gamma}_t' \Gamma_t^{-1} \boldsymbol{\gamma}_t, \tag{2.14}$$

which can be recursively computed using the Durbin-Levinson recursion as

$$MSE_{t+1}^t = MSE_t^{t-1}(1 - \phi(t,t)^2),$$

with $MSE_1^0 = \gamma(0)$.

Similarly, the h-step-ahead prediction of y_{t+h} based on $y_{1:t}$ is given by

$$y_{t+h}^t = \phi^{(h)}(t,1)y_t + \cdots + \phi^{(h)}(t,t)y_1, \tag{2.15}$$

with $\boldsymbol{\phi}_t^{(h)} = (\phi^{(h)}(t,1), \ldots, \phi^{(h)}(t,t))'$ the solution of $\Gamma_t \boldsymbol{\phi}_t^{(h)} = \boldsymbol{\gamma}_t^{(h)}$, where $\boldsymbol{\gamma}_t^{(h)} = (\gamma(h), \gamma(h+1), \ldots, \gamma(t+h-1))'$. The mean square error associated with the h-step-ahead prediction is given by

$$MSE_{t+h}^t = E(y_{t+h} - y_{t+h}^t)^2 = \gamma(0) - \boldsymbol{\gamma}_t'^{(h)} \Gamma_t^{-1} \boldsymbol{\gamma}_t^{(h)}. \tag{2.16}$$

It is also possible to compute the forecasts and mean square errors using the innovations algorithm proposed by Brockwell and Davis (1991) as follows. The one-step-ahead predictor and its associated mean squared error can be found iteratively via

$$y_{t+1}^t = \sum_{j=1}^{t} b_{t,j}(y_{t+1-j} - y_{t-j+1}^{t-j}), \tag{2.17}$$

$$MSE_{t+1}^t = \gamma(0) - \sum_{j=0}^{t-1} b_{t,t-j}^2 MSE_{j+1}^j, \tag{2.18}$$

for $t \geq 1$ where, for $j = 0 : (t-1)$,

$$b_{t,t-j} = \frac{\gamma(t-j) - \sum_{l=0}^{j-1} b_{j,j-l} b_{t,t-l} MSE_{l+1}^l}{MSE_{j+1}^j}.$$

The algorithm is initialized at $y_1^0 = 0$. Similarly, the h-step-ahead prediction and its corresponding mean squared error are given by

$$y_{t+h}^{t} = \sum_{j=h}^{t+h-1} b_{t+h-1,j}(y_{t+h-j} - y_{t+h-j}^{t+h-j-1}), \quad (2.19)$$

$$MSE_{t+h}^{t} = \gamma(0) - \sum_{j=h}^{t+h-1} b_{t+h-1,j}^{2} MSE_{t+h-j}^{t}. \quad (2.20)$$

For AR(p) models with $t > p$, the previous equations provide the exact one-step-ahead and h-step-ahead predictions. In particular, it is possible to see (e.g., Chapter 3 of Shumway and Stoffer 2017) that, if y_t is a stationary AR(p) process, then

$$y_{t+1}^{t} = \phi_1 y_t + \phi_2 y_{t-1} + \cdots + \phi_p y_{t-p+1}. \quad (2.21)$$

So far we have written the forecasting equations assuming that the parameters are known. If the parameters are unknown and need to be estimated, which is usually the case in practice, then it is necessary to substitute the parameter values by the estimated values in the previous equations. When a Bayesian analysis of the time series model is performed, the forecasts are obtained directly from the model equations. For instance, if we are dealing with an AR(p), the h-step-ahead predictions can be computed using either posterior estimates of the model parameters, or samples from the posterior distributions of such parameters. This will be discussed in detail in the next section.

2.3 Estimation in AR Models

2.3.1 Yule-Walker and Maximum Likelihood

Post-multiplying (2.1) by y_{t-h} and taking expected values, we obtain

$$\Gamma_p(h)\boldsymbol{\phi}_p = \boldsymbol{\gamma}_p(h), \quad \text{for } h > 0, \quad (2.22)$$

$$\sigma^2 = \gamma(0) - \phi_1\gamma(1) - \cdots - \phi_p\gamma(p), \quad (2.23)$$

where $\boldsymbol{\phi}_p = (\phi_1, \ldots, \phi_p)'$ and $\Gamma_p(h)$ and $\boldsymbol{\gamma}_p(h)$ are, respectively, a $p \times p$ matrix and a p-dimensional vector given by

$$\Gamma_p(h) = \begin{pmatrix} \gamma(h-1) & \gamma(h-2) & \cdots & \gamma(h-p+1) & \gamma(h-p) \\ \gamma(h) & \gamma(h-1) & \cdots & \gamma(h-p+2) & \gamma(h-p+1) \\ \vdots & \vdots & \ddots & \vdots & \vdots \\ \gamma(h+p-2) & \gamma(h+p-3) & \cdots & \gamma(h) & \gamma(h-1) \end{pmatrix}$$

and $\boldsymbol{\gamma}_p(h) = (\gamma(h), \ldots, \gamma(h+p-1))'$. We see that Γ_p and $\boldsymbol{\gamma}_p$ in (2.12) correspond to $\Gamma_p = \Gamma_p(1)$ and $\boldsymbol{\gamma}_p = \boldsymbol{\gamma}_p(1)$. Then, if autocovariances are substituted by sample autocovariances and a set of initial conditions is used, we obtain the Yule-Walker estimates $\hat{\boldsymbol{\phi}}_p$ and $\hat{v}$, such that

$$\hat{\Gamma}_p \hat{\boldsymbol{\phi}}_p = \hat{\boldsymbol{\gamma}}_p, \quad \hat{v} = \hat{\gamma}(0) - \hat{\boldsymbol{\gamma}}_p' \hat{\Gamma}_p^{-1} \hat{\boldsymbol{\gamma}}_p, \tag{2.24}$$

assuming that $\hat{\Gamma}_p$ is a nonsingular matrix. These estimates can also be computed via the Durbin-Levinson recursion (see Brockwell and Davis 1991 for details). The advantage of using the Durbin-Levinson recursion is that it avoids explicit computation of $\hat{\Gamma}_p^{-1}$. It is also possible to show (e.g., see Shumway and Stoffer 2017) that in the case of stationary AR processes, the Yule-Walker estimates are such that

$$\sqrt{T}(\hat{\boldsymbol{\phi}}_p - \boldsymbol{\phi}_p) \approx N(\mathbf{0}, v\Gamma_p^{-1}),$$

and that $\hat{v}$ is close to v when the sample size T is large. These results can be used to obtain confidence regions about $\hat{\boldsymbol{\phi}}_p$.

Maximum likelihood estimation (MLE) in AR(p) models can be achieved by maximizing the conditional likelihood given in (2.3). It is also possible to work with the unconditional likelihood. This will be discussed later when the MLE method for general autoregressive moving average (ARMA) models is described.

2.3.2 Basic Bayesian Inference for AR Models

Return to the basic model (2.1) and the conditional sampling density (2.3), and suppose that the data $y_{1:T}$ are observed. Given a model order p, Equation (2.3) defines the resulting likelihood function of $(\boldsymbol{\phi}, v)$, with $\boldsymbol{\phi} = (\phi_1, \ldots, \phi_p)'$. This is a conditional likelihood function—it is conditional on the assumed initial values $y_{1:p}$—and so the resulting inferences, reference posterior inferences, or otherwise, are also explicitly conditional on these initial values. More on dealing with this later. For now, we have a linear model $p(\mathbf{y}|\mathbf{F}, \boldsymbol{\phi}, v) = N(\mathbf{y}|\mathbf{F}'\boldsymbol{\phi}, v\mathbf{I}_{T-p})$ and we can apply standard theory. In particular, the reference posterior analysis described in Chapter 1 can be used to obtain baseline inferences for $(\boldsymbol{\phi}, v)$.

Example 2.2 *EEG data analysis.* Figure 2.2 displays recordings of an electroencephalogram (EEG). The data displayed represent variations in scalp potentials in microvolts during a seizure, the time intervals being just less than one fortieth of a second. The original data were sampled at 256 observations per second, and the 400 points in the plot were obtained by selecting every sixth observation from a midseizure section.

The sample autocorrelations (not shown) have an apparent damped sinusoidal form, indicative of the periodic behavior evident from the time series plot, with a period around 12 to 14 time units. The damping toward

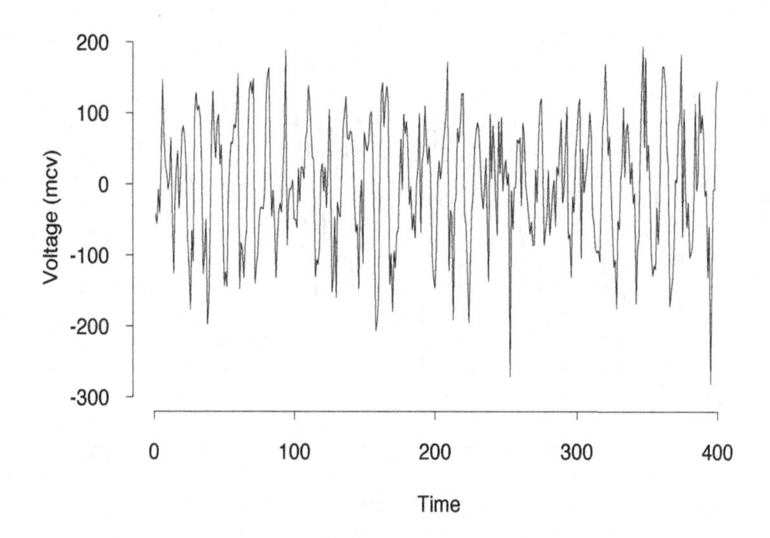

Figure 2.2 *A section of an EEG trace.*

zero shown in the sample autocorrelations is consistent with stationary autoregressive components with complex roots. The sample partial autocorrelations are strongly negative at lags between 2 and 7 or 8, but appear to drop off thereafter, suggesting an autoregression of order $p = 7$ or $p = 8$.

An AR(8) model is explored as an initial model for these data; $p = 8$ and $y_{9:400}$ represent the final $n = 392$ observations, the first 8 being conditioned upon for initial values. The posterior multivariate Student-t distribution has 384 degrees of freedom, and so it is practically indistinguishable from a normal; it has mean

$$\hat{\boldsymbol{\phi}} = (0.27, 0.07, -0.13, -0.15, -0.11, -0.15, -0.23, -0.14)'$$

and approximately common standard deviations at 0.05. This illustrates quite typical variation. The innovations standard deviation has posterior estimate $s = 61.52$.

We fix $\boldsymbol{\phi} = \hat{\boldsymbol{\phi}}$ to explore the model based on this point estimate of the parameter vector. The corresponding autoregressive polynomial equation $\Phi(u) = 0$ has four pairs of complex conjugate roots, whose corresponding moduli and wavelength pairs (r_j, λ_j) are, in order of decreasing modulus,

$$(0.97, 12.73); \quad (0.81, 5.10); \quad (0.72, 2.99); \quad (0.66, 2.23).$$

The first term here represents the apparent cyclical pattern of wavelength around 12 to 13 time units, and has a damping factor close to unity, indicating a rather persistent waveform; the half-life is about $h = 23$, i.e., 0.97^h

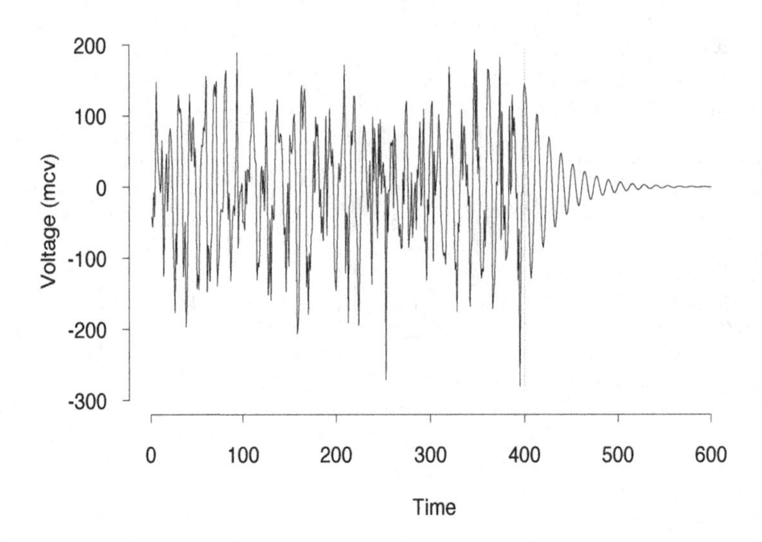

Figure 2.3 *EEG trace and forecast function from end of series.*

decays to about 0.5 at $h = 23$, so that, with zero future innovations, the amplitude of this waveform is expected to decay to half the starting level in about two full cycles. By comparison, the three other, higher frequency components have much faster decay rates. The pattern here is quite typical of quasicyclical series. The high frequency terms, close to the Nyquist frequency limit, capture very short run oscillations of very low magnitude, essentially tailoring the model to low level noise features in the data, rather than representing meaningful cyclical components. At time $T = 400$, or $t = n = 392$, the current state vector $\mathbf{x}_t$, together with the estimated parameter vector $\hat{\phi}$, imply a forecast function of the form given in (2.8), in which the components, four damped sinusoids, have relative amplitudes $2a_{tj}$ of approximately 157.0, 6.9, 18.0, and 7.0. So, the first component of wavelength around 12.73 is quite dominant at this time epoch (as it is over the full span of the data) both in terms of the initial amplitude and in terms of a much lower decay rate. Thus, the description of the series as close to a time-varying sine wave is reinforced.

Figure 2.3 displays the data and the forecast function from the end of the series over the next $h = 200$ times based on the estimated value $\hat{\phi}$. Figure 2.4 represents a more useful extrapolation, displaying a single "sampled future" based on estimated parameter values. This is generated simply by successively simulating future values $y_{T+h} = \sum_{j=1}^{p} \hat{\phi}_j y_{T+h-j} + \epsilon_{T+h}$ over $h = 1, 2, \ldots,$ where the ϵ_{T+h}s are drawn from $N(\cdot|0, s^2)$, and substituting sampled values as regressors for the future. This gives some flavor of likely

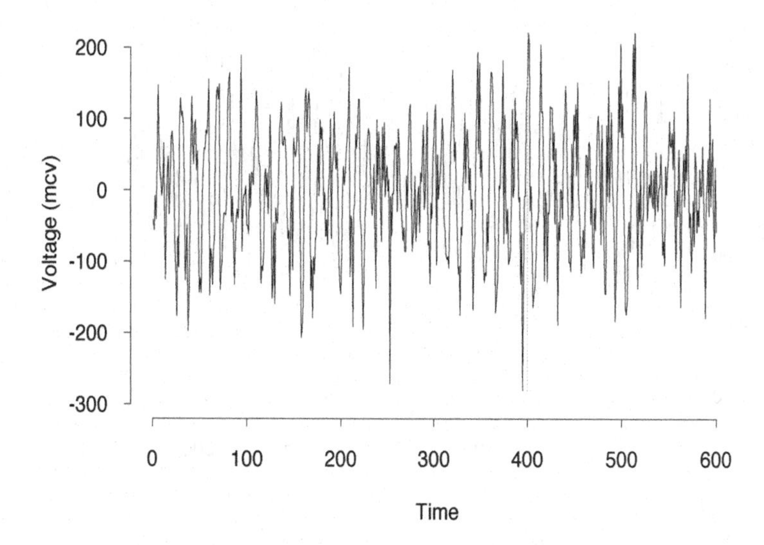

Figure 2.4 *EEG trace and sampled future conditional on parameter estimates* $(\hat{\phi}, s^2)$.

development, and the form is apparently similar to that of the historical data, suggesting a reasonable model description. These forecasts do not account for uncertainties about the estimated parameters $(\hat{\boldsymbol{\phi}}, s^2)$, so they do not represent formal predictive distributions, though are quite close approximations. This point is explored further below. Additional insight into the nature of the likely development, and also of aspects of model fit, are often gleaned by repeating this exercise, generating and comparing small sets of possible futures.

2.3.3 Simulation of Posterior Distributions

Inferences for other functions of model parameters and formal forecast distributions may be explored via simulation. Suppose interest lies in more formal inference about, for example, the period λ_1 of the dominant cyclical component in the above analysis of the EEG series, and other features of the structure of the roots of the AR polynomial. Though the posterior for $(\boldsymbol{\phi}, v)$ is analytically manageable, that for the reciprocal roots of the implied characteristic polynomial is not; posterior simulation may be used to explore these analytically intractable distributions. Similarly, sampled futures incorporating posterior uncertainties about $(\boldsymbol{\phi}, v)$ may be easily computed.

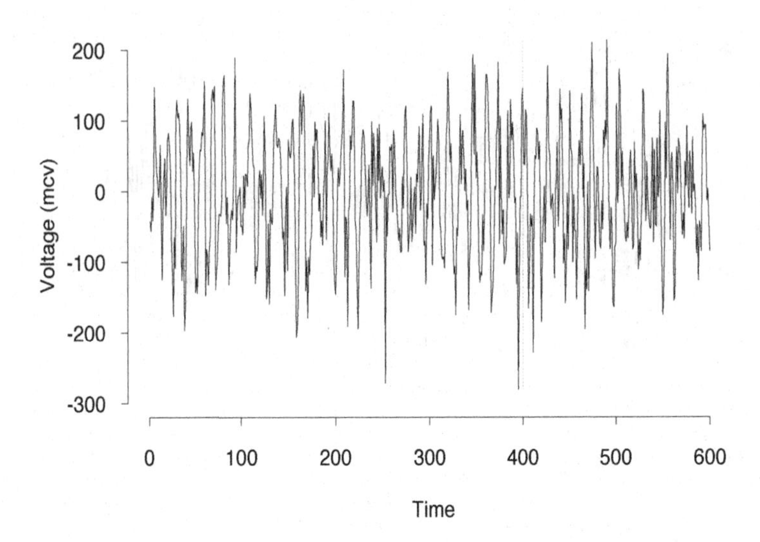

Figure 2.5 *EEG trace and sampled future from full posterior predictive distribution.*

Example 2.3 *EEG data analysis (continued).* A total number of 5,000 draws were made from the normal/inverse-gamma posterior distribution for $(\boldsymbol{\phi}, v)$. For each such draw, a sampled future $y_{T+1}, \ldots, y_{T+h}$, for any horizon h, was obtained as described before, but now based on the simulated values $(\boldsymbol{\phi}, v)$ at each sample, rather than the estimates $(\hat{\boldsymbol{\phi}}, s^2)$. This delivers a sample of size 5,000 from the full joint posterior predictive distribution for $(y_{T+1}, \ldots, y_{T+h})$. Averaging values across samples leads to a Monte Carlo approximation to the forecast function, which would provide a graph similar to that in Figure 2.3. Exploring sampled future values leads to graphs like Figure 2.5, where the sampled future values were computed based on one of the 5,000 draws from the posterior of $(\boldsymbol{\phi}, v)$. In this analysis, the additional uncertainties are small and have slight effects; other applications may be different.

Turn now to the inference on the AR polynomial roots $\boldsymbol{\alpha}$. Each posterior draw $(\boldsymbol{\phi}, v)$ delivers a corresponding root vector $\boldsymbol{\alpha}$ which represents a random sample from the full posterior $p(\boldsymbol{\alpha}|\mathbf{y})$. Various features of this posterior sample for $\boldsymbol{\alpha}$ may be summarized. Note first the inherent identification issue, that the roots are unidentifiable as the AR model is unchanged under permutations of the subscripts on the α_i. One way around this difficulty is to consider inference on roots ordered by modulus or frequency (note that the case of real roots formally corresponds to zero frequency). For example, the dominant component of the EEG model has been identified as that cor-

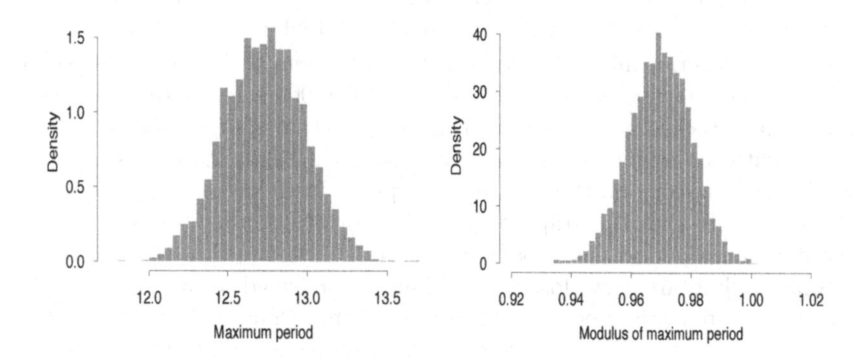

Figure 2.6 *Posterior for maximum period of sinusoidal components of the EEG series (left panel) and posterior for modulus of the damped sinusoidal component of maximum period in the EEG analysis (right panel).*

responding to the complex conjugate roots with the largest period around 12 to 13 time units. Ordering the complex values of each sampled set of roots leads to those with the largest period representing a sample from the posterior distribution for the period of the dominant component, and similarly for the samples of the corresponding modulus. The left and right panels in Figure 2.6 display the corresponding histograms in this analysis.

Note that no mention of stationarity has been made in this analysis. The reference posterior for $\boldsymbol{\phi}$, a multivariate Student-t distribution, is unconstrained and does not theoretically respect a constraint such as stationarity. In some applications, it may be physically meaningful and desirable to impose such an assumption, and the analysis should then be modified; theoretically, the prior for $(\boldsymbol{\phi}, v)$ should be defined as zero outside the stationarity region, whatever the form inside. In a simulation context, the simplest approach is to proceed as in the unconstrained analysis, but to simply reject sampled $(\boldsymbol{\phi}, v)$ values if the $\boldsymbol{\phi}$ vector lies outside the stationarity region, a condition that is trivially checked by evaluating the roots of the implied AR polynomial. In cases where the data/model match really supports a stationary series, the rejection rate will be low, providing a reasonable and efficient approximation to the analysis. In other cases, evidence of nonstationary features may lead to higher rejection rates and inefficient analyses; other methods are then needed. Some references below indicate work along these lines. Of course, an overriding consideration is the suitability of a strict stationarity assumption to begin with; if the series, conditional on the appropriateness of the assumed model, is really consistent with stationarity, this should be evidenced automatically in the posterior for the AR parameters, whose mass should be concentrated on values consistent with stationarity. This is true in the unconstrained EEG data analysis. Here the estimated AR polynomial root structure (at the reference posterior mean $\hat{\phi}$) has all reciprocal roots with moduli less than unity, suggesting stationarity. In addition, the 5,000 samples from the posterior can be checked similarly; in fact, the actual sample drawn has no values with roots violating stationarity, indicating high posterior probability (probability one on the Monte Carlo posterior sample) on stationarity. However, note that the graph of posterior distribution for the modulus of the maximum period (right panel in Figure 2.6) shows that some of these sampled values are very close to one. This is confirmed by the AR analysis with structured priors presented later in Example 2.5.

In other applications, sampling the posterior may give some values outside the stationary region; whatever the values, this provides a Monte Carlo approach to evaluating the posterior probability of a stationary series, conditional on the assumed AR model form.

2.3.4 Order Assessment

Analysis may be repeated for different values of model order p, it being useful and traditional to explore variations in inferences and predictions across a range of increasing values. Larger values of p are limited by the sample size, and fitting high order models to only moderate data sets produces meaningless reference posterior inferences; a large number of parameters, relative to sample size, can be entertained only with informed and proper prior distributions for those parameters, such as smoothness priors and others mentioned below. Otherwise, increasing p runs into the usual regression problems of overfitting and collinearity.

Simply proceeding to sequentially increase p and exploring fitted residuals, changes in posterior parameter estimates, and so forth is a very valuable exercise. Various numerical summaries may be easily computed as adjunct to this, the two most widely known and used being the so-called Akaike's information criterion, or AIC, and the Bayesian information criterion, or BIC (Akaike 1969; Akaike 1974; Schwarz 1978). The AIC and BIC are now described together with a more formal, reference Bayesian measure of model fit. As we are comparing models with differing numbers of parameters, we do so based on a common sample size; thus, we fix a maximum order p^* and, when comparing models of various orders $p \leq p^*$, we do so in conditional reference analyses using the latter $n = T - p^*$ of the full T observations in the series.

For a chosen model order p, explicit dependence on p is made by writing $\hat{\phi}_p$

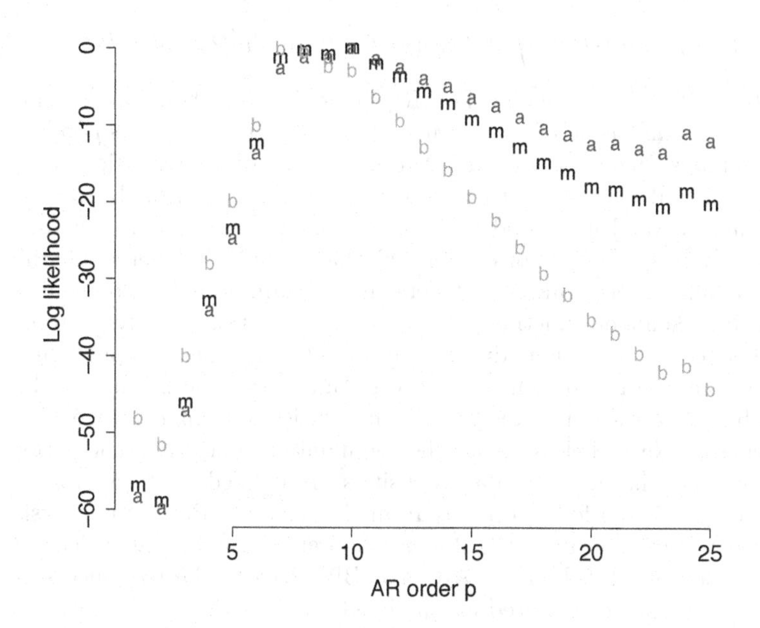

Figure 2.7 *Log-likelihood function for AR model order, computed from marginal data densities (labeled m), together with -AIC/2 criterion (labeled a) and -BIC/2 criterion (labeled b).*

for the MLE of the AR parameters, and s_p^2 for the corresponding estimate of innovations variance, i.e., the residual sum of squares divided by $n - p$. For our purposes, the AIC measure of model fit is taken as $2p + n\log(s_p^2)$, while the BIC is taken as $\log(n)p + n\log(s_p^2)$. Values of p leading to small AIC and BIC values are taken as indicative of relatively good model fits, within the class of AR models so explored (they may, of course, be poor models compared with other classes). Larger values of p will tend to give smaller variance estimates which decreases the second term in both expressions here, but this decrease is penalized for parameter dimension by the first term. BIC tends to choose simpler models than AIC. For the EEG series, negated AIC and BIC values, normalized to zero at the maximum, appear in Figure 2.7, based on $p^* = 25$. Also displayed there is a plot of the corresponding log-likelihood function for model order, computed as follows.

In a formal Bayesian analysis, the order p is viewed as an uncertain parameter, and so any prior over p is updated via a likelihood function proportional to the marginal data density

$$p(y_{(p^*+1):T}|y_{1:p^*}, p) = \int p(y_{(p^*+1):T}|\boldsymbol{\phi}_p, v, y_{1:p^*})p(\boldsymbol{\phi}_p, v)d\boldsymbol{\phi}_p dv,$$

where $p(\boldsymbol{\phi}_p, v)$ is the prior under the AR(p) model, and it should be remembered that the dimension of $\boldsymbol{\phi}_p$ depends on p. Given proper priors $p(\boldsymbol{\phi}_p, v)$ across the interesting range of order values $p \leq p^*$, a direct numerical measure of relative fit is available through this collection of marginal densities which defines a valid likelihood function for the model order. Doing this, however, requires a proper prior $p(\boldsymbol{\phi}_p, v)$ that naturally depends on the parameter dimension p, and this dependency is important in determining the resulting likelihood function. The use of the traditional reference prior invalidates these calculations due to impropriety. Alternative approaches to constructing proper but, in some senses, uninformative priors may be pursued but the critical need for priors to be consistent as model dimension varies remains. Nevertheless, under the commonly assumed reference prior $p(\boldsymbol{\phi}_p, v) \propto 1/v$, the marginal data densities are defined up to a proportionality constant and follow directly from the reference Bayesian analysis of the linear regression model (1.22) in Chapter 1. The marginal density values are closely related to the AIC and BIC values. The reference log-likelihood function so computed for the EEG series, with $p^* = 25$, appears in Figure 2.7. This reference log-likelihood function and the usual AIC and BIC criteria suggest orders between 8 and 10 as preferable, hence the earlier analysis was based on $p = 8$. Various alternatives based on different priors give similar results, at least in terms of identifying $p = 8$ or 9 as most appropriate. We note also that formal computation of, for example, predictive inferences involving averaging over p with respect to computed posterior probabilities on model order is possible, in contexts where proper priors for $(\boldsymbol{\phi}_p, v)$ are defined across models.

2.3.5 Initial values and Missing Data

The above analysis partitions the full data series $y_{1:T}$ into the p initial values $y_{1:p}$ and the final $n = T - p$ values $y_{(p+1):T}$, and is then conditional on $y_{1:p}$. Turn now to the unconditional analysis, in which the full likelihood function for $(\boldsymbol{\phi}, v)$ is

$$\begin{aligned} p(y_{1:T}|\boldsymbol{\phi}, v) &= p(y_{(p+1):T}|\boldsymbol{\phi}, v, y_{1:p})p(y_{1:p}|\boldsymbol{\phi}, v) \\ &= p(y_{(p+1):T}|\boldsymbol{\phi}, v, \mathbf{x}_p)p(\mathbf{x}_p|\boldsymbol{\phi}, v), \end{aligned} \quad (2.25)$$

with $\mathbf{x}_p = (y_p, \ldots, y_1)'$. The conditional analysis simply ignores the second component in (2.25). Whether or not this is justifiable or sensible depends on context, as follows.

In some applications, it is appropriate to assume some form of distribution for the initial values $\mathbf{x}_p$ that does not, in fact, depend on $(\boldsymbol{\phi}, v)$ at all. For example, it is perfectly reasonable to specify a model in which, say, the distribution $N(\mathbf{x}_p|\mathbf{0}, \mathbf{A})$ is assumed for some specified variance matrix $\mathbf{A}$. In such cases, (2.25) reduces to the first component alone, and the conditional analysis is exact.

Otherwise, when $p(\mathbf{x}_p|\boldsymbol{\phi}, v)$ actually depends on $(\boldsymbol{\phi}, v)$, there will be a contribution to the likelihood from the initial values, and the conditional analysis is only approximate. Note however that, as the series length T increases, the first term of the likelihood, based on $n = T - p$ observations, becomes more and more dominant; the effect of the initial values in the second likelihood factor is fixed based on these values, and does not change with n. On a log-likelihood scale, the first factor behaves in expectation as $o(n)$, and so the conditional and unconditional analyses are asymptotically the same. In real problems with finite n, but in which p is usually low compared to n, experience indicates that the agreement is typically close even with rather moderate sample sizes. It is therefore common practice, and completely justifiable in applications with reasonable data sample sizes, to adopt the conditional analysis.

The situation has been much studied under the stationarity assumption, and a variation of the reference Bayesian analysis is explored here. Under stationarity, any subset of the data will have a marginal multivariate normal distribution, with zero mean and a variance matrix whose elements are determined by the model parameters. In particular, the initial values follow $N(\mathbf{x}_p|\mathbf{0}, v\mathbf{A}(\boldsymbol{\phi}))$ where the $p \times p$ matrix $\mathbf{A}(\boldsymbol{\phi})$ depends (only) on $\boldsymbol{\phi}$ through the defining equations for autocorrelations in AR models. So (2.25), as a function of $(\boldsymbol{\phi}, v)$, is

$$p(y_{1:T}|\boldsymbol{\phi}, v) \propto v^{-T/2}|\mathbf{A}(\boldsymbol{\phi})|^{-1/2} \exp\left(-Q(y_{1:T}, \boldsymbol{\phi})/2v\right), \tag{2.26}$$

where $Q(y_{1:T}, \boldsymbol{\phi}) = \sum_{t=p+1}^{T}(y_t - \mathbf{f}_t'\boldsymbol{\phi})^2 + \mathbf{x}_p'\mathbf{A}(\boldsymbol{\phi})^{-1}\mathbf{x}_p$. As developed in Box, Jenkins, Reinsel, and Ljung (2015) Chapter 7, this reduces to a quadratic form $Q(y_{1:T}, \boldsymbol{\phi}) = a - 2\mathbf{b}'\boldsymbol{\phi} + \boldsymbol{\phi}'\mathbf{C}\boldsymbol{\phi}$, where the quantities $a, \mathbf{b}, \mathbf{C}$ are easily calculable, as follows. Define the symmetric $(p+1) \times (p+1)$ matrix $\mathbf{D} = \{D_{ij}\}$ by elements $D_{ij} = \sum_{r=0}^{T+1-j-i} y_{i+r}y_{j+r}$; then $\mathbf{D}$ is partitioned as

$$\mathbf{D} = \begin{pmatrix} a & -\mathbf{b}' \\ -\mathbf{b} & \mathbf{C} \end{pmatrix}.$$

One immediate consequence of this is that, if we ignore the determinant factor $|\mathbf{A}(\boldsymbol{\phi})|$, the likelihood function is of standard linear model form. The traditional reference prior $p(\boldsymbol{\phi}, v) \propto v^{-1}$ induces a normal/inverse-gamma posterior, for example; other normal/inverse-gamma priors might be used similarly. In the reference case, full details of the posterior analysis can be worked through by the reader. The posterior mode for $\boldsymbol{\phi}$ is now $\hat{\boldsymbol{\phi}}^* = \mathbf{C}^{-1}\mathbf{b}$. For the EEG series, the calculations lead to

$$\hat{\boldsymbol{\phi}}^* = (0.273, 0.064, -0.128, -0.149, -0.109, -0.149, -0.229, -0.138)'$$

to three decimal places. The approximate value based on the conditional analysis is

$$\hat{\boldsymbol{\phi}} = (0.272, 0.068, -0.130, -0.148, -0.108, -0.148, -0.226, -0.136)',$$

earlier quoted to only two decimal places in light of the corresponding posterior standard deviations around 0.05 in each case. The differences, in the third decimal place in each case, are negligible, entirely so in the context of spread of the posterior. Here we are in the (common) context where T is large enough compared to p, and so the effect of the initial values in (2.25) is really negligible. Repeating the analysis with just the first $T = 100$ EEG observations, the elements of $\hat{\boldsymbol{\phi}}$ and $\hat{\boldsymbol{\phi}}^*$ differ by only about 0.01, whereas the associated posterior standard errors are around 0.1; the effects become more marked with smaller sample sizes, though are still well within the limits of posterior standard deviations with much smaller values of T. In other applications, the effects may be more substantial.

Ignoring the determinant factor can be justified by the same asymptotic reasoning. Another justification is based on the use of an alternative reference prior: that based on Jeffreys' rule (Jeffreys 1961). Jeffreys' rule consists on using the density $p(\boldsymbol{\theta}) \propto \sqrt{|\mathcal{I}(\boldsymbol{\theta}|\mathbf{y})|}$ as a prior for $\boldsymbol{\theta}$, where $\mathcal{I}(\boldsymbol{\theta}|\mathbf{y})$ is the information matrix whose (i,j)-th element is given by

$$(\mathcal{I}(\boldsymbol{\theta}|\mathbf{y}))_{i,j} = -E[\partial^2 \log\{p(\mathbf{y}|\boldsymbol{\theta})\}/\partial\theta_i\partial\theta_j],$$

with $p(\mathbf{y}|\boldsymbol{\theta})$ the likelihood function for $\boldsymbol{\theta}$. The key feature of this prior is that it is invariant under reparameterization of $\boldsymbol{\theta}$. In this case, as shown in Box, Jenkins, Reinsel, and Ljung (2015), Jeffreys' prior is approximately $p(\boldsymbol{\phi}, v) \propto |\mathbf{A}(\boldsymbol{\phi})|^{1/2} v^{-1/2}$; this results in cancellation of the determinant factor so the above analysis is exact. Otherwise, under different prior distributions, the exact posterior involves the factor $|\mathbf{A}(\boldsymbol{\phi})|$, a complicated polynomial function of $\boldsymbol{\phi}$. However, $|\mathbf{A}(\boldsymbol{\phi})|$ can be evaluated at any specified $\boldsymbol{\phi}$ value, and numerical methods can be used to analyze the complete posterior. Numerical evaluation of the exact MLE is now a standard feature in some software packages. Bayesian analysis using Monte Carlo methods is also easy to implement in this framework.

2.3.6 Imputing Initial Values via Simulation

Introduce the truly uncertain initial values $\mathbf{x}_0 = (y_0, y_{-1}, \ldots, y_{-(p-1)})'$. Adjust the earlier conditional analysis to be based on all T observations $y_{1:T}$ and now to be conditional on these (imaginary) initial values $\mathbf{x}_0$. Then, whatever the prior, we have the posterior $p(\boldsymbol{\phi}, v|y_{1:T}, \mathbf{x}_0)$. In the reference analysis, we have a normal/inverse-gamma posterior now based on all T observations rather than just the last $n = T - p$, with obvious modifications. Note that this posterior can be simulated to deliver draws for $(\boldsymbol{\phi}, v)$ conditional on any specific initial vector $\mathbf{x}_0$. This can be embedded in an iterative simulation of the full joint posterior $p(\boldsymbol{\phi}, v, \mathbf{x}_0|y_{1:T})$ if, in addition, we can sample $\mathbf{x}_0$ vectors from the conditional posterior $p(\mathbf{x}_0|\boldsymbol{\phi}, v, y_{1:T})$ for any specified $(\boldsymbol{\phi}, v)$ parameters.

In the case of a stationary series, stationarity and the linear model form imply reversibility with respect to time; that is, the basic AR model holds backward, as well as forward, in time. Hence, conditional on $(\boldsymbol{\phi}, v)$ and future series values $y_{t+1}, y_{t+2}, \ldots$, the current value y_t follows the distribution $N(y_t|\mathbf{g}_t'\boldsymbol{\phi}, v)$ where $\mathbf{g}_t = rev(\mathbf{x}_{t+p}) = (y_{t+1}, \ldots, y_{t+p})'$; here the operator $rev(\cdot)$ simply reverses the elements of its vector argument. Applying this to the initial values at $t = 0, -1, \ldots$, leads to

$$p(\mathbf{x}_0|\boldsymbol{\phi}, v, y_{1:T}) = \prod_{t=0}^{-(p-1)} N(y_t|\mathbf{g}_t'\boldsymbol{\phi}, v).$$

Hence, given $(\boldsymbol{\phi}, v)$, a vector $\mathbf{x}_0$ is simulated by sequentially sampling the individual component normal distributions in this product: first draw y_0 given the known data $\mathbf{x}_p$ and the parameters; then substitute the sampled value y_0 as the first element of the otherwise known data vector $\mathbf{x}_{p-1}$, and draw y_{-1}; continue this way down to $y_{-(p-1)}$. This is technically similar to the process of simulating a future of the series illustrated earlier; now we are simulating the past.

This approach is both trivially implemented and practically satisfying as it provides, modulo the Monte Carlo simulation, exact analysis. Further extensions of basic AR models to incorporate various practically relevant additional features lead to Markov chain simulations as natural, and typically necessary, approaches to analysis, so that dealing with the starting value issue in this framework makes good sense.

It should also be clear that the same principle applies to problems of missing data. For any set of indices t such that the values y_t are missing (at random, that is, the reasons for missing data do not have a bearing on the values of the model parameters), then iterative simulation analysis can be extended and modified to incorporate the missing values as additional uncertain quantities to be estimated. Further details can be worked out in the framework here, as with the missing initial values above, and details are left to the reader. We revisit missing values later in the context of general state-space models.

2.4 Further Issues in Bayesian Inference for AR Models

2.4.1 Sensitivity to the Choice of Prior Distributions

Additional analyses explore inferences based on longer order AR models with various proper priors for the AR coefficients. One interest is in exploring the sensitivity of the earlier, reference inferences under ranges of proper and perhaps more plausible prior assumptions. In each case the model is based on (a maximum lag) $p = 25$, assuming that higher order models would have negligible additional coefficients and that, in any case, the higher order coefficients in the model are likely to decay. The two priors for $\boldsymbol{\phi}$ are centered around zero, inducing shrinkage of the posterior distributions towards the prior means of zero for all parameters. In each case, the first p values of $y_{1:T}$ are fixed to provide conditional analyses comparable to that earlier discussed at length.

2.4.1.1 Analysis Based on Normal Priors

A first analysis assumes a traditional prior with the coefficients i.i.d. normal; the joint prior is $N(\boldsymbol{\phi}|\mathbf{0}, w\mathbf{I}_p)$ for some scalar variance w, and so it induces shrinkage of the posterior toward the prior mean of zero for all parameters. The hyperparameter w will be estimated together with the primary parameters $(\boldsymbol{\phi}, v)$ via Gibbs sampling to simulate the full posterior for $(\boldsymbol{\phi}, v, w)$. We assume prior independence of v and w and adopt noninformative priors proportional to v^{-1} and w^{-1}, respectively. Posterior simulations draw sequentially from the following conditional posterior distributions, easily deduced from the likelihood form $N(\mathbf{y}|\mathbf{F}'\boldsymbol{\phi}, v\mathbf{I}_{T-p})$, with $\mathbf{y} = (y_T, \ldots, y_{p+1})'$, and $\mathbf{F} = [\mathbf{f}_T, \ldots, \mathbf{f}_{p+1}]$, where $\mathbf{f}_t = (y_{t-1}, \ldots, y_{t-p})'$, the prior, and the normal linear model theory reviewed in Chapter 1:

- Given (v, w), posterior for $\boldsymbol{\phi}$ is $N(\boldsymbol{\phi}|\hat{\boldsymbol{\phi}}, \mathbf{B})$ where $\mathbf{B}^{-1} = w^{-1}\mathbf{I}_p + v^{-1}\mathbf{F}\mathbf{F}'$ and $\hat{\boldsymbol{\phi}} = \mathbf{B}v^{-1}\mathbf{F}\mathbf{y}$.
- Given $(\boldsymbol{\phi}, w)$, posterior for v^{-1} is $G(v^{-1}|(T-p)/2, \mathbf{e}'\mathbf{e}/2)$ based on residual vector $\mathbf{e} = \mathbf{y} - \mathbf{F}'\boldsymbol{\phi}$.
- Given $(\boldsymbol{\phi}, v)$, posterior for w^{-1} is $G(w^{-1}|p/2, \boldsymbol{\phi}'\boldsymbol{\phi}/2)$.

For the EEG series, Figure 2.8 graphs the approximate posterior means of the ϕ_js based on a Monte Carlo sample of size 5,000 from the simulation analysis so specified. This sample is saved following burn-in of 500 iterations. Also plotted are the reference posterior means with two posterior standard deviation intervals, for comparison. Some shrinkage of the coefficients is evident, though apparently not dramatic in extent, and the posterior means are not incomparable with the reference values, indicating

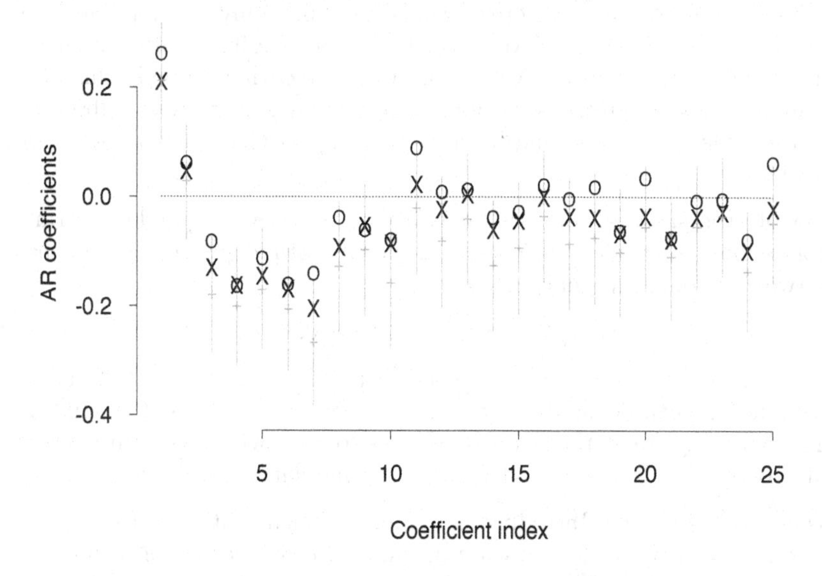

Figure 2.8 *Estimates of ϕ in EEG analyses. The vertical bars indicate approximate 95% posterior intervals for the ϕ_j from the reference analysis, centered about reference posterior means. The symbols X indicate approximate posterior means from the analysis based on independent normal priors. Symbols O indicate approximate posterior means from the analysis based on the two-component, normal mixture priors.*

some robustness to prior specification. Inferences and forecasts based on the normal prior will not differ substantially from those based on the reference prior. In this analysis, the posterior for the shrinkage parameter $\sqrt{w}$ is unimodal, centered around 0.12 with mass predominantly concentrated in the range 0.08–0.16.

2.4.1.2 Discrete Normal Mixture Prior and Subset Models

Further analysis illustrates priors inducing differential shrinkage effects across the ϕ_j parameters; some of the ϕ_js may indeed be close to zero, others are quite clearly distinct from zero, and a prior view that this may be the case can be embodied in standard modifications of the above analysis. One such approach uses independent priors conditional on individual scale factors, namely $N(\phi_j|0, w/\delta_j)$, where each weight δ_j is a random quantity to be estimated. For example, a model in which only one or two of the ϕ_js are really significant is induced by weights δ_j close to unity for those parameters, the other weights being relatively large resulting in priors and posteriors concentrated around zero for the negligible weights. This links to the concept of subset autoregressions, in which only a few parameters at specific lags are really relevant, the others, at possibly intervening lags, being zero or close to zero. A class of priors for $\boldsymbol{\phi}$ that embody this kind of qualitative view provides for automatic inference on relevant subsets of nonnegligible parameters and, effectively, addresses the variable selection question.

Probably the simplest approach extends the case of independent normal priors above, in which each $\delta_j = 1$, to the case of independent priors that are two-component normals, namely

$$\pi N(\phi_j|0, w) + (1 - \pi)N(\phi_j|0, w/L),$$

where π is a probability and L a specified precision factor. If $L >> 1$, the second normal component is very concentrated around zero, so this mixture prior effectively states that each ϕ_j is close to zero, with probability $1 - \pi$, and is otherwise drawn from the earlier normal with variance w.

Assume L is specified. Introduce indicators u_j such that $u_j = 1$ or $u_j = 0$ according to whether ϕ_j is drawn from the first or the second of the normal mixture components. These u_js are latent variables that may be introduced to enable the simulation analysis. Write $\mathbf{u} = (u_1, \ldots, u_p)$ and, for any set of values $\mathbf{u}$, write $\delta_j = u_j + (1 - u_j)L$, so that $\delta_j = 1$ or L; also, define the matrix $\Delta = \text{diag}(\delta_1, \ldots, \delta_p)$. Further, write $k = \sum_{j=1}^{p} u_j$ for the number of coefficients drawn from the first normal component; k can be viewed as the number of nonnegligible coefficients, the others being close to zero. Note that, given π, k has a prior binomial distribution with success probability π.

For completeness and robustness, π is usually viewed as uncertain too; in the analysis below, π is assigned a beta prior, $Be(\pi|a, b)$, independently of the other random quantities in the model. This implies, among other things, a beta-binomial marginal prior for the number k of significant coefficients, namely

$$p(k) = \binom{p}{k} \frac{B(a + k, b + p - k)}{B(a, b)},$$

over $k = 0 : p$, where $B(\cdot, \cdot)$ is the beta function (see Appendix). Finally, we assume priors of the form $p(v) \propto 1/v$ and $p(w) \propto 1/w$ on v and w.

Under this model and prior specification, the various conditional posterior distributions to be used in Gibbs sampling of the full posterior for $(\boldsymbol{\phi}, v, w, \mathbf{u}, \pi)$ are as follows:

- Given $(v, w, \mathbf{u}, \pi)$, posterior for $\boldsymbol{\phi}$ is $N(\boldsymbol{\phi}|\mathbf{b}, \mathbf{B})$ where $\mathbf{B}^{-1} = w^{-1}\Delta + v^{-1}\mathbf{F}\mathbf{F}'$ and $\mathbf{b} = \mathbf{B}v^{-1}\mathbf{F}\mathbf{y}$.

- Given $(\boldsymbol{\phi}, w, \mathbf{u}, \pi)$, posterior for v^{-1} is $G(v^{-1}|(T-p)/2, \mathbf{e}'\mathbf{e}/2)$ based on residual vector $\mathbf{e} = \mathbf{y} - \mathbf{F}'\boldsymbol{\phi}$.
- Given $(\boldsymbol{\phi}, v, \mathbf{u}, \pi)$, posterior for w^{-1} is $G(w^{-1}|p/2, q/2)$ with scale factor defined by $q = \sum_{j=1}^{p} \phi_j^2 \delta_j$.
- Given $(\boldsymbol{\phi}, v, w, \pi)$, the u_j are independent with conditional posterior probabilities $\pi_j = Pr(u_j = 1|\boldsymbol{\phi}, v, w, \pi)$ given, in odds form, by

$$\frac{\pi_j}{1-\pi_j} = \frac{\pi}{1-\pi} \exp\left(-(1-L)\phi_j^2/2w\right)/\sqrt{L}.$$

- Given $(\boldsymbol{\phi}, v, w, \mathbf{u})$, posterior for π is beta, namely $Be(\pi|a+k, b+p-k)$, where $k = \sum_{j=1}^{p} u_j$.

Iterative sampling of these conditional distributions provides samples of $\boldsymbol{\phi}, v, w, \mathbf{u}$, and π for inference. The additional symbols in Figure 2.8 indicate the posterior means for the ϕ_js from such an analysis, again based on a simulation sample size of 5,000 from the full posterior; the analysis adopts $a = 1, b = 4$, and $L = 25$. We note little difference in posterior means relative to the earlier analyses, again indicating robustness to prior specifications as there is a good deal of data here.

The implied beta-binomial prior for k appears in Figure 2.9, indicating mild support for smaller values consistent with the view that, though there is much prior uncertainty, many of the AR coefficients are likely to be negligible. The posterior simulation analysis provides posterior samples of k, and the relative frequencies estimate the posterior distribution, as plotted in Figure 2.9. This indicates a shift to favoring values in the 5–15 range based on the data analysis under this specific prior structure; there is much uncertainty about k represented under this posterior, though the indication of evidence for more than just a few coefficients is strong. Additional information is available in the full posterior sample; it carries, for instance, Monte Carlo estimates of the posterior probabilities that individual coefficients ϕ_j are drawn from the first or second mixture component, simply the approximate posterior means of the corresponding indicators u_j. This information can be used to assess subsets of significant coefficients, as adjunct to exploring posterior estimates and uncertainties about the coefficients, as in Figure 2.8.

2.4.2 Alternative Prior Distributions

2.4.2.1 Scale-mixtures and Smoothness Priors

Analyses based on alternative priors may be similarly explored; some examples are mentioned here, and may be explored by the reader. For instance,

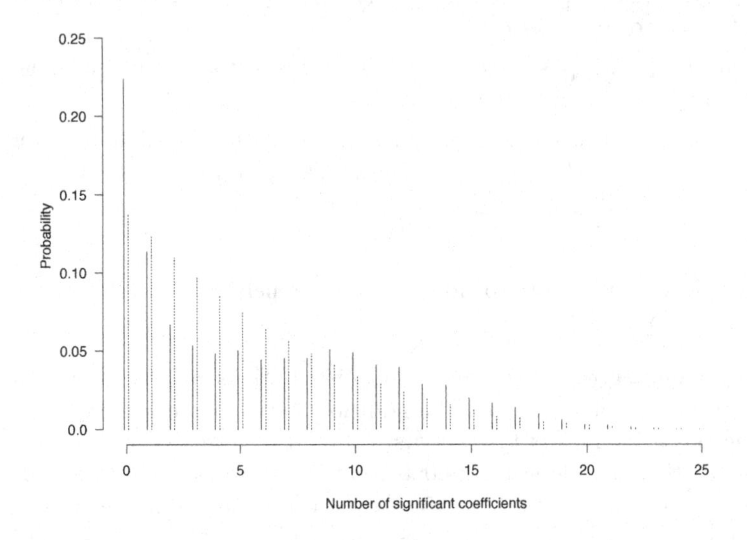

Figure 2.9 *Prior and approximate posterior distributions (dotted and solid lines, respectively) for the number of nonnegligible AR coefficients, out of the total* $p = 25$, *in the EEG analysis under the two-component mixture prior.*

the second analysis is an example of a prior constructed via scale-mixtures of a basic normal prior for the individual coefficients. The mixing distribution in that case is discrete, placing mass of π at $\delta_j = 1$ and $\delta_j = 25$. Other mixing distributions are common in applied Bayesian work, a key example being the class of gamma distributions. For instance, take the weights δ_j to be independently drawn from a gamma distribution with shape and scale equal to $k/2$ for some $k > 0$; this implies that the resulting marginal prior for each ϕ_j is a Student-t distribution with k degrees of freedom, mode at zero, and scale factor $\sqrt{w}$. This is, in some senses, a natural heavy-tailed alternative to the normal prior, assigning greater prior probabilities to ϕ_j values further from the prior location at zero. This can result in differential shrinkage, as in the case of the discrete normal mixture in the example.

Another class of priors incorporates the view that AR coefficients are unlikely to be large at higher lags, and ultimately decay toward zero. This kind of qualitative information may be important in contexts where p is large relative to expected sample sizes. This can be incorporated in the earlier normal prior framework, for example, by generalizing to independent priors $N(\phi_j|0, w/\delta_j)$, where the weights are now fixed constants that concentrate the priors around zero for larger lags j; an example would be $\delta_j = j^2$. Note that this may be combined with additional, random weights to develop decaying effects within a normal mixture prior, and is trivially implemented.

Traditional smoothness priors operate on differences of parameters at successive lags, so that priors for $|\phi_{j+1} - \phi_j|$ are also centered around zero to induce a smooth form of behavior of ϕ_j as a function of lag j, a traditional "distributed lag" concept; a smooth form of decay of the effects of lagged values of the series is often naturally anticipated. This is again a useful concept in contexts where long order models are being used. One example of a smoothness prior is given by generalizing the normal prior structure as follows. Take $N(\phi_1|0, w/\delta_1)$ and, for $j > 1$, assume conditional priors $N(\phi_j|\phi_{j-1}, w/\delta_j)$; here the δ_j weights are assumed to increase with lag j to help induce smoothness at higher lags. This specification induces a multivariate normal prior (conditional on the δ_j and w), $p(\boldsymbol{\phi}) = p(\phi_1) \prod_{j=2}^{p} p(\phi_j|\phi_{j-1}) = N(\boldsymbol{\phi}|\mathbf{0}, \mathbf{A}^{-1}w)$ (see Problem 13.), where the precision matrix $\mathbf{A} = \mathbf{H}'\Delta\mathbf{H}$ is defined by $\Delta = \text{diag}(\delta_1, \ldots, \delta_p)$ and

$$\mathbf{H} = \begin{pmatrix} 1 & 0 & 0 & \cdots & 0 & 0 \\ -1 & 1 & 0 & \cdots & 0 & 0 \\ 0 & -1 & 1 & \cdots & 0 & 0 \\ \vdots & \vdots & \cdots & \cdots & \vdots & \vdots \\ 0 & 0 & 0 & \cdots & -1 & 1 \end{pmatrix}.$$

Again, the δ_js may be either specified or random, or a mix of the two. Posterior inferences follow easily using iterative simulation, via straightforward modifications of the analyses above.

2.4.2.2 Priors Based on AR Latent Structure

Consider again the AR(p) model whose characteristic polynomial is given by $\Phi(u) = 1 - \phi_1 u - \cdots - \phi_p u^p$. The process is stable and stationary if the reciprocal roots of this polynomial have moduli less than unity. Now, consider the case in which there is a maximum number of C pairs of complex-valued reciprocal roots and a maximum number of R real-valued reciprocal roots, with $p = 2C + R$. The complex roots appear in pairs of complex conjugates, each pair having modulus r_j and wavelength λ_j—or equivalently, frequency $\omega_j = 2\pi/\lambda_j$—for $j = 1 : C$. Each real reciprocal root has modulus r_j, for $j = (C+1) : (C+R)$. Following Huerta and West (1999b), the prior structure given below can be assumed on the real reciprocal roots, i.e.,

$$\begin{aligned} r_j &\sim \pi_{r,-1} I_{(-1)}(r_j) + \pi_{c,0} I_0(r_j) + \pi_{r,1} I_1(r_j) \\ &\quad + (1 - \pi_{r,0} - \pi_{r,-1} - \pi_{r,1}) g_r(r_j), \end{aligned} \tag{2.27}$$

where $I(\cdot)$ denotes the indicator function, $g_r(\cdot)$ is a continuous distribution over $(-1, 1)$, and $\pi_{r,\cdot}$ are prior probabilities. The point masses at $r_j = \pm 1$

allow us to consider nonstationary unit roots. The point mass at $r_j = 0$ handles the uncertainty in the number of real roots, since this number may reduce below the prespecified maximum R. The default option for $g_r(\cdot)$ is the uniform $g_r(\cdot) = U(\cdot| - 1, 1)$, i.e., the reference prior for a component AR(1) coefficient r_j truncated to the stationary region. Similarly, for the complex reciprocal roots the following prior can be assumed:

$$\begin{aligned} r_j &\sim \pi_{c,0} I_0(r_j) + \pi_{c,1} I_1(r_j) + (1 - \pi_{c,1} - \pi_{c,0}) g_c(r_j), \\ \lambda_j &\sim h(\lambda_j), \end{aligned} \tag{2.28}$$

with $g_c(r_j)$ a continuous distribution on $0 < r_j < 1$ and $h(\lambda_j)$ a continuous distribution on $2 < \lambda_j < \lambda_u$, for $j = 1 : C$. The value of λ_u is fixed and by default it could be set to $T/2$. In addition, a so called "component reference prior" (Huerta and West 1999b) is induced by assuming a uniform prior on the implied AR(2) coefficients $2r_j \cos(2\pi/\lambda_j)$ and $-r_j^2$, but restricted to the finite support of λ_j for propriety. This is defined by $g_c(r_j) \propto r_j^2$, so that the marginal for r_j is $Be(\cdot|3, 1)$, and $h(\lambda_j) \propto \sin(2\pi/\lambda_j)/\lambda_j^2$ on $2 < \lambda_j < \lambda_u$. The probabilities $\pi_{c,0}$ and $\pi_{c,1}$ handle the uncertainty in the number of complex components and nonstationary unit roots, respectively. Uniform Dirichlet distributions are the default choice for the probabilities $\pi_{r,\cdot}$ and $\pi_{c,\cdot}$; this is

$$Dir(\pi_{r,-1}, \pi_{r,0}, \pi_{r,1}|1, 1, 1), \quad Dir(\pi_{c,0}, \pi_{c,1}|1, 1),$$

and an inverse-gamma prior is assumed for v, i.e., $IG(v|a, b)$.

A Markov chain Monte Carlo (MCMC) sampling scheme can be implemented to obtain samples from the posterior distribution of the model parameters

$$\boldsymbol{\theta} = \{(r_1, \lambda_1), \ldots, (r_C, \lambda_C), r_{(C+1):(C+R)}, \pi_{r,-1}, \pi_{r,0}, \pi_{r,1}, \pi_{c,0}, \pi_{c,1}, v, \mathbf{x}_0\},$$

with $\mathbf{x}_0 = (y_0, \ldots, y_{-(p-1)})'$, the p initial values. Specifically, if for any subset $\boldsymbol{\theta}^*$ of elements of $\boldsymbol{\theta}$, $\boldsymbol{\theta}\backslash\boldsymbol{\theta}^*$ denotes all the elements of $\boldsymbol{\theta}$ with the subset $\boldsymbol{\theta}^*$ removed, the MCMC algorithm can be summarized as follows:

- For each $j = (C+1) : (C+R)$, sample the real roots from the conditional marginal posterior $p(r_j|\boldsymbol{\theta}\backslash r_j, \mathbf{x}_0, y_{1:T})$. As detailed in Huerta and West (1999b), the conditional likelihood function for r_j provides a normal kernel in r_j, and so obtaining draws for each r_j reduces to sampling from a mixture posterior with four components, which can be easily done.
- For each $j = 1 : C$, sample the complex roots from the conditional marginal posterior $p(r_j, \lambda_j|\boldsymbol{\theta}\backslash(r_j, \lambda_j), \mathbf{x}_0, y_{1:T})$. Sampling from this conditional posterior directly is difficult, and so a reversible jump Markov chain Monte Carlo step is necessary. The reversible jump MCMC (RJMCMC) method introduced in Green (1995) permits jumps between pa-

rameter subspaces of different dimensions at each iteration. The method consists on creating a random sweep Metropolis-Hastings algorithm that adapts to changes in dimensionality. The RJMCMC algorithm is described in the Appendix.

- Sample $(\pi_{r,-1}, \pi_{r,0}, \pi_{r,1})$ and $(\pi_{c,0}, \pi_{c,1})$ from conditionally independent Dirichlet posteriors as detailed in Huerta and West (1999b).
- Sample v from an inverse-gamma distribution.
- Sample $\mathbf{x}_0$. Huerta and West (1999b) show the time reversibility property for AR models with unit roots, and so it is possible to sample the initial values $\mathbf{x}_0$ in a similar way to that described in Section 2.3.6.

Example 2.4 *RJMCMC for an AR(4) model with structured priors.* We consider the analysis of 100 observations simulated from an AR(2) process with a single pair of complex roots with modulus $r = 0.9$ and wavelength $\lambda = 8$. We fit an AR(4) to these data using the structured priors previously described. We set $C = 2$ and $R = 0$ and so two RJMCMC steps are needed to sample (r_1, λ_1) and (r_2, λ_2). Each RJMCMC step has a certain number of moves. For instance, if the chain is currently at $r_j = 0$, the following moves can be considered, each with probability 1/3:

- Remain at the origin.
- Jump at new values of the form $(1, \omega_j^*)$.
- Jump at new values of the form (r_j^*, ω_j^*).

The RJMCMC algorithm for the general AR(p) case is discussed in Huerta (1998). The analysis is implemented in the `ARcomp` public domain software* used for analysis here. This analysis leads to summaries for posterior inference in AR models with structured priors, and used to fit an AR(4) with structured priors to the simulated data.

In this analysis, two of the posterior summaries obtained are $Pr(p = 2|y_{1:T}) > 0.8$ and $Pr(C = 1|y_{1:T}) > 0.8$, indicating that the model adequately captures the quasiperiodic AR(2) structure in the data. Figure 2.10 displays the posterior distribution of (r_1, λ_1) (top panels) and (r_2, λ_2) (bottom panels). The histograms show the posterior distributions conditional on $r_1 \neq 0, 1$ and $r_2 \neq 0, 1$. We obtain $Pr(r_1 = 0|y_{1:T}) = 0$ and $Pr(r_2 = 0|y_{1:T}) = 0.98$, which are consistent with the fact that the data were simulated from an AR(2) process. In addition, the marginal posterior distributions for r_1 and λ_1 are concentrated around the true values $r = 0.9$ and $\lambda = 8$.

Note that the roots are not identified in the mathematical sense, since the

* `www.stat.duke.edu/research/software/west/`

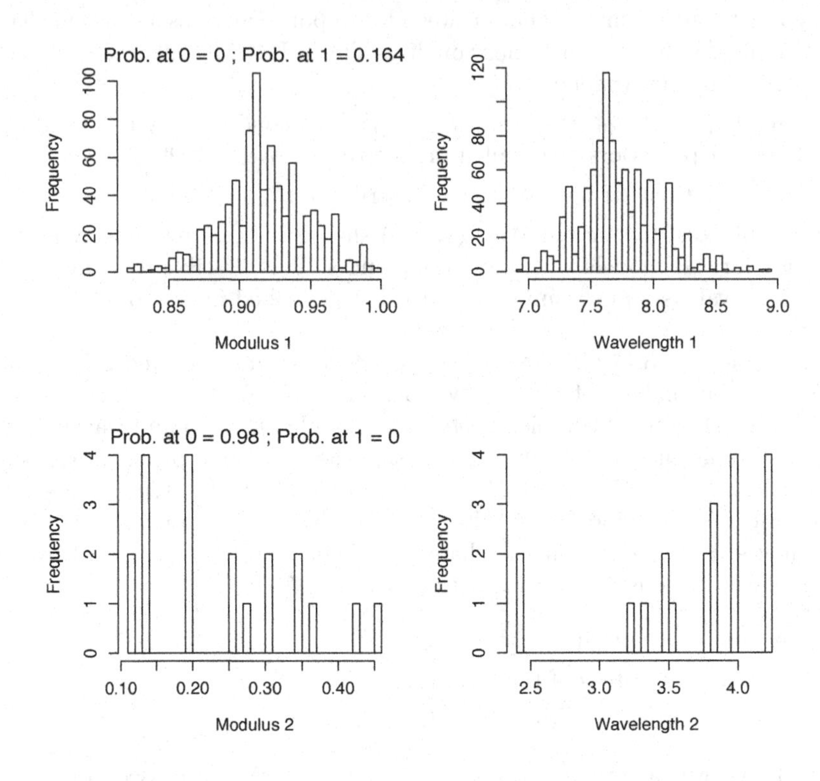

Figure 2.10 *Posterior distributions of* (r_1, λ_1) *and* (r_2, λ_2) *for the simulated data.*

AR(4) model is unchanged under permutations of the root index. For identifiability, posterior inferences are explored by ordering the roots by moduli, i.e., the roots were ordered by moduli before displaying the graphs in Figure 2.10, and so the bottom pictures correspond to the posterior distributions of the modulus and the wavelength of the root with the lowest modulus, while those pictures at the top correspond to the posterior distributions of the modulus and the wavelength of the root with the largest modulus.

Example 2.5 *Analysis of the EEG data with structured priors.* We now consider an analysis of the EEG data shown in Figure 2.2 using structured priors. In this example we set $C = R = 6$, and so the maximum model order is $p_{\max} = 2 * 6 + 6 = 18$. Figure 2.11 shows the posterior distributions of p, C, and R. This analysis gives highest posterior probability to a model with four pairs of characteristic complex roots and three real roots, or equivalently, a model with $p = 11$. However, there is considerable

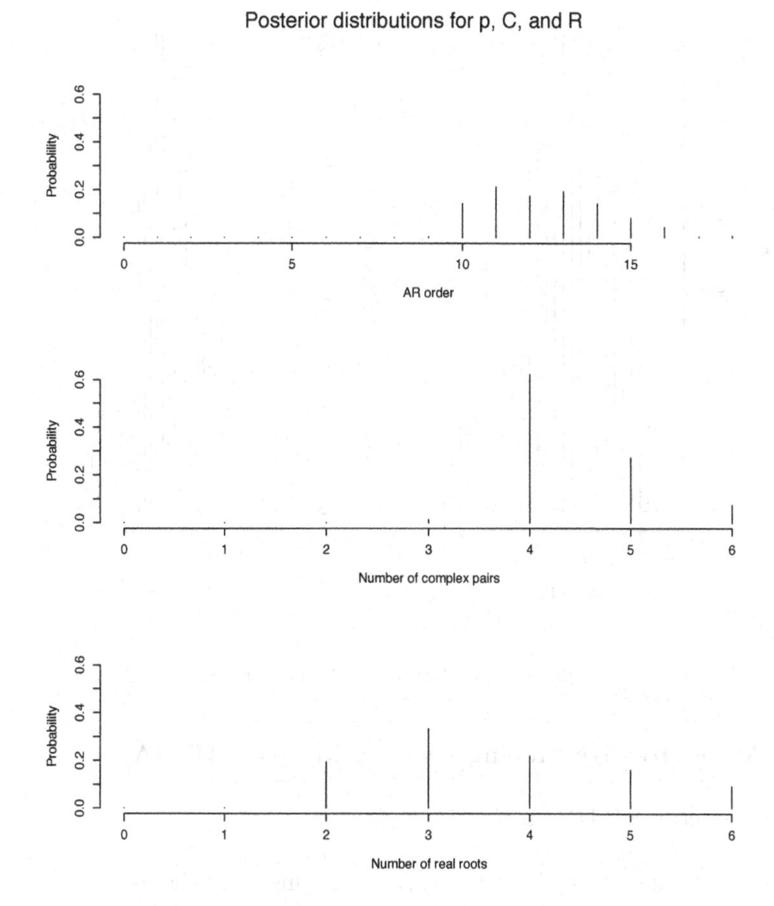

Figure 2.11 *Posterior distributions of the model order, C and R for the EEG data.*

uncertainty in the number of real and complex roots, and so models with $10 \leq p \leq 16$ get significant posterior probabilities. Figure 2.12 displays the marginal posterior distributions of r_1 and λ_1, i.e., the marginals for the modulus and wavelength of the component with the highest modulus. Note that these pictures are consistent with the results obtained from the reference analysis of an AR(8) model presented previously.

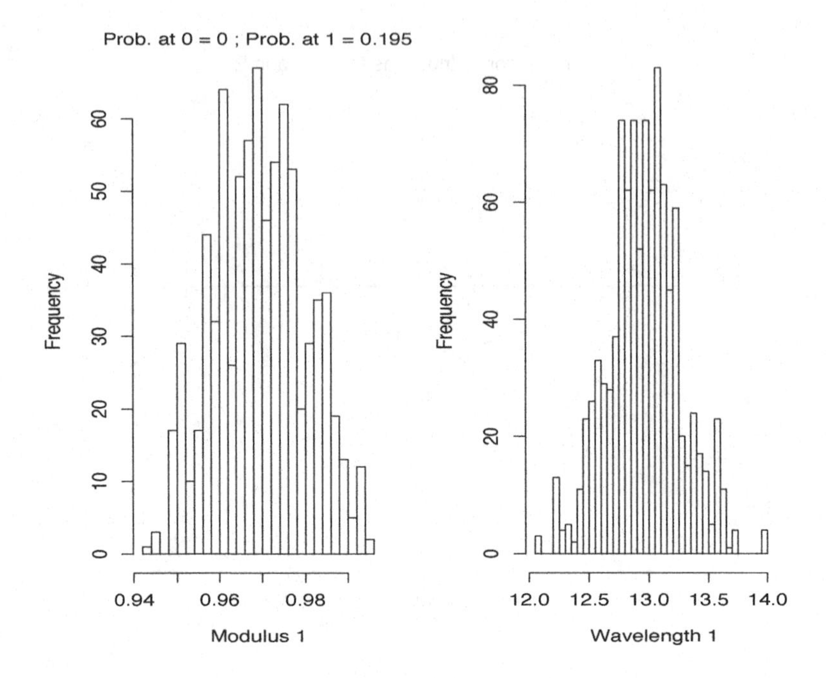

Figure 2.12 *Posterior distributions of* (r_1, λ_1) *for the EEG data.*

2.5 Autoregressive Moving Average Models (ARMA)

2.5.1 Structure of ARMA Models

Consider a time series y_t, for $t = 1, 2, \ldots,$ arising from the model

$$y_t = \sum_{i=1}^{p} \phi_i y_{t-i} + \sum_{j=1}^{q} \theta_j \epsilon_{t-j} + \epsilon_t, \tag{2.29}$$

with $\epsilon_t \sim N(0, v)$. Then, $\{y_t\}$ follows an autoregressive moving average model, or ARMA(p, q), where p and q are the orders of the autoregressive and moving average parts, respectively. When $p = 0$, $\{y_t\}$ is said to be a moving average process of order q or MA(q). Similarly, when $q = 0$, $\{y_t\}$ is an autoregressive process of order p or AR(p).

Example 2.6 *MA*(1) *process.* If $\{y_t\}$ follows a MA(1) process, $y_t = \theta\epsilon_{t-1} + \epsilon_t$, the process is stationary for all the values of θ. In addition, it is easy to see that the autocorrelation function has the following form

$$\rho(h) = \begin{cases} 1 & h = 0 \\ \frac{\theta}{(1+\theta^2)} & h = 1 \\ 0 & \text{otherwise.} \end{cases}$$

Now, if we consider a MA(1) process with coefficient $\frac{1}{\theta}$ instead of θ, we would obtain the same autocorrelation function, and so it would be impossible to determine which of the two processes generated the data. Therefore, it is necessary to impose identifiability conditions on θ. In particular, $|\theta| < 1$ is the identifiability condition for a MA(1), which is also known as the invertibility condition, given that it implies that the MA process can be "inverted" into an infinite order AR process.

If $\{y_t\}$ follows an ARMA(p, q), we can write $\Phi(B)y_t = \Theta(B)\epsilon_t$, with

$$\Phi(B) = 1 - \phi_1 B - \cdots - \phi_p B^p \quad \text{and} \quad \Theta(B) = 1 + \theta_1 B + \cdots + \theta_q B^q,$$

where B is the backshift operator.

In general, a MA(q) process is identifiable or invertible only when the roots of the MA characteristic polynomial $\Theta(u) = 1 + \theta_1 u + \ldots + \theta_q u^q$ lie outside the unit circle. In this case it is possible to write the MA process as an infinite order AR process. For an ARMA(p, q) process, the stability condition, which implies stationarity, is given in terms of the AR coefficients, i.e., the process is stable when the roots of the AR characteristic polynomial $\Phi(u) = 1 - \phi_1 u - \cdots - \phi_p u^p$ lie outside the unit circle. The ARMA process is invertible only when the roots of the MA characteristic polynomial lie outside the unit circle. So, when both conditions hold, i.e., when the roots of the AR polynomial and the roots of the MA polynomial lie outside the unit circle, the ARMA process can be written either as a purely AR process of infinite order, or as a purely MA process of infinite order.

If all the roots of the AR characteristic polynomial are outside the unit circle, then we can write the ARMA process as a purely MA process of infinite order

$$y_t = \Phi^{-1}(B)\Theta(B)\epsilon_t = \Psi(B)\epsilon_t = \sum_{j=0}^{\infty} \psi_j \epsilon_{t-j},$$

with $\Psi(B)$ such that $\Phi(B)\Psi(B) = \Theta(B)$. The ψ_j values can be found by solving the homogeneous difference equations given by

$$\psi_j - \sum_{h=1}^{p} \phi_h \psi_{j-h} = 0, \quad j \geq \max(p, q+1), \tag{2.30}$$

with initial conditions

$$\psi_j - \sum_{h=1}^{j} \phi_h \psi_{j-h} = \theta_j, \quad 0 \leq j < \max(p, q+1), \tag{2.31}$$

and $\theta_0 = 1$. The general solution to the Equations (2.30) and (2.31) is given by

$$\psi_j = \alpha_1^j p_1(j) + \cdots + \alpha_r^j p_r(j), \tag{2.32}$$

where $\alpha_1, \ldots, \alpha_r$ are the reciprocal roots of the characteristic polynomial $\Phi(u) = 0$, with multiplicities $m_1, \ldots, m_r$, respectively, and each $p_i(j)$ is a polynomial of degree $m_i - 1$.

2.5.2 *Autocorrelation and Partial Autocorrelation Functions*

If $\{y_t\}$ follows a MA(q) process, it is possible to show (see Shumway and Stoffer 2017) that its ACF is given by

$$\rho(h) = \begin{cases} 1 & h = 0 \\ \frac{\sum_{j=0}^{q-h} \theta_j \theta_{j+h}}{1+\sum_{j=1}^{q} \theta_j^2} & h = 1 : q \\ 0 & h > q, \end{cases} \tag{2.33}$$

and so, from a practical viewpoint it is possible to identify purely MA processes by looking at sample ACF plots, since the estimated ACF coefficients should drop after the q-th lag.

For general ARMA processes, the autocovariance function can be written in terms of the general homogeneous equations

$$\gamma(h) - \phi_1 \gamma(h-1) - \cdots - \phi_p \gamma(h-p) = 0, \quad h \geq \max(p, q+1), \tag{2.34}$$

with initial conditions given by

$$\gamma(h) - \sum_{j=1}^{p} \phi_j \gamma(h-j) = v \sum_{j=h}^{q} \theta_j \psi_{j-h}, \quad 0 \leq h < \max(p, q+1). \tag{2.35}$$

The ACF of an ARMA is obtained dividing (2.34) and (2.35) by $\gamma(0)$.

The PACF can be computed using any of the methods described in Section 2.1.5. The partial autocorrelation coefficients of a MA(q) process are never zero, as opposed to the partial autocorrelation coefficients of an AR(p) process which are zero after lag p. Similarly, for an invertible ARMA model, the partial autocorrelation coefficients will never drop to zero since the process can be written as an infinite order AR.

2.5.3 Inversion of AR Components

In contexts where the time series has a reasonable length, we can fit long order AR models rather than ARMA or other, more complex forms. One key reason is that the statistical analysis, at least the conditional analysis based on fixed initial values, is much easier. The reference analysis for AR(p) processes described previously, for example, is essentially trivial compared with the numerical analysis required to produce samples from posterior distributions in ARMA models (see next sections). Another driving motivation is that long order AR models will closely approximate ARMA forms. The proliferation of parameters is an issue, though with long series and possible use of smoothness priors or other constraints, this is not an overriding consideration.

If this view is adopted in a given problem, it may be informative to use the results of an AR analysis to explore possible MA component structure using the device of inversion, or partial inversion, of the AR model. This is described here. Assume that $\{y_t\}$ follows an AR(p) model with parameter vector $\boldsymbol{\phi} = (\phi_1, \ldots, \phi_p)'$, so we can write

$$\Phi(B)y_t = \prod_{i=1}^{p}(1 - \alpha_i B)y_t = \epsilon_t,$$

where the α_is are the autoregressive characteristic reciprocal roots. Often there will be subsets of pairs of complex conjugate roots corresponding to quasiperiodic components, perhaps with several real roots.

For some positive integer $r < p$, suppose that the final $p - r$ reciprocal roots are identified as having moduli less than unity; some or all of the first r roots may also represent stationary components, though that is not necessary for the following development. Then, we can rewrite the model as

$$\prod_{i=1}^{r}(1 - \alpha_i B)y_t = \prod_{i=r+1}^{p}(1 - \alpha_i B)^{-1}\epsilon_t = \Psi^*(B)\epsilon_t,$$

where the (implicitly) infinite order MA component has the coefficients of the infinite order polynomial $\Psi^*(u) = 1 + \sum_{j=1}^{\infty}\psi_j^* u^j$, defined by

$$1 = \Psi^*(u)\prod_{i=r+1}^{p}(1 - \alpha_i u).$$

So we have the representation

$$y_t = \sum_{j=1}^{r}\phi_j^* y_{t-j} + \epsilon_t + \sum_{j=1}^{\infty}\psi_j^*\epsilon_{t-j},$$

where the r new AR coefficients ϕ_j^*, for $j = 1 : r$, are defined by the characteristic equation $\Phi^*(u) = \prod_{i=1}^{r}(1-\alpha_i u) = 0$. The MA terms ψ_j^* can be easily calculated recursively, up to some appropriate upper bound on their number, say q. Explicitly, they are recursively computed as follows.

1. Initialize the algorithm by setting $\psi_i^* = 0$ for all $i = 1 : q$.
2. For $i = (r+1) : p$, update $\psi_1^* = \psi_1^* + \alpha_i$, and then,
   - for $j = 2 : q$, update $\psi_j^* = \psi_j^* + \alpha_i \psi_{j-1}^*$.

Suppose $\boldsymbol{\phi}$ is set at some estimate, such as a posterior mean, in the AR(p) model analysis. The above calculations can be performed for any specified value of r to compute the corresponding MA coefficients in an inversion to the approximating ARMA(r, q) model. If the posterior for $\boldsymbol{\phi}$ is sampled in the AR analysis, the above computations can be performed repeatedly for all sampled $\boldsymbol{\phi}$ vectors, so producing corresponding samples of the ARMA parameters $\boldsymbol{\phi}^*$ and $\boldsymbol{\psi}^*$. Thus, inference in various relevant ARMA models can be directly, and quite easily, deduced by inversion of longer order AR models. Typically, various values of r will be explored. Guidance is derived from the estimated amplitudes and, in the case of complex roots, periods of the roots of the AR model. Analyses in which some components are persistent suggest that these components should be retained in the AR description. The remaining roots, typically corresponding to high-frequency characteristics in the data with lower moduli, are then the candidates for inversion to what will often be a relatively low order MA component. The calculations can be repeated, sequentially increasing q and exploring inferences about the MA parameters, to assess a relevant approximating order.

Example 2.7 *Exploring ARMA structure in the EEG data.* It is of interest to determine whether or not the residual noise structure in the EEG series may be adequately described by alternative moving average structure with, perhaps, fewer parameters than the above eight or more in the AR description. This can be initiated directly from the AR analysis by exploring inversions of components of the autoregressive characteristic polynomial, as follows.

For any AR parameter vector $\boldsymbol{\phi}$, we have the model

$$\Phi(B)y_t = \prod_{i=1}^{8}(1-\alpha_i B)y_t = \epsilon_t,$$

where, by convention, the roots appear in order of decreasing moduli. In our AR(8) reference analysis there is a dominant component describing the major cyclical features that has modulus close to unity; the first two roots are complex conjugates corresponding to this component, the reference estimate of $\boldsymbol{\phi}$ produces an estimated modulus of 0.97 and frequency of 0.494. Identifying this as the key determinant of the AR structure, we can write the model as

$$\prod_{i=1}^{2}(1-\alpha_i B)y_t = \prod_{i=3}^{8}(1-\alpha_i B)^{-1}\epsilon_t = \Psi^*(B)\epsilon_t,$$

where the infinite order MA component is defined via

$$1 = \Psi^*(u)\prod_{i=3}^{8}(1-\alpha_i u),$$

leading to the representation

$$y_t = \phi_1^* y_{t-1} + \phi_2^* y_{t-2} + \epsilon_t + \sum_{j=1}^{\infty}\psi_j^*\epsilon_{t-j},$$

where $\phi_1^* = 2r_1\cos(\omega_1)$ and $\phi_2^* = -r_1^2$, and with (r_1, ω_1) being the modulus and frequency of the dominant cycle. In our case, the reference posterior mean from the fitted AR(8) model indicates values close to $\phi_1^* = 1.71$ and $\phi_2^* = -0.94$. The MA terms ψ_j^* can be easily calculated recursively, as detailed above. This can be done for any specified AR(8) vector $\boldsymbol{\phi}$. Note that the roots typically are complex, though the resulting ψ_j^* must be real-valued. Note also that the ψ_j^* will decay rapidly so that q in the recursive algorithm is often rather moderate. Figure 2.13 displays a summary of such calculations based on the existing AR(8) reference analysis. Here $q = 8$ is chosen, so that the approximating ARMA model is ARMA$(2, 8)$, but with the view that the MA term is possibly overfitting. The above computations are performed in parallel for each of the 5,000 $\boldsymbol{\phi}$ vectors sampled from the reference posterior. This provides a Monte Carlo sample of size 5,000 from the posterior for the MA parameters obtained via this inversion technique. For each j, the sample distribution of values of ψ_j^* is summarized in Figure 2.13. Note the expected feature that only rather few, in this case really only two, of the MA coefficients are nonnegligible; as a result, the inversion method suggests that the longer order AR model is an approximation to a perhaps more parsimonious ARMA$(2, 2)$ form with AR parameters near 1.71 and -0.94, and with MA parameters around -1.45 and 0.65.

This analysis is supported by an exploratory search across ARMA(p, q) models for p and q taking values between one and eight. This can be done simply to produce rough guidelines to determine model orders using the conditional and approximate log-likelihood and BIC computations, for example. Using the `arima` function in R (R Core Team 2018) with the conditional sum of squares maximum likelihood method and setting the maximum AR and MA orders both to 8, we find that the optimal model is an ARMA$(2, 2)$. The approximate MLEs of the ARMA$(2, 2)$ parameters, based on this conditional analysis in R are 1.72 (0.03) and -0.93 (0.03) for the AR component, and -1.46 (0.06) and 0.57 (0.08) for the MA compo-

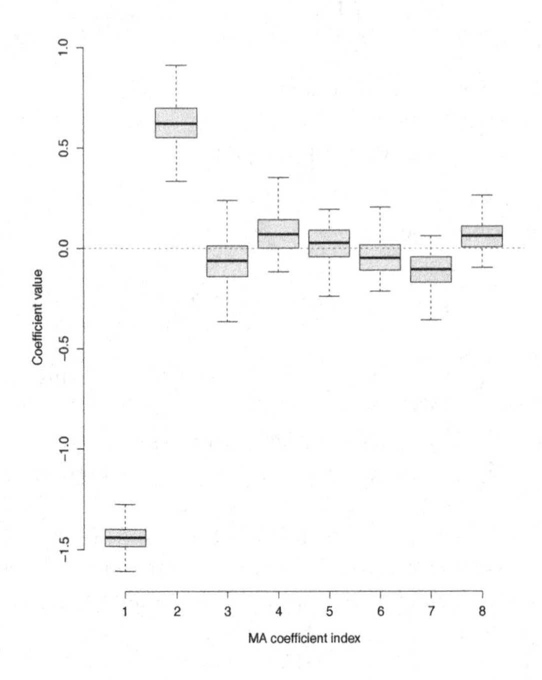

Figure 2.13 *Approximate posterior distributions for the first eight MA coefficients from a partial inversion of the reference AR(8) analysis of the EEG series.*

nent. These agree well with the inversion of the Bayesian AR(8) reference analysis. Note that the inversion approach directly supplies full posterior inferences, through easily implemented posterior simulations, in contrast to likelihood approaches. Note that this analysis could be repeated for higher order AR models. Proceeding to AR(10) or AR(12) produces models more tailored-to-minor noise features of the data. Subsequent inversion suggests possible higher order refinements, e.g., an ARMA(3, 3) model, though the global improvements in data fit and description are minor. Overall, though some additional insights are gleaned from exploring the MA structure, this particular segment of the EEG series is best described by the AR(8) and further analysis should be based on that. In other contexts, however, an ARMA structure may often be preferred.

2.5.4 Forecasting and Estimation of ARMA Processes

We now summarize some aspects of forecasting and estimation of autoregressive moving average processes. In particular, we consider maximum likelihood, least squares, and Bayesian approaches for parameter estimation. For more detailed developments and further discussion on some of these topics see for example Shumway and Stoffer (2017) and Box, Jenkins, Reinsel, and Ljung (2015).

2.5.4.1 Forecasting ARMA Models

Consider a stable and invertible ARMA process with parameters $\phi_1, \ldots, \phi_p$ and $\theta_1, \ldots, \theta_q$. Then, it is possible to write the process as a purely AR process of infinite order, and so

$$y_{t+h} = \sum_{j=1}^{\infty} \phi_j^* y_{t+h-j} + \epsilon_{t+h}. \tag{2.36}$$

Alternatively, it can also be written as an infinite order MA process

$$y_{t+h} = \sum_{j=1}^{\infty} \theta_j^* \epsilon_{t+h-j} + \epsilon_{t+h}. \tag{2.37}$$

Let $y_{t+h}^{-\infty}$ be the minimum mean square predictor of y_{t+h} based on y_t, y_{t-1}, $\ldots$, y_1, y_0, $y_{-1}, \ldots$, which we denote as $y_{-\infty:t}$. In other words, $y_{t+h}^{-\infty} = E(y_{t+h}|y_{-\infty:t})$. Then, it is possible to show that (see Problem 16.)

$$y_{t+h} - y_{t+h}^{-\infty} = \sum_{j=0}^{h-1} \theta_j^* \epsilon_{t+h-j}, \tag{2.38}$$

with $\theta_0^* = 1$, and so the mean square prediction error is given by

$$\text{MSE}_{t+h}^{-\infty} = E(y_{t+h} - y_{t+h}^{-\infty})^2 = v \sum_{j=0}^{h-1} (\theta_j^*)^2. \tag{2.39}$$

For a given sample size T, only the observations $y_{1:T}$ are available, and so the following truncated predictor is used as an approximation:

$$y_{T+h}^{-\infty,T} = \sum_{j=1}^{h-1} \phi_j^* y_{T+h-j}^{-\infty,T} + \sum_{j=h}^{T+h-1} \phi_j^* y_{T+h-j}. \tag{2.40}$$

This predictor is computed recursively for $h = 1, 2, \ldots$, and the mean square prediction error is given approximately by (2.39).

In the AR(p) case, if $T > p$, the predictor y_{T+1}^T computed via (2.13) and given by

$$y_{T+1}^T = \phi_1 y_T + \phi_2 y_{T-1} + \ldots + \phi_p y_{T-p+1} \tag{2.41}$$

yields to the exact predictor. This is true in general for any h, in other words, $y^T_{T+h} = y^{-\infty}_{T+h} = y^{-\infty,T}_{T+h}$, and so there is no need for approximations.

For general ARMA(p,q) models, the truncated predictor in (2.40) is

$$y^{-\infty,T}_{T+h} = \sum_{j=1}^{p} \phi_j y^{-\infty,T}_{T+h-j} + \sum_{j=1}^{q} \theta_j \epsilon^T_{T+h-j}, \quad (2.42)$$

where $y^{-\infty,T}_t = y_t$ for $1 \le t \le T$, $y^{-\infty,T}_t = 0$ for $t \le 0$, and the truncated prediction errors are given by $\epsilon^T_t = 0$, for $t \le 0$ or $t > T$, and

$$\epsilon^T_t = \phi(B) y^{-\infty,T}_t - \theta_1 \epsilon^T_{t-1} - \ldots - \theta_q \epsilon^T_{t-q}$$

for $1 \le t \le T$.

2.5.4.2 MLE and Least Squares Estimation

For an ARMA(p,q) model we need to estimate the parameters $\boldsymbol{\beta}$ and v where $\boldsymbol{\beta} = (\phi_1, \ldots, \phi_p, \theta_1, \ldots, \theta_q)'$. The likelihood function can be written as follows:

$$p(y_{1:T}|\boldsymbol{\beta}, v) = \prod_{t=1}^{T} p(y_t|y_{1:(t-1)}, \boldsymbol{\beta}, v). \quad (2.43)$$

Now, assuming that the conditional distribution of y_t given $y_{1:(t-1)}$ is Gaussian with mean y^{t-1}_t and variance $V^{t-1}_t = v r^{t-1}_t$, we can write

$$\begin{aligned} -2\log\left[p(y_{1:T}|\boldsymbol{\beta}, v)\right] &= T\log(2\pi v) + \\ &+ \sum_{t=1}^{T}\left[\log(r^{t-1}_t) + \frac{(y_t - y^{t-1}_t)^2}{v r^{t-1}_t}\right], \end{aligned} \quad (2.44)$$

where y^{t-1}_t and r^{t-1}_t are functions of $\boldsymbol{\beta}$, and so the MLEs of $\boldsymbol{\beta}$ and v are computed by minimizing the expression (2.44) with respect to $\boldsymbol{\beta}$ and v. Equation (2.44) is usually a nonlinear function of the parameters, and so the minimization has to be done using a nonlinear optimization algorithm such as the Newton-Raphson algorithm described in Chapter 1.

Least squares (LS) estimation can be performed by minimizing the expression

$$S(\boldsymbol{\beta}) = \sum_{t=1}^{T} \frac{(y_t - y^{t-1}_t)^2}{r^{t-1}_t},$$

with respect to $\boldsymbol{\beta}$. Similarly, conditional least squares estimation is performed by conditioning on the first p values of the series $y_{1:p}$ and assuming that $\epsilon_p = \epsilon_{p-1} = \cdots = \epsilon_{p-(q-1)} = 0$. In this case we can minimize the conditional sum of squares given by

$$S_c(\boldsymbol{\beta}) = \sum_{t=p+1}^{T} \epsilon_t^2(\boldsymbol{\beta}), \tag{2.45}$$

where $\epsilon_t(\boldsymbol{\beta}) = y_t - \sum_{i=1}^{p} \phi_i y_{t-i} - \sum_{j=1}^{q} \theta_j \epsilon_{t-j}(\boldsymbol{\beta})$. When $q = 0$ this reduces to a linear regression problem, and so no numerical minimization technique is required. When the number of observations T is not very large, conditioning on the first initial values will have an influence on the parameter estimates. In such cases working with the unconditional sum of squares might be preferable. Several methodologies have been proposed to handle unconditional least squares estimation. In particular, Box, Jenkins, Reinsel, and Ljung (2015, Appendix A7.3), showed that an approximation to the unconditional sum of squares $S(\boldsymbol{\beta})$ is

$$\hat{S}(\boldsymbol{\beta}) = \sum_{t=-M}^{T} \hat{\epsilon}_t^2(\boldsymbol{\beta}), \tag{2.46}$$

with $\hat{\epsilon}_t(\boldsymbol{\beta}) = E(\epsilon_t | y_{1:T})$ and, if $t \leq 0$, these values are obtained by backcasting. Here M is chosen to be such that $\sum_{t=-\infty}^{-M} \hat{\epsilon}_t^2(\boldsymbol{\beta}) \approx 0$.

A Gauss-Newton procedure (see Shumway and Stoffer 2017, Section 3.5 and references therein) can be used to obtain an estimate of $\boldsymbol{\beta}$, say $\hat{\boldsymbol{\beta}}$, that minimizes $S(\boldsymbol{\beta})$ or $S_c(\boldsymbol{\beta})$. For instance, in order to find an estimate of $\boldsymbol{\beta}$ that minimizes the conditional sum of squares in (2.45), the following algorithm is repeated by computing $\boldsymbol{\beta}^{(j)}$ at each iteration $j = 1, 2, \ldots,$ until convergence is reached, with

$$\boldsymbol{\beta}^{(j)} = \boldsymbol{\beta}^{(j-1)} + \Delta(\boldsymbol{\beta}^{(j-1)}).$$

Here

$$\Delta(\boldsymbol{\beta}) = \frac{\sum_{t=p+1}^{T} \mathbf{z}_t(\boldsymbol{\beta})\epsilon_t(\boldsymbol{\beta})}{\sum_{t=p+1}^{T} \mathbf{z}_t'(\boldsymbol{\beta})\mathbf{z}_t(\boldsymbol{\beta})}$$

and

$$\mathbf{z}_t(\boldsymbol{\beta}) = \left(-\frac{\partial \epsilon_t(\boldsymbol{\beta})}{\partial \beta_1}, \ldots, -\frac{\partial \epsilon_t(\boldsymbol{\beta})}{\partial \beta_{p+q}} \right)'. \tag{2.47}$$

Convergence is considered to be achieved when $|\boldsymbol{\beta}^{(j+1)} - \boldsymbol{\beta}^{(j)}| < \delta_{\boldsymbol{\beta}}$, or when $|Q_c(\boldsymbol{\beta}^{(j+1)}) - Q_c(\boldsymbol{\beta}^{(j)})| < \delta_Q$, for $\delta_{\boldsymbol{\beta}}$ and δ_Q set to some fixed small values. Here, $Q_c(\boldsymbol{\beta})$ is a linear approximation of $S_c(\boldsymbol{\beta})$ given by

$$Q_c(\boldsymbol{\beta}) = \sum_{t=p+1}^{T} \left[\epsilon_t(\boldsymbol{\beta}^{(0)}) - (\boldsymbol{\beta} - \boldsymbol{\beta}^{(0)})' \mathbf{z}_t(\boldsymbol{\beta}^{(0)}) \right]^2$$

and $\boldsymbol{\beta}^{(0)}$ is an initial estimate of $\boldsymbol{\beta}$.

Example 2.8 *Conditional least squares estimation of the parameters of an ARMA*(1, 1). Consider a stable and invertible ARMA(1, 1) process described by

$$y_t = \phi_1 y_{t-1} + \theta_1 \epsilon_{t-1} + \epsilon_t,$$

with $\epsilon_t \sim N(0, v)$. Then, we can write $\epsilon_t(\boldsymbol{\beta}) = y_t - \phi_1 y_{t-1} - \theta_1 \epsilon_{t-1}(\boldsymbol{\beta})$, with $\boldsymbol{\beta} = (\phi_1, \theta_1)'$. Additionally, we condition on $\epsilon_1(\boldsymbol{\beta}) = 0$ and y_1. Now, using the expression (2.47) we have that $\mathbf{z}_t(\boldsymbol{\beta}) = (z_{t,1}(\boldsymbol{\beta}), z_{t,2}(\boldsymbol{\beta}))'$ with $z_{t,1}(\boldsymbol{\beta}) = y_{t-1} - \theta_1 z_{t-1,1}(\boldsymbol{\beta})$ and $z_{t,2}(\boldsymbol{\beta}) = \epsilon_{t-1}(\boldsymbol{\beta}) - \theta_1 z_{t-1,2}(\boldsymbol{\beta})$, with $\mathbf{z}_0(\boldsymbol{\beta}) = \mathbf{0}$. The Gauss-Newton algorithm starts with some initial value of $\boldsymbol{\beta}^{(0)} = (\phi_1^{(0)}, \theta_1^{(0)})'$ and then, at each iteration $j = 1, 2, \dots,$ we have

$$\boldsymbol{\beta}^{(j)} = \boldsymbol{\beta}^{(j-1)} + \frac{\sum_{t=2}^{T} \mathbf{z}_t(\boldsymbol{\beta}^{(j-1)})\epsilon_t(\boldsymbol{\beta}^{(j-1)})}{\sum_{t=2}^{T} \mathbf{z}'_t(\boldsymbol{\beta}^{(j-1)})\mathbf{z}_t(\boldsymbol{\beta}^{(j-1)})}.$$

2.5.4.3 State-space Representation

Due to the computational burden of maximizing the exact likelihood given in (2.43), many of the existing methods for parameter estimation in the ARMA modeling framework consider approximations to the exact likelihood, such as the backcasting method of Box, Jenkins, Reinsel, and Ljung (2015). There are also approaches that allow computation of the exact likelihood function. Some of these approaches involve rewriting the ARMA model in state-space or dynamic linear model (DLM) form, and then applying the Kalman filter to achieve parameter estimation (see for example Kohn and Ansley 1985; Harvey 1981, 1991).

A state-space model or DLM is usually defined in terms of two equations, one that describes the evolution of the time series at the observational level, and another equation that describes the evolution of the system over time. One of the most useful ways of representing the ARMA(p, q) model given in (2.29) is by writing it in the state-space or DLM form given by the following equations,

$$\begin{aligned} y_t &= \mathbf{E}'_m \boldsymbol{\gamma}_t \\ \boldsymbol{\gamma}_t &= \mathbf{G}\boldsymbol{\gamma}_{t-1} + \boldsymbol{\omega}_t, \end{aligned} \qquad (2.48)$$

where $\mathbf{E}_m = (1, 0, \dots, 0)'$ is a vector of dimension m, with $m = \max(p, q+1)$; $\boldsymbol{\omega}_t$ is also a vector of dimension m with $\boldsymbol{\omega}_t = (1, \theta_1, \dots, \theta_{m-1})'\epsilon_t$ and $\mathbf{G}$

is an $m \times m$ matrix given by

$$\mathbf{G} = \begin{pmatrix} \phi_1 & 1 & 0 & \dots & 0 \\ \phi_2 & 0 & 1 & \dots & 0 \\ \vdots & \vdots & \vdots & \ddots & \vdots \\ \phi_{m-1} & 0 & 0 & \dots & 1 \\ \phi_m & 0 & 0 & \cdots & 0 \end{pmatrix}.$$

Here $\phi_r = 0$ for all $r > p$ and $\theta_r = 0$ for all $r > q$. The evolution noise has a variance-covariance matrix $\mathbf{U} = v(1, \theta_1, \dots, \theta_{m-1})'(1, \theta_1, \dots, \theta_{m-1})$.

Using this representation it is possible to perform parameter estimation for general ARMA(p, q) models. We will revisit this topic after developing the theory of DLMs in Chapter 4.

2.5.4.4 Bayesian Estimation of ARMA Processes

There are several approaches to Bayesian estimation of general ARMA models, e.g., Monahan (1983), Marriott and Smith (1992), Chib and Greenberg (1994), Box, Jenkins, Reinsel, and Ljung (2015), Zellner (1996), Marriott, Ravishanker, Gelfand, and Pai (1996), and Barnett, Kohn, and Sheather (1997), among others.

We briefly outline the approach of Marriott, Ravishanker, Gelfand, and Pai (1996) and discuss some aspects related to alternative ways of performing Bayesian estimation in ARMA models. This approach leads to MCMC parameter estimation of ARMA(p, q) models by reparameterizing the ARMA parameters in terms of partial autocorrelation coefficients. Specifically, let $p(y_{1:T}|\boldsymbol{\psi}^*)$ be the likelihood for the T observations given the vector of parameters $\boldsymbol{\psi}^* = (\boldsymbol{\phi}', \boldsymbol{\theta}', v, \mathbf{x}_0', \boldsymbol{\epsilon}_0')'$, with $\boldsymbol{\epsilon}_0 = (\epsilon_0, \epsilon_{-1}, \dots, \epsilon_{1-q})'$. This likelihood function is given by

$$p(y_{1:T}|\boldsymbol{\psi}^*) = (2\pi v)^{-T/2} \exp\left\{ -\frac{1}{2v} \sum_{t=1}^{T} (y_t - \mu_t)^2 \right\}, \tag{2.49}$$

where

$$\begin{aligned} \mu_1 &= \sum_{i=1}^{p} \phi_i y_{1-i} + \sum_{i=1}^{q} \theta_i \epsilon_{1-i}, \\ \mu_t &= \sum_{i=1}^{p} \phi_i y_{t-i} + \sum_{i=1}^{t-1} \theta_i (y_{t-i} - \mu_{t-i}) + \sum_{i=t}^{q} \theta_i \epsilon_{t-i}, \quad t = 2:q, \\ \mu_t &= \sum_{i=1}^{p} \phi_i y_{t-i} + \sum_{i=1}^{q} \theta_i (y_{t-i} - \mu_{t-i}), \quad t = (q+1):T. \end{aligned}$$

The prior specification is as follows,

$$\pi(\boldsymbol{\psi}^*) = \pi(\mathbf{x}_0, \boldsymbol{\epsilon}_0|\boldsymbol{\phi}, \boldsymbol{\theta}, v)\pi(v)\pi(\boldsymbol{\phi}, \boldsymbol{\theta}),$$

with $\pi(\mathbf{x}_0, \boldsymbol{\epsilon}_0|\boldsymbol{\phi}, \boldsymbol{\theta}, v) = N(\mathbf{0}, v\Omega)$, $\pi(v) \propto 1/v$, and $\pi(\boldsymbol{\phi}, \boldsymbol{\theta})$ a uniform distribution in the stationary and invertibility regions of the ARMA process denoted by $\mathcal{C}_p$ and $\mathcal{C}_q$, respectively. The matrix $v\Omega$ is the covariance matrix of $(\mathbf{x}_0, \boldsymbol{\epsilon}_0)'$, which can be easily computed for any ARMA(p, q) model. Therefore, the joint posterior for $\boldsymbol{\psi}^*$ is given by

$$p(\boldsymbol{\psi}^*|y_{1:T}) \propto (v)^{-(T+2)/2} \exp\left\{-\frac{1}{2v}\sum_{t=1}^{T}(y_t - \mu_t)^2\right\} \times \tag{2.50}$$

$$N((\mathbf{x}_0', \boldsymbol{\epsilon}_0')'|\mathbf{0}, v\Omega). \tag{2.51}$$

The MCMC algorithm can be summarized in terms of the following steps.

- Sample $(v|\boldsymbol{\phi}, \boldsymbol{\theta}, \mathbf{x}_0, \boldsymbol{\epsilon}_0, y_{1:T})$. This is done by sampling v from the inverse-gamma full conditional distribution with the following form:

$$IG\left(\frac{T+p+q}{2}, \frac{1}{2}\left[\begin{pmatrix}\mathbf{x}_0\\ \boldsymbol{\epsilon}_0\end{pmatrix}'\Omega^{-1}\begin{pmatrix}\mathbf{x}_0\\ \boldsymbol{\epsilon}_0\end{pmatrix} + \sum_{t=1}^{T}(y_t - \mu_t)^2\right]\right).$$

- Sample $(\mathbf{x}_0, \boldsymbol{\epsilon}_0|\boldsymbol{\phi}, \boldsymbol{\theta}, v, y_{1:T})$. The full conditional distribution of $(\mathbf{x}_0', \boldsymbol{\epsilon}_0')$ is a multivariate normal; however, it is computationally simpler to use a Metropolis step with Gaussian proposal distributions.
- Sample $(\boldsymbol{\phi}, \boldsymbol{\theta}|v, \mathbf{x}_0, \boldsymbol{\epsilon}_0, y_{1:T})$. In order to sample $\boldsymbol{\phi}$ and $\boldsymbol{\theta}$, successive transformations for $\mathcal{C}_p$ and $\mathcal{C}_q$ to p-dimensional and q-dimensional hypercubes and then to R^p and R^q, respectively, are considered. The transformations of $\mathcal{C}_p$ and $\mathcal{C}_q$ to the p-dimensional and q-dimensional hypercubes were proposed by Monahan (1984), extending the work of Barndorff-Nielsen and Schou (1973). Specifically, the transformation for the AR parameters is given by

$$\phi(i,h) = \phi(i,h-1) - \phi(h,h)\phi(h-i,h-1), \;\; i = 1:(h-1),$$

where $\phi(h,h)$ is the partial autocorrelation coefficient and $\phi(j,p) = \phi_j$ is the jth coefficient from the AR(p) process defined by the characteristic polynomial $\Phi(u) = 1 - \phi_1 u - \cdots - \phi_p u^p$. The inverse transformation in iterative form is given by

$$\phi(i,h-1) = [\phi(i,h) + \phi(h,h)\phi(h,h-i)]/[1 - \phi^2(h,h)], \;\; i = 1:(h-1),$$

and the Jacobian of the transformation is

$$J = \prod_{h=1}^{p}(1 - \phi^2(h,h))^{[(h-1)/2]} \prod_{j=1}^{[p/2]}(1 - \phi(2j,2j)).$$

Now, the stationarity condition on $\boldsymbol{\phi}$ can be written in terms of the partial autocorrelation coefficients as $|\phi(h,h)| < 1$ for all $h = 1 : p$. Marriott, Ravishanker, Gelfand, and Pai (1996) propose a transformation from $\mathbf{r}_\phi = (\phi(1,1), \ldots, \phi(p,p))'$ to $\mathbf{r}^*_\phi = (\phi^*(1,1), \ldots, \phi^*(p,p))'$, with $\mathbf{r}^*_\phi \in R^p$. The $\phi^*(j,j)$ elements are given by

$$\phi^*(j,j) = \log\left(\frac{1+\phi(j,j)}{1-\phi(j,j)}\right).$$

Similarly, a transformation from $\boldsymbol{\theta}$ to $\mathbf{r}^*_\theta \in R^q$ can be defined using the previous two steps replacing $\boldsymbol{\phi}$ by $\boldsymbol{\theta}$. Then, instead of sampling $\boldsymbol{\phi}$ and $\boldsymbol{\theta}$ from the constrained full conditional distributions, we can sample unconstrained full conditional distributions for $\mathbf{r}^*_\phi$ and $\mathbf{r}^*_\theta$ on R^p and R^q, respectively. Marriott, Ravishanker, Gelfand, and Pai (1996) suggest using a Metropolis step as follows. First, compute MLE estimates of $\boldsymbol{\phi}$ and $\boldsymbol{\theta}$, say $(\hat{\boldsymbol{\phi}}, \hat{\boldsymbol{\theta}})$, with its asymptotic variance-covariance matrix $\Sigma_{(\hat{\phi},\hat{\theta})}$. Use the transformations described above to obtain $(\hat{\mathbf{r}}^*_{\hat{\phi}}, \hat{\mathbf{r}}^*_{\hat{\theta}})$ and a corresponding variance-covariance matrix Σ^* (computed via the delta method). Let $g_{p+q}(\mathbf{r}^*_\phi, \mathbf{r}^*_\theta)$ be the $p+q$-dimensional multivariate normal distribution with mean $(\hat{\mathbf{r}}^*_{\hat{\phi}}, \hat{\mathbf{r}}^*_{\hat{\theta}})$ and variance-covariance matrix Σ^*. Take g_{p+q} to be the proposal density in the Metropolis step build to sample $\mathbf{r}^*_\phi$ and $\mathbf{r}^*_\theta$.

Example 2.9 *Bayesian estimation in the ARMA(1,1) model.* Consider an ARMA(1,1) model described by $y_t = \phi y_{t-1} + \theta\epsilon_{t-1} + \epsilon_t$, with $N(\epsilon_t|0,v)$. In this case $\mathbf{x}_0 = y_0$, $\boldsymbol{\epsilon}_0 = \epsilon_0$, $\mathbf{r}_\phi = \phi$, $\mathbf{r}_\theta = \theta$, $\mathbf{r}^*_\phi = \phi^*$, $\mathbf{r}^*_\theta = \theta^*$,

$$\Omega = \begin{pmatrix} \frac{(1+\theta^2+2\phi\theta)}{(1-\phi^2)} & 1 \\ 1 & 1 \end{pmatrix}, \quad \phi^* = \log\left(\frac{1+\phi}{1-\phi}\right), \quad \theta^* = \log\left(\frac{1+\theta}{1-\theta}\right),$$

and the inverse of the determinant of the Jacobian of the transformation is given by $(1-\phi^2)(1-\theta^2)/4$.

2.6 Other Models

Extensions to ARMA models can be considered to account for nonstationary time series. A class of nonstationary models often used in practice is that of autoregressive integrated moving average models, or ARIMA models. Such models assume that the d-th difference of the process has a stationary ARMA structure. In other words, a process $\{y_t\}$ is an ARIMA(p,d,q) process if

$$(1 - \phi_1 B - \cdots - \phi_p B^p)(1-B)^d y_t = (1 + \theta_1 B + \cdots + \theta_q B^q)\epsilon_t,$$

with $\epsilon_t \sim N(0,v)$.

Often in practical settings the observed time series may display a strong correlation with past values that occur at multiples of some seasonal lag s. For example, the median home price in the month of April of a given year typically shows a stronger correlation with the median home price of April of the previous year than with the median home prices in the months preceding April of that same year. Seasonal ARMA, or SARMA, models can be used to capture such seasonal behavior. A process y_t is a seasonal ARMA process, or $\text{SARMA}_s(p, q)$, if

$$(1 - \phi_1 B^s - \phi_2 B^{2s} - \cdots - \phi_p B^{ps}) y_t = (1 + \theta_1 B^s + \theta_2 B^{2s} + \cdots + \theta_q B^{qs}) \epsilon_t,$$

with $\epsilon_t \sim N(0, v)$.

Seasonal and nonseasonal operators can also be combined into multiplicative seasonal autoregressive moving average models. Processes that follow such models can be represented as

$$\Phi_p(B) \Phi_P(B^s)(1 - B)^d y_t = \Theta_q(B) \Theta_Q(B^s) \epsilon_t,$$

with $\epsilon_t \sim N(0, v)$. In this case p and q are the orders of the nonseasonal ARMA components, P and Q are the orders of the seasonal ARMA components, while s is the period.

For an extensive treatment of ARIMA and seasonal ARIMA models, as well as their multiplicative versions, see Box, Jenkins, Reinsel, and Ljung (2015). Examples involving the analysis of real data with these models can also be found in Shumway and Stoffer (2017).

Other extensions include ARMA models whose coefficients vary periodically in time (Troutman 1979), ARMA models with time-varying coefficients (Chapters 4 and 5), fractionally integrated ARMA models (Chapter 3), and ARMA models for multivariate time series (Chapter 9) among others.

2.7 Appendix

2.7.1 The Reversible Jump MCMC Algorithm

In general the reversible jump MCMC (RJMCMC) method can be described as follows (see Green 1995 for details). Assume that $\boldsymbol{\theta}$ is a vector of parameters to be estimated and $\pi(d\boldsymbol{\theta})$ is the target probability measure, which often is a mixture of densities, or a mixture with continuous and discrete parts. Suppose that $m = 1, 2, \ldots,$ indexes all the possible dimensions of the model. If the current state of the Markov chain is $\boldsymbol{\theta}$ and a move of type m and destination $\boldsymbol{\theta}^*$ is proposed from a proposal measure $q_m(\boldsymbol{\theta}, d\boldsymbol{\theta}^*)$, the move is accepted probability

$$\alpha_m(\boldsymbol{\theta}, \boldsymbol{\theta}^*) = \min\left\{1, \frac{\pi(d\boldsymbol{\theta}^*) q_m(\boldsymbol{\theta}^*, d\boldsymbol{\theta})}{\pi(d\boldsymbol{\theta}) q_m(\boldsymbol{\theta}, d\boldsymbol{\theta}^*)}\right\}.$$

For cases in which the move type does not change the dimension of the parameter, the expression above reduces to the Metropolis-Hastings acceptance probability,

$$\alpha_m(\boldsymbol{\theta}, \boldsymbol{\theta}^*) = \min\left\{1, \frac{p(\boldsymbol{\theta}^*|y_{1:T})q_m(\boldsymbol{\theta}^*|\boldsymbol{\theta})}{p(\boldsymbol{\theta}|y_{1:T})q_m(\boldsymbol{\theta}|\boldsymbol{\theta}^*)}\right\},$$

where $p(\cdot|y_{1:T})$ denotes the target density or posterior density in our case. If $\boldsymbol{\theta}$ is a parameter vector of dimension m_1 and $\boldsymbol{\theta}^*$ a parameter vector of dimension m_2, with $m_1 \neq m_2$, the transition between $\boldsymbol{\theta}$ and $\boldsymbol{\theta}^*$ is done by generating $\mathbf{u}_1$ of dimension n_1 from a density $q_{1,m}(\mathbf{u}_1|\boldsymbol{\theta})$, and $\mathbf{u}_2$ of dimension n_2 from a density $q_{2,m}(\mathbf{u}_2|\boldsymbol{\theta}^*)$, such that $m_1 + n_1 = m_2 + n_2$. Now, if $J(m, m^*)$ denotes the probability of a move of type m^* given that the chain is at m, the acceptance probability is

$$\alpha_m(\boldsymbol{\theta}, \boldsymbol{\theta}^*) = \min\left\{1, \frac{p(\boldsymbol{\theta}^*, m_2|y_{1:T})J(m_1, m_2)q_{2,m}(\mathbf{u}_2|\boldsymbol{\theta}^*)}{p(\boldsymbol{\theta}, m_1|y_{1:T})J(m_2, m_1)q_{1,m}(\mathbf{u}_1|\boldsymbol{\theta})}\left|\frac{\partial(\boldsymbol{\theta}^*, \mathbf{u}_2)}{\partial(\boldsymbol{\theta}, \mathbf{u}_1)}\right|\right\}.$$

2.7.2 *The Binomial Distribution*

A random variable x follows a binomial distribution with parameters n and p, $x \sim Bin(n, p)$, if its probability function is given by

$$p(x) = \binom{n}{x} p^x (1-p)^{n-x}, \quad x = 0, 1, \ldots, n.$$

In addition, $E(x) = np$ and $V(x) = np(1-p)$.

2.7.3 *The Beta Distribution*

A random variable x follows a beta distribution with parameters $\alpha > 0$ and $\beta > 0$, $x \sim Be(\alpha, \beta)$, if its density is given by

$$p(x) = \frac{1}{B(\alpha, \beta)} x^{\alpha-1}(1-x)^{\beta-1}, \quad x \in [0, 1],$$

where $B(\alpha, \beta)$ is the beta function given by

$$B(\alpha, \beta) = \frac{\Gamma(\alpha)\Gamma(\beta)}{\Gamma(\alpha+\beta)}.$$

We also have that $E(x) = \alpha/(\alpha+\beta)$ and $V(x) = \alpha\beta/[(\alpha+\beta)^2(\alpha+\beta+1)]$.

2.7.4 *The Dirichlet Distribution*

A k-dimensional random variable $\mathbf{x}$ follows a Dirichlet distribution with parameters $\alpha_1, \ldots, \alpha_k$, $\mathbf{x} \sim Dir(\alpha_1, \ldots, \alpha_k)$, if its density function is given by

$$p(\mathbf{x}) = \frac{\Gamma(\alpha_1 + \cdots + \alpha_k)}{\Gamma(\alpha_1) \times \cdots \times \Gamma(\alpha_k)} \theta_1^{\alpha_1 - 1} \times \cdots \times \theta_k^{\alpha_k - 1},$$

with $\theta_1, \ldots, \theta_k \geq 0$, $\sum_{j=1}^k \theta_j = 1$, and $\alpha_j > 0$ for all j. In addition, $E(x_j) = \alpha_j/\alpha$, $V(x_j) = \alpha_j(\alpha - \alpha_j)/[\alpha^2(\alpha+1)]$ and $Cov(x_i, x_j) = -\alpha_i\alpha_j/[\alpha^2(\alpha+1)]$, where $\alpha = \sum_{j=1}^k \alpha_j$.

2.7.5 The Beta-binomial Distribution

A random variable x follows a beta-binomial distribution with parameters n, $\alpha > 0$ and $\beta > 0$, $x \sim Be-Bin(n, \alpha, \beta)$, if its probability mass function is given by

$$p(x) = \frac{\Gamma(n+1)}{\Gamma(x+1)\Gamma(n-x+1)} \times \frac{\Gamma(\alpha+x)\Gamma(n+\beta-x)}{\Gamma(\alpha+\beta+n)} \times \frac{\Gamma(\alpha+\beta)}{\Gamma(\alpha)\Gamma(\beta)},$$

with $x = 0, 1, 2, \ldots, n$. In this case we have $E(x) = n\alpha/(\alpha+\beta)$ and $V(x) = n\alpha\beta(\alpha+\beta+n)/[(\alpha+\beta)^2(\alpha+\beta+1)]$.

2.8 Problems

1. Consider the AR(1) process $y_t = \phi y_{t-1} + \epsilon_t$, with $\epsilon_t \sim N(0, v)$. If $|\phi| < 1$ then $y_t = \sum_{j=0}^{\infty} \phi^j \epsilon_{t-j}$. Use this fact to prove that $y_1 \sim N(0, v/(1-\phi^2))$ and that, as a consequence, the likelihood function has the form (1.17).
2. Consider the AR(1) process $y_t = \phi y_{t-1} + \epsilon_t$, with $\epsilon_t \sim N(0, v)$. Show that the process is nonstationary when $\phi = \pm 1$.
3. Suppose y_t follows a stationary AR(1) model with AR parameter ϕ and innovation variance v. Define $\mathbf{x} = (y_1, \ldots, y_n)'$. We know that $\mathbf{x} \sim N(\mathbf{0}, s\mathbf{\Phi}_n)$ where $s = v/(1-\phi^2)$ is the marginal variance of the y_t process and the correlation matrix $\mathbf{\Phi}_n$ has (i, j) element $\phi^{|i-j|}$, viz.

$$\mathbf{\Phi}_n = \begin{pmatrix} 1 & \phi & \phi^2 & \cdots & \phi^{n-1} \\ \phi & 1 & \phi & \cdots & \phi^{n-2} \\ \phi^2 & \phi & 1 & \cdots & \phi^{n-3} \\ \vdots & \vdots & \vdots & \ddots & \vdots \\ \phi^{n-1} & \phi^{n-2} & \phi^{n-3} & \cdots & 1 \end{pmatrix}.$$

   Find the precision matrix $\mathbf{K}_n = s^{-1}\mathbf{\Phi}_n^{-1}$ and comment on its form. *Hint: One way to find this is "brute-force" matrix inversion using induction; but, that is just linear algebra that—in particular—ignores the probability model that defines $\mathbf{\Phi}_n$. There is a simpler and more instructive way to identify $\mathbf{K}_n$ based on reflecting on the probability model.*

4. Consider an AR(2) process with AR coefficients $\phi = (\phi_1, \phi_2)'$.
   (a) Show that the process is stationary for parameter values lying in the region $-1 < \phi_2 < 1$, $\phi_1 < 1 - \phi_2$, and $\phi_1 > \phi_2 - 1$.
   (b) Show that the partial autocorrelation function of this process is $\phi_1/(1-\phi_2)$ for the first lag, ϕ_2 for the second lag, and equal to zero for any lag h with $h \geq 3$.

5. This question concerns a time series model for continuous and positive outcomes y_t. Suppose a series x_t follows a stationary AR(1) model with parameters ϕ, v and the usual normal innovations. Define a transformed time series $y_t = \exp(\mu + x_t)$ for each t for some known constant μ.
   (a) Show that y_t a first-order Markov process.
   (b) Is y_t a stationary process?
   (c) Find $E(y_t|y_{t-1})$ as a function of y_{t-1} and show that it has the form $E(y_t|y_{t-1}) = ay_{t-1}^{\phi}$ for some positive constant a. Give an expression for a in terms of μ, ϕ, v.
   (d) Can you imagine applied time series contexts that might utilize this simple model as a component? Comment on potential uses.

6. Show that the eigenvalues of the matrix $\mathbf{G}$ given by (2.7) correspond to the reciprocal roots of the AR(p) characteristic polynomial.

7. Consider the AR(2) series $y_t = \phi_1 y_{t-1} + \phi_2 y_{t-2} + \epsilon_t$ with $\epsilon_t \sim N(0, v)$. Following Section 2.1.2, rewrite the model in the standard DLM form $y_t = \mathbf{F}'\mathbf{x}_t$ and $\mathbf{x}_t = \mathbf{G}\mathbf{x}_{t-1} + \mathbf{F}\epsilon_t$ where

$$\mathbf{F} = \begin{pmatrix} 1 \\ 0 \end{pmatrix}, \quad \mathbf{x}_t = \begin{pmatrix} y_t \\ y_{t-1} \end{pmatrix}, \quad \mathbf{G} = \begin{pmatrix} \phi_1 & \phi_2 \\ 1 & 0 \end{pmatrix}.$$

   We know that this implies that, for any given t and over $k \geq 0$, the forecast function is $E(y_{t+k}|\mathbf{x}_t) = \mathbf{F}'\mathbf{G}^k\mathbf{x}_t$.
   (a) Show that the eigenvalues of $\mathbf{G}$ denoted by λ_1 and λ_2 are the roots of the quadratic in λ given by $\lambda^2 - \phi_1\lambda - \phi_2 = 0$. Deduce that $\phi_1 = \lambda_1 + \lambda_2$ and $\phi_2 = -\lambda_1\lambda_2$.
   (b) Suppose that the eigenvalues λ_1, λ_2 are distinct, whether they be real or a pair of complex conjugates. Define

$$\mathbf{\Lambda} = \begin{pmatrix} \lambda_1 & 0 \\ 0 & \lambda_2 \end{pmatrix} \quad \text{and} \quad \mathbf{E} = \begin{pmatrix} \lambda_1 & \lambda_2 \\ 1 & 1 \end{pmatrix} \tau$$

   for any nonzero τ. Note that $\mathbf{E}$ is nonsingular since $\lambda_1 \neq \lambda_2$. Verify that $\mathbf{GE} = \mathbf{E\Lambda}$, so that $\mathbf{G} = \mathbf{E\Lambda E}^{-1}$, that is, $\mathbf{E}$ has columns that are eigenvectors of $\mathbf{G}$ corresponding to eigenvalues (λ_1, λ_2).
   (c) We can take $\tau = 1$ with no loss of generality as τ cancels in the identity $\mathbf{G} = \mathbf{E\Lambda E}^{-1}$; do so from here on. Show that

$$\mathbf{\Lambda}^k\mathbf{E}^{-1} = \frac{1}{(\lambda_1 - \lambda_2)} \begin{pmatrix} \lambda_1^k & -\lambda_1^k\lambda_2 \\ -\lambda_2^k & \lambda_1\lambda_2^k \end{pmatrix}.$$

(d) Deduce that $E(y_{t+k}|\mathbf{x}_t) = a_k y_t + b_k y_{t-1}$ with lagged coefficients

$$a_k = \frac{(\lambda_1^{k+1} - \lambda_2^{k+1})}{(\lambda_1 - \lambda_2)} \quad \text{and} \quad b_k = \frac{(-\lambda_1^{k+1}\lambda_2 + \lambda_1\lambda_2^{k+1})}{(\lambda_1 - \lambda_2)}.$$

(e) Verify that this resulting expression $E(y_{t+k}|\mathbf{x}_t) = a_k y_t + b_k y_{t-1}$ gives the known results in terms of ϕ_1, ϕ_2 when $k = 0$ and $k = 1$.

(f) Consider now the special case of complex eigenvalues $\lambda_1 = re^{i\omega}$ and $\lambda_2 = re^{-i\omega}$ for some real-valued modulus $r > 0$ and argument $\omega > 0$. Show that the lagged coefficients a_k, b_k become

$$a_k = r^k \sin((k+1)\omega)/\sin(\omega) \quad \text{and} \quad b_k = -r^{k+1}\sin(k\omega)/\sin(\omega).$$

(g) Continuing in the case of complex eigenvalues, use simple trigonometric identities to show that the forecast function can be reduced to

$$E(y_{t+k}|\mathbf{x}_t) = r^k h_t \cos(k\omega + g_t), \qquad k = 0, 1, \ldots,$$

a damped cosine form in k (in stationary models with $0 < r < 1$). Give explicit expressions for the time-dependent amplitude $h_t > 0$ and phase g_t in terms of ω and y_{t-1}, y_t.

8. Show that the general solution of the homogeneous difference Equation (2.9) has the form (2.10).
9. Show that, when the characteristic roots are all different, the forecast function of an AR(p) process has the representation given in (2.8).
10. Show that if an AR(2) process has a pair of complex roots given by $r\exp(\pm i\omega)$, they can be written in terms of the AR coefficients as $r = \sqrt{-\phi_2}$ and $\cos(\omega) = \phi_1/2r$.
11. Plot the corresponding forecast functions for the AR(2) processes considered in Example 2.1.
12. Verify that the expressions for the conditional posterior distributions in Section 2.4.1 are correct.
13. Show that a prior on the vector of AR(p) coefficients $\boldsymbol{\phi}$ of the form $N(\phi_1|0, w/\delta_1)$ and $N(\phi_j|\phi_{j-1}, w/\delta_j)$ for $1 < j \leq p$ can be written as $p(\boldsymbol{\phi}) = N(\boldsymbol{\phi}|\mathbf{0}, \mathbf{A}^{-1}w)$, where $\mathbf{A} = \mathbf{H}'\Delta\mathbf{H}$ with $\mathbf{H}$ and Δ defined in Section 2.4.2.
14. Verify the ACF of a MA(q) process given in (2.33).
15. Find the ACF of a general ARMA(1,1) process.
16. Show that Equations (2.38) and (2.39) hold by taking expected values in (2.36) and (2.37) with respect to the whole past history $y_{-\infty,t}$.
17. Consider the AR(1) model given by

$$(1 - \phi B)(y_t - \mu) = \epsilon_t,$$

where $\epsilon_t \sim N(0, v)$.

(a) Find the MLEs for ϕ and μ when $\mu \neq 0$.

(b) Assume that v is known, $\mu = 0$, and that the prior distribution for ϕ is $U(\phi|0, 1)$. Find an expression for the posterior distribution of ϕ.

18. Suppose you observe $y_t = x_t + \nu_t$ where:
    - x_t follows a stationary AR(1) process with AR parameter ϕ and innovation variance v, i.e., $x_t = \phi x_{t-1} + \epsilon_t$ with independent innovations $\epsilon_t \sim N(0, v)$;
    - The ν_t are independent measurement errors with $\nu_t \sim N(0, w)$;
    - The ϵ_t and ν_t series are mutually independent.

    It easily follows that $q = V(y_t) = s + w$ where $s = V(x_t) = v/(1 - \phi^2)$.

    (a) Show that $y_t = \phi y_{t-1} + \eta_t$ where $\eta_t = \epsilon_t + \nu_t - \phi\nu_{t-1}$.

    (b) Show that the lag-1 correlation in the η_t sequence is given by the expression $-\phi w/(w(1 + \phi^2) + v)$.

    (c) Find an expression for the lag$-k$ autocorrelation of the y_t process in terms of k, ϕ, and the signal to noise ratio s/q. Comment on this result.

    (d) Is y_t an AR(1) process? Is it Markov? Discuss and provide theoretical rationalization.

19. You observe $y_t = x_t + \mu$, $t = 1, 2, \ldots,$ where x_t follows a stationary AR(1) process with AR parameter ϕ and innovation variance v, i.e., $x_t = \phi x_{t-1} + \epsilon_t$ with independent innovations $\epsilon_t \sim N(0, v)$. Assume all parameters (μ, ϕ, v) are known.

    (a) Identify the ACF and PACF of y_t, and comment of comparisons with those of x_t.

    (b) What is the marginal distribution of y_t?

    (c) What is the distribution of $(y_t|y_{t-1})$?

    (d) What is the distribution of $(y_t|y_1, \ldots, y_{t-1})$?

    (e) Now consider μ as a parameter to be estimated. As a function of μ and conditioning on the initial value y_1, what is the likelihood function $p(y_2, \ldots, y_{T+1}|y_1, \mu)$?

    (f) Assume ϕ, v are known. Under the reference prior $p(\mu) \propto$ constant, show that the resulting posterior for μ based on the conditional likelihood above is normal with precision $(1 - \phi)^2 T/v$, and give an expression for the mean of this posterior.

    (g) Show that, for large T, the reference posterior mean above is approximately the sample mean of the y_t data.

    (h) If $\phi = 0$, we have the usual normal random sampling problem. For nonzero values of ϕ, the above posterior for the mean of the normal data y_t depends on ϕ in the posterior variance. Comment on how the posterior changes with ϕ and why this makes sense.

20. Consider the ARMA(1,1) model described by

$$y_t = 0.95y_{t-1} + 0.8\epsilon_{t-1} + \epsilon_t,$$

with $\epsilon_t \sim N(0,1)$ for all t.

(a) Show that the one-step-ahead truncated forecast is given by $y_{t+1}^{t,-\infty} = 0.95y_t + 0.8\epsilon_t^{t,-\infty}$, with $\epsilon_t^{t,-\infty}$ computed recursively via $\epsilon_j^{t,-\infty} = y_j - 0.95y_{j-1} - 0.8\epsilon_{j-1}^{t,-\infty}$, for $j = 1:t$ with $\epsilon_0^{t,-\infty} = 0$ and $y_0 = 0$.

(b) Show that the approximate mean square prediction error is

$$MSE_{t+h}^{t,-\infty} = v\left[1 + \frac{(\phi+\theta)^2(1-\phi^{2(h-1)})}{(1-\phi^2)}\right].$$

21. Consider a MA(2) process.

(a) Find its ACF.

(b) Use the innovations algorithm to obtain the one-step-ahead predictor and its mean square error.

22. Let x_t be an AR(p) process with characteristic polynomial $\Phi_x(u)$ and y_t be an AR(q) process with characteristic polynomial $\Phi_y(u)$. What is the structure of the process z_t with $z_t = x_t + y_t$?

23. Consider the infinite order MA process defined by

$$y_t = \epsilon_t + a(\epsilon_{t-1} + \epsilon_{t-2} + \cdots),$$

where a is a constant and the ϵ_ts are i.i.d. N$(0, v)$ random variables.

(a) Show that y_t is nonstationary.

(b) Consider the series of first differences $z_t = y_t - y_{t-1}$. Show that z_t is a MA(1) process. What is the MA coefficient of this process?

(c) For which values of a is z_t invertible?

(d) Find the ACF of z_t.

24. Consider the AR(2) process $y_t = \phi_1 y_{t-1} + \phi_2 y_{t-2} + \epsilon_t$ with $\epsilon_t \sim N(0, v)$ independent with $\phi_1 = 0.9$, and $\phi_2 = -0.9$. Is this process stable? If so write the process as an infinite order MA process, $y_t = \sum_{j=0}^{\infty} \psi_j \epsilon_{t-j}$. Find ψ_j for all j.

25. Consider a process of the form

$$y_t = -2t + \epsilon_t + 0.5\epsilon_{t-1}, \quad \epsilon_t^{i.i.d.} \sim N(0, v).$$

(a) Find the ACF of this process.

(b) Now define $z_t = y_t - y_{t-1} + 2$. What kind of process is this? Find its ACF.

26. Figure 2.14 plots the monthly changes in the US S&P stock market index over 1965 to 2016. Consider an AR(1) model as a very simple exploratory model—for understanding local dependencies but not for

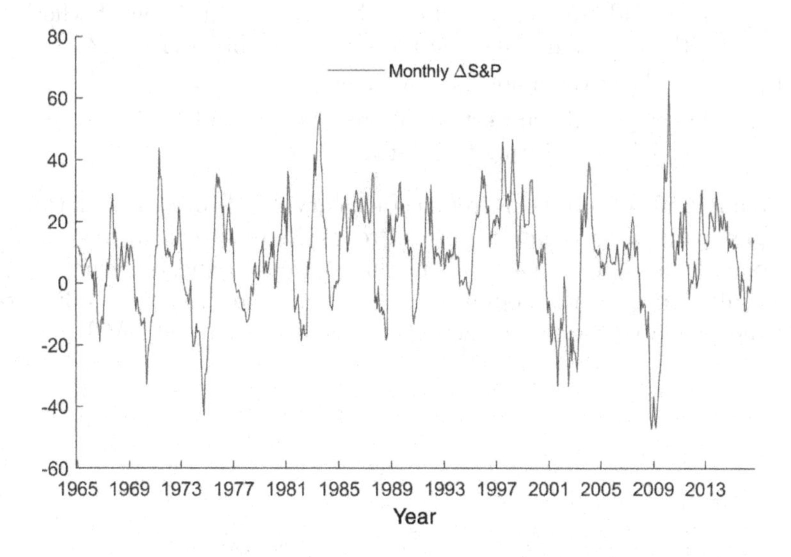

Figure 2.14 *Monthly changes in the US S&P index over 1965–2015.*

forecasting more than a month or two ahead. We know there is a great deal of variation across the years in the market economy and that we might expect "change" that an AR(1) model does not capture. To explore this, we can simply fit the AR(1) model to shorter sections of the data and examine the resulting inferences on parameters to see if they seem to vary across time. Do this as follows. The full series has $T = 621$ months of data; look at many separate time series by selecting a month m and taking some number k months either side; for example, you might take $k = 84$ and for any month m analyze the data over the "windowed period" from $m - k$ to $m + k$ inclusive. Repeat this for each month m running from $m = k + 1$ to $m = T - k$. These repeated analyses will define a "trajectory" of AR(1) analyses over time, one for each sub-series.

For each sub-series, subtract the sub-series mean (to roughly center the sub-series series about zero) and then compute the summaries of the reference posterior for an AR(1) model to just those $2k+1$ time points—just treating each selected sub-series separately. Using the theoretical posterior T distribution for the ϕ parameter, compute and compare (graphically) the exact posterior 90% credible intervals.

(a) Comment on what you see in the plot and comparison, and what you might conclude in terms of changes over time.

(b) Do you believe that short-term changes in S&P have shown real changes in month-month dependencies since 1965?

(c) How would you suggest also addressing the question of whether or not the underlying mean of the series is stable over time?

(d) What about the innovations variance?

(e) What does this suggest for more general models that might do a better job of imitating this data?

27. Sample 500 observations from a stationary AR(4) process with two complex pairs of conjugate reciprocal roots. More specifically, assume that one of the complex pairs has modulus $r_1 = 0.9$ and frequency $\omega_1 = 5$, while the other has modulus $r_2 = 0.75$ and frequency $\omega_2 = 1.35$. Graph the simulated series and their corresponding ACF and PACF.

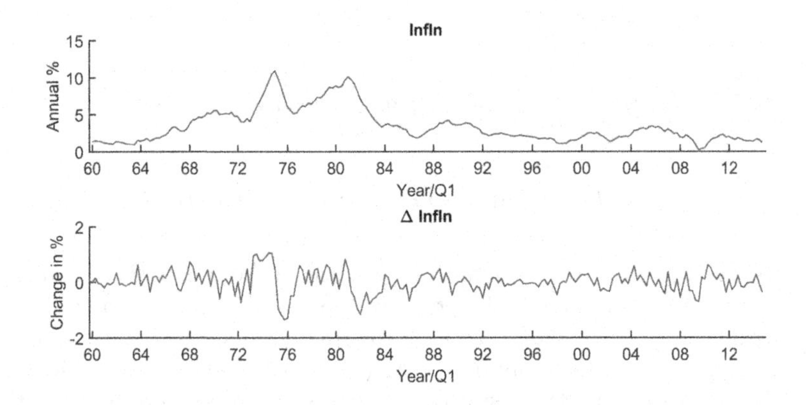

Figure 2.15 *Quarterly US inflation and* Δ*Infln.*

28. Consider the quarterly US macro-economic data in Figure 2.15. Let y_t be the implied series of quarterly changes (i.e., difference values of quarterly actual inflation levels). Fit the reference analysis of an AR(8) model to the y_t data and address the following.

(a) If the AR(8) is accepted as a good model for this data, do you think the data-model match supports stationarity? Give full numerical support for this based on the reference posterior.

(b) Assuming that there is some indication of quasi-periodic behavior under this posterior, summarize inferences on the *maximum wavelength* of (quasi-)periodic components.

(c) Explore and discuss aspects of inference on the implied decomposition of the series into underlying components implied by the eigenstructure of the AR model.

(d) Produce and display graphical summaries—in terms of (Monte Carlo based) posterior medians, upper and lower quartiles, and 10% and

90% points of the predictive distributions of *actual inflation* over the 12 quarters following the end of the data series.

(e) Assuming an AR(p) model is agreeable, do you think $p = 8$ makes sense for the differenced inflation series? Consider features of the fitted residuals from the model as well as numerical measures of model fit including marginal likelihood, AIC and BIC.

29. For the EEG data discussed in Section 2.3, perform the AR(8) Bayesian reference analysis as described there.

(a) Draw histograms of the marginal posterior distributions of the model coefficients ϕ_j for $j = 1:8$.

(b) Draw histograms of the marginal posterior distributions of the moduli and wavelengths of the complex reciprocal roots ordered by decreasing wavelength. In addition, compute a 95% posterior probability interval for each of these moduli and wavelengths.

(c) According to this analysis, what is the approximate posterior probability that the process is nonstationary?

(d) Obtain samples for some of the model error terms and draw histograms of these samples. What is your assessment about the underlying normality assumption?

(e) Repeat the analysis described in Example 2.7 that keeps 2 AR components and inverts the remaining 6 components. Compare your results with Figure 2.13.

30. Consider an ARMA(1, 1) process with AR parameter ϕ, MA parameter θ, and variance v.

(a) Simulate 400 observations from a process with $\phi = 0.9$, $\theta = 0.6$, and $v = 1$.

(b) Compute the conditional least squares estimates of ϕ and θ based on the 400 observations simulated above.

(c) Implement a MCMC algorithm to obtain samples from the posterior distribution of ϕ, θ, and v under the conditional likelihood. Assume a uniform prior distribution in the stationary and invertibility regions for ϕ and θ, and a prior of the form $\pi(v) \propto 1/v$ on the variance parameter. Summarize your posterior inference and forecasting (for up to 100 steps ahead) under this model.

31. Consider the detrended oxygen isotope data analyzed in Chapter 5 (see also Aguilar, Huerta, Prado, and West 1999).

(a) Under AR models use the AIC/BIC criteria to obtain the model order p that is the most compatible with the data.

(b) Fit an AR model with the value of p obtained in (a) via maximum likelihood and compute the residuals. Graph the ACF, PACF, and Q-Q plot of the residuals. Are these plots consistent with the assumption of normality?

(c) Using your AR MLE fit, find point estimates for the moduli and wavelengths of the complex reciprocal roots ordering the roots by decreasing wavelength.

(d) Compute MSE forecasts for the next 100 observations. Plot these forecasts along with 95% prediction intervals.

32. This question concerns the alternative state-space representation of an AR(p) model that arises as a special case of the state-space representation of ARMA(p, q) models when $q = 0$. This is also easily seen to be defined by a direct linear transformation of state vectors in the standard representation, as follows.

Begin with the standard state-space representation of the AR(p) model; the state vector is $\mathbf{x}_t = (y_t, y_{t-1}, \ldots, y_{t-p+1})'$ and the model equations are $y_t = \mathbf{F}'\mathbf{x}_t$ and $\mathbf{x}_t = \mathbf{G}\mathbf{x}_{t-1} + \mathbf{F}\epsilon_t$ where

$$\mathbf{F} = \begin{pmatrix} 1 \\ 0 \\ 0 \\ \vdots \\ 0 \end{pmatrix} \quad \text{and} \quad \mathbf{G} = \begin{pmatrix} \phi_1 & \phi_2 & \cdots & \phi_{p-1} & \phi_p \\ 1 & 0 & \cdots & 0 & 0 \\ \vdots & \ddots & \ddots & & \vdots \\ 0 & 0 & \ddots & 0 & 0 \\ 0 & 0 & \cdots & 1 & 0 \end{pmatrix}$$

with AR parameters $\boldsymbol{\phi} = (\phi_1, \ldots, \phi_p)'$ and innovations $\epsilon_t \sim N(0, v)$. Define the $p \times p$ symmetric matrix $\mathbf{A}$ by

$$\mathbf{A} = \begin{pmatrix} 1 & 0 & 0 & 0 & \cdots & 0 & 0 \\ 0 & \phi_2 & \phi_3 & \phi_4 & \cdots & \phi_{p-1} & \phi_p \\ 0 & \phi_3 & \phi_4 & \phi_5 & \cdots & \phi_p & 0 \\ \vdots & \vdots & \vdots & & \iddots & \iddots & \vdots \\ \vdots & \vdots & \vdots & \iddots & \iddots & & \vdots \\ 0 & \phi_{p-1} & \phi_p & 0 & \cdots & \cdots & 0 \\ 0 & \phi_p & 0 & 0 & \cdots & \cdots & 0 \end{pmatrix}$$

(a) Verify that the matrix product $\mathbf{AG}$ is given by

$$\mathbf{AG} = \begin{pmatrix} \phi_1 & \phi_2 & \phi_3 & \phi_4 & \cdots & \phi_{p-1} & \phi_p \\ \phi_2 & \phi_3 & \phi_4 & \phi_5 & \cdots & \phi_p & 0 \\ \phi_3 & \phi_4 & \phi_5 & \phi_5 & \cdots & 0 & 0 \\ \vdots & \vdots & \vdots & & \iddots & \iddots & \vdots \\ \vdots & \vdots & \vdots & \iddots & \iddots & & \vdots \\ \phi_{p-1} & \phi_p & 0 & 0 & \cdots & \cdots & 0 \\ \phi_p & 0 & 0 & 0 & \cdots & \cdots & 0 \end{pmatrix}$$

noting that this is also symmetric.

(b) Show or deduce that:

i. For a proper AR(p) model in which $\phi_p \neq 0$, then $|\mathbf{A}| \neq 0$ so that $\mathbf{A}$ is nonsingular.

ii. $\mathbf{AGA}^{-1} = \mathbf{G}'$.

iii. $\mathbf{AF} = \mathbf{F}$ and, as a result, $\mathbf{F}' = \mathbf{F}'\mathbf{A}^{-1}$.

(c) Hence show that an equivalent state-space AR(p) form is given by $y_t = \mathbf{F}'\mathbf{z}_t$ and $\mathbf{z}_t = \mathbf{G}'\mathbf{z}_{t-1} + \mathbf{F}\epsilon_t$ based on a new $p \times 1$ state vector $\mathbf{z}_t = \mathbf{A}\mathbf{x}_t$ and where the state evolution matrix is $\mathbf{G}'$, i.e.,

$$\mathbf{G}' = \begin{pmatrix} \phi_1 & 1 & 0 & \cdots & 0 \\ \phi_2 & 0 & 1 & \cdots & 0 \\ \vdots & \vdots & & \ddots & \vdots \\ \phi_{p-1} & 0 & 0 & \cdots & 1 \\ \phi_p & 0 & 0 & \cdots & 0 \end{pmatrix}$$

(d) What is the interpretation of the elements of the transformed state vector $\mathbf{z}_t$?

33. Two univariate time series y_t, z_t follow the *coupled* dynamic models over $t = 1, \ldots$ given by

$$y_t = \phi y_{t-1} + \gamma z_t + \nu_t \qquad \text{and} \qquad z_t = \theta z_{t-1} + \epsilon_t$$

where $\nu_t \sim N(0, v)$ and $\epsilon_t \sim N(0, u)$ are independent and mutually independent innovations sequences. In backshift operator notation,

$$\phi(B) y_t = \gamma z_t + \nu_t \qquad \text{and} \qquad \theta(B) z_t = \epsilon_t$$

where $\phi(B) = 1 - \phi B$ and $\theta(B) = 1 - \theta B$. The two AR(1) coefficients have values such that $|\phi| < 1$ and $|\theta| < 1$. Now suppose that you do not observe z_t. The model equations imply a *marginal* model for the y_t series alone when z_t is not observed.

(a) Show that this marginal model for y_t is

$$y_t = \alpha_1 y_{t-1} + \alpha_2 y_{t-2} + \eta_t$$

where $\alpha_1 = \phi + \theta$ and $\alpha_2 = -\phi\theta$, and η_t is a zero-mean, normal random quantity.

(b) Comment on values of the α_1, α_2 coefficients in cases when ϕ, θ are positive and close to 1. As part of this, discuss the kinds of behavior you would expect to see in the y_t series.

(c) Is this marginal process y_t Markovian? Is it an AR process?

(d) Consider a context in macro-economics where y_t is quarterly inflation and z_t is, for example, a quarterly measure of national money supply, perhaps slightly lagged. Two econometricians Chas and Dave are discussing predicting inflation. They agree that z_t looks like an AR(1) process. For y_t, Chas wants to use z_t as a predictor of y_t in a model as presented above. Dave says he will ignore z_t and will just use a higher-order AR model, such as an AR(6).

i. Are they really building different models?
ii. If you look at predicting multiple quarters ahead, what complications arise for each approach?

(e) Discuss predictive models to potentially improve (i) the description of the US inflation time series and (ii) its predictability. Any ideas and comments should be specific to the above discussion of simple (y_t, z_t) models/relationships.

34. Suppose y_t follows a stationary AR(1) process with AR parameter ϕ and innovation variance v with (ϕ, v) uncertain. At any time t write $\mathcal{D}_t$ for the past data and information, including all past observations. If no additional information arises over the time interval $(t-1, t]$, then $\mathcal{D}_t$ sequentially updates as the new observation is made via—simply—$\mathcal{D}_t = \{\mathcal{D}_{t-1}, y_t\}$.

Now suppose you are standing at the end of time interval $t-1$ so that you have current information set $\mathcal{D}_{t-1}$. The current posterior for (ϕ, v) based on this information has a conjugate normal-inverse gamma form written as

$$\begin{aligned}(\phi|v, \mathcal{D}_{t-1}) &\sim N(m_{t-1}, C_{t-1}(v/s_{t-1})),\\ (v^{-1}|\mathcal{D}_{t-1}) &\sim G(n_{t-1}/2, n_{t-1}s_{t-1}/2)\end{aligned}$$

with known defining parameters. This would be the case, for example, of a reference posterior based on the first $t-1$ observations. Here m_{t-1} and $s_{t-1} > 0$ are natural point estimates of ϕ and v respectively, while $C_{t-1} > 0$ and $n_{t-1} > 0$ relate to uncertainty.

(a) What is the current marginal posterior for ϕ, namely $p(\phi|\mathcal{D}_{t-1})$?

(b) Show that, conditional on v and marginalizing over ϕ, the implied 1−step ahead forecast distribution for y_t given v is

$$(y_t|v, \mathcal{D}_{t-1}) \sim N(f_t, q_t v/s_{t-1})$$

with $f_t = m_{t-1}y_{t-1}$ and $q_t = s_{t-1} + C_{t-1}y_{t-1}^2$.

(c) Now marginalize also over v to find the implied 1−step ahead forecast distribution for y_t, namely $p(y_t|\mathcal{D}_{t-1})$, i.e., the distribution you will use in practice to predict y_t 1−step ahead. What is this distribution?

(d) Now move to time t and observe the outcome y_t. Show that the time t posterior $p(\phi, v|\mathcal{D}_t)$ is also normal-inverse gamma, having the same form as in at time $t-1$ above but now with $t-1$ updated to t and updated defining parameters $\{m_t, C_t, n_t, s_t\}$ that can be written in the following forms:

- $m_t = m_{t-1} + A_t e_t$,
- $C_t = r_t(C_{t-1} - A_t^2 q_t)$,
- $n_t = n_{t-1} + 1$,
- $s_t = r_t s_{t-1}$ with $r_t = (n_{t-1} + e_t^2/q_t)/n_t$,

where

- $e_t = y_t - f_t$ is the realized 1−step ahead (point) forecast error, and
- $A_t = C_{t-1} y_{t-1}/q_t$ is the adaptive coefficient.

(e) Comment on these expressions, giving particular attention to the following:

i. How (m_t, C_t) depend on the new data y_t relative to the prior values (m_{t-1}, C_{t-1}).
ii. The role of the adaptive coefficient in the update of (m_{t-1}, C_{t-1}) to (m_t, C_t).
iii. The updates for the degrees of freedom n_t and point estimate s_t and how they depend on y_t.

(f) Consider an example in which the forecast error is very large relative to expectation, resulting in a value of e_t^2/q_t much greater than 1. Comment on how the posterior for (ϕ, v) responds.

Chapter 3

The frequency domain

Harmonic regression provides the basic background and introduction to methods of cyclical time series modeling and spectral theory for stationary time series analysis. These topics are covered here. Spectral analysis, particularly that based on traditional nonparametric statistical approaches, is widely applied to time series data processing in the physical and engineering sciences. Spectral theory, nonparametric methods, and relationships with parametric time series models are also discussed in this chapter.

3.1 Harmonic Regression

Harmonic regression refers to—usually linear—models describing periodicities in data by sinusoids. The simplest case of a single sinusoid is most illuminating.

3.1.1 The One-component Model

Consider a time series with mean zero, observed at possibly unequally spaced times $t_1, \ldots, t_T$, in a context in which the series has a suspected periodic and sinusoidal component. The simplest model is the single-component harmonic regression given by

$$y_{t_i} = \rho \cos(\omega t_i + \eta) + \epsilon_{t_i}, \tag{3.1}$$

where ϵ_{t_i} is a noise series for $i = 1 : T$. Some terminology, notation, and general considerations are now introduced.

- ω is the angular frequency, measured in radians. The frequency in cycles per unit time is $\omega/2\pi$, and the corresponding wavelength or period is $\lambda = 2\pi/\omega$.

• The phase η lies between zero and 2π.

• The angular frequency ω is usually restricted to lie between zero and π for identification. In fact, note that given any value of ω, the model is unchanged at angular frequencies $\omega \pm k\pi$ for integer k, suggesting a restriction to $\omega < 2\pi$; also, for $0 < \omega < \pi$, we have the same model with a sign change of $\rho \to -\rho$ at angular frequency $2\pi - \omega$, and so the restriction $0 < \omega < \pi$ is added. This implies a period $\lambda > 2$. The highest possible frequency $\omega = \pi$, at which $\lambda = 2$, is the Nyquist frequency, this period being the smallest detectable in cases of equally spaced data one time unit apart, i.e., $t_i = i$.

• The origin and scale of measurement on the time axis are essentially arbitrary. Transforming the time scale to $u = (t - x)/s$, for any x and $s > 0$, changes the model form to have angular frequency $s\omega$, phase $\eta + x\omega$, and unchanged amplitude on the new time scale.

• We can rewrite the model as

$$y_{t_i} = \alpha_1 \cos(\omega t_i) + \alpha_2 \sin(\omega t_i) + \epsilon_{t_i}, \tag{3.2}$$

where $\alpha_1 = \rho \cos(\eta)$ and $\alpha_2 = -\rho \sin(\eta)$, so that $\rho^2 = \alpha_1^2 + \alpha_2^2$ and $\eta = \tan^{-1}(-\alpha_2/\alpha_1)$.

Interest lies in estimating the frequency, amplitude, and phase (ω, ρ, η), or equivalently, the frequency plus the two harmonic coefficients $(\omega, \alpha_1, \alpha_2)$, together with characteristics of the noise series. The most basic model assumes that the ϵ_{t_i}s are i.i.d., usually with $N(\epsilon_{t_i}|0, v)$. The reference analysis under this assumption is now detailed.

3.1.1.1 Reference Analysis

Denote $p(\omega, \alpha_1, \alpha_2, v)$ the joint prior density for the four model parameters, resulting in the posterior density

$$p(\omega, \alpha_1, \alpha_2, v|y_{t_1:t_T}) \propto p(\omega, \alpha_1, \alpha_2, v) \prod_{i=1}^{T} N(y_{t_i}|\mathbf{f}'_{t_i}\boldsymbol{\beta}, v),$$

where $\boldsymbol{\beta} = (\alpha_1, \alpha_2)'$ and $\mathbf{f}_{t_i} = (\cos(\omega t_i), \sin(\omega t_i))'$. Thus, conditional on any specified value of ω,

$$p(a, b, v|\omega, y_{t_1:t_T}) \propto p(a, b, v|\omega) \prod_{i=1}^{T} N(y_{t_i}|\mathbf{f}'_{t_i}\boldsymbol{\beta}, v),$$

and the likelihood function here is that from the simple linear regression $y_{t_i} = \mathbf{f}'_{t_i}\boldsymbol{\beta} + \epsilon_{t_i}$ with ω fixed. Analysis under any prior, or class of priors, may now proceed; we detail the traditional reference analysis in which $p(\boldsymbol{\beta}, v|\omega) \propto v^{-1}$. Notice that this distribution implies no dependence a priori between the harmonic coefficients and the frequency ω, in addition to the usual assumption of a noninformative form. The standard linear model theory applies (see Chapter 1); we use this to evaluate the conditional posterior $p(\boldsymbol{\beta}, v|\omega, y_{t_1:t_T})$. Then, the posterior for ω follows via $p(\omega|y_{t_1:t_T}) \propto p(\omega)p(y_{t_1:t_T}|\omega)$, where $p(y_{t_1:t_T}|\omega)$ is the marginal data density under the linear regression at the specified ω value. As a result, we have the following ingredients. First condition on a value of ω. Quantities $\mathbf{f}_{t_i}$ and others below are implicitly dependent on ω, though this is not explicitly recognized in the notation. With this understood, write $\mathbf{F}'$ for the $T \times 2$ design matrix with rows $\mathbf{f}'_{t_i}$. Recall (again, see Chapter 1) that $\hat{\boldsymbol{\beta}} = (\mathbf{F}\mathbf{F}')^{-1}\mathbf{F}\mathbf{y}$, with $\mathbf{y} = (y_{t_1}, \ldots, y_{t_T})'$, and the residual sum of squares is given by $R = \mathbf{e}'\mathbf{e} = \mathbf{y}'\mathbf{y} - \hat{\boldsymbol{\beta}}'\mathbf{F}\mathbf{F}'\hat{\boldsymbol{\beta}}$, the term $\hat{\boldsymbol{\beta}}'\mathbf{F}\mathbf{F}'\hat{\boldsymbol{\beta}}$ being the usual sum of squares explained by the regression. Then, we have the following:

- $p(\boldsymbol{\beta}|v, \omega, \mathbf{y})$ is $N(\boldsymbol{\beta}|\hat{\boldsymbol{\beta}}, v(\mathbf{F}\mathbf{F}')^{-1})$ and $p(\boldsymbol{\beta}|\omega, \mathbf{y})$ is $T_{T-2}(\boldsymbol{\beta}|\hat{\boldsymbol{\beta}}, s^2(\mathbf{F}\mathbf{F}')^{-1})$, with $s^2 = R/(T-2)$. For large T, the multivariate Student-t distribution is roughly $N(\boldsymbol{\beta}|\hat{\boldsymbol{\beta}}, s^2(\mathbf{F}\mathbf{F}')^{-1})$.
- The marginal data density at the assumed ω value is
$$\begin{aligned} p(y_{t_1:t_T}|\omega) &\propto |\mathbf{F}\mathbf{F}'|^{-1/2}R^{-(T-2)/2} \\ &\propto |\mathbf{F}\mathbf{F}'|^{-1/2}\{1 - \hat{\boldsymbol{\beta}}'\mathbf{F}\mathbf{F}'\hat{\boldsymbol{\beta}}/(\mathbf{y}'\mathbf{y})\}^{(2-T)/2}, \end{aligned} \tag{3.3}$$
where the proportionality constants do not depend on ω.

Performing this analysis for various values of ω produces the likelihood function $p(y_{t_1:t_T}|\omega)$, resulting in a marginal posterior
$$\begin{aligned} p(\omega|y_{t_1:t_T}) &\propto p(\omega)p(y_{t_1:t_T}|\omega) \\ &\propto p(\omega)|\mathbf{F}\mathbf{F}'|^{-1/2}\{1 - \hat{\boldsymbol{\beta}}'\mathbf{F}\mathbf{F}'\hat{\boldsymbol{\beta}}/(\mathbf{y}'\mathbf{y})\}^{(2-T)/2}, \end{aligned} \tag{3.4}$$
where the dependence of $\mathbf{F}$ and $\hat{\boldsymbol{\beta}}$ on ω is noted. This can be evaluated across a range of ω values and the resulting density summarized to infer ω, or, via transformation, the uncertain period $\lambda = 2\pi/\omega$. Note that, when the error variance v is assumed known, the above development would simplify to give
$$p(\omega|y_{t_1:t_T}) \propto p(\omega)|\mathbf{F}\mathbf{F}'|^{-1/2}\exp(\hat{\boldsymbol{\beta}}'\mathbf{F}\mathbf{F}'\hat{\boldsymbol{\beta}}/2v).$$
Note further that priors $p(\boldsymbol{\beta}, v|\omega)$ other than the reference prior used here would, naturally, lead to different posterior distributions.

Useful insights and connections with developments in spectral analysis arise if we have equally spaced data, i.e., when $t_i = i$ for all i, and if we restrict the interest to values of ω at the so-called Fourier frequencies for the data set, namely $\omega_k = 2\pi k/T$ across integers $1 \leq k < T/2$. For any such ω, it is easily verified (using the simple trigonometric identities given in the Appendix) that the linear regression models are orthogonal. At $\omega = \omega_k$, this results in $\mathbf{FF}' = (T/2)\mathbf{I}_2$ in each case, and the MLEs $\hat{\boldsymbol{\beta}}_k = (\hat{\alpha}_1(\omega_k), \hat{\alpha}_2(\omega_k))'$ given by

$$\begin{aligned} \hat{\alpha}_1(\omega_k) &= (2/T)\sum_{i=1}^{T} y_i \cos(\omega_k i), \\ \hat{\alpha}_2(\omega_k) &= (2/T)\sum_{i=1}^{T} y_i \sin(\omega_k i), \end{aligned}$$

where we now make the dependence on the chosen frequency quite explicit in the notation.

It then follows that the sum of squares explained by regression is just

$$\hat{\boldsymbol{\beta}}_k' \mathbf{FF}' \hat{\boldsymbol{\beta}}_k = I(\omega_k) \equiv \frac{T[\hat{\alpha}_1^2(\omega_k) + \hat{\alpha}_2^2(\omega_k)]}{2},$$

and we deduce that

$$p(\omega|y_{1:T}) \propto p(\omega)\{1 - I(\omega)/\mathbf{y}'\mathbf{y}\}^{(2-T)/2} \tag{3.5}$$

for the Fourier frequencies $\omega_k = 2\pi k/T$ for $1 \leq k < T/2$. In the case of a known error variance v, this result is modified as

$$p(\omega|y_{1:T}) \propto p(\omega)\exp(I(\omega)/2v).$$

In addition, we have the following:

- In many cases of unequally spaced observations, assuming that T is large and ω not too small, the exact marginal posterior (3.4) is closely approximated by the special form (3.5) for varying ω, not just at the Fourier frequencies.
- The functions $p(y_{1:T}|\omega)$ are volatile, especially as ω decreases toward zero, being typically highly multimodal. Often a global mode is apparent and clearly dominant, though sometimes this is not the case.
- Various numerical integration methods can be used to compute posterior means, probabilities, etc., as the numerical problem is effectively in one dimension. For instance, discretization of the range or rejection methods may be used to draw samples from $p(\omega|y_{1:T})$.

3.1.2 The Periodogram

The function

$$I(\omega) = \frac{T}{2}(\hat{\alpha}_1^2(\omega) + \hat{\alpha}_2^2(\omega)) \tag{3.6}$$

is referred to as the periodogram in traditional spectral analysis. Other authors have used various scalar multiples of $I(\omega)$, however, this is not an issue, as the relative values are the ones that determine the importance of different ωs. Traditionally, the view is that if a frequency ω is related to a large value of $I(\omega)$ then such frequency should be important.

From the definitions of $\hat{\alpha}_1(\omega)$ and $\hat{\alpha}_2(\omega)$ in the orthogonal case, it follows that

$$I(\omega) = \frac{2}{T}|\sum_{i=1}^{T} y_{t_i} e^{-i\omega t_i}|^2,$$

and so the periodogram is efficiently computed—particularly for very large T—via the fast Fourier transform. Note that this is an extension of the usual definition of the periodogram which is restricted to evaluation at the Fourier frequencies based on equally spaced data.

3.1.3 Some Data Analyses

The basic, single harmonic regression model is rarely an adequate global model for time series in practical work, and the interpretation of the periodogram in connection with this model must be treated carefully. A periodogram, or the corresponding likelihood function, may take high values over certain frequency ranges, and may appear to heavily support a single frequency above all others, even though the model fit is globally very poor and the model may be inappropriate for the series in terms of explanation and prediction. Keeping this point in mind, these kinds of computations often provide informative preliminary analyses that may serve as a prelude to further analyses with more refined models. A couple of examples elaborate these points.

Example 3.1 *Electroencephalogram (EEG) series.* First, revisit the EEG series from Chapter 2. The above computations produce the log-likelihood function for period $\lambda = 2\pi/\omega$ displayed in Figure 3.1; the assumption of a uniform prior implies that this is the log-posterior density for the period of a single harmonic regression model. The range of likely periods agrees with previous autoregressive analyses, but note the variability and erratic behavior exhibited. On the likelihood or posterior density scale, the global mode around $\lambda = 13$ dominates, i.e., the function visually appears unimodal after exponentiation. This is a case in which the series is, as we know, adequately represented by a stationary autoregressive model; the autocorrelations decay with lag and the autoregressive models fitted earlier seem appropriate. These models indicated dominant, quasicyclical components of period around 12 to 13, though with substantial variation in

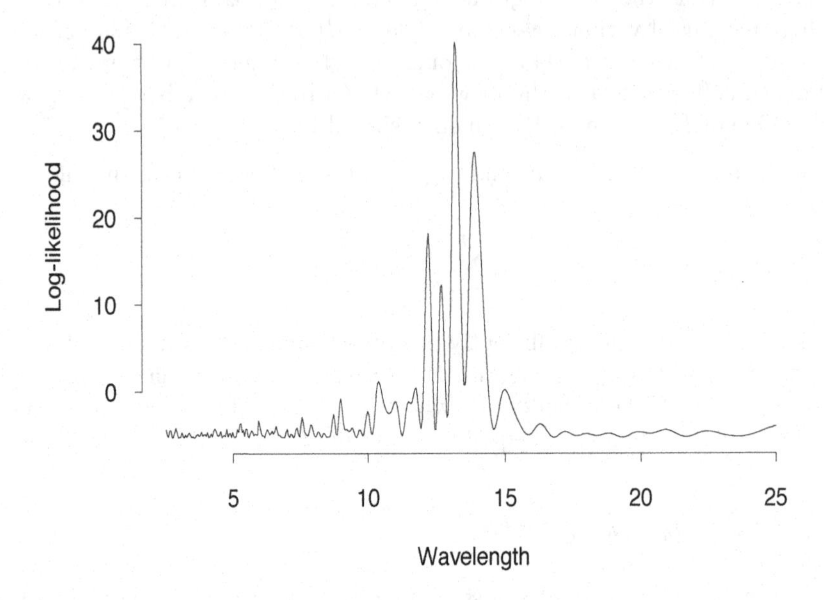

Figure 3.1 *Log-likelihood function for period in a single harmonic regression model of EEG series.*

amplitude and phase characteristics. This is reflected in Figure 3.1, though the sustained and time-invariant sinusoidal variations predicted by a simple harmonic regression are not so appropriate. Note also the spread of larger values of the log-likelihood function in this wavelength region, which is characteristic of such functions when the series is adequately modeled by a stationary process with quasicyclical components, and so the computations do give some guide to aspects of the time series structure.

Example 3.2 *Mauna Loa series.* The data displayed in Figure 3.2 are monthly measures of ground level carbon dioxide concentrations at Mauna Loa, Hawaii, during the period from January 1959 to December 1975. This data set is available in `R` with `data(co2)`. Figure 3.3 displays the first differences of the series and a series constructed by subtracting a lowess (lowess stands for locally weighted scatterplot smoother, see Cleveland and Devlin 1988) estimate of trend. There are apparent differences induced by these two crude methods of detrending. Subtracting a smooth trend line reduces the series to an apparently sinusoidal pattern, while differencing induces obvious departures around the peaks and troughs, resulting in a series that is not symmetric about the time axis. Periodic and decaying autocorrelations, in each case, support annual periodicities, a key wavelength of $\lambda = 12$, as is

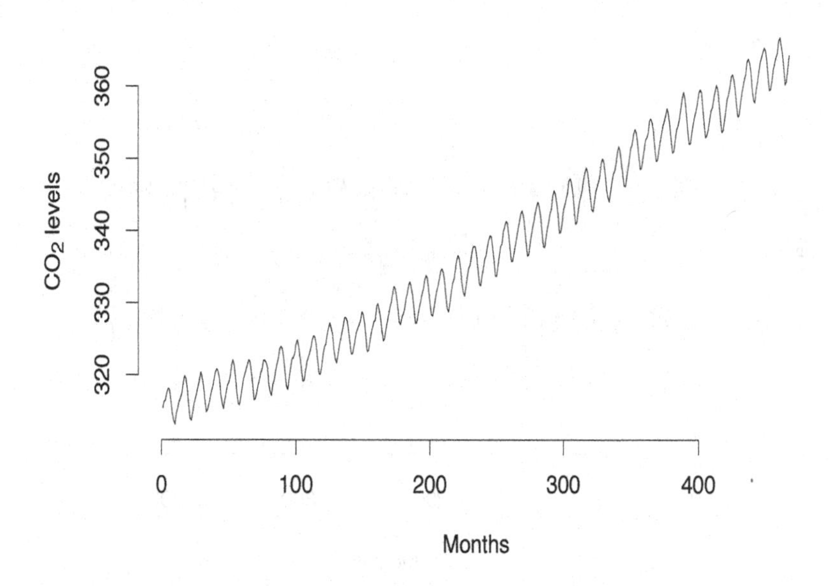

Figure 3.2 *Mauna Loa carbon dioxide series.*

naturally expected and supported by the periodogram/log-likelihood plots in Figure 3.4.

The tradition in displaying periodograms is to plot $\log_{10}(I(\omega))$ versus ω, which, though we plot here as a function of the wavelength $\lambda = 2\pi/\omega$ rather than the frequency, has the effect of substantially amplifying the subsidiary peaks. Applied here, this enhances the small subsidiary peaks near $\lambda = 6$, and makes evident another spike around $\lambda = 4$ not visually apparent in the log-likelihood plot. Similar pictures arise in plotting the logarithm of the log-likelihood (adding a constant to ensure positivity of the log-likelihood). Generally, such peaks are often interpreted as suggestive of subsidiary structure, though this example typifies periodic phenomena whose forms are not strictly sinusoidal; the frequencies $2\pi/6$ and $2\pi/4$ are the higher order harmonics of the fundamental frequency $2\pi/12$, and their relevance in describing periodic behavior of period 12 is evident from standard theory of Fourier representations, as follows in the next section. Note, as an aside, that the secondary peak is more marked for the differenced data than for the directly detrended data, consistent with the fact that former methods of detrending lead to a series rather less well described by the single sinusoid; this supports the view that it is often better to model trends (and other components) of a series directly, and then either adjust as here or, better, estimate together with the periodic components, than to

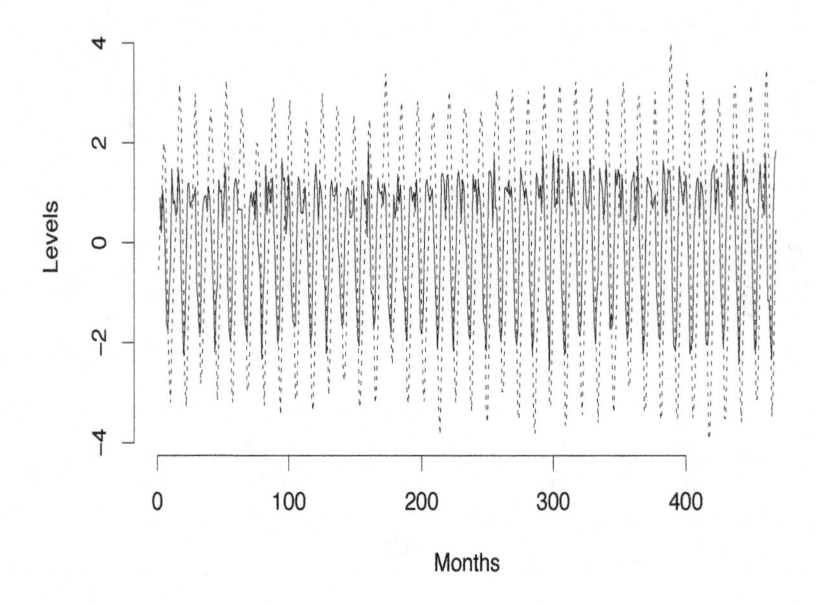

Figure 3.3 *Differenced (solid line) and smooth-detrended (dashed line) carbon dioxide series.*

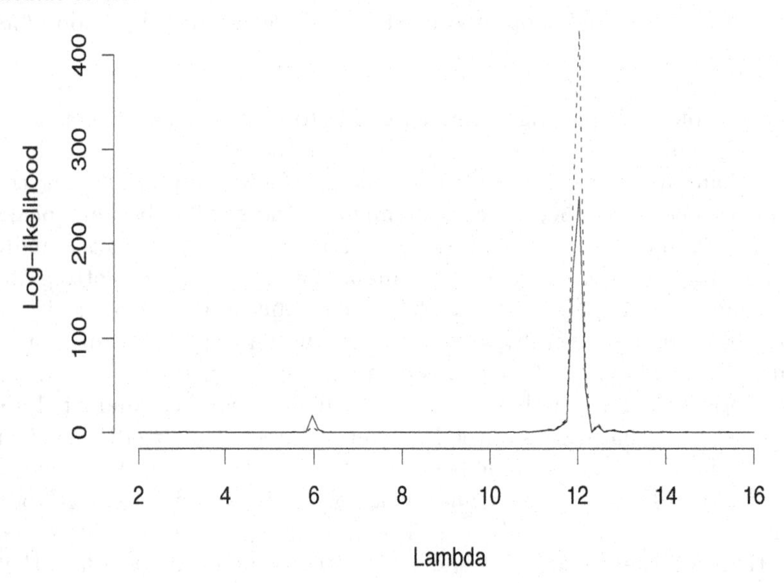

Figure 3.4 *Log-likelihood function for period in single harmonic regression models of differenced (solid line) and detrended (dashed line) carbon dioxide series.*

transform the series using differencing operations, as the latter naturally induce irregularities such as those observed here.

3.1.4 Several Uncertain Frequency Components

The foregoing is directly extensible to several frequencies, as may be of interest in exploring data for evidence of quite distinct periodic components. A two-component model, for example, is $y_{t_i} = \mathbf{f}_{t_i}'\boldsymbol{\beta} + \epsilon_{t_i}$ with $\boldsymbol{\beta} = (\alpha_1(\omega_1), \alpha_2(\omega_1), \alpha_1(\omega_2), \alpha_2(\omega_2))'$, and

$$\mathbf{f}_{t_i}' = (\cos(\omega_1 t_i), \sin(\omega_1 t_i), \cos(\omega_2 t_i), \sin(\omega_2 t_i)).$$

Viewing ω_1 and ω_2 as distinct frequencies, we may proceed to analyze this model via the same approach used in the one-cycle model. Then, under a conditional reference prior specification with

$$p(\boldsymbol{\beta}, v|\omega_1, \omega_2) = p(\boldsymbol{\beta}, v) \propto v^{-1},$$

the key results are as follows, all analogous to the earlier development in the one-cycle model. First, the likelihood function for the two frequencies is given by

$$p(y_{t_1:t_T}|\omega_1, \omega_2) \quad \propto \quad |\mathbf{F}\mathbf{F}'|^{-1/2}\{1 - \hat{\boldsymbol{\beta}}'\mathbf{F}\mathbf{F}'\hat{\boldsymbol{\beta}}'/(\mathbf{y}'\mathbf{y})\}^{(p-T)/2}. \quad (3.7)$$

Here the number of parameters is $p = 4$. This can be evaluated and plotted over a grid of values of (ω_1, ω_2) or, alternatively, in terms of the wavelengths (λ_1, λ_2), with $\lambda_j = 2\pi/\omega_j$. Bretthorst (1988) displays contour plots in some examples. Further analysis requires numerical or analytic approximation to this bivariate posterior.

In the special case of equally spaced observations, the regression is orthogonal at any of the Fourier frequencies $\omega_{1,k} = 2\pi k/T$ and $\omega_{2,l} = 2\pi l/T$ for any integers k, l between zero and $T/2$. Then the estimated coefficients are again Fourier transform-based, namely,

$$\begin{aligned}
\hat{\alpha}_1(\omega_{1,k}) &= (2/T)\sum_{i=1}^{T} y_i \cos(\omega_{1,k} i), \\
\hat{\alpha}_2(\omega_{1,k}) &= (2/T)\sum_{i=1}^{T} y_i \sin(\omega_{1,k} i), \\
\hat{\alpha}_1(\omega_{2,l}) &= (2/T)\sum_{i=1}^{T} y_i \cos(\omega_{2,l} i), \\
\hat{\alpha}_2(\omega_{2,l}) &= (2/T)\sum_{i=1}^{T} y_i \sin(\omega_{2,l} i),
\end{aligned}$$

and the fitted sum of squares is

$$\begin{aligned}\hat{\boldsymbol{\beta}}_{k,l}'\mathbf{F}\mathbf{F}'\hat{\boldsymbol{\beta}}_{k,l} &= \frac{T}{2}[\hat{\alpha}_1^2(\omega_{1,k})+\hat{\alpha}_2^2(\omega_{1,k})+\hat{\alpha}_1^2(\omega_{2,l})+\hat{\alpha}_2^2(\omega_{2,l})]\\ &= I(\omega_{1,k})+I(\omega_{2,l}),\end{aligned}$$

so that

$$p(y_{1:T}|\omega_1,\omega_2)\propto\{1-[I(\omega_1)+I(\omega_2)]/\mathbf{y}'\mathbf{y}\}^{(4-T)/2}$$

closely approximates (3.7) in the case of unequally spaced observations when T is large if ω_1 and ω_2 are not too small. In the case of known error variance v, this is modified to

$$p(y_{1:T}|\omega_1,\omega_2)\propto\exp\{[I(\omega_1)+I(\omega_2)]/2v\}.$$

Either way, the resulting likelihood function leads to a corresponding posterior for the two frequencies or wavelengths. In some cases, the data support well-separated frequencies in which case the likelihood is essentially orthogonal and similar results are obtained by sequentially fitting first one cycle and then, on the estimated residuals, a second. The likelihood is a monotonic function of the sum of the two periodogram ordinates, and so the periodogram will have its peaks at the maximum likelihood value, being appreciable in other regions favored by the likelihood function. Further development appears in Bretthorst (1988).

The next section discusses models in which harmonic components of a base frequency, i.e., sinusoids with integer multiples of the base frequency, are represented. Note that the above development can be pursued with the modification that each of the two (or more) distinct periodic components may be better represented by a collection of harmonics. This suggests inclusion of some or all of the harmonics of each of the ω_1 and ω_2 frequencies.

3.1.5 Harmonic Component Models of Known Period

Standard Fourier representation of periodic functions proves useful in modeling series with persistent periodic patterns having identifiable integer periods, especially in dealing with seasonal phenomena. The basic theory of transformation to Fourier or harmonic regression coefficients is as follows. Given an integer period p, the numbers $y_1,\ldots,y_p$ have the exact representations

$$y_t = a_0+\sum_{k=1}^{m}\{a_k\cos(2\pi kt/p)+b_k\sin(2\pi kt/p)\},\qquad (t=1:p),$$

where $m = \lfloor p/2 \rfloor$, the integer part of $p/2$, and the Fourier coefficients are defined as follows: $a_0 = (1/p)\sum_{t=1}^{p} y_t$, and for $1 \le k < m$,

$$\begin{aligned} a_k &= (2/p)\sum_{t=1}^{p} y_t \cos(2\pi kt/p), \\ b_k &= (2/p)\sum_{t=1}^{p} y_t \sin(2\pi kt/p). \end{aligned}$$

In addition, $b_m = 0$ and $a_m = 0$ if p is odd. Otherwise, $a_m = a_{p/2} = (1/p)\sum_{t=1}^{p}(-1)^{t-1}y_t$, and $b_m = b_{p/2} = 0$.

For a zero-mean time series y_t assumed to vary as $y_t = \mu(t)+\epsilon_t$ for $t = 1 : T$, where $\mu(t)$ is periodic with integer period p, it is now evident that an appropriate model is the harmonic component form

$$y_t = \sum_{k=1}^{m}\{\alpha_{1,k}\cos(2\pi kt/p) + \alpha_{2,k}\sin(2\pi kt/p)\} + \epsilon_t.$$

Fitting a single harmonic, as in the previous section, will tend to produce a periodogram with a peak at the fundamental frequency $2\pi/p$ and subsidiary peaks at the higher harmonic frequencies $2\pi k/p$ for $k > 1$. The full harmonic component description above defines a linear regression model conditional on the specified period p. It is easily verified that, when T is an integer multiple of p, the regression is orthogonal; for other values of T, orthogonality is approximately achieved and is close for large T. In such cases, the reference posterior means/maximum likelihood estimators of the orthogonal regression parameters are given by

$$\hat{\alpha}_{1,k} = (2/T)\sum_{t=1}^{T} y_t \cos(2\pi kt/p),$$

and

$$\hat{\alpha}_{2,k} = (2/T)\sum_{t=1}^{T} y_t \sin(2\pi kt/p),$$

for $k < m$, and with $\hat{\alpha}_{1,p/2} = (1/T)\sum_{t=1}^{T}(-1)^{t-1}y_t$, and $\hat{\alpha}_{2,p/2} = 0$ in the case of even p. It also easily follows that the fitted sum of squares partitions as the sum of harmonic contributions

$$\frac{T}{2}\sum_{k=1}^{m-1}(\hat{\alpha}_{1,k}^2 + \hat{\alpha}_{2,k}^2) + T\hat{\alpha}_{1,m}^2 = \sum_{k=1}^{m} I(\omega_k),$$

where $I(\cdot)$ is the periodogram previously introduced at harmonic frequencies $\omega_k = 2\pi k/p$ for $k < m$, with a slightly modified form at the Nyquist frequency.

Those harmonics with largest amplitudes $\alpha_{1,k}^2 + \alpha_{2,k}^2$ will tend to have larger estimated amplitudes and hence, larger values of periodogram ordinates. The periodogram therefore indicates relative contributions of harmonics to the composition of the periodic function $\mu(\cdot)$. Harmonics with low amplitudes can sometimes be dropped from the representation with little loss in the approximation of the series. This can be assessed for each harmonic using the posterior Student-t distribution for the regression coefficients, by evaluating the support for nonzero coefficients. We can write the model above in the linear model setting discussed in Section 1.5 as $\mathbf{y} = \mathbf{F}'\boldsymbol{\beta} + \boldsymbol{\epsilon}$, with $\mathbf{y} = (y_1, \ldots, y_T)'$, $\boldsymbol{\beta} = (\alpha_{1,1}, \alpha_{2,1}, \ldots, \alpha_{1,m}, \alpha_{2,m})'$, $\mathbf{F}$ with columns $\mathbf{f}_i = (\cos(2\pi i/p), \sin(2\pi i/p), \ldots, \cos(2\pi im/p), \sin(2\pi im/p))'$, and $\boldsymbol{\epsilon} = (\epsilon_1, \ldots, \epsilon_T)'$, for $i = 1 : T$. Then, using the results in Chapter 1 we see that the joint posterior for all coefficients is a multivariate Student-t, so for any harmonic k but the Nyquist, the coefficients $(\alpha_{1,k}, \alpha_{2,k})'$ have a bivariate Student-t distribution. The degrees of freedom parameter is $\nu = T - p + 1$ (there are $p - 1$ parameters in the regression; ν will differ in models with an intercept and/or other regression terms). If harmonic k has nonnegligible coefficients, then zero is an unlikely value under this posterior. Given (at least approximate if not exact) posterior orthogonality, the bivariate Student-t distribution has mode at $\hat{\boldsymbol{\beta}}_k = (\hat{\alpha}_{1,k}, \hat{\alpha}_{2,k})'$, and scale matrix $(2s^2/T) \times \mathbf{I}_2$ where $s^2 = \mathbf{e}'\mathbf{e}/\nu$, where $\mathbf{e} = (\mathbf{y} - \mathbf{F}'\hat{\boldsymbol{\beta}})$. The posterior density contour running through zero has probability content determined by the F distribution, namely $p_k = Pr(F_{2,\nu} \leq z_k)$ where $F_{2,\nu}$ represents the standard F distribution, and $z_k = (\hat{\alpha}_{1,k}^2 + \hat{\alpha}_{2,k}^2)T/4s = I(\omega_k)/2s$. In the case of even p, the Nyquist harmonic, as usual, is different in detail; the univariate posterior Student-t distribution for $\alpha_{1,p/2}$ leads to $p_{p/2} = Pr(F_{1,\nu} \leq z_{p/2})$ with $z_{p/2} = \hat{\alpha}_{1,p/2}^2 T/s = I(\pi)/s$. A standard testing approach views the harmonic k as significant if $1 - p_k$ is small.

Example 3.3 *UK natural gas consumption series.* Figure 3.5 displays the logged values of monthly estimates of UK inland natural gas consumption data over the period October 1979 to September 1984 inclusive, 60 observations in all. The units of measurement are log millions of tons of coal equivalent. The data, deriving from the Central Statistical Office Monthly Digest, appears in West and Harrison (1997), Table 8.1. The annual seasonality is obvious, as are clear departures from simple sinusoidal forms. The data are analyzed after subtracting the sample mean (though this could

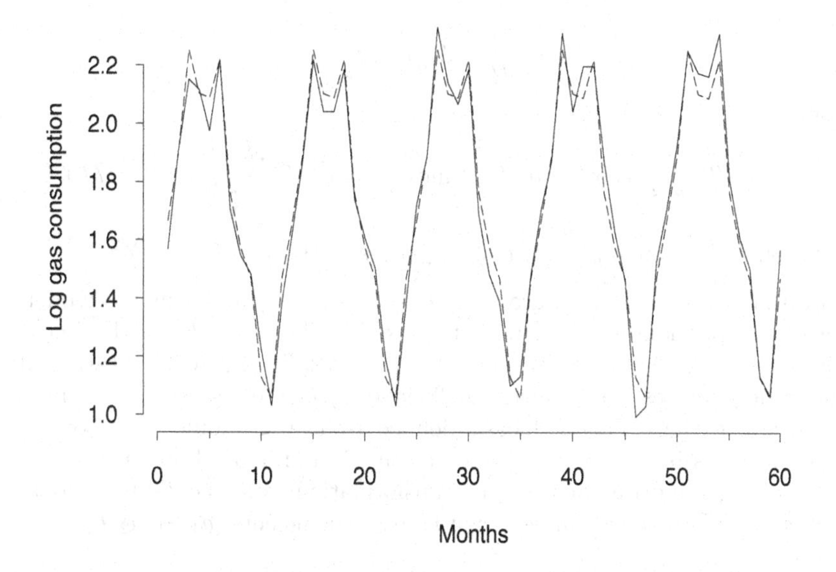

Figure 3.5 *UK natural gas consumption series. The months index runs from October 1979 to September 1984 inclusive, and the consumption is measured in millions of tons of coal equivalent. The dashed line represents fitted values based on a simple harmonic component description.*

be added as a parameter too). So $p = 12$, $T = 60$, and $\nu = 49$. The above calculations lead to posterior hpd (highest probability density) regions for harmonic coefficients with probability levels $1 - p_k$, for $k = 1 : 6$, given—up to two decimal places—by 0, 0, 0.07, 0, 0.08, and 0.37. This indicates the importance of the fundamental, second, and fourth harmonics in modeling the seasonal pattern, the apparent though less significant contributions of the third and fifth harmonics, and the negligible contribution of Nyquist term. The significant higher order harmonics are needed to model the asymmetries between peaks and troughs, and the interesting dips in midwinter induced by industrial close down during vacations, apparent in the fitted values appearing in the graph. Further exploration includes fitting models that do not have some of the less important harmonics (see Problem 2). Related discussion appears in West and Harrison (1997), Section 8.6.

3.1.6 *The Periodogram (revisited)*

We have already introduced the periodogram, defined by Equation (3.6). We revisit it here to emphasize some connections with traditional spectral theory. At Fourier frequencies $\omega_j = 2\pi j/T$, for $j = 1 : m = \lfloor T/2 \rfloor$, we have that, if $j < m$

$$I(\omega_j) = \frac{T}{2}(a_j^2 + b_j^2),$$

where

$$a_j = (2/T)\sum_{t=1}^{T} y_t \cos(2\pi jt/T) \quad \text{and} \quad b_j = (2/T)\sum_{t=1}^{T} y_t \sin(2\pi jt/T).$$

For even T, $b_m = 0$, $a_m = (1/T)\sum_{t=1}^{T} y_t \cos(\pi jt)$, and $I(\pi) = Ta_m^2$.

Note that these equations are just those arising in the harmonic modeling of the previous section, with the fixed period $p = T$. In the linear regression context, this corresponds to an exact fit and so the statistical theory degenerates. The Fourier coefficients a_j, b_j are derived by mapping the series to the orthogonal basis defined by the harmonic regressors at all frequencies in a Fourier representation of the full T observations; i.e., by inverting the orthogonal linear transformation from the T-dimensional vector $\mathbf{y}$, to T-dimensional vector of Fourier coefficients $(a_1, b_1; a_2, b_2; \ldots)'$, given by

$$y_t = \sum_{j=1}^{m} \{a_j \cos(2\pi jt/T) + b_j \sin(2\pi jt/T)\}, \tag{3.8}$$

for $t = 1 : T$. The periodogram is widely used in exploring series for important frequency components. Evaluating and plotting the periodogram based on data with evident periodicities of some fixed period p will typically lead to peaks at p and at the harmonics of this fundamental period. More widely, the interpretation and statistical evaluation of the periodogram is linked to its theoretical relationship with specific hypothesized forms of underlying structure in the time series, based on developments in spectral theory summarized below.

3.2 Some Spectral Theory

The mathematical foundation for spectral theory of stationary time series connects with the foregoing on harmonic analysis through the following heuristics. Consider the representation (3.8) for large T; the separation between Fourier frequencies is $2\pi/T$, suggesting an integral limiting form of the summation as $T \to \infty$. Assuming that y_t is a realization of a stationary process for all t, then such an integral representation exists. The technical development requires the introduction of limiting versions of the

Fourier coefficients—that are themselves realizations of underlying random processes of intricate structure—and stochastic integrals.

A brief summary overview and details of relevant concepts and theory are now discussed. A full, detailed theoretical development is beyond our scope and well beyond the needs to underpin our later methodological and applied developments; the theory is, in any case, presented and abundantly developed in many other classical texts of time series theory. In particular, readers interested in delving deeper into spectral theory might consult the seminal work of Priestley (1994). Additional references include Percival and Walden (1993) and Brockwell and Davis (1991). Our summary covers key aspects and provides heuristic development and motivation.

3.2.1 Spectral Representation of a Time Series Process

One version of the fundamental spectral representation is as follows. Suppose the real-valued, discrete time stationary stochastic process $\{y_t\}$, for $t = 0, \pm 1, \pm 2, \ldots,$ has mean zero and finite variance. It then has the representation

$$y_t = \int_{-\pi}^{\pi} e^{i\omega t} dU(\omega), \tag{3.9}$$

where $dU(\cdot)$ is a complex, orthogonal increments process (see Appendix), with some variance function $E(|dU(\omega)^2|) = dF(\omega)$. Here $F(\cdot)$ is a nondecreasing function in $-\pi < \omega < \pi$ with $dF(-\omega) = -dF(\omega)$ so that the nonnegative function $f(\omega)$ given by $dF(\omega) = f(\omega)d\omega$ is symmetric about zero.

Since $\{y_t\}$ is a real process we can write

$$y_t = \int_0^{\pi} \{dA(\omega)\cos(\omega t) + dB(\omega)\sin(\omega t)\}, \tag{3.10}$$

where the formal notation involves functions $A(\cdot)$ and $B(\cdot)$ that are real-valued stochastic processes, now discussed. Also, this form is a stochastic integral defined in the mean square sense (e.g., Priestley 1994).

The representation in (3.10) is based on underlying random functions $A(\omega)$ and $B(\omega)$, with properties of orthogonal increments processes. Let us focus on $A(\omega)$. Then the random quantity $A(\omega + \delta) - A(\omega)$ is uncorrelated with $A(\omega^* + \delta^*) - A(\omega^*)$ if the intervals $(\omega, \omega + \delta)$ and $(\omega^*, \omega^* + \delta^*)$ are disjoint. The increments of $A(\omega)$ are zero-mean, so that $E(A(\omega + \delta) - A(\omega)) = 0$ for all ω and δ. Further, there exists a nonnegative, nondecreasing function $F(\omega)$ such that

$$V(A(\omega + \delta) - A(\omega)) = F(\omega + \delta) - F(\omega),$$

where F has the properties of a positive multiple of a probability distribution over ω.

The process $A(\omega)$ is not differentiable anywhere; the notation $dA(\omega)$ in (3.10) is therefore purely formal, the standard notation for stochastic integration. F may be differentiable, having derivative $f(\omega)$, in which case the limiting version $dA(\omega)$ of $A(\omega+\delta) - A(\omega)$ as $\delta \to 0$ has variance $f(\omega)\delta$; so, though $A(\omega)$ is nowhere differentiable, the notation captures the intuitive view that $dA(\omega)$ is like a zero-mean random variable with variance $f(\omega)\delta$ as $\delta \to 0$. This discussion applies also to the function $B(\omega)$, and it is also true that increments of A and B are uncorrelated. Hence, the spectral representation has the heuristic interpretation as an infinite sum of cosine/sine terms whose coefficients are zero-mean, uncorrelated random quantities, the coefficients of terms at frequency ω having variance proportional to a vanishingly small increment multiplied by the derivative of a function $F(\omega)$. As a result, rapidly varying values of F lead to large variances of these coefficients, and so the sinusoids at those frequencies could potentially have a large impact in the spectral representation of the series, with high power at these frequencies. We see that $dF(\omega) \approx f(\omega)\delta$ represents the contribution of frequencies near ω to the variation in y_t; the total variation is distributed across frequencies according to the spectral distribution or density function. If f is continuous and positive, then all frequencies contribute to the variation in the stochastic process. If F is a completely discrete distribution, then only a discrete set of frequencies are important. Low density implies less important frequencies.

$F(\omega)$ is the spectral distribution function over frequencies in $(-\pi, \pi)$. F, and hence f, may be discrete, continuous, or mixed. The increments of F, or, in the case of differentiability, its derivatives, define the spectral density function $f(\omega)$. Some key results about this function are reviewed in the following section.

Before proceeding, note that the spectral representation does not fully describe the process $\{y_t\}$, as the distribution of the processes A and B are not completely specified. If they are Gaussian processes, then $\{y_t\}$ is a stationary Gaussian time series process; in this case, the lack of correlation in the A and B processes implies independence. Generally, this latter result does not hold; for non-Gaussian series the A and B processes will have dependent, though uncorrelated, increments.

3.2.2 Representation of Autocorrelation Functions

The autocovariances $\gamma(h)$, and hence autocorrelations $\rho(h)$, of the stationary process $\{y_t\}$, are related to the spectral distribution via

$$\gamma(h) = \int_{-\pi}^{\pi} e^{i\omega h} dF(\omega) = \int_{-\pi}^{\pi} \cos(\omega h) dF(\omega). \tag{3.11}$$

Notice that $V(y_t) = \gamma(0) = \int_{-\pi}^{\pi} dF(\omega)$ so that the scaled function $F(\omega)/\gamma(0)$ is actually a probability distribution function over $-\pi < \omega < \pi$. By the symmetry of the autocovariance function, we can write

$$\gamma(h) = 2\int_{0}^{\pi} \cos(\omega h) dF(\omega),$$

restricting our attention to $h \geq 0$. Note also that if F is everywhere continuous and differentiable, with $f(\omega) = dF(\omega)/d\omega$, (3.11) can be written as

$$\gamma(h) = \int_{-\pi}^{\pi} e^{i\omega h} f(\omega) d\omega,$$

where $f(\omega)$ is the spectral density. Under mild conditions, the above Fourier cosine transform inverts to give the spectral density function as

$$\begin{aligned} f(\omega) &= \frac{1}{2\pi}\{\gamma(0) + 2\sum_{h=1}^{\infty} \gamma(h)\cos(\omega h)\} \\ &= \frac{\gamma(0)}{2\pi}\{1 + 2\sum_{h=1}^{\infty} \rho(h)\cos(\omega h)\}. \end{aligned} \tag{3.12}$$

Hence, the process $\{y_t\}$ is summarized by either the autocovariance function or the spectral density function; they are essentially equivalent. Knowing $f(\omega)$ provides insight into the frequency components of the process.

One further result, often useful for computational purposes, relates spectral densities to the autocovariance generating function, $G(\cdot)$, with $G(x) = \sum_{h=-\infty}^{\infty} \gamma(h)x^h$. That is,

$$f(\omega) = \frac{1}{2\pi} G(e^{-i\omega}).$$

3.2.3 Other Facts and Examples

If $\{z_t\}$ is a process with spectral distribution $F_z(\omega)$, it is possible to write $F_z(\omega)$ as

$$F_z(\omega) = aF_x(\omega) + bF_y(\omega)$$

(see, e.g., Priestley 1994), where $F_x(\omega)$ is an absolutely continuous spectral distribution function and $F_y(\omega)$ is a purely discontinuous (or step) spectral distribution function, with $a \geq 0, b \geq 0$, and $a + b = 1$. When $a = 1$ and $b = 0$, z_t has a purely continuous spectrum. Examples of processes with purely continuous spectra include autoregressive moving average (ARMA) processes and general linear processes. The spectral densities of some of these processes are illustrated in this section. The so called *line spectrum* below is an example of a purely discrete spectrum (i.e., in this case $a = 0$ and $b = 1$).

Also, the spectral density of the sum of independent stochastic processes is the sum of their spectral densities. In other words, if $\{z_t\}$ is a stationary time series process such that $z_t = x_t + y_t$, with $\{x_t\}$ and $\{y_t\}$ independent time series processes, then $dF_z(\omega) = dF_x(\omega) + dF_y(\omega)$.

Example 3.4 *Line spectrum.* Suppose that $F(\omega)$ is discrete, with masses $\gamma(0)p_j$ at points ω_j, for $j = 1 : J$ and $\sum_j p_j = 1$. Then, $dF(\omega) = 0$ except at frequencies ω_j. In the spectral representation of y_t, the processes $dA(\cdot)$ and $dB(\cdot)$ have zero variance everywhere except at these frequencies, giving

$$y_t = \sum_{j=1}^{J} \{a_j \cos(\omega_j t) + b_j \sin(\omega_j t)\},$$

a harmonic process with random coefficients a_j, b_j; these have zero mean, are uncorrelated, and have variances $V(a_j) = V(b_j) = \gamma(0)p_j$. Also

$$\gamma(h) = 2\gamma(0) \sum_{j=1}^{J} p_j \cos(\omega_j h).$$

Hence, an estimate of the spectral density exhibiting spikes at various frequencies is indicative of sustained and constant form cyclical components.

Example 3.5 *White noise.* Suppose that $\gamma(h) = 0$ for $h > 0$, so that $\{y_t\}$ is an independent noise process. Then $f(\omega) = \gamma(0)/2\pi$, constant over all frequencies.

Example 3.6 *Cycles plus noise.* As a result of the previous examples, we see that a process exhibiting a spectral density with spikes at a number of frequencies and being roughly flat elsewhere will be represented by the linear combination of a collection of harmonic components and a noise series. More generally, F may have jumps at any number of frequencies, and be continuous or differentiable elsewhere. In such case we have a spectral density that is a mixture of line spectra and a purely continuous spectrum, and so $\{y_t\}$ is a linear combination of a mixture of exactly harmonic components and stationary but completely aperiodic components.

Example 3.7 *AR(1) process.* Assume that $y_t = \phi y_{t-1} + \epsilon_t$, with ϵ_t having

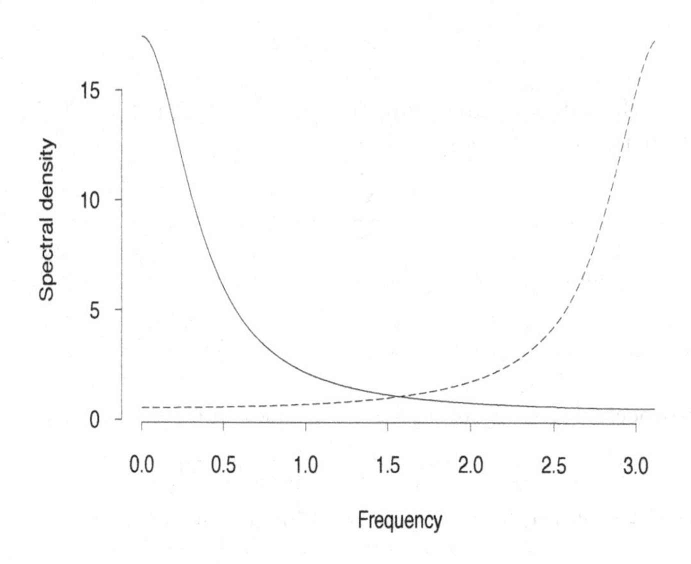

Figure 3.6 *Spectra for two AR(1) processes: $\phi = 0.7$ (full line) and $\phi = -0.7$ (dashed line).*

zero mean and variance v, and $|\phi| < 1$. Then $\gamma(0) = v/(1 - \phi^2)$ and $\gamma(h) = \gamma(0)\phi^h$. Direct calculation gives

$$f(\omega) = \frac{v}{2\pi}[1 + \phi^2 - 2\phi \cos(\omega)]^{-1}. \tag{3.13}$$

For $\phi > 0$, the process is positively correlated so low frequencies are evident, whereas high power at high frequencies corresponds to $\phi < 0$; see Figure 3.6.

Example 3.8 *MA(1) process.* Suppose that $y_t = \epsilon_t - \theta\epsilon_{t-1}$ with $|\theta| < 1$ and ϵ_t a sequence of uncorrelated zero-mean innovations with variance v. Then, $\rho(h) = 0$ for $h > 1$ and $\rho(1) = -\theta/(1 + \theta^2)$. Direct calculation gives

$$f(\omega) = \frac{v}{2\pi}[1 + \theta^2 - 2\theta \cos(\omega)]. \tag{3.14}$$

This is, in some sense, a reciprocal version of the AR case.

Example 3.9 *General linear processes.* Any stationary Gaussian process has a general linear representation as an infinite order MA process, $y_t = \sum_{j=0}^{\infty} \psi_j \epsilon_{t-j}$, for a white noise sequence ϵ_t with variance v, where the sequence of coefficients $\{\psi_j\}$ are absolutely summable and $\psi_0 = 1$.

Let $\Gamma(x)$ denote the autocovariance generating function of the process $\{y_t\}$

with

$$\Gamma(x) = \sum_{k=-\infty}^{\infty} \gamma(k)x^k,$$

where $\gamma(\cdot)$ is the autocovariance function. Now, writing $\psi(u) = \sum_{j=0}^{\infty} \psi_j u^j$, we have that the autocovariances are

$$\gamma(h) = v\sum_{j=0}^{\infty} \psi_j\psi_{j+h},$$

leading to the following form of the autocovariance generating function

$$\Gamma(x) = v\{\sum_{j=0}^{\infty} \psi_j x^j)\}\{\sum_{i=0}^{\infty} \psi_i x^{-i})\} = v\psi(x)\psi(x^{-1}).$$

Hence the spectral density for this general linear process is

$$f(\omega) = \frac{v}{2\pi}\psi(e^{-i\omega})\psi(e^{i\omega}) = \frac{v}{2\pi}|\psi(e^{-i\omega})|^2. \tag{3.15}$$

This result is useful in the following additional examples.

Example 3.10 *AR(p) processes.* Let us begin with the AR(2) case, $y_t = \phi_1 y_{t-1} + \phi_2 y_{t-2} + \epsilon_t$, which has the general linear process form with $\psi(u) = 1/(1 - \phi_1 u - \phi_2 u^2)$. Hence

$$f(\omega) = \frac{v}{2\pi}|(1 - \phi_1 e^{-i\omega} - \phi_2 e^{-2i\omega})|^{-2}.$$

This can be expanded to give

$$\frac{2\pi f(\omega)}{v} = \frac{1}{[1 + \phi_1^2 + 2\phi_2 + \phi_2^2 + 2(\phi_1\phi_2 - \phi_1)\cos(\omega) - 4\phi_2\cos^2(\omega)]}.$$

Note that, for stationarity, the roots of the characteristic polynomial $\phi(u) = 1 - \phi_1 u - \phi_2 u^2$ must have moduli greater than one, so we need to constrain the ϕ_js to the stationary region. If the roots are real, then $f(\omega)$ has a mode at either zero or π; otherwise, the roots are complex conjugates and $f(\omega)$ is unimodal at $\omega = \arccos[-\phi_1(1 - \phi_2)/4\phi_2]$ lying strictly between zero and π. This frequency is that of the harmonic component of the forecast function of the process, corresponding period $2\pi/\omega$.

In other AR(p) cases, the spectral density function may be computed using the representation in (3.15). This implies that the spectrum of an AR(p) process is given by

$$f(\omega) = \frac{v}{2\pi|\Phi(e^{-i\omega})|^2} = \frac{v}{2\pi|(1 - \phi_1 e^{-i\omega} - \cdots - \phi_p e^{-ip\omega})|^2}.$$

As an example, consider again the EEG series of Chapter 1. In any fitted AR model, we may simply use the reference posterior mean $\hat{\beta}$ and posterior innovations variance s^2 as estimates of ϕ and v, and compute the

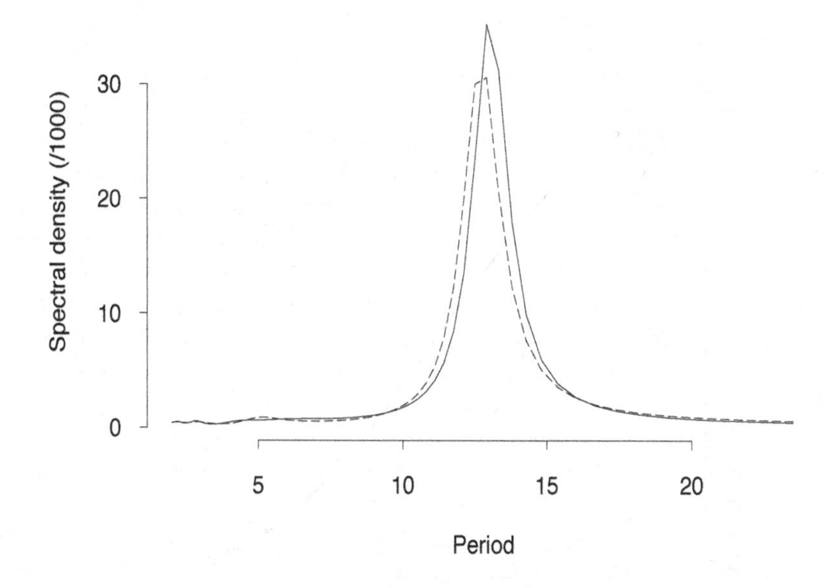

Figure 3.7 *Reference posterior estimates of the AR(8) (dashed lines) and AR(10) (full line) spectral density functions for the EEG series.*

corresponding estimate of the spectral density function. This is done here in the two cases $p = 8$ and $p = 10$, conditioning on the first 10 observations in each case. The resulting spectral density estimates are plotted in Figure 3.7 as a function of period $\lambda = 2\pi/\omega$, restricting to the range 0 to 25 to compare with Figure 3.1.

Example 3.11 *Linear filtering.* Suppose a process $\{y_t\}$ is defined as the output or response to an input or innovations process $\{x_t\}$ via the linear filter $y_t = \sum_{j=-\infty}^{\infty} c_j x_{t-j}$, for some filter coefficients c_j. Assuming spectral densities exist, they are related via

$$f_y(\omega) = f_x(\omega)|c(e^{i\omega})|^2,$$

where $c(\cdot)$ is defined by $c(u) = \sum_{j=-\infty}^{\infty} c_j u^j$. The function $c(u)$ is the *filter transfer function.* $|c(e^{i\omega})|^2$ is the *power transfer function* and its choice may be viewed as a way of transforming an input process of known spectral characteristics into one, or close to one, with certain specified spectral properties.

For example, a one-sided moving average has weights $c_j = 1/m$, for $j = 0 : (m-1)$, and zero otherwise, for some integer m; then $|c(e^{i\omega})|^2 = (1 - \cos(m\omega))/[m^2(1 - \cos(\omega))]$. Figure 3.8 shows the spectra for $m = 2, 3, 4$

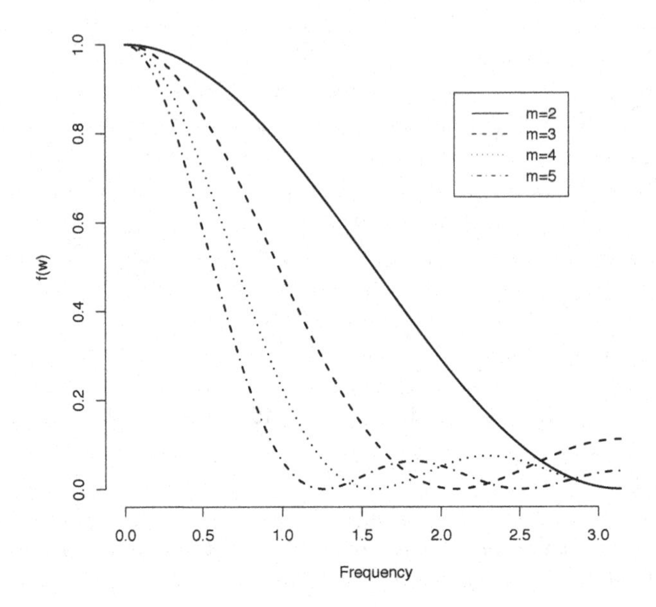

Figure 3.8 *Spectra for moving average filters.*

and $m = 5$. The larger the value of m the closer to zero is the spectrum for larger frequencies. This illustrates how moving averages act to preserve low frequencies (e.g., trends) and dampen high frequencies (e.g., noise), and so they are referred to as *low-pass filters.*

The differencing operation $y_t = x_t - x_{t-1}$ has $c(u) = 1 - u$ and so $f(\omega) = |c(e^{i\omega})|^2 = (1-e^{i\omega})(1-e^{-i\omega}) = 2(1-\cos(\omega))$. In this case the filter enhances high frequencies, damps low frequencies, and is known as a *detrending* or *high-pass filter.*

Example 3.12 *ARMA processes.* Using again (3.15), we have that if $\{y_t\}$ is a stable ARMA(p, q) process defined by $\Phi(B)y_t = \Theta(B)\epsilon_t$, with $\epsilon_t \sim N(0, v)$, its spectral density is given by

$$f(\omega) = \frac{v}{2\pi} \frac{|\Theta(e^{-i\omega})|^2}{|\Phi(e^{-i\omega})|^2}. \tag{3.16}$$

It can also be shown (see Fan and Yao 2003) that (3.16) can be written as

$$f(\omega) = \frac{v}{2\pi} \frac{1 + \sum_{j=1}^{q} \theta_j^2 + 2\sum_{k=1}^{q}(\sum_{j=k}^{q} \theta_j \theta_{t-j}) \cos(k\omega)}{1 + \sum_{j=1}^{p} \phi_j^2 + 2\sum_{k=1}^{p}(\sum_{j=k}^{p} \phi_j^* \phi_{j-k}^*) \cos(k\omega)},$$

with $\theta_0 = \phi_0^* = 1$ and $\phi_j^* = -\phi_j$ for $1 \leq j \leq p$.

3.2.4 Traditional Nonparametric Spectral Analysis

The periodogram is the basic tool in traditional nonparametric estimation of spectral density functions. As described above, the periodogram indicates the relative contributions of collections of Fourier frequencies in a standard Fourier representation of any given set of T consecutive observations, whatever their genesis. Making the further assumption that the observations arise as realizations of an underlying stationary process, the periodogram may also be interpreted as a natural estimate of the underlying spectral density function, as follows (for details see, for example, Brockwell and Davis 1991).

From Section 3.1, we have that the periodogram is given by

$$I(\omega_j) = \frac{T}{2}(a_j^2 + b_j^2) = \frac{2}{T}|\sum_{j=1}^{T} y_j e^{-i\omega_j}|^2$$

at all Fourier frequencies $\omega_j = 2\pi j/T$.

The traditional asymptotic theory of spectral estimation may be simply motivated as follows. Assuming zero-mean and stationarity of the $\{y_t\}$ process, the implied sampling distribution of the Fourier coefficients has zero mean, the coefficients are uncorrelated as a result of orthogonality of the transforms, and it can be shown that

$$V(a_j) = V(b_j) = T^{-1}\{2\gamma(0) + (4/T)\sum_{k=1}^{T-1}(T-k)\gamma(k)\cos(k\omega_j)\},$$

for $1 < j < T/2$, excluding zero and Nyquist frequencies. For large T, using the autocovariance representation of the spectral density function, we have that

$$V(a_j) = V(b_j) \approx T^{-1}\{2\gamma(0) + 4\sum_{h=1}^{\infty}\gamma(h)\cos(h\omega_j)\} \equiv \frac{4\pi f(\omega)}{T}.$$

Furthermore, the asymptotic distribution of each a_j and b_j is normal as a result of the central limit theorem. Therefore, as $T \rightarrow \infty$, a_j and b_j are independent $N(\cdot|0, T^{-1}4\pi f(\omega_j))$ and the coefficients are approximately asymptotically mutually independent for ω_j away from zero. Notice further that the normality here is exact for all finite T if it is known that the y_t series is Gaussian.

The resulting asymptotic approximation to the joint density of the Fourier coefficients (excluding the Nyquist) is therefore proportional to

$$\prod_{j=1}^{m} N(a_j|0, T^{-1}4\pi f(\omega_j))N(b_j|0, T^{-1}4\pi f(\omega_j))$$
$$\propto \textstyle\prod_{j=1}^{m} f(\omega_j)^{-1}\exp\left\{-\frac{I(\omega_j)}{4\pi f(\omega_j)}\right\}, \tag{3.17}$$

where $m = \lfloor (T-1)/2 \rfloor$. From a likelihood or Bayesian approach, viewing the spectral density function as uncertain, the above equation provides the likelihood function for the collection of unknown values $f(\omega_j)$ at the Fourier frequencies. Basing this likelihood function on the sampling density of the Fourier transforms of the data is sufficient, as they are one-to-one transforms of the original data. Note further that the same likelihood function is obtained by using, instead, the corresponding asymptotic distribution of the periodogram ordinates, $I(\omega_j) = (T/2)(a_j^2 + b_j^2)$. We see that $E[I(\omega_j)] = (T/2)(V(a_j) + V(b_j))$ which, as $T \to \infty$, converges to

$$E[I(\omega_j)] = 4\pi f(\omega_j).$$

Further, the asymptotic normality of the Fourier coefficients implies that the $I(\omega_j)$s follow asymptotically independent, scaled chi-square distributions with two degrees of freedom; specifically, $I(\omega_j)/(4\pi f(\omega_j))$ is standard exponential, and the above likelihood function arises from the corresponding joint density of periodogram ordinates. This latter result is the basis for sampling theoretic inference based on the raw sample spectrum; $I(\omega_j)/4\pi$ is used as a natural estimate of $f(\omega_j)$ and, by extrapolation between Fourier frequencies, the so-called sample spectrum $I(\omega)/4\pi$ estimates $f(\omega)$.

Figure 3.9 displays the sample spectrum of the EEG series on the decibel scale ($10\log_{10}$). The smooth spectra from the autoregressions previously illustrated are superimposed for comparison.

3.3 Discussion and Extensions

For likelihood approaches based on (3.17), note the factorization into unrelated components involving the $f(\omega_j)$ individually. Thus, the data provide essentially independent packets of information on the spectral density ordinates at the Fourier frequencies, and only this information. The sample spectrum $I(\omega)/4\pi$ provides, at each ω_j, the MLE of $f(\omega_j)$. A Bayesian analysis in which the $f(\omega_j)$s are independent under the prior, or which uses a prior distribution that is essentially uniform relative to the likelihood function, results in independent posterior distributions for the $f(\omega_j)$s

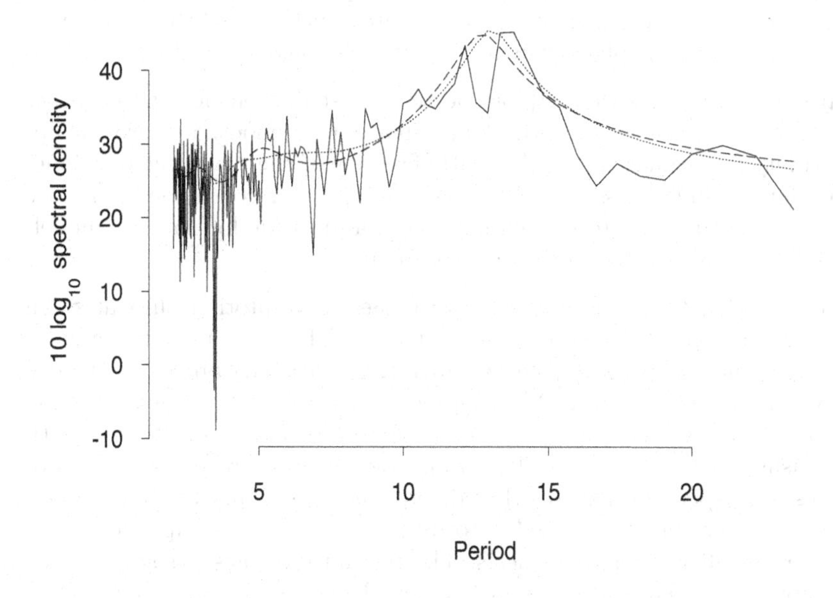

Figure 3.9 *Logged sample spectrum (full line) and AR spectra for EEG series.*

and with marginal posterior densities essentially proportional to the likelihood components in (3.17). Notice that this typically leads to quite erratic estimates as we move through Fourier frequencies; there is no tie between the sample spectrum at differing values. Critically, there is no asymptotic concentration of the likelihood components as T increases, i.e., the information content of the observations for the ordinate of the spectral density at any frequency is the same as T increases and so, the erratic nature of sample spectral density estimates does not disappear as T increases. In other words, the periodogram is not a consistent estimator of the spectral density.

There are various traditional, data analytic, and sampling theoretic approaches to inducing smoothness in spectral estimates. Kernel methods of periodogram smoothing estimate $f(\omega_j)$ via weighted averages of values of the sample spectrum at ω_j and at neighboring Fourier frequencies; this is analogous to local averaging methods of density and regression estimation, notably kernel regression methods. In addition to smoothing otherwise erratic sample spectra, such techniques enjoy the desirable sampling theoretic property of consistency. A class of related transformations of sample spectra involve the concept of tapering. Spectral tapering is based on weighting the raw data y_t in the computation of modified Fourier coefficients in order to give lower weight to data values at the extremes, near $t = 1$ and $t = T$.

A key objective of tapering methods is to (hopefully) enhance and sharpen the appearance of peaks in spectral estimates, such peaks being sometimes obscured due to the phenomenon of spectral leakage.

Bayesian approaches that explore posterior distributions for collections of values of $f(\cdot)$ based on priors over $f(\cdot)$ that induce smoothness provide opportunity for regularizing such erratic features of the raw likelihood function; for example, classes of prior distributions over continuous functions would constrain $f(\omega)$ to be well-behaved, and provide for natural, model-based smoothing of the raw sample spectrum.

Note also that there are difficulties with these asymptotic results at small frequencies in cases when the autocovariances of the series are not summable (though square-summable). This characterizes stationary processes which—unlike usual cases, such as ARMA processes, with exponentially decaying autocovariances—have autocovariances that decay only geometrically with increasing lag, referred to as long range dependence processes. Such processes have spectral densities that diverge as $\omega \to 0$, and this impacts on the regularity conditions needed to deduce the above asymptotic results for very small Fourier frequencies. The relevance of these issues for both asymptotic inference need exploration, and here, as in the more regular cases, there are needs for exact likelihood-based calculations for finite sample sizes.

Some relevant references regarding traditional nonparametric approaches to spectral estimation include Bloomfield (2000), Brockwell and Davis (1991), Percival and Walden (1993), and Shumway and Stoffer (2017), among others. As mentioned before, a key reference for a Bayesian approach to spectral analysis is Bretthorst (1988).

Several Bayesian model-based approaches based on the likelihood approximation discussed above have also been considered. In particular, Carter and Kohn (1997) consider a model of the form

$$\log(I(\omega_j)) = \log(f(\omega_j)) + \epsilon_j,$$

and approximate the distribution of ϵ_j with a mixture of normal distributions while assigning a smoothness prior on $\log(f(\omega))$. Choudhuri, Ghosal, and Roy (2004) describe a Bayesian approach to estimating the spectral density of a stationary process using a nonparametric prior described through Bernstein polynomials. This framework was extended by Macaro and Prado (2014) to consider spectral decompositions of multiple time series in factorial experiments. Rosen and Stoffer (2007) also make use of the likelihood approximation to the spectral density in the multivariate context, and use smoothing splines and a Markov chain Monte Carlo scheme to obtain posterior inference on the spectral density. All these references use the likelihood approximation for the log-periodogram observations, often referred to as Whittle's likelihood approximation, and then place different types of priors on the log-spectral density. More recently, Cadonna, Kottas, and Prado (2017) and Cadonna, Kottas, and Prado (2019) model the distribution of the log-periodogram with a mixture of Gaussian distributions. This results in an implied mixture model on the log-spectral density. Cadonna, Kottas, and Prado (2017) focuses on univariate spectral analysis, while Cadonna, Kottas, and Prado (2019) develops and illustrates models and spectral inference for multiple time series. It is worth noticing that Whittle's likelihood approximation is exact only for Gaussian white noise but, as shown in Contreras-Cristán, Gutiérrez-Peña, and Walker (2006) among others, this approximation losses efficiency in cases where there are small samples and the autocorrelation of the observed process is high. In order to relax the assumption of asymptotic independence between the periodogram ordinates at different Fourier frequencies that is used in the likelihood approximation, Kirch, Edwards, Meier, and Meyer (2017) propose a semi-parametric Bayesian approach that uses a nonparametric correction of a parametric likelihood. This approach is illustrated in the context of estimating univariate spectral densities of gravitational wave data. The R package `beyondWhittle` (Meier, Kirch, Edwards, and Meyer 2018) implements these methods.

In relation to spectral analysis and parametric methods, Huerta and West (1999a) studied the implications of AR-structured priors with unitary roots for spectral estimation of *poles* for an astronomy time series data with predominant frequency behavior and harmonics. McCoy and Stephens (2004) proposed parameterizing the spectrum of an ARMA process in terms of its reciprocal characteristic roots and then, using an asymptotic approximation like that in (3.17), they performed Bayesian inference directly on the frequency domain as an alternative to identify periodic behavior. Many other references illustrate how Bayesian spectral analysis can be applied in different areas, such as econometrics, hydrology, and signal processing (see, e.g., Bretthorst 1990, Rao and Tirtotjondro 1996, Cadonna, Kottas, and Prado 2019 and Chen 1997).

Approaches for spectral analysis of nonstationary time series that consider time-varying spectral densities have also been proposed. For example, in the Bayesian context, Rosen, Stoffer, and Wood (2009) model the time-varying log spectral density using a mixture of a finite but unknown number of individual log spectra with time-varying mixture weights. More specifically, Rosen, Stoffer, and Wood (2009) estimate the time-varying log spectral density using a Bayesian mixture of splines. Rosen, Wood, and Stoffer (2012) consider an alternative approach in which the time series are divided adaptively into segments of variable length and the local spectra are estimated using smoothing splines. The R package `BayesSpec` (Rosen, Wood, and Stoffer 2017) implements several methods for spectral analysis within the Bayesian framework, including those in Rosen, Wood, and Stoffer (2012).

In addition, many of the spectral theoretical aspects summarized in this chapter can be extended to the multivariate framework. This topic is discussed later in Chapter 8, in the context of analyzing multiple time series with a common underlying structure. References to Bayesian multivariate spectral analysis include, among other, Rosen and Stoffer (2007) and Li and Krafty (2018).

3.3.1 Long Memory Time Series Models

These models are used to describe time series processes for which the autocorrelation functions decay very slowly in time, as opposed to processes for which such functions decay exponentially fast. Beran (1994) covers in detail several theoretical aspects of long memory processes as well as statistical methods for the analysis of time series with long range dependence.

Long memory processes can be defined in terms of their spectral density $f(\omega)$ for $\omega \in (0, \pi]$. Specifically, we say that a weakly stationary time series process $\{y_t\}$ is a long memory process with parameter $d \in (0, 1)$ if

$$\lim_{\omega \to 0} \frac{f(\omega)}{c\omega^{-d}} = 1,$$

for some real constant c. In other words, $f(\omega) \sim \omega^{-d}$ as $\omega \to 0$. Many time series data sets in geophysical and climatological applications have shown this type of behavior (see for example Mandelbrot and Wallis 1968). The fractional Gaussian noise and the fractionally differenced white noise are examples of long memory processes. The first type of processes were used by Mandelbrot and coauthors in the study of the so called *Hurst* effect (Hurst 1951; Mandelbrot and van Ness 1968; Mandelbrot and Wallis 1968). Such processes are characterized by the Hurst parameter H, where $1/2 < H < 1$ indicates the presence of long range dependence. The relationship between d and H is given by $d = 2H - 1$. The second type of processes were introduced by Granger and Joyeux (1980) and Hosking (1981), and are briefly described below.

A process $\{y_t\}$ is called a *fractionally differenced white noise* with parameter $d^* \in (-0.5, 0.5)$ if, for some white noise process $\{\epsilon_t\}$, we can write $y_t = D^{-d^*}\epsilon_t$ for all t, where $D = (1 - B)$ and B is the backshift operator. The difference operator of order d^* can be formally defined as

$$D^{d^*} = \sum_{k=0}^{\infty} \binom{d^*}{k} (-B)^k.$$

Equivalently, we can write that $D^{d^*} y_t = \epsilon_t$, and so $\{y_t\}$ would have an ARIMA$(0, d^*, 0)$, or autoregressive moving average, representation. Therefore, a generalization of these processes is given by the class of *fractionally integrated* ARMA processes, also referred to as ARFIMA(p, d^*, q) or FARIMA(p, d^*, q) processes (see Beran 1994). $\{y_t\}$ is an ARFIMA(p, d^*, q) process if it satisfies the relation $\Phi(B)D^{d^*} y_t = \Theta(B)\epsilon_t$, for all t and $d^* \in (-0.5, 0.5)$. The relationship between d^* and the Hurst parameter is given by $d^* = H - 1/2$, and so the ARFIMA processes are long memory processes when $d^* > 0$.

Several methods have been developed to make inferences in various types of long memory processes. Geweke and Porter-Hudak (1983) implemented an approximate semiparametric method to obtain an estimate of the parameter d^* as follows. The spectral density of an ARFIMA(p, d^*, q) can be written as

$$f(\omega) = |1 - e^{-i\omega}|^{-2d^*} \frac{v}{2\pi} \frac{|\Theta(e^{-i\omega})|^2}{|\Phi(e^{-i\omega})|^2}.$$

Taking logarithms and replacing ω by the Fourier frequencies $\omega_j = 2\pi j/T$ for $j = 1 : m$ with $m = \lfloor T/2 \rfloor$, $\omega_j \leq \omega^*$, and ω^* small, we obtain

$$y_j \approx \beta_0 + \beta_1 x_j + \epsilon_j,$$

where $y_j = \log I(\omega_j)$, $x_j = \log(|1 - e^{-i\omega_j}|^2)$, $\epsilon_j = \log[I(\omega_j)/f(\omega_j)]$, $\beta_0 = \log f_U(0)$, with $f_U(\omega) = v|\Theta(e^{-i\omega})|^2/2\pi|\Phi(e^{-i\omega})|^2$, and $\beta_1 = -d^*$. Here $I(\omega_j)$ are the periodogram coordinates evaluated at the Fourier frequencies. Using asymptotic theory Geweke and Porter-Hudak (1983) considered the ϵ_js to be independent and identically distributed with a $\log(\frac{1}{2}\chi_2^2)$ distribution, and estimated $d^* = -\beta_1$ by least squares.

Petris (1997) develops Bayesian nonparametric time series models for data that may exhibit structured trends and long range dependencies. This approach assumes that the spectral density may be written as $f(\omega) = \omega^{-d} \exp(g(\omega))$, where $d \in [0, 1)$ and $g(\omega)$ is a continuous and bounded function on $[0, \pi)$. Then, if $d > 0$, the spectral density function has a long memory component, while $d = 0$ implies that the spectral density is that of an essentially arbitrary short memory process. More specifically, taking $s = [T/2] - 1$,

$$y_j = -d \log(\omega_j) + g_j + \epsilon_j, \quad j = 1 : s, \tag{3.18}$$

with $y_j = \log(I(\omega_j))$ and where the ϵ_js are independent and distributed as $\log(\frac{1}{2}\chi_2^2)$ under asymptotic theory, and $g_j = g(\omega_j)$. Following Carter and Kohn (1997), Petris (1997) approximates $\log(\frac{1}{2}\chi_2^2)$ with a mixture of normal distributions. That is, $\log(\frac{1}{2}\chi_2^2) \approx \sum_{i=1}^{K} p_i N(\mu_i, V_i)$ for some constant values of μ_i, V_i, and fixed K. In addition, in order to proceed with a fully Bayesian approach to inference, Petris (1997) completes the model in (3.18) with the following prior distributions. First, a smoothness prior is assumed for g, with $g_j = \tilde{g}_j + l$, l a constant, and $\tilde{g}$ a centered stationary Gaussian process such that $(1 - \gamma B)^p \tilde{g} = \eta$, with $\gamma \in (0, 1)$, p a positive integer, and $\eta \sim N(0, v_\eta)$. Hyperprior distributions are set on γ, p, and v_η. Finally, a mixture with weights π and $(1 - \pi)$, of a point mass at zero and a continuous distribution on $(0, 1)$ is assigned to d. This mixture prior on d allows practitioners to compute $Pr(d = 0|y_{1:T})$ and determine if there is a strong evidence of long memory in the data. Petris (1997) and Petris and West (1998) use a Markov chain Monte Carlo scheme to obtain samples from the posterior distribution of the parameters. These models are then applied to simulated and real data such as monthly land and sea average temperature values in the Southern Hemisphere during 1854–1989 (see also Petris and West 1996).

Liseo, Marinucci, and Petrella (2001) considered a Bayesian semiparametric approach for the analysis of stationary long memory time series that partitions the parameter space into two regions: a region where prior information on the actual form of the spectral density can be specified and a region where vague prior beliefs are adopted. Markov chain Monte Carlo methods are used to achieve posterior inference in this framework. More recently, Holan, McElroy, and Chakraborty (2009) consider fractional exponential models to obtain posterior inference on the long memory parameter d of a long-memory stationary Gaussian time series.

3.4 Appendix

3.4.1 The F Distribution

If $x \sim \chi^2_m$ and $y \sim \chi^2_n$ are two independent random variables, then

$$z = \frac{x/m}{y/n} \sim F(m, n).$$

That is, z has an F distribution with m degrees of freedom in the numerator and n degrees of freedom in the denominator. In addition, $1/z \sim F(n, m)$.

3.4.2 Distributions of Quadratic Forms

If $\mathbf{x}$ and $\mathbf{y}$ are independent random variables with $\mathbf{x} \sim N(\mathbf{0}, \mathbf{I}_m)$ and $\mathbf{y} \sim N(\mathbf{0}, \mathbf{I}_n)$, then

$$\frac{\mathbf{x}'\mathbf{x}/m}{\mathbf{y}'\mathbf{y}/n} \sim F(m, n).$$

If $\mathbf{x} \sim N(\mathbf{0}, \mathbf{I}_k)$ and $\mathbf{A}$ and $\mathbf{B}$ are symmetric, idempotent $k \times k$ matrices with $\text{rank}(\mathbf{A}) = m$, $\text{rank}(\mathbf{B}) = n$ such that $\mathbf{AB} = \mathbf{0}_k$, then

$$\frac{\mathbf{x}'\mathbf{A}\mathbf{x}}{\mathbf{x}'\mathbf{B}\mathbf{x}} \sim F(m, n).$$

3.4.3 Orthogonality of Harmonics

For any integer T and $\omega = 2\pi/T$, and for all $j, k = 1, \ldots, \lfloor T/2 \rfloor$, we have the following identities:

- $\sum_{t=1}^{T} \cos(\omega jt) = \sum_{t=1}^{n} \sin(\omega jt) = 0.$
- $\sum_{t=1}^{T} \cos(\omega jt) \sin(\omega kt) = 0.$
- $\sum_{t=1}^{T} \cos(\omega jt) \cos(\omega kt) = \sum_{t=1}^{T} \sin(\omega jt) \sin(\omega kt) = 0$ if $j \neq k$.
- If $j = 0$ or $j = T/2$, then

$$\sum_{t=1}^{T} \cos^2(\omega jt) = T, \quad \sum_{t=1}^{T} \sin^2(\omega jt) = 0.$$

- Else, if $j \neq 0$ and $j \neq T/2$, then

$$\sum_{t=1}^{T} \cos^2(\omega jt) = \sum_{t=1}^{T} \sin^2(\omega jt) = T/2.$$

3.4.4 Complex Valued Random Variables

Suppose x and y are two real-valued random variables. Then, $z = x + iy$ is a complex random variable. $E(z) = E(x) + iE(y)$ and $V(z) = E(|z|) = E(zz^*)$, where $z*$ is the conjugate of z, i.e., $z* = x - iy$. For instance, if $E(z) = 0$, $V(z) = V(x) + V(y)$.

3.4.5 Orthogonal Increments Processes

3.4.5.1 Real-valued Orthogonal Increments Processes

$dU(\omega)$ is a real-valued orthogonal increments (OI) process if, for each ω, $dU(\omega)$ is a real-valued random quantity and the following are true:

1. At any point ω and any increment $d\omega > 0$, the random variable $dU(\omega) = \{U(\omega + d\omega) - U(\omega)\}$ for $d\omega > 0$ has zero mean and variance $\sigma^2(\omega)d\omega$ for some specified positive function $\sigma^2(\omega)$.
2. For any two distinct points ω_1 and ω_2, the random variables $dU(\omega_1)$ and $dU(\omega_2)$ are uncorrelated, i.e., $dU(\omega_1) = \{U(\omega_1 + d\omega_1) - U(\omega_1)\}$ and $dU(\omega_2) = \{U(\omega_2 + d\omega_2) - U(\omega_2)\}$ are uncorrelated.

If, in addition, $dU(\omega)$ is a Gaussian process, then we have an independent increments process.

3.4.5.2 Complex-valued Orthogonal Increments Processes

$dU(\omega) = dA(\omega)+idB(\omega)$ is a complex valued orthogonal increments process if $dA(\omega)$ and $dB(\omega)$ are real-valued orthogonal increment processes and if $E(dA(\omega_1), dB(\omega_2)) = 0$ for all distinct points ω_1 and ω_2.

3.5 Problems

1. Show that the reference analysis of the model in (3.2) leads to the expressions of $p(\boldsymbol{\beta}|v, \omega, \mathbf{y})$, $p(\boldsymbol{\beta}|\omega, \mathbf{y})$, and $p(\omega|\mathbf{y})$ given in Section 3.1.1.
2. Consider the UK gas consumption series `ukgasconsumption.dat` analyzed in Example 3.3. Analyze these series with models that include only the significant harmonics for $p = 12$. Compare the fitted values obtained from such analyses to those obtained with the model used in Example 3.3.
3. Consider the Southern Oscillation Index series (`soi.dat`) shown in Figure 1.7 (a). Perform a Bayesian spectral analysis of such series.
4. Consider the three time series of concentrations of luteinizing hormone in blood samples from Diggle (1990). Perform a Bayesian spectral analysis of such series based on the single-harmonic regression model presented in Section 3.1.1.
5. Show that the spectral densities of AR(1) and MA(1) processes are given by (3.13) and (3.14), respectively.
6. Let $y_t = \phi_1 y_{t-1} + \phi_2 y_{t-2} + \epsilon_t$, with $\epsilon_t \sim N(0, 1)$. Plot the spectra of y_t in the following cases:
   (a) When the AR(2) characteristic polynomial has two real reciprocal roots given by $r_1 = 0.9$ and $r_2 = -0.95$.
   (b) When the AR(2) characteristic polynomial has a pair of complex reciprocal roots with modulus $r = 0.95$ and frequency $2\pi/8$.
   (c) When the AR(2) characteristic polynomial has a pair of complex reciprocal roots with modulus $r = 0.5$ and frequency $2\pi/8$.
7. Let $\{x_t\}$ be a stationary AR(1) process given by $x_t = \phi x_{t-1} + \epsilon_t^x$, with $\epsilon_t^x \sim N(0, v_x)$. Let $y_t = x_t + \epsilon_t^y$, with ϵ_t^y uncorrelated with the process x_t, and with $\epsilon_t^y \sim N(0, v_y)$.
   (a) Find the spectrum of $\{y_t\}$.
   (b) Simulate 500 observations from the process above, y_t, $t = 1 : 500$ using parameters $\phi = 0.9$, $v_x = v_y = 1$.

(c) Draw the periodogram of the 500 time series data points you simulated in part (b).

(d) Fit a Bayesian autoregressive model to the simulated data $y_{1:500}$ using a reference analysis as explained in Chapter 2 (you can pick the optimal model order using BIC). Obtain samples from the posterior distribution of the AR model parameters and the corresponding observational variance in this AR setting. Finally, draw samples of the resulting posterior spectral densities computed using posterior samples of the AR model parameters. Compare these to the true spectral density.

Chapter 4

Dynamic linear models

Dynamic linear models (DLMs) arise via state-space formulation of standard time series models as already illustrated in Chapter 2 and also, as natural structures for modeling time series with nonstationary components. A review of the structure and statistical theory of basic normal DLMs is given here, with various special cases exemplified, followed by development of simulation methods for routine time series analysis within the DLM class. Most of the methods summarized here are based on the theory of West and Harrison (1997). Markov chain Monte Carlo methods for filtering in conditionally Gaussian dynamic linear models are also summarized and illustrated.

4.1 General Linear Model Structures

We begin by describing the class of normal DLMs for univariate time series of equally spaced observations. Specifically, assume that y_t is modeled over time by the equations

$$y_t = \mathbf{F}_t' \boldsymbol{\theta}_t + \nu_t, \tag{4.1}$$

$$\boldsymbol{\theta}_t = \mathbf{G}_t \boldsymbol{\theta}_{t-1} + \mathbf{w}_t, \tag{4.2}$$

with the following components and assumptions:

- $\boldsymbol{\theta}_t = (\theta_{t,1}, \dots, \theta_{t,p})'$ is the $p \times 1$ state vector at time t.
- $\mathbf{F}_t$ is a p-dimensional vector of known constants or regressors at time t.
- ν_t is the observation noise, with $N(\nu_t|0, v_t)$.
- $\mathbf{G}_t$ is a known $p \times p$ matrix, usually referred to as the state evolution matrix at time t.

- $\mathbf{w}_t$ is the state evolution noise, or innovation, at time t, distributed as $N(\mathbf{w}_t|\mathbf{0}, \mathbf{W}_t)$.
- The noise sequences ν_t and $\mathbf{w}_t$ are independent and mutually independent.

We note there are generalizations of the basic DLM assumptions to allow for known, nonzero means for innovations and for dependencies between ν_t and $\mathbf{w}_t$ terms.

The form of a process adequately represented by a specified DLM can be exhibited, in part, through the forecast function defined by the model, as introduced in Chapter 2. For a given value of the state vector $\boldsymbol{\theta}_t$ at an arbitrary origin time t, the expected development of the series into the future up to $h > 0$ steps ahead is

$$E(y_{t+h}|\boldsymbol{\theta}_t) = \mathbf{F}'_{t+h}\mathbf{G}_{t+h}\mathbf{G}_{t+h-1}\cdots\mathbf{G}_{t+1}\boldsymbol{\theta}_t.$$

Generally, $\boldsymbol{\theta}_t$ is unobservable, but data up to time t, and any other available information at time t, are used to estimate $\boldsymbol{\theta}_t$. Write such information as $\mathcal{D}_t$. Then, the h-step-ahead forecast function from time t is given by

$$f_t(h) = E(y_{t+h}|\mathcal{D}_t) = \mathbf{F}'_{t+h}\mathbf{G}_{t+h}\mathbf{G}_{t+h-1}\cdots\mathbf{G}_{t+1}E(\boldsymbol{\theta}_t|\mathcal{D}_t). \tag{4.3}$$

Whatever $\mathcal{D}_t$ and the estimate $E(\boldsymbol{\theta}_t|\mathcal{D}_t)$ may be, this forecast function has a form in h essentially determined by the state evolution matrices. Often, key special cases have constant matrices, $\mathbf{G}_t = \mathbf{G}$ for all time t, and so

$$f_t(h) = \mathbf{F}'_{t+h}\mathbf{G}^h E(\boldsymbol{\theta}_t|\mathcal{D}_t). \tag{4.4}$$

A shorthand notation for the structure described above is given by the quadruple $\{\mathbf{F}_t, \mathbf{G}_t, v_t, \mathbf{W}_t\}$, on the understanding that this represents a sequence of defining quadruples over time.

Prior to reviewing basic theory, some simple examples illustrate the scope of this model class.

Example 4.1 *Regressions.* Take $\mathbf{G}_t = \mathbf{I}_p$, the $p \times p$ identity matrix, to give a linear regression of y_t on regressors in $\mathbf{F}_t$, for all t, but in which the regression parameters are now time-varying according to a random walk, i.e., $\boldsymbol{\theta}_t = \boldsymbol{\theta}_{t-1} + \mathbf{w}_t$. Traditional static regression is the special case in which $\mathbf{w}_t = \mathbf{0}$ for all t, arising (with probability one) by specifying $\mathbf{W}_t = \mathbf{0}$ for all t.

Example 4.2 *Autoregressions.* A particular class of static regressions is the class of autoregressions (AR), in which $\mathbf{F}'_t = (y_{t-1}, \ldots, y_{t-p})$. This basic DLM representation of AR models was noted in Chapter 2. It is useful for various purposes, one such being that it immediately suggests extensions to models with time-varying AR coefficients when $\mathbf{W}_t \neq \mathbf{0}$.

Alternative DLM forms of AR models are useful in other contexts. Various representations of static AR models have constant components $\mathbf{F}_t = \mathbf{F}, \mathbf{G}_t = \mathbf{G}$, and $\mathbf{W}_t = \mathbf{W}$. One common alternative is specified by $\mathbf{F} = (1, 0, \ldots, 0)'$ and

$$\mathbf{G} = \begin{pmatrix} \phi_1 & 1 & 0 & \cdots & 0 \\ \phi_2 & 0 & 1 & \cdots & 0 \\ \vdots & & & \ddots & \vdots \\ \phi_{p-1} & 0 & 0 & \cdots & 1 \\ \phi_p & 0 & 0 & \cdots & 0 \end{pmatrix},$$

with $v_t = 0$ for all t, and with $\mathbf{W}$ having entries all zero except $\mathbf{W}_{1,1} = w > 0$, the variance of the AR innovations. Note the trivial extension to time dependent variances, w_t at time t.

Example 4.3 *Autoregressive moving average (ARMA) models.* A zero-mean ARMA model described by $y_t = \sum_{j=1}^{p} \phi_j y_{t-j} + \sum_{j=1}^{q} \theta_j \epsilon_{t-j} + \epsilon_t$, with $N(\epsilon_t|0, w_t)$, has a representation like that above. Set $m = \max(p, q+1)$, extend the ARMA coefficients to $\phi_j = 0$ for $j > p$ and $\theta_j = 0$ for $j > q$, and write $\mathbf{u} = (1, \theta_1, \ldots, \theta_{m-1})'$. Then, the DLM form holds with $\mathbf{F}' = (1, 0, \ldots, 0)$, $v_t = 0$, $\mathbf{W}_t = w_t \mathbf{u}\mathbf{u}'$, and $\mathbf{G}$ as above but with dimension $m \times m$ rather than $p \times p$. Of course, the AR model is a special case when $q = 0$. Note that in this representation the variance-covariance matrix of the innovations term, $\mathbf{W}_t$, is of rank one.

Example 4.4 *Nonstationary polynomial trend models.* Local polynomial models are often relevant components of larger models that provide flexible and adaptive estimation of underlying, and often nonstationary, trends in time series. These are special cases of models in which $\mathbf{F}_t = \mathbf{F}$ and $\mathbf{G}_t = \mathbf{G}$, referred to as time series models (West and Harrison 1997).

The simplest of these models is the first order polynomial model. This model has a scalar state vector θ_t that represents the expected level of the series at time t, which changes over time according to a simple random walk. Then, we have

$$y_t = \theta_t + \nu_t \quad \text{and} \quad \theta_t = \theta_{t-1} + w_t,$$

where the innovation w_t represents the stochastic change in level between times $t-1$ and t. This model therefore has $\mathbf{F} = \mathbf{G} = 1$.

The class of second order polynomials has $\mathbf{F} = (1, 0)'$ and

$$\mathbf{G} = \begin{pmatrix} 1 & 1 \\ 0 & 1 \end{pmatrix}.$$

The state vector $\boldsymbol{\theta}_t$ has two elements, the first representing the expected level of the series and the second representing expected change in level, each subject to stochastic change through the innovation vector $\mathbf{w}_t$. The change in level parameter wanders through time according to a random walk, and the underlying level of the series is represented as a random walk with a time-varying drift. The model is sometimes referred to as locally linear for this reason, or, alternatively, as a linear growth model.

Higher order polynomial DLMs are defined by extension; the class of pth order polynomial models have p-dimensional state vectors, $\mathbf{F} = (1, 0, \ldots, 0)'$ and $\mathbf{G}$ matrix given by the $p \times p$ Jordan form with diagonal and super-diagonal entries of unity, and all other entries being zero. This is

$$\mathbf{G} = \mathbf{J}_p(1) = \begin{pmatrix} 1 & 1 & 0 & \cdots & 0 \\ 0 & 1 & 1 & \cdots & 0 \\ 0 & 0 & 1 & \cdots & 0 \\ \vdots & \vdots & \vdots & \ddots & \vdots \\ 0 & 0 & 0 & \cdots & 1 \end{pmatrix}.$$

Then, the element $\theta_{t,r}$ of the state vector represents the $(r-1)$-th difference, or derivative, of the trend in the series at time t, and is subject to stochastic changes over time based on the corresponding elements of the innovations vector $\mathbf{w}_t$.

The p-th order polynomial model has a forecast function of the form

$$f_t(h) = \sum_{j=0}^{p-1} a_{t,j} h^j,$$

i.e., it has the form of a $(p-1)$-th degree polynomial predictor into the future, but with time-varying coefficients depending on the past of the series, and any other information available at the forecast time t.

Cyclical forms. We now consider three representations for models that incorporate cyclical components.

Example 4.5 *Cyclical forms: seasonal factor representation.* This is the most basic representation for processes exhibiting periodic behavior with a fixed and integer period p. If $\boldsymbol{\theta}_t$ represents the p seasonal levels, with the first element representing the level at the current time, we have a DLM in which $\mathbf{F} = (1, 0, \ldots, 0)'$ and

$$\mathbf{G} = \begin{pmatrix} 0 & 1 & 0 & \cdots & 0 \\ 0 & 0 & 1 & \cdots & 0 \\ \vdots & \vdots & \vdots & & \vdots \\ 0 & 0 & 0 & \cdots & 1 \\ 1 & 0 & 0 & \cdots & 0 \end{pmatrix}.$$

This evolution matrix acts to permute the seasonal factors prior to the additions of stochastic changes in the factors through the innovations term $\mathbf{w}_t$. Other seasonal factor representations are possible (see West and Harrison 1997).

Example 4.6 *Cyclical forms: Fourier representation.* Fourier form dynamic linear models simply adapt standard Fourier representations of periodic functions to permit time variation in Fourier coefficients, thus providing for variation over time in the observed amplitudes and phases of harmonic components. For a zero-mean series, the full Fourier description of a seasonal DLM of integer period p is as follows.

If p is odd, $p = 2m - 1$ for some integer m, the state vector has dimension $2m - 2$, and the DLM has defining components $\mathbf{F}' = (1, 0, 1, 0, \cdots, 1, 0)$ and $\mathbf{G} = \text{blockdiag}(\mathbf{G}_1, \cdots, \mathbf{G}_{m-1})$ where, for $j = 1 : (m - 1)$,

$$\mathbf{G}_j = \begin{pmatrix} \cos(\omega_j) & \sin(\omega_j) \\ -\sin(\omega_j) & \cos(\omega_j) \end{pmatrix},$$

with Fourier frequencies $\omega_j = 2\pi j/p$.

If p is even, $p = 2m$ for some integer m, and so the DLM form is as above with the modifications that $\mathbf{F}$ is extended by an additional one, and $\mathbf{G}$ has a final diagonal entry given by $\cos(\omega_m) = \cos(\pi) = -1$, representing the Nyquist frequency. Thus, components $1, 3, \ldots,$ represent the current expected harmonic components of the series, and vary over time in response to the elements of the innovations vector $\mathbf{w}_t$. The forecast function has the form

$$f_t(h) = \sum_{j=1}^{m-1} a_{t,j} \cos(\omega_j h + b_{t,j}) + (-1)^h a_{t,m},$$

where $a_{t,m} = 0$ if p is odd, and the amplitudes and phases of harmonics depend on data and information available up to the forecast time point t.

Note that, as with other DLMs, the special case of zero innovations variance implies a static model form, in this case a standard Fourier description of a fixed, time-invariant periodic form.

Example 4.7 *Cyclical forms: autoregressive component representation.* An alternative representation of periodic behavior is based on autoregressive components, as follows. Consider the scalar difference equation $\theta_t = \beta\theta_{t-1} - \theta_{t-2}$ in which $|\beta| < 2$. This equation has solution $\theta_t = a\cos(\omega t + b)$ for some a, b, and ω such that $\beta = 2\cos(\omega)$. Extending this to a stochastic version we observe that the DLM with $\mathbf{F} = (1, 0)'$ and

$$\mathbf{G} = \begin{pmatrix} \beta & -1 \\ 0 & 1 \end{pmatrix}$$

has the cyclical forecast function $f_t(h) = a_t \cos(\omega h + b_t)$. This is simply a state-space representation of an AR(2) model in which the autoregressive polynomial roots are complex and have unit modulus, i.e., they lie on the boundary of the stationary region, generating a sustained sinusoidal solution rather than one that is exponentially damped. As a result, a process exhibiting harmonic behavior at several distinct and fixed frequencies, but with time-varying amplitude and phase patterns, may be represented in DLM form using $\mathbf{F}' = (1, 0, 1, 0, \cdots, 1, 0)$ and $\mathbf{G} = \text{blockdiag}(\mathbf{G}_1, \cdots, \mathbf{G}_m)$ where, in this case,

$$\mathbf{G}_j = \begin{pmatrix} \beta_j & -1 \\ 0 & 1 \end{pmatrix}$$

and $\beta_j = 2\cos(\omega_j)$ for $j = 1 : m$. This permits varying degrees of stochasticity in individual components, and also allows for extensions to include full harmonic descriptions for each or any of the individual periodicities.

4.2 Forecast Functions and Model Forms

As we have seen in the previous examples, the basic, qualitative structure of a DLM is essentially determined by the form of its forecast function. We now review some key general theories regarding forecast functions and implied model forms.

4.2.1 Superposition of Models

DLMs are additive, as are all linear models. Suppose that a set of $m > 1$ models is defined by individual model quadruples,

$$\{\mathbf{F}_{i,t}, \mathbf{G}_{i,t}, v_{i,t}, \mathbf{W}_{i,t}\},$$

for all t and $i = 1 : m$; write $y_{i,t}$ for the observation on the ith series at time t. Suppose we add the series following these models, resulting in $y_t = \sum_{i=1}^{m} y_{i,t}$. Then, y_t has a DLM representation given by $\{\mathbf{F}_t, \mathbf{G}_t, v_t, \mathbf{W}_t\}$, characterized by the following features:

- the regression vector $\mathbf{F}_t$, the state vector $\boldsymbol{\theta}_t$, and the evolution innovation vector $\mathbf{w}_t$ are obtained via the concatenation of the corresponding elements of the individual models; thus, $\mathbf{F}_t = (\mathbf{F}'_{1,t}, \cdots, \mathbf{F}'_{m,t})'$, with a similar form for $\boldsymbol{\theta}_t$ and $\mathbf{w}_t$ in the overall model;
- $v_t = \sum_{i=1}^{m} v_{i,t}$;
- $\mathbf{G}_t = \text{blockdiag}(\mathbf{G}_1, \cdots, \mathbf{G}_m)$; and
- $\mathbf{W}_t = \text{blockdiag}(\mathbf{W}_1, \cdots, \mathbf{W}_m)$.

This new DLM results from the superposition of the m models. The forecast function for the superposition of the m models is given by $f_t(h) = \sum_{i=1}^{m} f_{i,t}(h)$, with $f_{i,t}(h)$ the forecast function of model i.

4.2.2 Time Series Models

These are models in which $\mathbf{F}_t = \mathbf{F}$ and $\mathbf{G}_t = \mathbf{G}$ for all t. We explore the form of the forecast function (4.4) in this case.

First, assume that the $p \times p$ state matrix $\mathbf{G}$ has p distinct eigenvalues. Let $\mathbf{A} = \text{diag}(\alpha_1, \cdots, \alpha_p)$ be the diagonal matrix of these eigenvalues and let $\mathbf{E}$ be the matrix whose columns are the corresponding eigenvectors, so that $\mathbf{G}$ is similar to $\mathbf{A}$ with similarity transform $\mathbf{G} = \mathbf{EAE}^{-1}$. This leads to

$$f_t(h) = \mathbf{F}'\mathbf{E}\mathbf{A}^h\mathbf{E}^{-1}E(\boldsymbol{\theta}_t|\mathcal{D}_t) = \sum_{i=1}^{p} c_{t,i}\alpha_i^h, \tag{4.5}$$

where each $c_{t,i}$ depends on $\mathbf{F}, \mathbf{E}$, and $\mathcal{D}_t$, but not on h.

The forecast function is written in terms of the real and complex eigenvalues of $\mathbf{G}$, as has been explored in the case of autoregressions in Chapter 2. Real eigenvalues introduce exponentially decaying (usually) or exploding (rarely) functions of h. Complex eigenvalues appear in conjugate pairs and introduce exponentially damped (usually) or exploding (rarely) multiples of cosines of fixed wavelength and time dependent amplitudes and phases.

This additive decomposition of the forecast function relates to the principle of superposition, as discussed above; the model is decomposed into p component models, each having a scalar state vector effectively defined via quadruples $\{1, \alpha_i, (\cdot), (\cdot)\}$, where the $(\cdot)$ terms indicate arbitrary values of the variance components. One immediate implication is that a forecast function of the same form is obtained from any model whose state evolution matrix has the same eigenstructure; we call any two such models *similar models*, reflecting the similarity of their evolution matrices. For all practical purposes, any two similar models are exchangeable, in the sense that they will provide essentially equivalent analyses of any observed series. This implies the need for selection of specific models within any class of similar models. Models with the most basic forms are generally preferred from the viewpoint of simplicity and parsimony, with exceptions being made in cases where this would imply working with complex valued $\mathbf{G}$ matrices and state vectors.

Examples include the autoregressive DLMs noted in Chapter 2 and the various cyclical component models discussed above, with some generalizations. Suppose, for instance, that $p = 2$ and that a model with $\mathbf{F} = (1, 0)'$ has a $\mathbf{G}$ matrix with complex conjugate eigenvalues $r \exp(\pm i\omega)$ for some angular frequency ω and modulus $r = 1$. This model is similar to all others with the same $\mathbf{F}$ vector and $\mathbf{G}$ matrices selected from

$$\begin{pmatrix} e^{i\omega} & 0 \\ 0 & e^{-i\omega} \end{pmatrix}, \quad \begin{pmatrix} \cos(\omega) & \sin(\omega) \\ -\sin(\omega) & \cos(\omega) \end{pmatrix}, \quad \begin{pmatrix} \beta & -1 \\ 0 & 1 \end{pmatrix},$$

with $\beta = 2\cos(\omega)$. In practice, we would work with either of the real-valued $\mathbf{G}$ matrices, as already illustrated.

In cases when $\mathbf{G}$ has repeated eigenvalues, the above development is modified somewhat. The $\mathbf{G}$ matrix is no longer diagonalizable, but is formally similar to matrices of Jordan form. The main effect of this is to introduce polynomial functions of h as multipliers of components of the forecast function (4.4) corresponding to eigenvalues of multiplicity greater than unity. Key examples are the polynomial DLMs introduced above. In a pth order polynomial model the Jordan form of $\mathbf{G}$ has all p eigenvalues of unity, resulting in a polynomial forecast function. Detailed theoretical developments and many more examples appear in West and Harrison (1997).

Example 4.8 *Google trends data*

Figure 4.1 displays data corresponding to monthly Google search volume index values for the words "time series" from January 2014 to December 2020 in the United States. The series presents a clear seasonal pattern with the highest google search volume index values occurring in November. There is also an increase in the average search volume index starting around 2010. The seasonal pattern varies over time, with higher peaks in the months of November after 2012. In particular, November 2016 shows the largest search volume index. Note that November is election month in the US. US Presidential elections are held every 4 years, with 2004, 2008, 2012, and 2016 being Presidential election years. Elections for federal offices (Presidential and US Congress) occur in even-numbered years. Elections to the House of Representatives and the Senate are held every two years. Elections for state, local government offices and local propositions, occur in odd-numbered years.

A DLM to describe these data needs to incorporate at least two components, a linear trend component, and a cyclical component with fundamental period $p = 12$. We can use a seasonal factor representation or a Fourier representation for the cyclical component. A Fourier representation allows us to consider a model that includes a component for the fundamental period $p = 12$, as well as the components for all the harmonics of this period, or a reduced representation that includes a component for the fundamental period and additional components for a subset of the harmonics. For instance, for a DLM that includes a linear trend and only one Fourier component with period $p = 12$ we can obtain $\mathbf{F}$ and $\mathbf{G}$ using the superposition

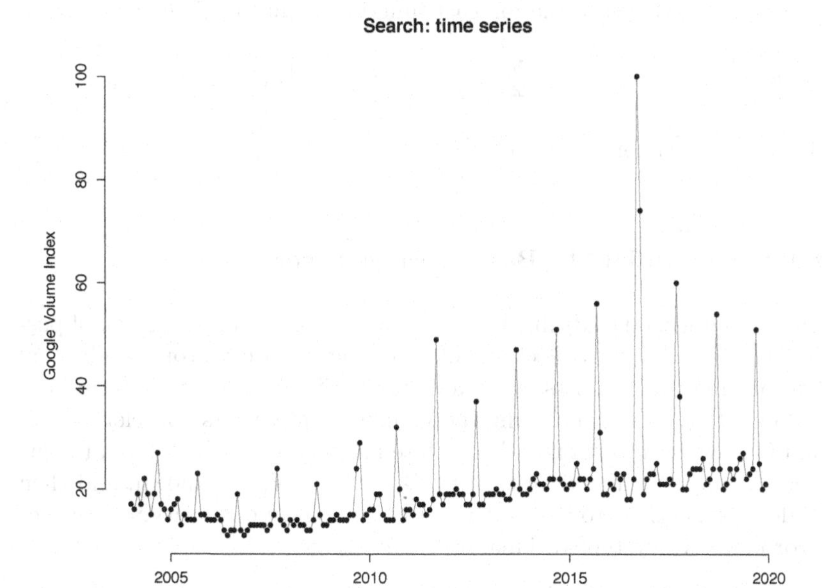

Figure 4.1 *Monthly Google search volume index for "time series" from January 2014 to December 2019.*

principle as $\mathbf{F} = (1, 0, 1, 0)'$ and

$$\mathbf{G} = \begin{pmatrix} 1 & 1 & 0 & 0 \\ 0 & 1 & 0 & 0 \\ 0 & 0 & \cos(2\pi/12) & \sin(2\pi/12) \\ 0 & 0 & -\sin(2\pi/12) & \cos(2\pi/12) \end{pmatrix}.$$

This model would have a forecast function of the form

$$f_t(h) = (a_{t,1} + a_{t,2}h) + a_{t,3}\cos(2\pi h/12) + a_{t,4}\sin(2\pi h/12),$$

where $a_{t,i}$, $i = 1:4$, depending on $\mathbf{F}, \mathbf{G}$, and $\mathcal{D}_t$ but not on h.

Similarly, a model with a linear trend component plus the full Fourier representation for $p = 12$ would have $\mathbf{F} = (1, 0, \mathbf{E}_1', \dots, \mathbf{E}_5', 1)'$, and $\mathbf{G} = \text{blockdiag}(\mathbf{G}_1, \mathbf{G}_2)$ where $\mathbf{E}_i = (1, 0)$, for $i = 1:5$,

$$\mathbf{G}_1 = \begin{pmatrix} 1 & 1 \\ 0 & 1 \end{pmatrix},$$

$\mathbf{G}_2 = \text{blockdiag}(\mathbf{G}_{2,1}, \dots, \mathbf{G}_{2,5}, -1)$, and

$$\mathbf{G}_{2,r} = \begin{pmatrix} \cos(2\pi r/12) & \sin(2\pi r/12) \\ -\sin(2\pi r/12) & \cos(2\pi r/12) \end{pmatrix},$$

for $r = 1:5$. In this case the forecast function would be of the form

$$f_t(h) = (a_{t,1} + a_{t,2}h) + + \sum_{r=1}^{5} a_{t,2+r} \cos(\omega_r h + b_{t,r}) + (-1)^h a_{t,8},$$

with $\omega_r = 2\pi r/12$ for $r = 1:5$.

4.3 Inference in DLMs: Basic Normal Theory

Inference in normal DLMs involves computation and summarization of posterior distributions for collections of component state vectors conditional on various information sets, as well as predictive inferences for future observations. Parameters including the variance components and elements of $\mathbf{F}$ and $\mathbf{G}$ may be uncertain and therefore included in analysis too, at some stage. We begin by assuming that $\mathbf{F}$, $\mathbf{G}$, v_t, and $\mathbf{W}_t$ are known, and then provide modifications to deal with cases of unknown observation variances and/or unknown system variances.

A sequential updating analysis is based on the normal linear model structure and the additional assumption that, at an arbitrary time origin $t = 0$, the state vector $\boldsymbol{\theta}_0$ is assigned a normal initial distribution with specified moments, $(\boldsymbol{\theta}_0|\mathcal{D}_0) \sim N(\boldsymbol{\theta}_0|\mathbf{m}_0, \mathbf{C}_0)$, where $\mathcal{D}_0$ represents all initial prior information. Then, proceeding through time, information sets are sequentially updated by received observations as $\mathcal{D}_t = \{\mathcal{D}_{t-1}, y_t\}$. Though additional information may be used, we now restrict to this specific form of information. Then, from the linear normal structure of the model, the results that follow are relevant to sequential learning about sequences of state vectors.

At any time t we have various distributions available, and the analysis implies specific forms of change to these distributions over time. Specifically, we will be interested in the following:

- $p(\boldsymbol{\theta}_t|\mathcal{D}_{t-1})$: the prior density for the state vector at time t given information up to the preceding time;
- $p(y_t|\mathcal{D}_{t-1})$: the one-step-ahead predictive density for the next observation;
- $p(\boldsymbol{\theta}_t|\mathcal{D}_t)$: the posterior density for the state vector at time t given $\mathcal{D}_{t-1}$ and y_t;
- the h-step-ahead forecasts $p(y_{t+h}|\mathcal{D}_t)$ and $p(\boldsymbol{\theta}_{t+h}|\mathcal{D}_t)$;
- $p(\boldsymbol{\theta}_t|\mathcal{D}_T)$: the smoothing density for $\boldsymbol{\theta}_t$ where $T > t$.

4.3.1 Sequential Updating: Filtering

For each $t > 0$, we have the following distributions (West and Harrison 1997):

- Based on the information up to time $t-1$, the prior for the state vector at time t is $N(\boldsymbol{\theta}_t|\mathbf{a}_t, \mathbf{R}_t)$ with moments

$$\mathbf{a}_t = \mathbf{G}_t \mathbf{m}_{t-1} \quad \text{and} \quad \mathbf{R}_t = \mathbf{G}_t \mathbf{C}_{t-1} \mathbf{G}_t' + \mathbf{W}_t. \tag{4.6}$$

- At time $t-1$, the one-step-ahead predictive distribution is given by $(y_t|\mathcal{D}_{t-1}) \sim N(y_t|f_t, q_t)$, where

$$f_t = \mathbf{F}_t' \mathbf{a}_t \quad \text{and} \quad q_t = \mathbf{F}_t' \mathbf{R}_t \mathbf{F}_t + v_t. \tag{4.7}$$

  Observing y_t produces the forecast error $e_t = y_t - f_t$.
- The posterior distribution for the state vector $\boldsymbol{\theta}_t$ given the current information set $\mathcal{D}_t$ is $N(\boldsymbol{\theta}_t|\mathbf{m}_t, \mathbf{C}_t)$, for moments $\mathbf{m}_t$ and $\mathbf{C}_t$ given by

$$\mathbf{m}_t = \mathbf{a}_t + \mathbf{A}_t e_t \quad \text{and} \quad \mathbf{C}_t = \mathbf{R}_t - \mathbf{A}_t \mathbf{A}_t' q_t, \tag{4.8}$$

  where $\mathbf{A}_t$ is the adaptive coefficient vector $\mathbf{A}_t = \mathbf{R}_t \mathbf{F}_t / q_t$.

Equations (4.6) to (4.8) are often referred to as the Kalman filtering equations (Kalman 1960; Harvey 1991). Proceeding through time, sequentially derived inferences on the time trajectory of the state vector are based on the sequence of posterior normal distributions so derived. Computations are typically quite straightforward, though, especially in cases of higher dimensional state vectors, numerical problems may arise, particularly when $\mathbf{C}_t$ is updated, due to collinearities between state vector elements at adjacent times.

In time series models with constant $\mathbf{F}$ and $\mathbf{G}$, and assuming that the variance components are constant too, i.e., $v_t = v$ and $\mathbf{W}_t = \mathbf{W}$, rather general results exist that relate to convergence of components of these updating equations. In particular, under certain general conditions, the defined sequences of components $\mathbf{C}_t, \mathbf{R}_t, \mathbf{A}_t$, and q_t tend to rapidly converge to stable limiting values, say $\mathbf{C}, \mathbf{R}, \mathbf{A}$, and q, implying that the limiting form of the updating equations reduces in essence to the pair of equations $\mathbf{a}_t = \mathbf{G}\mathbf{m}_{t-1}$ and $\mathbf{m}_t = \mathbf{a}_t + \mathbf{A}(y_t - \mathbf{F}'\mathbf{a}_t)$. This is interesting for two main reasons. First, the limiting forms provide connections with methods of filtering and smoothing based on autoregressive integrated moving average (ARIMA) and exponential regression techniques (see West and Harrison 1997, Chapter 5). Second, using limiting values for some components can substantially simplify the calculations.

4.3.2 Learning a Constant Observation Variance

Modifications to include inference on an uncertain and constant observation variance are based on quite standard conjugate linear model concepts (West and Harrison 1997, Chapter 4). The DLM structure is updated with the observational variance now given by $v_t = k_t v$, where k_t is a fixed, known constant multiplier of the uncertain variance v; in many cases, $k_t = 1$ though the generality allows variance weights. At each time $t-1$, the equations for evolution and updating of distributions to time t are augmented by learning on v with some changes to state and forecast distributions now detailed.

At each time t, the uncertain variance v has a marginal inverse-gamma posterior, corresponding to a gamma posterior for the precision $\phi = 1/v$, viz.,

$$(v|\mathcal{D}_t) \sim IG(n_t/2, d_t/2) \;\; \text{and} \;\; (\phi|\mathcal{D}_t) \sim G(n_t/2, d_t/2)$$

with $d_t = n_t s_t$ where $s_t = d_t/n_t = E(v^{-1}|\mathcal{D}_t)^{-1}$ is the common point estimate of v, the posterior harmonic mean. In the sequential updating analysis, this estimate and the accompanying degrees of freedom n_t are updated. In moving from time $t-1$ to time t, the updates to the gamma parameters are

$$d_t = d_{t-1} + s_{t-1}e_t^2/q_t \;\; \text{and} \;\; n_t = n_{t-1} + 1,$$

with the corresponding implied update to the point estimate being

$$s_t = s_{t-1} + \frac{s_{t-1}}{n_t}\left(\frac{e_t^2}{q_t} - 1\right),$$

based on specified initial values n_0, d_0 with $s_0 = d_0/n_0$. Here the forecast scale is now $q_t = \mathbf{F}_t'\mathbf{R}_t\mathbf{F}_t + k_t s_{t-1}$ where the time $t-1$ estimate s_{t-1} of v has been substituted for in the expression for conditional observation variance $v_t = k_t v$.

The corresponding modifications to the above summaries of prior and posterior distributions for state vectors are as follows. First, in addition to the above change to the forecast variance term q_t, the updating equation for the variance matrix $\mathbf{C}_t$ involves a scale change to reflect the revised estimate of v; that is,

$$\mathbf{C}_t = \frac{s_t}{s_{t-1}}(\mathbf{R}_t - \mathbf{A}_t\mathbf{A}_t' q_t).$$

Finally, all prior and posterior normal distributions for state vectors, which are exactly normal only conditional on v, become unconditional Student-t distributions. For example, the prior for the state vector at time t is $T_{n_{t-1}}(\boldsymbol{\theta}_t|\mathbf{a}_t, \mathbf{R}_t)$ and is updated to the posterior $T_{n_t}(\boldsymbol{\theta}_t|\mathbf{m}_t, \mathbf{C}_t)$. Similarly, the one-step-ahead predictive distribution is $t_{n_{t-1}}(y_t|f_t, q_t)$.

In the following subsections, we discuss extensions and modifications to the basic DLM formulation to handle missing data, forecasting, and retrospective smoothing. We also discuss the idea of discounting for dealing with models that assume unknown system variances as well as extensions to time-varying observational variances.

4.3.3 Missing and Unequally Spaced Data

Suppose observation y_t is missing for some time point t. Under the assumption that the reasons for a missing value do not provide information relevant to inference on the series, i.e., that the missing data mechanism is "noninformative," then the corresponding information set update is vacuous, and so $\mathcal{D}_t = \mathcal{D}_{t-1}$. This implies no update of prior to posterior distributions at the time, or, in other words, no data means posterior equals prior. Formally, the equations reduce to $\mathbf{m}_t = \mathbf{a}_t, \mathbf{C}_t = \mathbf{R}_t, s_t = s_{t-1}$, and $n_t = n_{t-1}$.

Sometimes missing data are not so noninformative. A recorded observation may be identified as suspect, for some external reasons, and to be deleted from the analysis as an effective outlier, while considering additional effects of the background circumstances on the expected behavior of the series into the future. For example, an observation may be so extreme as to call into question the current model form and numerical values of the parameters, and some form of external intervention to adapt the model is made. Such interventions are often critical in live forecasting systems, as responses to events unforeseen by the model. A wild observation may be a simple data recording or processing error, or it may indicate that the current estimate $\mathbf{a}_t$ of the state vector is, relative to estimated uncertainties $\mathbf{R}_t$, inappropriate. One conservative intervention technique consists on modifying the prior by increasing uncertainties to reflect the view that some additional changes in the state vector require modeling and, as a result, should lead to increased adaptation to future observations. This is often naturally achieved by remodeling the innovations term $\mathbf{w}_t$, at that time point alone, with a more diffuse distribution, i.e., larger variance elements in $\mathbf{W}_t$. The end result with this and other forms of intervention is that there is a change of information set, $\mathcal{D}_t = \{\mathcal{D}_{t-1}, \mathcal{I}_t\}$, say, where $\mathcal{I}_t$ represents the intervention information; the posterior $p(\boldsymbol{\theta}_t|\mathcal{D}_t)$ will now reflect this information, differing from the prior $p(\boldsymbol{\theta}_t|\mathcal{D}_{t-1})$, though ignoring the missing observation. We revisit the topic of intervention later in this chapter.

Cases of unequally spaced observations can often be handled by adapting the randomly missing data feature. This is true in cases where observations arise at times that are exact integer multiples of an underlying baseline timing interval, and when specifying an equally spaced DLM on that underlying time scale makes sense. For instance, suppose the series is a nominally monthly record of average daytime temperatures at some location, but data are observed only sporadically; we may have observation times $t_1 = 1, t_2 = 2, t_3 = 3$, $t_4 = 7, t_5 = 9, t_6 = 10$, etc., with no observations made in months 4, 5, 6, 8, and so forth. Clearly this can be handled by building a DLM for the monthly series, and treating the unequal spacings as cases of randomly missing data.

Though many problems of unequally spaced data can be handled this way, some cannot. For example, it may be that the timings are really arbitrary on an effectively continuous time scale, and no minimum, baseline interval can be identified in advance. In such cases discretized versions of underlying continuous time models can sometimes be developed. Another kind of difficulty arises when the data represent flows or aggregates, e.g., cumulative totals of underlying stock variables.

4.3.4 Forecasting

Distributions for future values of the state vector and, consequently, future observations, are available at any time t as follows. Conditional on $\mathcal{D}_t$, the h-step-ahead forecast distribution for the state vector and the corresponding h-step-ahead predictive distribution are, respectively,

$$(\boldsymbol{\theta}_{t+h}|\mathcal{D}_t) \sim N(\boldsymbol{\theta}_{t+h}|\mathbf{a}_t(h), \mathbf{R}_t(h)), \text{ and } (y_{t+h}|\mathcal{D}_t) \sim N(y_{t+h}|f_t(h), q_t(h)),$$

where the moments of the distribution of the state vector are obtained sequentially by extrapolation from time t, as

$$\mathbf{a}_t(h) = \mathbf{G}_{t+h}\mathbf{a}_t(h-1) \text{ and } \mathbf{R}_t(h) = \mathbf{G}_{t+h}\mathbf{R}_t(h-1)\mathbf{G}'_{t+h} + \mathbf{W}_{t+h}, \quad (4.9)$$

for $h = 1, 2, \ldots$, with initial values $\mathbf{a}_t(0) = \mathbf{m}_t$ and $\mathbf{R}_t(0) = \mathbf{C}_t$. Similarly, the moments of the h-step-ahead forecast predictive distribution are

$$f_t(h) = \mathbf{F}'_{t+h}\mathbf{a}_t(h) \quad \text{and} \quad q_t(h) = \mathbf{F}'_{t+h}\mathbf{R}_t(h)\mathbf{F}_{t+h} + v_{t+h}.$$

Note that these calculations require that the future values of the variance components v_{t+h} and $\mathbf{W}_{t+h}$ be known or estimated up to the forecast horizon, together with the values of future regression vectors and state evolution matrices. Note also that joint distributions of future state vectors and observations may be similarly computed; thus, for example, the joint forecast distribution of y_{t+1} and y_{t+2} may be of interest in a particular application. If $v_t = v$ is estimated, the only modifications involved are to substitute v by the current estimate s_t in defining the forecast variances $q_t(h)$, and so normal distributions become Student-t distributions on the current n_t degrees of freedom.

4.3.5 Retrospective Updating: Smoothing

Analogous to extrapolating forward for prediction is the activity known as retrospective filtering or smoothing, i.e., extrapolating into the past. Recall that the sequential analysis provides, at each time t, a current summary of past information $\mathcal{D}_t$ about the current state vector $\boldsymbol{\theta}_t$, and that this is a sufficient summary, in the formal statistical sense, of historical information for inference about the future. Proceeding to receive new observations, sequentially updated information sets also provide information about the past; e.g., the observed value of y_{t+1} provides information about $\boldsymbol{\theta}_t$ as well as about $\boldsymbol{\theta}_{t+1}$. Sometimes the inferences made "on-line" can be radically updated retrospectively in the light of future observations, through a retrospective analysis. Generally, we are interested in posteriors for past state vectors given all information currently available. The central technical issues in retrospective analysis involve computing smoothing equations to "transmit" recent information back over time. These are summarized.

Standing at an arbitrary time T, consider extrapolating back over time to compute posterior distributions $p(\boldsymbol{\theta}_t|\mathcal{D}_T)$ for all $t < T$. We are sometimes also interested in joint distributions, such as $p(\boldsymbol{\theta}_t, \boldsymbol{\theta}_{t-1}|\mathcal{D}_T)$, though details are left to the reader (see Problem 1).

One way to efficiently compute such distributions is recursively as follows. Start with the known distribution $N(\boldsymbol{\theta}_T|\mathbf{m}_T, \mathbf{C}_T)$; recall that $\mathbf{a}_T(0) = \mathbf{m}_T$ and $\mathbf{R}_T(0) = \mathbf{C}_T$ from the forecasting section above. Then, for $t < T$, we have that

$$(\boldsymbol{\theta}_t|\mathcal{D}_T) \sim N(\boldsymbol{\theta}_t|\mathbf{a}_T(t-T), \mathbf{R}_T(t-T))$$

where the notation now extends the definitions of $\mathbf{a}_T(\cdot)$ and $\mathbf{R}_T(\cdot)$ to negative arguments backward over time. These are computed recursively for $t = T-1, T-2, \ldots$, via

$$\mathbf{a}_T(t-T) = \mathbf{m}_t - \mathbf{B}_t[\mathbf{a}_{t+1} - \mathbf{a}_T(t-T+1)] \quad (4.10)$$

$$\mathbf{R}_T(t-T) = \mathbf{C}_t - \mathbf{B}_t[\mathbf{R}_{t+1} - \mathbf{R}_T(t-T+1)]\mathbf{B}_t', \quad (4.11)$$

where $\mathbf{B}_t = \mathbf{C}_t\mathbf{G}_{t+1}'\mathbf{R}_{t+1}^{-1}$.

Plotting components of the full posterior means given by $\mathbf{a}_T(t-T) = E(\boldsymbol{\theta}_t|\mathcal{D}_T)$, over times $t = 1:T$, provides a way of visually exploring the nature of changes in the corresponding elements of the state vector over time, as estimated based on the model and all observed data. The elements of $\mathbf{R}_T(t-T)$ measure uncertainty about these so-called retrospective trajectories of state parameters.

If $v_t = v$ is estimated, the modifications involved are simple. All normal distributions become Student-t distributions on the final degrees of freedom n_T, and the state variance equations involve a simple change of scale to reflect the update error variance estimate, i.e., the expression for $\mathbf{R}_T(t-T)$ above is multiplied by s_T/s_t.

In connection with the development of simulation methods in DLMs to follow, we note the following additional component distributions. For any time t and any fixed number of observations $T > t$, suppose we are interested in the conditional distributions $p(\boldsymbol{\theta}_t|\boldsymbol{\theta}_{t+1},\ldots,\boldsymbol{\theta}_T,\mathcal{D}_T)$, i.e., the distribution of a past state vector conditional on all future state vectors and information up to time T. The Markovian structure of the model implies that

$$p(\boldsymbol{\theta}_t|\boldsymbol{\theta}_{t+1},\ldots,\boldsymbol{\theta}_T,\mathcal{D}_T) = p(\boldsymbol{\theta}_t|\boldsymbol{\theta}_{t+1},\mathcal{D}_t),$$

and so state vectors other than the next one, and all future data, are irrelevant to determine the distribution of $\boldsymbol{\theta}_t$. Furthermore, this is easily seen to be normal with moments computed as

$$\mathbf{m}_t^* \equiv E(\boldsymbol{\theta}_t|\boldsymbol{\theta}_{t+1},\mathcal{D}_t) = \mathbf{m}_t + \mathbf{B}_t(\boldsymbol{\theta}_{t+1} - \mathbf{a}_{t+1}), \tag{4.12}$$

$$\mathbf{C}_t^* \equiv V(\boldsymbol{\theta}_t|\boldsymbol{\theta}_{t+1},\mathcal{D}_t) = \mathbf{C}_t - \mathbf{B}_t\mathbf{R}_{t+1}\mathbf{B}_t'. \tag{4.13}$$

4.3.6 Discounting for DLM State Evolution Variances

The model specification and inference process detailed above considered two cases. We first dealt with models in which the observational variance v_t and the system variance $\mathbf{W}_t$ were assumed known for all t. We then considered models in which $v_t = v$ for all t, with v unknown, and the matrices $\mathbf{W}_t$ were assumed known for all t. We now introduce the central and practically critical idea and method of discounting to handle models with unknown $\mathbf{W}_t$. The key idea builds on the use of component models and component discounting for $\mathbf{W}_t$.

Recall that the prior variance of the state vector at time t, denoted by $\mathbf{R}_t$, is obtained as

$$\mathbf{R}_t = V(\boldsymbol{\theta}_t|\mathcal{D}_{t-1}) = \mathbf{P}_t + \mathbf{W}_t, \tag{4.14}$$

where $\mathbf{P}_t = \mathbf{G}_t\mathbf{C}_{t-1}\mathbf{G}_t'$. The matrix $\mathbf{P}_t$ can be seen as the prior variance in a DLM with no evolution error at time t, i.e., a model with $\mathbf{W}_t = 0$. In other words, $\mathbf{P}_t$ corresponds to the prior variance in an ideal scenario in which the state vector is stable and requires no stochastic variation. This is not a realistic assumption in most practical scenarios, however, it can be assumed that $\mathbf{R}_t = \mathbf{P}_t/\delta$, for $\delta \in (0,1]$, and so the prior variance at time t is that of a model with no system stochastic variation, times a correction factor that inflates such variance. When $\delta = 1$ we have a static model. Combining $\mathbf{R}_t = \mathbf{P}_t/\delta$ with (4.14) we have that

$$\mathbf{W}_t = \frac{(1-\delta)}{\delta}\mathbf{P}_t$$

and so, given δ and $\mathbf{C}_0$, the whole sequence $\mathbf{W}_t$ for $t = 1, 2, \ldots$, is identified. Low values of the discount factor are consistent with high variability in the $\boldsymbol{\theta}_t$ sequence, while high values, with $\delta \geq 0.9$, are typically relevant in practice.

Discount factors can be chosen by maximizing joint log-likelihood functions defined in terms of the observed predictive density. In other words, we could choose the value of δ that maximizes the function

$$\log L(\delta) \equiv \log[p(y_{1:T}|\mathcal{D}_0, \delta)] = \sum_{t=1}^{T} \log[p(y_t|\mathcal{D}_{t-1}, \delta)],$$

where $p(y_t|\mathcal{D}_{t-1}, \delta)$ are the one-step-ahead univariate densities. Similarly, other criteria could be used to select the optimal value of δ, such as choosing the optimal δ value that minimizes the mean squared error (MSE) or the mean absolute deviation (MAD).

It is also possible to show that, if $\mathbf{G}_t$ is nonsingular, the smoothing equations in (4.10) and (4.11) can be written in a recursive way as

$$\begin{aligned} \mathbf{a}_T(t-T) &= (1-\delta)\mathbf{m}_t + \delta \mathbf{G}_{t+1}^{-1}\mathbf{a}_T(t-T+1) && (4.15) \\ \mathbf{R}_T(t-T) &= (1-\delta)\mathbf{C}_t + \delta^2 \mathbf{G}_{t+1}^{-1}\mathbf{R}_T(t-T+1)(\mathbf{G}_{t+1}')^{-1}, && (4.16) \end{aligned}$$

using the fact that $\mathbf{B}_t = \mathbf{C}_t\mathbf{G}_{t+1}'\mathbf{R}_{t+1}^{-1} = \delta \mathbf{G}_{t+1}^{-1}$ (Problem 2).

When several component models $\{\mathbf{F}_{i,t}, \mathbf{G}_{i,t}, v_{i,t}, \mathbf{W}_{i,t}\}$, for $i = 1 : m$, are superposed, a component discount DLM can be considered by defining the evolution matrices $\mathbf{W}_{1,t}, \ldots, \mathbf{W}_{m,t}$ in terms of m discount factors $\delta_1, \ldots, \delta_m$ as

$$\mathbf{W}_{i,t} = \frac{(1-\delta_i)}{\delta_i}\mathbf{P}_{i,t}.$$

Further discussion on discount models and their limiting behavior when $t \to \infty$, $\mathbf{F}_t = \mathbf{F}$, $\mathbf{G}_t = \mathbf{G}$, and $v_t = v$ appear in Section 6 of West and Harrison (1997) and references therein.

4.3.7 Stochastic Variances and Discount Learning

In many practical applications the observational variance may change over time, and we consider here a core model of slow, steady random changing variances that engenders an ability to adapt to and track a randomly changing variance v_t. This induces robustness and protection against potential biases in estimation of the state vector due to both real changes in volatility of measurements around the state, and can also protect and ensure inferences against aspects of model misspecification. The discounted variance learning model here, involving a random-walk like stochastic beta-gamma evolution for the observational precision sequence $1/v_t$, is a first stochastic volatility model.

Assume that the observational variance v_t is unknown. Write $\phi_t = 1/v_t$ for the precision at time t—we can work interchangeably between variance and precision. Suppose that the sequence of variances follows a stochastic, Markov evolution model defined by

$$v_t = \beta v_{t-1}/\gamma_t \quad \text{or, equivalently,} \quad \phi_t = \phi_{t-1}\gamma_t/\beta \tag{4.17}$$

where γ_t is a time t random "shock," with

$$(\gamma_t|\mathcal{D}_{t-1}) \sim Be(\beta n_{t-1}/2, (1-\beta)n_{t-1}/2),$$

independently of v_{t-1}, and $\beta \in (0,1]$. The specified parameter β acts as a discount factor, that is, the larger the value of β is, the smaller is the random "shock" to the observational variance at each time, with $\beta = 1$ leading to the constant variance model with $v_t = v$ for all t.

It is easy to see that, based on a time $t-1$ posterior $(v_{t-1}|\mathcal{D}_{t-1}) \sim IG(n_{t-1}/2, d_{t-1}/2)$, the implied prior for v_t following the evolution of Equation (4.17) is $IG(\beta n_{t-1}/2, \beta d_{t-1}/2)$. Equivalently in terms of precision, the distribution evolves as

$$(\phi_{t-1}|\mathcal{D}_{t-1}) \sim G(n_{t-1}/2, d_{t-1}/2) \rightarrow (\phi_t|\mathcal{D}_{t-1}) \sim G(\beta n_{t-1}/2, \beta d_{t-1}/2) \tag{4.18}$$

at each time t. Note that the harmonic mean estimate of variance remains unchanged at $s_{t-1} = d_{t-1}/n_{t-1}$ through this evolution; the evolution decreases the degrees of freedom which translates into an increase in the spread of the distribution, reflecting the evolution noise but maintaining the same general location.

On observing y_t, we obtain the posterior $IG(n_t/2, d_t/2)$ via the equations

$$n_t = \beta n_{t-1} + 1 \quad \text{and} \quad d_t = \beta d_{t-1} + s_{t-1}e_t^2/q_t.$$

This clearly shows how past information is discounted to reflect changes in volatility, with the updated posterior distribution being more heavily weighted on the new observation than in the case of static variance when $\beta = 1$.

Finally, the beta-gamma evolution model leads to theory that can be used to modify the retrospective smoothing analysis in this model. In particular, we recursively compute retrospective estimates

$$E(\phi_t|\mathcal{D}_T) = (1-\beta)s_t^{-1} + \beta E(\phi_{t+1}|\mathcal{D}_T)$$

back over time $t = T-1, \ldots, 1$. Similarly, and practically most usefully, we also recursively simulate the past trajectory of precisions, hence variances, using a similar construction. Details are left to the reader in the exercises in Section 4.6 below.

We note that in this case the smoothing distributions are no longer Gamma but can be approximated using the following Gamma distributions:

$$p(\phi_t|\mathcal{D}_T) \approx G(\phi_t|n_T(t-T)/2, d_T(t-T)/2),$$

with $n_T(t-T) = (1-\beta)n_t + \beta n_T(t-T+1)$, $d_T(t-T) = n_T(t-T)s_T(t-T)$, where

$$[s_T(t-T)]^{-1} = (1-\beta)s_t^{-1} + \beta[s_T(t-T+1)]^{-1},$$

initialized at $n_T(0) = n_T$ and $s_T(0) = s_T$.

4.3.7.1 References and additional comments

The variance discount method above grew out of the historical use of more general discounting ideas pioneered by P.J. Harrison for many years, as far back as the early 1960s. The specific development of variance discounting as above stems from Ameen and Harrison (1985a, 1985b), and were quickly adopted in Bayesian forecasting with state-space models as exemplified in various extensions, examples, and applications in West and Harrison (1986), Harrison and West (1987), West and Harrison (1989), and Pole, West, and Harrison (1994). The methodology was developed further in the 1989 first edition of West and Harrison (1997) (Chapter 10), and led to multivariate volatility discount models in Quintana and West (1987, 1988). Shephard (1994) described the beta-Markov modeling foundation and developed connections with univariate stochastic volatility modeling. West and Harrison (1997) developed the retrospective analysis for smoothing.

Clearly, this model has broad utility though it is not a model for anticipating directional drifts in variances, being restricted to allowing for and adjusting to, or tracking changes. Note finally that a generalization allows different discount factors at any time; β can be replaced by specified β_t at time t. One use of this is to allow larger changes at a given time via intervention (West and Harrison 1989; Pole, West, and Harrison 1994), the topic of the next section.

Example 4.9 *Google trends data (continued)* As mentioned in Example 4.8 we consider DLMs with a linear trend component, and a seasonal component with a Fourier representation with fundamental period $p = 12$ to analyze these data. We first considered a model that incorporates the fundamental period $p = 12$ and all its harmonics, however, based on this analysis we found that the last two harmonic components, i.e., the one with frequency $2\pi * 5/12 = 5\pi/6$ and the one corresponding to the Nyquist at $2\pi * 6/12 = \pi$, are essentially negligible in terms of the proportion of

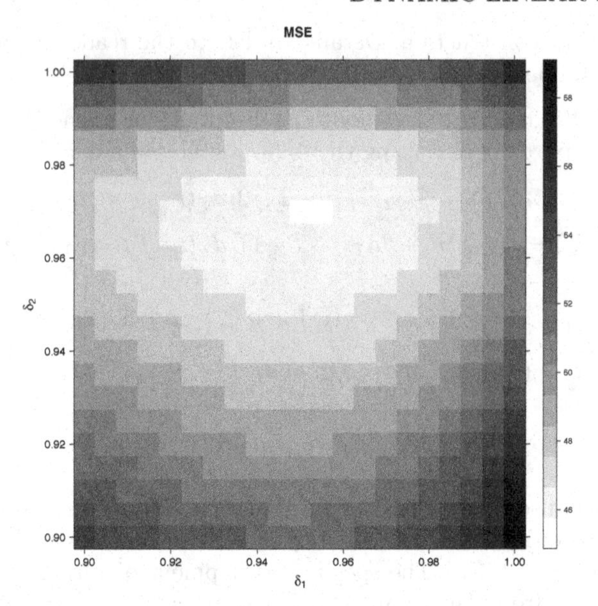

Figure 4.2 *MSE values for different combinations of discount factors* δ_1 *and* δ_2.

variation that they explain of the total seasonal amplitude variation for these data. Therefore, we fit a DLM with a linear trend component and reduce the number of seasonal components in the Fourier representation to include only frequencies $2\pi r/12$, for $r = 1:4$. This model has a state-space parameter vector of dimension 10, with 2 parameters for the linear trend component and 2 parameters for each of the 4 seasonal components.

We further assume that the observational variance is constant but unknown, i.e., $v_t = v$, and impose a Gamma prior on $1/v$ with parameters $n_0 = 1$ and $d_0 = 10$. Conditional on v we also assume a $N(\mathbf{m}_0, v\mathbf{C}_0)$ prior on the state vector with $\mathbf{m}_0 = (25, \mathbf{0}')'$ and $\mathbf{C}_0 = 10\mathbf{I}$. We use component discounting to specify the variance at the system level, with two discount factors, δ_1 and δ_2 to control the variation of polynomial trend component and the harmonic components, respectively. In order to determine the optimal discount factor values we did a grid search in $[0.9, 1] \times [0.9, 1]$ and chose the discount factor values that minimized the MSE. This resulted in optimal values of $\hat{\delta}_1 = 0.955$ and $\hat{\delta}_2 = 0.97$. Figure 4.2 shows the MSE values obtained for different combinations of discount factor factors δ_1 and δ_2 in this analysis.

Fixing the discount factors at their optimal values we then proceeded to obtain filtering and smoothing estimates for the model parameters as well as the one-step-ahead and 12-steps-ahead predictions. Figure 4.3 displays the

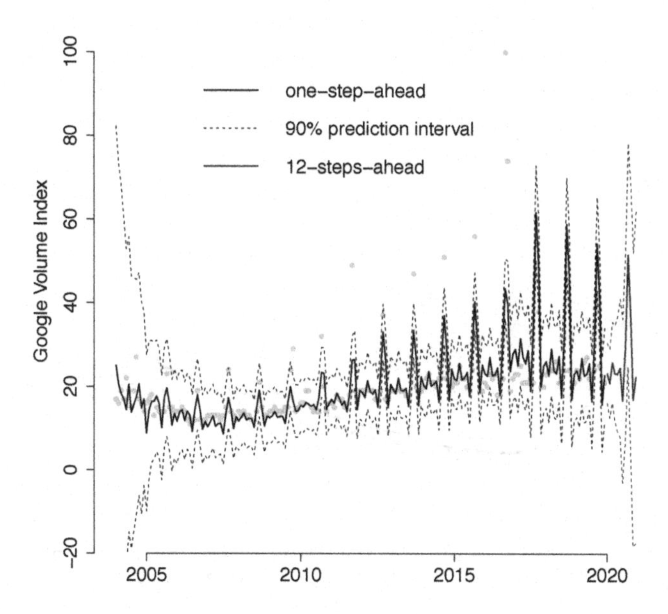

Figure 4.3 *One-step ahead and 12-steps-ahead distributions.*

mean of the one-step-ahead distribution over time along with corresponding 90% prediction intervals from January 2004 up to December 2019. The plot also displays the 12-steps ahead predictions and corresponding 90% prediction intervals for all the months of the 2020 year. We see that the model adequately captures the increase in trend and the time-varying seasonal behavior of the series. There is a clear change of pattern in the seasonal behavior after November 2011, with the seasonal amplitude increasing over time after this month, resulting in relatively low volume index values for non-November months and large volume index values for November months. The model does a reasonable job at short term prediction in terms of capturing the general behavior of the series for most months, but underestimates the observed values for the months of November from 2009 until 2018. Alternative models with error structures that allow for increased variation during the months of November or models with more general error structures that can accommodate extreme observations can be considered (e.g., see Example 4.10 for a model with a mixture structure on the obser-

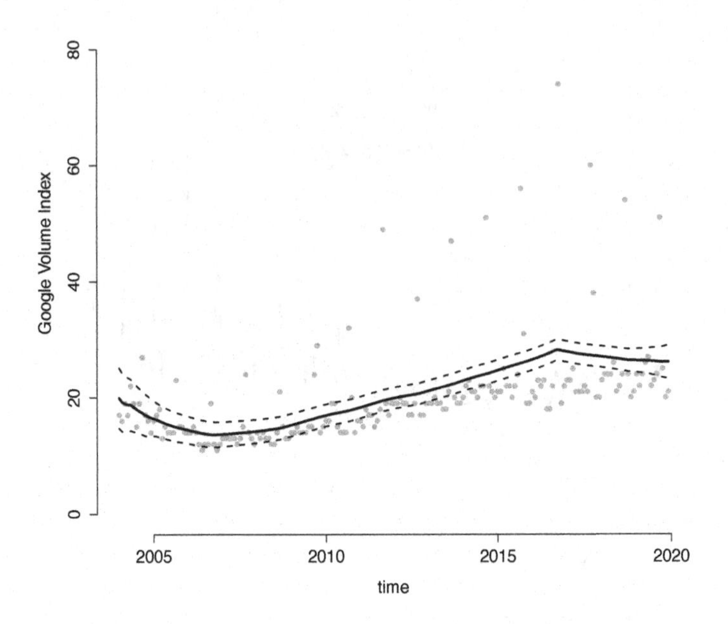

Figure 4.4 *Smoothing distribution for the linear trend component.*

vational errors). Note also that since the model assumes a Gaussian error structure the predictions are not restricted to be positive, which is not adequate for volume index data. Fitting DLMs on a transformation of the index such as the log volume index is a possibility (see Problem 5 in Section 4.6). Non-Gaussian dynamic models and dynamic generalized linear models (Section 4.4) can be also be considered and are typically used in practice.

Finally, Figures 4.4 and 4.5 show summaries of the smoothing distributions for the trend and seasonal components. There is an increasing trend from approximately 2008 up to 2016 and then a minor decrease. The seasonal components also show an increase in amplitude over time.

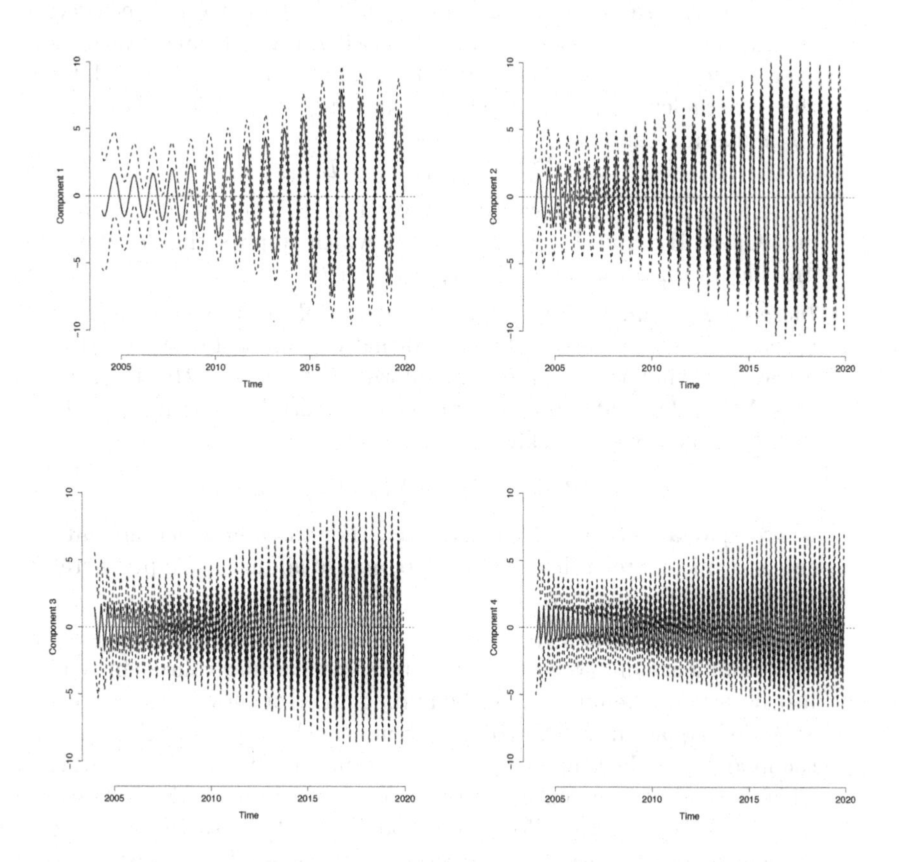

Figure 4.5 *Smoothing distributions for the seasonal components.*

4.3.8 Intervention, Monitoring, and Model Performance

4.3.8.1 Intervention

When information about possible immediate changes at the observational level is available, it should be incorporated into the model by means of a process called *intervention.*

There are various ways of making interventions, allowing the model to adapt to possible future changes. Choosing which way is more appropriate is not trivial, as it depends greatly on the type of information available. We summarize three intervention procedures below. For details and examples see West and Harrison (1997) , Chapter 11.

- *Treating y_t as an outlier.* This procedure is used for cases in which the immediate future observation does not follow the pattern forecasted by the model. The observation is considered of critical importance, however, since it does not offer relevant information for future updates at the system level, it is treated as a missing observation. Thus, $\mathcal{I}_t = \{y_t \text{ is missing}\}$ and

$$\mathcal{D}_t = \{\mathcal{D}_{t-1}, \mathcal{I}_t\} = \mathcal{D}_{t-1}.$$

More formally, the posterior distribution for $\boldsymbol{\theta}_t$ is updated via $\mathbf{m}_t = \mathbf{a}_t$ and $\mathbf{C}_t = \mathbf{R}_t$, and $\mathcal{I}_t = \{v_t^{-1} = 0\}$, which implies that y_t does not provide information about the state vector $\boldsymbol{\theta}_t$.

- *Increasing the uncertainty at the system level.* In this case the system variance $\mathbf{R}_t$ is increased. This is formally done as follows. At time t we assume that the intervention information is $\mathcal{I}_t = \{\mathbf{h}_t, \mathbf{H}_t\}$, where $\boldsymbol{\epsilon}_t \sim N(\mathbf{h}_t, \mathbf{H}_t)$, and $\boldsymbol{\epsilon}_t$ is a random vector uncorrelated with $(\boldsymbol{\theta}_{t-1}|\mathcal{D}_{t-1})$, $(\boldsymbol{\theta}_t|\mathcal{D}_{t-1})$, and with $\mathbf{w}_t$. Then,

$$(\boldsymbol{\theta}_t|\mathcal{D}_{t-1}, \mathcal{I}_t) \sim N(\mathbf{a}_t^*, \mathbf{R}_t^*),$$

  where $\mathbf{a}_t^* = \mathbf{a}_t + \mathbf{h}_t$ and $\mathbf{R}_t^* = \mathbf{R}_t + \mathbf{H}_t$. For cases in which no shifts in the prior mean are anticipated $\mathbf{h}_t = \mathbf{0}$, and so adding $\boldsymbol{\epsilon}_t$ simply increases the uncertainty about $\boldsymbol{\theta}_t$.

- *Arbitrary intervention.* In some cases, the modeler may wish to set the prior moments of the state vector to some specific values, instead of just increasing the uncertainty by adding an error term $\boldsymbol{\epsilon}_t$ to the system evolution equation. Then, postintervention values $\mathbf{a}_t^*$ and $\mathbf{R}_t^*$ are set by the modeler, and so the information at time t is given by $\mathcal{I}_t = \{\mathbf{a}_t^*, \mathbf{R}_t^*\}$. This method of intervention is the most general one, as it includes the previous cases. It also allows the modeler to decrease the variance of $\boldsymbol{\theta}_t$, which may be desirable in certain situations. This arbitrary form of the prior does not provide a coherent distribution for filtering and smoothing within the DLM structure described by (4.1) and (4.2). Then, if the new moments $\mathbf{a}_t^*$ and $\mathbf{R}_t^*$ need to be specified, it is necessary to define two new arrays, $\mathbf{K}_t$ and $\mathbf{h}_t$, and consider a DLM in which the evolution equation is given by $\boldsymbol{\theta}_t = \mathbf{G}_t^*\boldsymbol{\theta}_{t-1} + \mathbf{w}_t^*$, with $\mathbf{w}_t^* \sim N(\mathbf{h}_t, \mathbf{W}_t^*)$, where $\mathbf{w}_t^*$ is uncorrelated with $\boldsymbol{\theta}_{t-1}$ given $\mathcal{D}_{t-1}$ and $\mathcal{I}_t$, $\mathbf{G}_t^* = \mathbf{K}_t\mathbf{G}_t$, $\mathbf{w}_t^* = \mathbf{K}_t\mathbf{w}_t + \mathbf{h}_t$, and $\mathbf{W}_t^* = \mathbf{K}_t\mathbf{W}_t\mathbf{K}_t'$. $\mathbf{K}_t$ and $\mathbf{h}_t$ are given by

$$\mathbf{K}_t = \mathbf{L}_t\mathbf{Z}_t^{-1} \quad \text{and} \quad \mathbf{h}_t = \mathbf{a}_t^* - \mathbf{K}_t\mathbf{a}_t,$$

  where $\mathbf{L}_t$ and $\mathbf{Z}_t$ are, respectively, the unique lower-triangular and non-singular square matrices of $\mathbf{R}_t^*$ and $\mathbf{R}_t$, i.e., $\mathbf{R}_t^* = \mathbf{L}_t\mathbf{L}_t'$ and $\mathbf{R}_t = \mathbf{Z}_t\mathbf{Z}_t'$.

4.3.8.2 Model monitoring and performance

We now focus on how to assess model fitting and performance. We also briefly discuss automatic methods for sequentially monitoring the forecasts obtained from a given DLM. Comparing actual observations to the corresponding forecasts produced by the model is the central idea behind these methods.

Plots of the standardized forecast errors over time, given by $e_t/\sqrt{q_t}$, together with 95% probability limits from the one-step-ahead forecast distribution, are typically used as a model diagnostic tool. Forecast errors can also be used to compute basic measures of model predictive performance, such as the total absolute deviation, $\sum |e_t|$, and the mean absolute deviation (MAD); the total square error, $\sum e_t^2$, and the mean square error (MSE); and the model likelihood, which can be used to compare models with different structures.

For instance, if two DLMs, denoted by $\mathcal{M}_0$ and $\mathcal{M}_1$, have the same structure and differ only in their discount factor values, the relative likelihood of model $\mathcal{M}_0$ versus model $\mathcal{M}_1$ at time t is given by

$$H_t = \frac{p(y_t|\mathcal{D}_{t-1}, \mathcal{M}_0)}{p(y_t|\mathcal{D}_{t-1}, \mathcal{M}_1)}.$$

This is also called the *Bayes' factor* for $\mathcal{M}_0$ versus $\mathcal{M}_1$ based on y_t. It is also possible to aggregate the densities for the observations $y_t, \ldots, y_{t-h+1}$, to obtain the cumulative likelihood ratio, or cumulative Bayes' factor, based on h consecutive observations given as

$$H_t(h) = \prod_{r=t-h+1}^{t} H_r = \frac{p(y_t, y_{t-1}, \ldots, y_{t-h+1}|\mathcal{D}_{t-h}, \mathcal{M}_0)}{p(y_t, y_{t-1}, \ldots, y_{t-h+1}|\mathcal{D}_{t-h}, \mathcal{M}_1)}.$$

The Bayes' factor for $\mathcal{M}_1$ versus $\mathcal{M}_0$ is simply $H_t(h)^{-1}$. These quantities can also be computed recursively as $H_t(h) = H_t H_t(h-1)$, for $h = 2 : t$, with $H_t(1) = H_t$. On the logarithmic scale, the evidence against or in favor of model $\mathcal{M}_0$ is given by $\log[H_t(h)] = \log(H_t) + \log[H_{t-1}(h-1)]$. Log Bayes' factors of one (-1) indicate evidence in favor of model $\mathcal{M}_0$ $(\mathcal{M}_1)$ while values of two or more $(-2$ or less) point toward strong evidence. Plots of cumulative Bayes' factors over time are used to monitor the performance of $\mathcal{M}_0$ against $\mathcal{M}_1$ based on the most recent h observations. For a discussion on various monitoring methods based on these quantities, as well as decision theoretical approaches to model monitoring, see West and Harrison (1997).

4.4 Extensions: Non-Gaussian and Nonlinear Models

The DLMs previously discussed assume that the distributions of the errors at the observational and system levels are Gaussian, and that the observation and system equations are linear.

Relaxing the Gaussian assumption is key in many applied situations. For instance, the following model—in which the observational error ν_t is assumed to follow a mixture of normal distributions—has been used to handle outliers (see Box and Tiao 1973, Carter and Kohn 1997, and West and Harrison 1997):

$$\begin{aligned} y_t &= \mathbf{F}_t'\boldsymbol{\theta}_t + \nu_t, \\ \boldsymbol{\theta}_t &= \mathbf{G}\boldsymbol{\theta}_{t-1} + \mathbf{w}_t, \end{aligned}$$

where ν_t follows a mixture of two normals, i.e.,

$$\nu_t \sim \pi N(0, v) + (1 - \pi)N(0, \kappa^2 v),$$

and $\mathbf{w}_t \sim N(\mathbf{0}, \mathbf{W}_t)$. Assume that κ is known. This model is a non-Gaussian DLM that can be written as a conditionally Gaussian DLM (or CDLM) by considering auxiliary (latent) variables λ_t such that, when $\lambda_t = 1$ ν_t is normally distributed with zero mean and variance v, and when $\lambda_t = \kappa^2$ ν_t is normally distributed with zero mean and variance $\kappa^2 v$. Then, conditional on λ_t, the model above can be written as a Gaussian DLM. That is,

$$\begin{aligned} y_t &= \mathbf{F}'\boldsymbol{\theta}_t + \nu_t, \quad \nu_t \sim N(0, v_{\lambda_t}), \\ \boldsymbol{\theta}_t &= \mathbf{G}\boldsymbol{\theta}_{t-1} + \mathbf{w}_t, \quad \mathbf{w}_t \sim N(\mathbf{0}, \mathbf{W}_t), \end{aligned}$$

with $v_{\lambda_t} = v$ if $\lambda_t = 1$ and $v_{\lambda_t} = \kappa^2 v$ if $\lambda_t = \kappa^2$. Markov chain Monte Carlo methods for posterior estimation within the class of CDLMs were proposed in Frühwirth-Schnatter (1994) and Carter and Kohn (1994), and will be discussed and illustrated later in this chapter.

In many cases relaxing the assumption of normality does not lead to a CDLM structure. General state-space or dynamic models are specified in terms of an observation density $y_t \sim p(y_t|\boldsymbol{\theta}_t)$ and a state evolution density $\boldsymbol{\theta}_t \sim p(\boldsymbol{\theta}_t|\boldsymbol{\theta}_{t-1})$. Both densities may involve nonlinearities. In *dynamic generalized linear models* (DGLMs) (West, Harrison, and Migon 1985 and West and Harrison 1997), the observational density belongs to the exponential family, and so

$$p(y_t|\eta_t) \propto \exp\left\{\frac{y_t\eta_t - a(\eta_t)}{v_t}\right\},$$

with $\mu_t = E(y_t|\eta_t) = a'(\eta_t)$. The mean μ_t is related to the state parameters via the link function $g(\eta_t) = \mathbf{F}_t'\boldsymbol{\theta}_t$, and the evolution equation is linear and Gaussian. Applications and algorithms for posterior estimation within the class of DGLMs appear in West and Harrison (1997) (see Chapter 14), Gamerman (1998), and Ferreira and Gamerman (2000). Markov chain Monte Carlo (MCMC) methods and sequential Monte Carlo methods for posterior inference in the class of general state-space models are discussed in Sections 4.5 and 6.2.

4.5 Posterior Simulation: MCMC Algorithms

Posterior inference and forecasting can be easily achieved in the normal DLM framework as described previously in this chapter. When more general models are considered, such as nonlinear and non-Gaussian dynamic models, other types of simulation-based algorithms are typically needed for posterior estimation. We briefly discuss MCMC posterior simulation schemes in general frameworks and then focus on algorithms specifically designed for the class of conditionally linear/Gaussian dynamic models.

In a general framework, a nonlinear/non-Gaussian dynamic model is defined by the densities $p(y_t|\boldsymbol{\theta}_t, \mathcal{D}_{t-1})$, $p(\boldsymbol{\theta}_t|\boldsymbol{\theta}_{t-1}, \mathcal{D}_{t-1})$ and the prior density $p(\boldsymbol{\theta}_0|\mathcal{D}_0)$. We are interested in obtaining samples from the filtering distribution $p(\boldsymbol{\theta}_t|\mathcal{D}_t)$ and the joint posterior distribution $p(\boldsymbol{\theta}_{0:T}|\mathcal{D}_T)$. In a Gibbs sampling framework, we would iteratively sample from the conditional posterior distributions $p(\boldsymbol{\theta}_t|\boldsymbol{\theta}_{(-t)}, \mathcal{D}_T)$—where $\boldsymbol{\theta}_{(-t)}$ consists of $\boldsymbol{\theta}_{0:T}$ except the tth element $\boldsymbol{\theta}_t$—sequencing through $t = 0 : T$ as follows:

1. Set initial values $\boldsymbol{\theta}_{0:T}^{(0)}$.
2. For each iteration m, sample $\boldsymbol{\theta}_{0:T}^{(m)}$ component by component, i.e., for each t, sample $\boldsymbol{\theta}_t^{(m)}$ from $p(\boldsymbol{\theta}_t|\boldsymbol{\theta}_{0:(t-1)}^{(m)}, \boldsymbol{\theta}_{(t+1):T}^{(m-1)}, \mathcal{D}_T)$.
3. Repeat the previous step until MCMC convergence.

Note that in the case of Gaussian densities at the observation and system evolution levels, the conditional posteriors $p(\boldsymbol{\theta}_t|\boldsymbol{\theta}_{0:(t-1)}^{(m)}, \boldsymbol{\theta}_{(t+1):T}^{(m-1)}, \mathcal{D}_T)$ are also Gaussian whose moments are easily obtained. However, this is rarely the case in general frameworks, and so alternative posterior simulation schemes such as those considering Metropolis-Hastings steps within the Gibbs iterations could be used. Carlin, Polson, and Stoffer (1992) discuss this in the context of nonlinear models as well as scale mixture of normal models.

We now restrict our attention to the class of conditionally linear and Gaussian dynamic models for which more efficient MCMC approaches can be designed. In particular, we summarize the steps of a Gibbs sampling algorithm referred to as the *forward filtering backward sampling* algorithm, or FFBS. This algorithm was introduced independently by Carter and Kohn (1994) and Frühwirth-Schnatter (1994). Details and examples can be found in these two references and in West and Harrison (1997). More specifically, we begin by assuming that the model defined by the quadruple $\{\mathbf{F}_t, \mathbf{G}_t, v_t, \mathbf{W}_t\}$ depends on some latent parameters $\boldsymbol{\lambda}_t$. In other words, we assume that the model has a standard normal DLM structure conditional on these latent parameters. Then, in order to achieve full posterior inference based on the observations $y_{1:T}$, we will obtain samples from the full posterior distribution given by $p(\boldsymbol{\theta}_{0:T}, \boldsymbol{\lambda}_{1:T}|\mathcal{D}_T)$, by iterating between the two conditional posteriors $p(\boldsymbol{\theta}_{0:T}|\boldsymbol{\lambda}_{1:T}, \mathcal{D}_T)$ and $p(\boldsymbol{\lambda}_{1:T}|\boldsymbol{\theta}_{0:T}, \mathcal{D}_T)$. The FFBS algorithm described below allows us to sample from $p(\boldsymbol{\theta}_{0:T}|\boldsymbol{\lambda}_{1:T}, \mathcal{D}_T)$. No general procedures are available to efficiently sample from $p(\boldsymbol{\lambda}_{1:T}|\boldsymbol{\theta}_{0:T}, \mathcal{D}_T)$, since this requires taking into account the model structure.

The FFBS algorithm takes into account the Markovian structure of the system Equation (4.2), and so a sample of $\boldsymbol{\theta}_{0:T}$ is obtained by noting that

$$p(\boldsymbol{\theta}_{0:T}|\boldsymbol{\lambda}_{1:T}, \mathcal{D}_T) = p(\boldsymbol{\theta}_T|\boldsymbol{\lambda}_{1:T}, \mathcal{D}_T) \prod_{t=0}^{T-1} p(\boldsymbol{\theta}_t|\boldsymbol{\theta}_{t+1}, \boldsymbol{\lambda}_{1:T}, \mathcal{D}_t),$$

where $p(\boldsymbol{\theta}_t|\boldsymbol{\theta}_{t+1}, \boldsymbol{\lambda}_{1:T}, \mathcal{D}_t)$ is easily obtained from $p(\boldsymbol{\theta}_{t+1}|\boldsymbol{\theta}_t, \boldsymbol{\lambda}_{1:T}, \mathcal{D}_t)$ and $p(\boldsymbol{\theta}_t|\boldsymbol{\lambda}_{1:T}, \mathcal{D}_t)$. Then, the algorithm is summarized as follows. For each MCMC iteration i, obtain $\boldsymbol{\theta}_{0:T}^{(i)}$ conditional on $\boldsymbol{\lambda}_{1:T}^{(i)}$ via the following steps:

1. Use the DLM filtering equations to compute $\mathbf{m}_t, \mathbf{a}_t, \mathbf{C}_t$, and $\mathbf{R}_t$ for $t = 1:T$.
2. At time $t = T$ sample $\boldsymbol{\theta}_T^{(i)}$ from $N(\boldsymbol{\theta}_T|\mathbf{m}_T, \mathbf{C}_T)$ and then,
3. For $t = (T-1):0$ sample $\boldsymbol{\theta}_t^{(i)}$ from $N(\boldsymbol{\theta}_t|\mathbf{m}_t^*, \mathbf{C}_t^*)$, with $\mathbf{m}_t^*$ and $\mathbf{C}_t^*$ given by (4.12) and (4.13).

Note that the moments of the Gaussian distributions above, $\mathbf{m}_T, \mathbf{C}_T$, $\mathbf{m}_t^*$, and $\mathbf{C}_t^*$, are typically functions of $\boldsymbol{\lambda}_{1:T}$, although this is not made explicit in the notation. It is also possible to obtain more computationally efficient versions of the FFBS algorithm by considering block sampling schemes that take advantage of fast inversion algorithms as detailed in Migon, Gamerman, Lopes, and Ferreira (2005).

4.5.1 Examples

Example 4.10 *AR*(1) *with normal mixture structure on observational errors.* Consider the model

$$y_t = \mu_t + \nu_t, \quad \mu_t = \phi\mu_{t-1} + w_t, \tag{4.19}$$

where ν_t has the following distribution,

$$\nu_t \sim \pi N(0, v) + (1-\pi)N(0, \kappa^2 v), \tag{4.20}$$

and $w_t \sim N(0, w)$. Here $\kappa > 1$ and $\pi \in (0, 1)$, with κ and π assumed known. This model can be written as a conditionally Gaussian DLM given by $\{\mathbf{F}_t, \mathbf{G}_t, v\lambda_t, w\}$, where λ_t is a latent variable that takes the values one or κ^2 with probabilities π and $(1-\pi)$, respectively.

Figure 4.6 shows 200 data points simulated from the model given by (4.19) and (4.20), with $\pi = 0.9$, $\kappa^2 = 4$, $\phi = 0.9$, and $v = w = 1$. The circles in

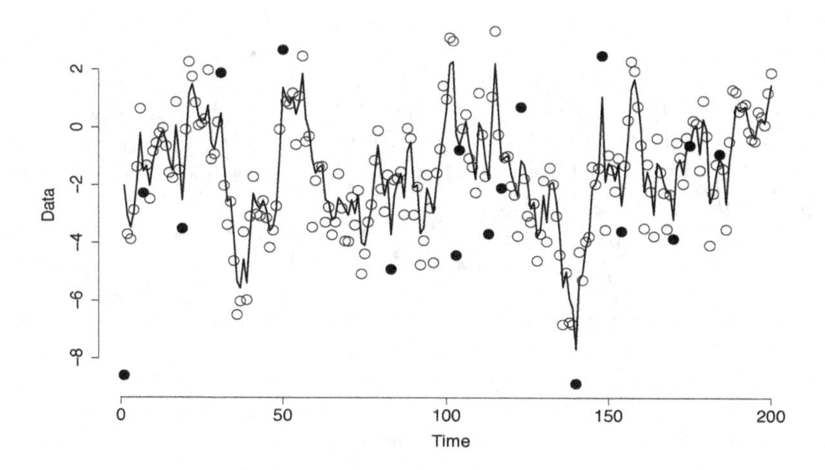

Figure 4.6 *Simulated series from an AR(1) with mixture observational errors (circles and solid circles). The solid line corresponds to the posterior mean of $(\mu_t|\mathcal{D}_t)$ over time obtained from a model that ignores the mixture structure in the observational errors.*

the figure correspond to y_t for $t = 1 : 200$. The solid circles are observations that were simulated under the $N(0, \kappa^2 v)$ mixture component.

We fitted two different models to the simulated data. We first fitted a model that ignored the mixture structure in the innovations at the observational level and assumed ϕ, v, and w known. In other words, we fitted a standard DLM described by $\{1, 0.9, 1, 1\}$. Figure 4.6 displays the posterior means of the filtering distributions, i.e., $E(\mu_t|\mathcal{D}_t)$, for $t = 1 : 200$. Figure 4.7 shows the posterior mean of $(\mu_t|\mathcal{D}_{200})$ and corresponding 95% posterior bands. As can be seen from these figures and from the residual plots and the p-values for the Ljung-Box statistic shown in Figure 4.8, the model does a reasonable job in describing the process underlying the series. However, since it was not designed to capture the mixture structure in the innovations at the observational level, it results in large residuals for some of the observations generated under the second mixture component (solid circles).

We then fitted the model given by (4.19) and (4.20) assuming that w, v, and ϕ were unknown. For this we implemented the Markov chain Monte Carlo algorithm summarized in the following steps.

- Sample $(v|w, \phi, \mu_{0:200}, \lambda_{1:200}, y_{1:200})$. This reduced to sample v from an inverse-gamma distribution, i.e., $\text{IG}(v|\alpha_v, \beta_v)$, with

$$\alpha_v = \alpha_{0,v} + T/2, \quad \beta_v = \beta_{0,v} + s_v^2/2.$$

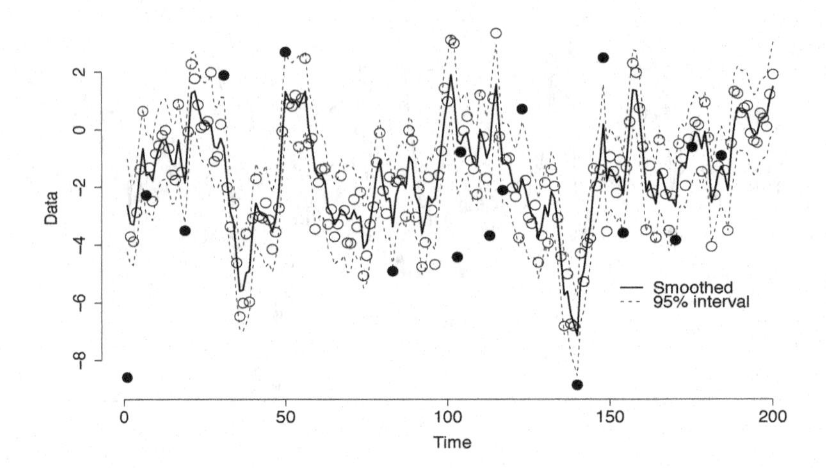

Figure 4.7 *Simulated series (circles and solid circles). The solid line corresponds to the posterior mean of the smoothing distribution over time* $(\mu_t|\mathcal{D}_{200})$ *obtained from a model that ignores the mixture structure in the observational errors. The dotted lines are 95% posterior bands for* $(\mu_t|\mathcal{D}_{200})$.

Here, $\alpha_{0,v}$ and $\beta_{0,v}$ are prior fixed values, $T = 200$, and s_v^2 is given by

$$s_v^2 = \sum_{\{\lambda_t=1\}} (y_t - \mu_t)^2 + \sum_{\{\lambda_t=\kappa^2\}} (y_t - \mu_t)^2/\kappa^2.$$

- Sample $(w|v, \phi, \mu_{0:200}, \lambda_{1:200}, y_{1:200})$. Again, this is a Gibbs step and so w is sampled from an inverse-gamma distribution $\text{IG}(w|\alpha_w, \beta_w)$ with

$$\alpha_w = \alpha_{0,w} + T/2, \quad \beta_w = \beta_{w,0} + \sum_{t=1}^{T}(\mu_t - \phi\mu_{t-1})^2/2,$$

where $\alpha_{0,w}$ and $\beta_{0,w}$ are prior fixed values.

- Sample $(\phi|v, w, \mu_{0:200}, y_{1:200})$. ϕ is sampled from $N(\phi|m_\phi, C_\phi)$, with

$$m_\phi = \left(\sum_{t=1}^{200} \mu_t\mu_{t-1}\right) / \sum_{t=1}^{200} \mu_{t-1}^2, \quad \text{and} \quad C_\phi = w/\sum_{t=1}^{200} \mu_{t-1}^2.$$

- Sample $(\mu_{0:200}|v, w, \phi, \lambda_{1:200}, y_{1:200})$. A forward filtering backward sampling algorithm was used to obtain a sample of $\mu_{0:200}$.
- Sample $(\lambda_{1:200}|v, w, \phi, \mu_{0:200}, y_{1:200})$. At each time t, λ_t is sampled from a discrete distribution, i.e., λ_t is set to 1 or κ^2 with probabilities defined

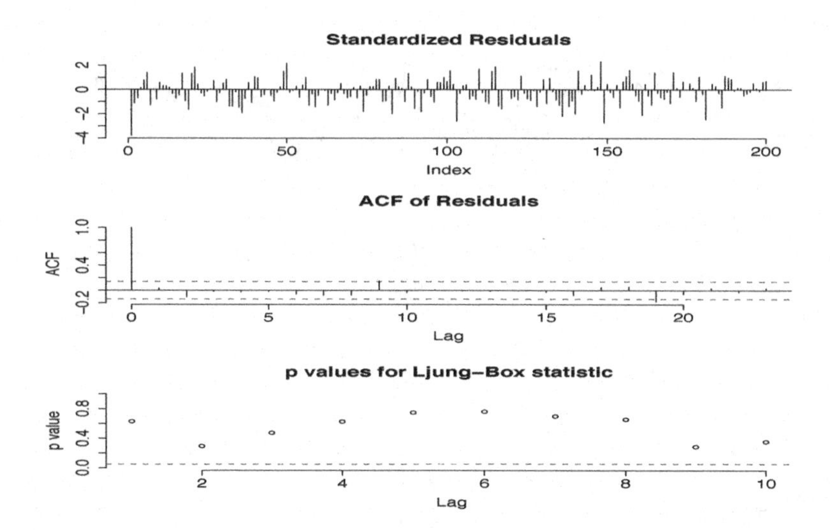

Figure 4.8 *Residual analysis obtained from fitting the model* $\{1, 0.9, 1, 1\}$ *to the simulated data shown in Figure 4.6. The top and middle plots display, respectively, the standardized residuals and the sample ACF of the residuals. The bottom plot shows the p-values for the Ljung-Box statistics at lags 1 to 10. The Ljung-Box statistics allow testing the "randomness" of the residuals and the p-values are computed as a function of the residual autocorrelations up to lag h (see Brockwell and Davis 2002 for details).*

in terms of the ratio

$$\frac{Pr(\lambda_t = 1|v, \mu_{0:200}, y_{1:200})}{Pr(\lambda_t = \kappa^2|v, \mu_{0:200}, y_{1:200})} = \frac{\pi\kappa}{(1-\pi)} \exp\left\{-(y_t - \mu_t)^2(1-\kappa^{-2})/2v\right\}.$$

The top graph in Figure 4.9 displays the simulated series together with the posterior mean for μ_t and the corresponding 95% posterior bands based on 1,000 MCMC iterations obtained after a burn-in period of 500 iterations. The bottom graph shows the posterior mean for the latent process λ_t. These graphs show that the model is effectively capturing the latent structure underlying the simulated series. The most extreme observations are correctly identified as observations generated from the $N(0, \kappa^2 v)$ mixture component. In addition, the posterior means and 95% posterior intervals for ϕ, v, and w are given below.

	ϕ	v	w
posterior mean	0.888	1.036	1.318
95% P.I.	$(0.820, 0.967)$	$(0.771, 1.319)$	$(0.789, 1.905)$

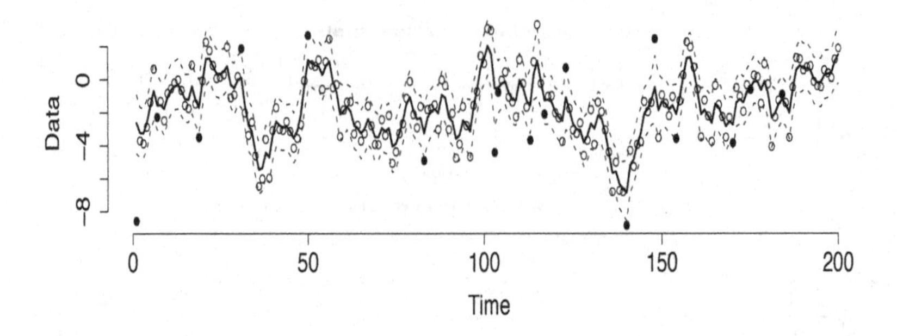

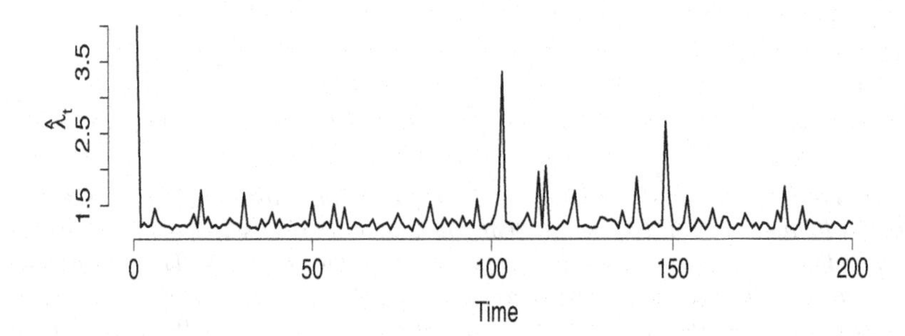

Figure 4.9 *Top: simulated series (circles and solid circles), posterior mean, and 95% posterior bands for* $(\mu_t|D_{200})$ *obtained from the model described by Equations (4.19) and (4.20). Bottom: posterior mean of the latent process* λ_t.

The models were fitted using the `R` software package `dlm` recently developed by Giovanni Petris (Petris, Petrone, and Campagnoli 2009). This package allows the user to easily implement the FFBS algorithm. The software is available online at `http://cran.r-project.org`.

The `BATS` package of Pole, West, and Harrison (1994) implements Gaussian dynamic linear models. DLMs with discount factors are handled by `BATS`. Various other libraries that implement Kalman filters and specific classes of DLMs are also available in `R` and `Matlab`©.

4.6 Problems

1. Assuming a DLM structure given by $\{\mathbf{F}_t, \mathbf{G}_t, v_t, \mathbf{W}_t\}$, find the distributions of $(\boldsymbol{\theta}_{t+k}, \boldsymbol{\theta}_{t+j}|\mathcal{D}_t)$, $(y_{t+k}, y_{t+j}|\mathcal{D}_t)$, $(\boldsymbol{\theta}_{t+k}, y_{t+j}|\mathcal{D}_t)$, $(y_{t+k}, \boldsymbol{\theta}_{t+j}|\mathcal{D}_t)$, and $(\boldsymbol{\theta}_{t-k-j}, \boldsymbol{\theta}_{t-k}|\mathcal{D}_t)$.
2. Show that the smoothing equations in (4.10) and (4.11) can be written

as (4.15) and (4.16) when a single discount factor $\delta \in (0, 1]$ is used to specify $\mathbf{W}_t$.

3. Consider a DLM with time t observation variance v_t known. At time $t-1$, we have the summary posterior $(\boldsymbol{\theta}_{t-1}|\mathcal{D}_{t-1}) \sim N(\mathbf{m}_{t-1}, \mathbf{C}_{t-1})$ and the state vector evolves through the state equation $\boldsymbol{\theta}_t = \mathbf{G}_t\boldsymbol{\theta}_{t-1} + \boldsymbol{\omega}_t$ where $\boldsymbol{\omega}_t \sim N(\mathbf{0}, \mathbf{W}_t)$ with $\boldsymbol{\theta}_{t-1}, \boldsymbol{\omega}_t$ are independent. Suppose now the special case in which:
   - the evolution is a random walk, i.e., $\mathbf{G}_t = \mathbf{I}$ for all t, and
   - $\mathbf{W}_t = \epsilon\mathbf{C}_{t-1}$ where $\epsilon = (1-\delta)/\delta$ for some discount factor $\delta \in (0, 1)$.

   (a) Show how the update equations for prior:posterior analysis at time t simplify in this special case.

   (b) Comment on the simplified structure and how it depends on the chosen/specified discount factor δ.

   (c) Comment on the computational implications of this simplified structure. As part of this, you might consider how the update for $\mathbf{C}_t$ can be rewritten in terms of how the precision matrix $\mathbf{C}_t^{-1}$ is updated from $\mathbf{C}_{t-1}^{-1}$.

4. For a univariate series y_t, consider the simple first-order polynomial (locally constant) DLM with local level μ_t at time t. The $p = 1$–dimensional state is $\boldsymbol{\theta}_t = \mu_t$, while $\mathbf{F}_t = 1$ and $\mathbf{G}_t = 1$ for all t. Also, assume a constant, known observation variance v.

   (a) Show that the usual updating equations for m_t, C_t can be written in the alternative forms
   $$m_t = C_t(R_t^{-1}m_{t-1} + v^{-1}y_t) \qquad \text{with} \qquad C_t^{-1} = R_t^{-1} + v^{-1}.$$

   (b) Suppose that $R_t = C_{t-1}/\delta$ for some discount factor $\delta \in (0, 1]$. Show that $C_t^{-1} = v^{-1} + \delta v^{-1} + \delta^2 v^{-1} + \cdots + \delta^t C_0^{-1}$.

   (c) Deduce that, as $t \to \infty$, the variance C_t has the limiting form $C_t \approx (1-\delta)v$. Comment on this result in connection with the amount of information arising for inference on the local level at t after observing the data for several time points.

   (d) Show that the implied limiting form of the usual updating equation for the posterior mean m_t is, as $t \to \infty$,
   $$m_t \approx \delta m_{t-1} + (1-\delta)y_t$$
   and comment on this form.

   (e) Assuming t is large enough so that this limiting form of m_t is accurate, what is the contribution of a past observation y_{t-k} to the value of m_t?

5. Consider a dynamic regression DLM for a univariate time series, namely $y_t = \mathbf{F}_t'\boldsymbol{\theta}_t + \nu_t$ with $\nu_t \sim N(0, v)$ and v known. Suppose a random walk

evolution for $\boldsymbol{\theta}_t$ so that $\mathbf{G} = \mathbf{I}$ and $\boldsymbol{\theta}_t = \boldsymbol{\theta}_{t-1} + \boldsymbol{\omega}_t$ and $\boldsymbol{\omega}_t \sim N(0, v\mathbf{W}_t)$ where $\mathbf{W}_t$ is defined by a single discount factor δ. With an initial prior $\theta_0|\mathcal{D}_0 \sim N(\mathbf{m}_0, v\mathbf{C}_0)$, it follows for all $t \geq 1$ that $\boldsymbol{\theta}_t|\mathcal{D}_t \sim N(\mathbf{m}_t, v\mathbf{C}_t)$ where $(\mathbf{m}_t, \mathbf{C}_t)$ are updated by the usual filtering equations.

(a) Show that the updating equations can be written in an alternative form using precision matrices as, for all $t > 0$,

$$\mathbf{m}_t = \mathbf{C}_t(\mathbf{R}_t^{-1}\mathbf{m}_{t-1} + \mathbf{F}_t y_t) \quad \text{and} \quad \mathbf{C}_t^{-1} = \mathbf{R}_t^{-1} + \mathbf{F}_t\mathbf{F}_t'$$

where $\mathbf{R}_t = \mathbf{C}_{t-1} + \mathbf{W}_t$.

(b) Show that $\mathbf{C}_t^{-1} = \delta^t \mathbf{C}_0^{-1} + \sum_{r=1}^t \delta^{t-r}\mathbf{F}_r\mathbf{F}_r'$.

(c) Show that $\mathbf{C}_t^{-1}\mathbf{m}_t = \delta^t \mathbf{C}_0^{-1}\mathbf{m}_0 + \sum_{r=1}^t \delta^{t-r}\mathbf{F}_r y_r$.

(d) Interpret these results in connection with the role and choice of the discount factor δ.

6. A DLM for the univariate series y_t is given by $y_t = \mathbf{F}'\boldsymbol{\theta}_t + \nu_t$ where $\nu_t \sim N(0, v)$, and $\boldsymbol{\theta}_t = \mathbf{G}\boldsymbol{\theta}_{t-1} + \boldsymbol{\omega}_t$ where $\boldsymbol{\omega}_t \sim N(\mathbf{0}, v\mathbf{W})$ with the usual conditional independence assumptions. All model parameters $\mathbf{F}, v, \mathbf{G}, \mathbf{W}$ are known and *constant over time*. The modeler specifies the model such that:

   - $\mathbf{G}$ has p *real and distinct* eigenvalues $\lambda_i,\ i = 1, \ldots, p$, with $|\lambda_i| < 1$ for each i; and
   - at $t = 0$, the state distribution $\boldsymbol{\theta}_0|\mathcal{D}_0 \sim N(\mathbf{m}_0, v\mathbf{C}_0)$ where $\mathbf{m}_0 = \mathbf{0}$ and $\mathbf{C}_0 \equiv \mathbf{C}$ satisfies the equation $\mathbf{C} = \mathbf{G}\mathbf{C}\mathbf{G}' + \mathbf{W}$. It can be shown that there is a unique variance matrix $\mathbf{C}$ satisfying this equation when $|\lambda_i| < 1$ as is true in this exercise.

(a) Show that the $t-$step ahead prior distribution for future state vectors $p(\boldsymbol{\theta}_t|\mathcal{D}_0)$ is given by $\boldsymbol{\theta}_t|\mathcal{D}_0 \sim N(\mathbf{0}, v\mathbf{C})$ for all $t \geq 0$.

(b) For any time point t and $k \geq 0$, show that $C(\boldsymbol{\theta}_{t+k}, \boldsymbol{\theta}_t|\mathcal{D}_0) = v\mathbf{G}^k\mathbf{C}$.

(c) Show that the $t-$step ahead forecast distribution $p(y_t|\mathcal{D}_0) = N(0, vs)$ for some constant $s > 0$, and give the expression for s in terms of $\mathbf{F}, \mathbf{G}, \mathbf{C}$.

(d) For any time point t and $k \geq 1$, show that $p(y_{t+k}, y_t|\mathcal{D}_0)$ is bivariate normal with covariance that depends on k but not t. Give an expression for this covariance in terms of k and model parameters.

(e) Deduce that y_t is a stationary time series.

(f) Describe the qualitative form of the implied autocorrelation function $\rho(k)$ as a function of lag k.

(g) Comment on the connections with a stationary AR(p) model for y_t.

7. A DLM has the forecast function defined– over $k = 0, 1, \ldots$ at any "current" time t– by

$$f_t(k) = a_{t,1} + a_{t,2}k + a_{t,3}r^k \cos(2\pi k/\mu + c_t)$$

for some positive wavelength μ and some positive number $r < 1$, and where the quantities $a_{t,1}, a_{t,2}, a_{t,3}, c_t$ are constants known at time t. Give real-valued and *constant* observation vector $\mathbf{F}$ and state evolution matrix $\mathbf{G}$ of *two different DLMs* with this forecast function.

8. Consider the three DLMs below, each with a 2−dimensional state vector $\boldsymbol{\theta}_t = (\theta_{t,1}, \theta_{t,2})'$. Each model is defined by the constant $\mathbf{F}, \mathbf{G}$ elements shown. For each of these DLMs,
   - give details of the implied form of the forecast function $f_t(k)$ over $k = 1, 2, \ldots,$ and
   - comment on the meaning/interpretation of the elements of the state vector.

   (a) The first DLM has
$$\mathbf{F} = \begin{pmatrix} 1 \\ 0 \end{pmatrix} \quad \text{and} \quad \mathbf{G} = \begin{pmatrix} 1 & 0.9 \\ 0 & 0.9 \end{pmatrix}.$$
   (b) The second DLM has
$$\mathbf{F} = \begin{pmatrix} 1 \\ 1 \end{pmatrix} \quad \text{and} \quad \mathbf{G} = \begin{pmatrix} 0.95 & 0 \\ 0 & 0.80 \end{pmatrix}.$$
   (c) The third DLM has
$$\mathbf{F} = \begin{pmatrix} 1 \\ 1 \end{pmatrix} \quad \text{and} \quad \mathbf{G} = \begin{pmatrix} 1 & 0 \\ 0 & 0 \end{pmatrix}.$$

9. Work through the key results of Section 4.3.5 to ensure understanding of the role of the Markovian structure of a DLM in retrospective analysis. Do this in a DLM which, for all time t, has known observation variance v_t. Given $\mathcal{D}_{t-1}$, the two consecutive state vectors $\boldsymbol{\theta}_t$ and $\boldsymbol{\theta}_{t-1}$ are related linearly with Gaussian error, and so the two state vectors have a joint normal distribution $p(\boldsymbol{\theta}_t, \boldsymbol{\theta}_{t-1}|\mathcal{D}_{t-1})$ with
$$E(\boldsymbol{\theta}_{t-1}|\mathcal{D}_{t-1}) = \mathbf{m}_{t-1}, \quad E(\boldsymbol{\theta}_t|\mathcal{D}_{t-1}) = \mathbf{a}_t,$$
$$V(\boldsymbol{\theta}_{t-1}|\mathcal{D}_{t-1}) = \mathbf{C}_{t-1}, \quad V(\boldsymbol{\theta}_t|\mathcal{D}_{t-1}) = \mathbf{R}_t.$$
   (a) Show that the covariance matrix $C(\boldsymbol{\theta}_t, \boldsymbol{\theta}_{t-1}|\mathcal{D}_{t-1}) = \mathbf{G}_t\mathbf{C}_{t-1}$, and hence that $C(\boldsymbol{\theta}_{t-1}, \boldsymbol{\theta}_t|\mathcal{D}_{t-1}) = \mathbf{C}_{t-1}\mathbf{G}_t'$.
   (b) Deduce that $p(\boldsymbol{\theta}_{t-1}|\boldsymbol{\theta}_t, \mathcal{D}_{t-1})$ is normal with mean vector $\mathbf{m}^*_{t-1}$ and variance matrix $\mathbf{C}^*_{t-1}$ as defined in Section 4.3.5.
   (c) For a time specified $T \geq t$, what is the distribution $p(\boldsymbol{\theta}_{t-1}|\boldsymbol{\theta}_t, \mathcal{D}_T)$?
   (d) Comment on the role of this theory in quantifying the retrospective distribution for a full trajectory of states $p(\boldsymbol{\theta}_1, \ldots, \boldsymbol{\theta}_T|\mathcal{D}_T)$.
   (e) Consider now a specific class of DLMs in which:
   - the evolution is a random walk, i.e., $\mathbf{G}_t = \mathbf{I}$ for all t, and

- $\mathbf{W}_t = \epsilon \mathbf{C}_{t-1}$ where $\epsilon = (1-\delta)/\delta$ for some discount factor $\delta \in (0,1)$.

Show how the above results simplify in these special cases, discussing both the role of δ as well as computational considerations.

10. The basic distribution theory in this question underlies the discount volatility model of Section 4.3.7 and the results to be shown below in Problem 11. Two positive scalar random quantities ϕ_0 and ϕ_1 have a joint distribution under which:
   - $\phi_0 \sim G(a,b)$ for some scalars $a>0, b>0$; and
   - $p(\phi_1|\phi_0)$ is implicitly defined by

   $$\phi_1 = \phi_0 \eta/\beta, \quad \text{where} \quad \eta \sim Be(\beta a, (1-\beta)a)$$

   with η independent of ϕ_0 and where $\beta \in (0,1)$ is a known, constant discount factor.

   (a) What is $E(\phi_1|\phi_0)$?
   (b) What are $E(\phi_0)$ and $E(\phi_1)$?
   (c) Starting with the joint density $p(\phi_0)p(\eta)$ (a product form since ϕ_0 and η are independent), make the bivariate transformation to (ϕ_0, ϕ_1) and show that

   $$p(\phi_0, \phi_1) = c\, \mathrm{e}^{-b\phi_0}\, \phi_1^{\beta a-1}\, (\phi_0 - \beta\phi_1)^{(1-\beta)a-1}, \quad \text{on } 0 < \phi_1 < \phi_0/\beta,$$

   being zero otherwise. Here c is a normalizing constant that does not depend on the conditioning value of ϕ_0.
   (d) Derive the p.d.f. $p(\phi_1)$ (up to a proportionality constant). Deduce that the marginal distribution of ϕ_1 is $\phi_1 \sim G(\beta a, \beta b)$.
   (e) Show that the reverse conditional $p(\phi_0|\phi_1)$ is implicitly defined by

   $$\phi_0 = \beta\phi_1 + \gamma \quad \text{where} \quad \gamma \sim G((1-\beta)a, b)$$

   with γ independent of ϕ_1.

11. Consider the observational variance discount model of Section 4.3.7. You may use the results from Problem 10.

   (a) Show that the time $t-1$ prior $(\phi_{t-1}|\mathcal{D}_{t-1}) \sim G(n_{t-1}/2, d_{t-1}/2)$ combined with the beta-gamma evolution model $\phi_t = \phi_{t-1}\gamma_t/\beta$ yields a conditional density $p(\phi_{t-1}|\phi_t, \mathcal{D}_{t-1})$ that can be expressed as $\phi_{t-1} = \beta\phi_t + v^*_{t-1}$, where

   $$(v^*_{t-1}|\mathcal{D}_{t-1}) \sim G((1-\beta)n_{t-1}/2, d_{t-1}/2)$$

   is independent of ϕ_t.
   (b) Show further that $p(\phi_{t-1}|\phi_t, \mathcal{D}_T) \equiv p(\phi_{t-1}|\phi_t, \mathcal{D}_{t-1})$ for all $T \geq t$.
   (c) Describe how this result can be used to recursively compute retrospective point estimates $E(\phi_t|\mathcal{D}_T)$ backward in time, beginning at $t = T$.

(d) Describe how this result can similarly be used to recursively simulate a full trajectory of values of $\phi_T, \phi_{T-1}, \ldots, \phi_1$ from the retrospective smoothed posterior conditional on $\mathcal{D}_T$.

12. Go to Google Trends and download the monthly data for the searches of the term "time series" in the U.S. and the rest of the world from January 2004 until December 2019. For each of these two time series fit the following DLMs and provide a summary of the filtering and smoothing distributions of the model parameters and also summaries of the one-step ahead predictions:

(a) DLMs with a second order polynomial and the first 4 harmonics of a Fourier representation with fundamental period $p = 12$. Use a single discount factor $\delta \in [0.9, 1.0]$ to determine the system variance choosing the optimal discount factor that minimizes the MSE.

(b) Now consider the same DLM structure above but fit the model to the log volume index data.

13. Consider the following model:

$$\begin{aligned} y_t &= \theta_t + \nu_t, \quad \nu_t \sim N(0, \sigma^2), \\ \theta_t &= \sum_{j=1}^{p} \phi_j \theta_{t-j} + \omega_t, \quad \omega_t \sim N(0, \tau^2). \end{aligned}$$

This model is a simplified version of that proposed in West (1997c). Develop the conditional distributions required to define an MCMC algorithm to obtain samples from $p(\theta_{1:T}, \boldsymbol{\phi}, \tau^2, \sigma^2 | y_{1:T})$, and implement the algorithm.

14. Derive the conditional distributions for posterior MCMC simulation in Example 4.10, verifying that the algorithm outlined there is correct.

15. Consider again the AR(1) model with mixture observational errors described in Example 4.10. Modify the MCMC algorithm in order to perform posterior inference when λ_t has the following Markovian structure:

$$Pr(\lambda_t = \kappa^2 | \lambda_{t-1} = \kappa^2) = Pr(\lambda_t = 1 | \lambda_{t-1} = 1) = p$$

and

$$Pr(\lambda_t = \kappa^2 | \lambda_{t-1} = 1) = Pr(\lambda_t = 1 | \lambda_{t-1} = \kappa^2) = (1 - p),$$

where p is known. For suggestions see, for instance, Carter and Kohn (1994).

16. Consider the dynamic trend model $\{\mathbf{F}, \mathbf{G}, v(\alpha_1), \mathbf{W}(\alpha_2, \alpha_3)\}$ introduced by Harrison and Stevens (1976) and revisited in Frühwirth-Schnatter (1994), where

$$\mathbf{F}' = (1, 0), \quad \mathbf{G} = \begin{pmatrix} 1 & 1 \\ 0 & 1 \end{pmatrix}, \quad v(\alpha_1) = \alpha_1,$$

and

$$\mathbf{W}(\alpha_2, \alpha_3) = \mathbf{G}\,\mathrm{diag}(\alpha_2, \alpha_3)\mathbf{G}' = \begin{pmatrix} \alpha_2 + \alpha_3 & \alpha_3 \\ \alpha_3 & \alpha_3 \end{pmatrix}.$$

Simulate a time series data set from this model. Propose and implement a MCMC algorithm for posterior simulation assuming that α_1, α_2, and α_3 are unknown, where each α_i is assumed to follow an inverse gamma prior distribution.

Chapter 5

State-space TVAR models

This chapter concerns the class of autoregressive models with time-varying parameters, or TVAR models, that are widely used in empirical analysis of time series and short-term forecasting. They extend AR models to a broader class useful in describing nonstationary time series. One main focus is on the underlying structure in time series that TVAR models can identify, especially related to quasiperiodic components. Such data arise in many areas of applications, such as biomedical monitoring, speech signal processing, climatological studies, and econometrics. TVAR decompositions allow us to partition a time series into a collection of processes that are often scientifically meaningful, defining a statistically sound approach to time-frequency analysis. TVAR models are special classes of DLMs, so the theory of Chapter 4 applies for estimation, forecasting, smoothing, and inference on latent structure in nonstationary time series.

5.1 Time-Varying Autoregressions and Decompositions

5.1.1 Basic DLM Decomposition

Consider a scalar time series y_t observed at times $t = 1 : T$, modeled with a DLM of the form

$$y_t = x_t + \nu_t, \quad x_t = \mathbf{F}'\boldsymbol{\theta}_t, \quad \boldsymbol{\theta}_t = \mathbf{G}_t\boldsymbol{\theta}_{t-1} + \mathbf{w}_t, \tag{5.1}$$

where x_t is a latent unobservable process driving the process y_t; ν_t is an observation error; $\boldsymbol{\theta}_t$ is the $p \times 1$ state vector; $\mathbf{F}$ is a p-vector of constants; $\mathbf{G}_t$ is a $p \times p$ state evolution matrix; and $\mathbf{w}_t$ is the p-vector of state innovations. The noise terms ν_t and $\mathbf{w}_t$ are zero-mean sequences, mutually independent and having Gaussian distributions with variance v_t for ν_t, and a variance-covariance matrix $\mathbf{W}_t$ for $\mathbf{w}_t$. As discussed previously in Chapter 4, more general modeling assumptions to deal with measurement error and outlier components include heavy-tailed error distributions and mixture error distributions (e.g., Carter and Kohn 1994, West 1997b, and Fonseca, Ferreira, and Migon 2008).

We are interested in decomposing x_t into relevant latent components. The decomposition results presented here are obtained by considering a linear transformation of the state parameter vector $\boldsymbol{\theta}_t$ that reparameterizes model (5.1) in terms of a new evolution matrix with a simpler structure than that of $\mathbf{G}_t$. This is related to the theory of similar and equivalent models briefly discussed in Chapter 4 and extensively developed in Chapter 5 of West and Harrison (1997).

More specifically, suppose that at each time t, $\mathbf{G}_t$ in (5.1) has p different eigenvalues $\lambda_{t,1}, \ldots, \lambda_{t,p}$ (and so, each eigenvalue has multiplicity $m_{t,i} = 1$). Some of these eigenvalues could be complex, and in such case they would appear in complex conjugate pairs. The number of complex and real eigenvalues may also vary over time but, for the sake of simplicity, let us assume that for all t, there are exactly c pairs of complex eigenvalues, denoted by $r_{t,j} \exp(\pm i\omega_{t,j})$ for $j = 1 : c$, and $r = p - 2c$ real and distinct eigenvalues denoted by $r_{t,j}$ for $j = (2c+1) : p$. Since the eigenvalues are distinct, the eigenvectors of $\mathbf{G}_t$ are unique up to a constant, and so $\mathbf{G}_t = \mathbf{B}_t\mathbf{A}_t\mathbf{B}_t^{-1}$, where $\mathbf{A}_t$ is the diagonal matrix of eigenvalues in arbitrary but fixed order, and $\mathbf{B}_t$ is a $p \times p$ eigenmatrix whose columns correspond to right eigenvectors of $\mathbf{G}_t$ appearing in the order given by the eigenvalues. Now, for each time t define $\mathbf{H}_t = \text{diag}(\mathbf{B}_t'\mathbf{F})\mathbf{B}_t^{-1}$ and reparameterize the model (5.1), linearly transforming $\boldsymbol{\theta}_t$ and $\mathbf{w}_t$ via $\boldsymbol{\gamma}_t = \mathbf{H}_t\boldsymbol{\theta}_t$ and $\boldsymbol{\delta}_t = \mathbf{H}_t\mathbf{w}_t$. The eigenmatrix $\mathbf{B}_t$ is not unique, but the transformation defined by $\mathbf{H}_t$ is unique. Then, we can rewrite (5.1) in terms of the new state and innovation vectors, $\boldsymbol{\gamma}_t$ and $\boldsymbol{\delta}_t$, as follows,

$$y_t = x_t + \nu_t, \quad x_t = \mathbf{1}'\boldsymbol{\gamma}_t, \quad \boldsymbol{\gamma}_t = \mathbf{A}_t\mathbf{K}_t\boldsymbol{\gamma}_{t-1} + \boldsymbol{\delta}_t, \tag{5.2}$$

where $\mathbf{1}' = (1, \ldots, 1)$ and $\mathbf{K}_t = \mathbf{H}_t\mathbf{H}_{t-1}^{-1}$. Therefore, the equations in (5.2) imply that x_t is the sum of the p components of $\boldsymbol{\gamma}_t = (\gamma_{t,1}, \ldots, \gamma_{t,p})'$. In other words, x_t can be written as a sum of p latent processes related to the p distinct eigenvalues of $\mathbf{G}_t$. The final r elements of $\boldsymbol{\gamma}_t$ are real, corresponding to the real eigenvalues $r_{t,j}$ at each time t. Rename these real-valued processes $x_{t,j}^{(2)}$, for $j = 1 : r$. The initial $2c$ elements of $\boldsymbol{\gamma}_t$ appear in complex pairs, and so $x_{t,j}^{(1)} = \gamma_{t,2j-1} + \gamma_{t,2j}$ is a real process for $j = 1 : c$. Each $x_{t,j}^{(1)}$ is related to the pair of complex eigenvalues $r_{t,j}\exp(\pm i\omega_{t,j})$ at time t.

The basic decomposition result for the class of models that can be expressed in the form (5.1) is simply $y_t = x_t + \nu_t$, where

$$x_t = \sum_{j=1}^{c} x_{t,j}^{(1)} + \sum_{j=1}^{r} x_{t,j}^{(2)}. \tag{5.3}$$

Given known, estimated, or simulated values of $\mathbf{F}$, $\mathbf{G}_t$, and $\boldsymbol{\theta}_t$ at each time t, the processes $x_{t,j}^{(1)}$ and $x_{t,j}^{(2)}$ can be evaluated over time by computing the eigenstructure of $\mathbf{G}_t$ and the transformations described above.

5.1.2 Latent Structure in TVAR Models

A time-varying autoregression of order p, or TVAR(p), is described by

$$x_t = \sum_{j=1}^{p} \phi_{t,j} x_{t-j} + \epsilon_t, \tag{5.4}$$

where $\boldsymbol{\phi}_t = (\phi_{t,1}, \ldots, \phi_{t,p})'$ is the time-varying parameter vector and ϵ_t are zero-mean independent innovations with variance v_t. This model has a DLM representation (5.1) with $\nu_t = 0$, $\boldsymbol{\theta}_t = (x_t, x_{t-1}, \ldots, x_{t-p+1})'$, $\mathbf{w}_t = (\epsilon_t, 0, \ldots, 0)'$, $\mathbf{F}' = (1, 0, \ldots, 0)$, and

$$\mathbf{G}_t \equiv \mathbf{G}(\boldsymbol{\phi}_t) = \begin{pmatrix} \phi_{t,1} & \phi_{t,2} & \ldots & \phi_{t,p-1} & \phi_{t,p} \\ 1 & 0 & \ldots & 0 & 0 \\ 0 & 1 & \ldots & 0 & 0 \\ \vdots & & \ddots & & \vdots \\ 0 & 0 & \ldots & 1 & 0 \end{pmatrix}. \tag{5.5}$$

Similarly to the DLM representation of the autoregressive model of order p, or AR(p), this DLM representation of the TVAR(p) model is such that the eigenvalues of $\mathbf{G}_t$ are precisely the reciprocal roots of the autoregressive characteristic equation at time t, $\Phi_t(u) = 0$, with $\Phi_t(u) = (1 - \phi_{t,1}u - \cdots - \phi_{t,p}u^p)$.

5.1.2.1 Decompositions for standard autoregressions

If $\mathbf{G}_t$ is constant over time we have a standard AR(p) process. Then, the eigenvalues of $\mathbf{G}$ are constant for all t, and so $r_{t,j} = r_j$ and $\omega_{t,j} = \omega_j$ for all t. Reparameterizing the model via $\mathbf{H}_t = \mathbf{H} = \text{diag}(\mathbf{B}'\mathbf{F})\mathbf{B}^{-1}$, where $\mathbf{B}$ is a matrix of eigenvectors, we obtain $\mathbf{K}_t = \mathbf{H}\mathbf{H}^{-1} = \mathbf{I}_p$, and the new evolution matrix is the diagonal matrix of eigenvalues $\mathbf{A}_t = \mathbf{A} = \text{diag}(\lambda_1, \ldots, \lambda_p)$. Once again, we are assuming that $\mathbf{G}$ has exactly p distinct eigenvalues. The final r real components of the transformed state vector $\gamma_{t,2c+j}$, renamed $x_{t,j}^{(2)}$, are AR(1) processes with coefficients r_j, namely

$$x_{t,j}^{(2)} = r_j x_{t-1,j}^{(2)} + \delta_{t,j}, \tag{5.6}$$

for $j = 1 : r$. The initial $2c$ elements of $\boldsymbol{\gamma}_t$ will generate the real processes $x_{t,j}^{(1)}$, with $x_{t,j}^{(1)} = \gamma_{t,2j-1} + \gamma_{t,2j} = 2Re(\gamma_{t,2j-1})$ for $j = 1 : c$. It is possible to show that $x_{t,j}^{(1)}$ is an autoregressive moving average process ARMA$(2, 1)$ with AR parameters $2r_j \cos(\omega_j)$ and $-r_j^2$ (see Problem 1), that is

$$x_{t,j}^{(1)} = 2r_j \cos(w_j) x_{t-1,j}^{(1)} - r_j^2 x_{t-2,j}^{(1)} + \eta_{t,j},$$

where $\eta_{t,j}$ is a real, zero-mean AR(1) process itself. The AR(2) component here is quasiperiodic with time-varying random amplitude and phase, and constant characteristic frequency and modulus ω_j and r_j, respectively. The innovations $\delta_{t,j}$ and $\eta_{t,j}$ are not independent, having conditional variances that are functions of the AR parameter vector $\boldsymbol{\phi}$; hence, $x_{t,j}^{(1)}$ and $x_{t,j}^{(2)}$ will be correlated for a fixed t.

5.1.2.2 Decompositions in the TVAR case

In general, the matrix $\mathbf{K}_t$ in (5.2) does not reduce to the identity. This implies that the structures of $x_{t,j}^{(1)}$ and $x_{t,j}^{(2)}$ over time are more complicated than those of the latent components in the standard AR case. Note that, if $\mathbf{K}_t = \mathbf{I}_p$ for all t, the decomposition result is exactly as in the standard AR(p) case, but now each $x_{t,j}^{(2)}$ is a TVAR(1) with time-varying AR coefficient $r_{t,j}$ at time t. Similarly, each $x_{t,j}^{(1)}$ follows a time-varying ARMA model, TVARMA$(2, 1)$, whose amplitude and phase change randomly over time, as in the static case, but now the frequencies $\omega_{t,j}$ and the moduli $r_{t,j}$ for each j are also time-varying. The series x_t is then decomposed as a sum of TVAR(1) and quasiperiodic TVARMA$(2, 1)$ processes. The spectrum of x_t is time-varying: it is a function of the instantaneous spectra of the $x_{t,j}^{(1)}$ and $x_{t,j}^{(2)}$ processes. The spectral density is peaked around the frequencies $\omega_{t,j}$s and the sharpness of each of these peaks is proportional to its corresponding modulus $r_{t,j}$.

If $\mathbf{G}_t$ is slowly varying over time the resulting $\mathbf{K}_t$ matrices are typically very close to the identity for each t. Therefore, x_t can be decomposed as the sum of processes that can be *approximately* represented by TVAR(1) and TVARMA$(2, 1)$. Alternatively, in those cases where the difference between the eigenmatrices at times t and $t - 1$ is not negligible, $x_{t,j}^{(1)}$ and $x_{t,j}^{(2)}$ are "mixed" over time through $\mathbf{K}_t$ and they will not generally follow TVAR(1) and TVARMA$(2, 1)$ processes (see Problem 2). Consequently, for general TVAR(p) models, even though the signal x_t can still be decomposed into latent processes associated with the complex and real characteristic reciprocal roots, it is not always straightforward to describe the structure of these latent processes.

The decomposition results for TVAR models summarized here were introduced in Prado (1998) and West, Prado, and Krystal (1999). We refer the reader to these references for further discussion and details on these results.

5.1.3 Interpreting Latent TVAR Structure

In many time series applications the TVAR coefficients vary very slowly in time implying that, at each time t, $\mathbf{G}_t$ and $\mathbf{G}_{t-1}$ have very similar eigenstructures. This results in $\mathbf{K}_t$ matrices that are almost equal to the identity for every t, and so the model in (5.2) can be approximated by another model whose state evolution matrix is $\mathbf{A}_t$ instead of $\mathbf{A}_t\mathbf{K}_t$. As mentioned above, the latent processes in the decomposition of x_t for models with evolution matrix $\mathbf{A}_t$ have a very specific structure: they are TVAR(1) and TVARMA(2,1) processes. Then, the question that arises in connection with the interpretability of the latent components is how close is the structure of these components in a general TVAR(p) model, with state evolution matrix $\mathbf{A}_t\mathbf{K}_t$, to the TVARMA$(2,1)$ and TVAR(1) structure obtained when $\mathbf{K}_t = \mathbf{I}_p$? Following the ideas of West and Harrison (1997), we can measure this by comparing the forecast function of the general TVAR(p) model with evolution matrix $\mathbf{A}_t\mathbf{K}_t$ to that of a model with evolution matrix $\mathbf{A}_t$. Specifically, consider the models $\mathcal{M}_1$ and $\mathcal{M}_2$ given by

$$\begin{array}{rrcl} \mathcal{M}_1: & y_t & = & x_t + \nu_t \\ & x_t & = & \mathbf{1}'\gamma_t \\ & \gamma_t & = & \mathbf{A}_t\mathbf{K}_t\gamma_{t-1} + \boldsymbol{\delta}_t \end{array} \qquad \begin{array}{rrcl} \mathcal{M}_2: & y_t & = & x_t + \nu_t \\ & x_t & = & \mathbf{1}'\gamma_t \\ & \gamma_t & = & \mathbf{A}_t\gamma_{t-1} + \boldsymbol{\delta}_t. \end{array} \tag{5.7}$$

The h-step-ahead forecast functions for $\mathcal{M}_1$ and $\mathcal{M}_2$ are, respectively,

$$f_t^{(1)}(h) = E(x_{t+h} \mid \gamma_t, \mathcal{M}_1) = \mathbf{1}'\mathbf{A}_{t+h}\mathbf{K}_{t+h}\mathbf{A}_{t+h-1}\mathbf{K}_{t+h-1}\cdots\mathbf{A}_{t+1}\mathbf{K}_{t+1}\gamma_t$$

and

$$f_t^{(2)}(h) = E(x_{t+h} \mid \gamma_t, \mathcal{M}_2) = \mathbf{1}'\mathbf{A}_{t+h}\mathbf{A}_{t+h-1}\cdots\mathbf{A}_{t+1}\gamma_t.$$

It can be shown (see Problem 3) that the relative difference between $f_t^{(1)}(h)$ and $f_t^{(2)}(h)$ can be bounded as follows,

$$\frac{|f_t^{(1)}(h) - f_t^{(2)}(h)|}{||\gamma_t||_\infty} \leq (\lambda^*)^h \times [(1+\epsilon^*)^h - 1], \tag{5.8}$$

with $\lambda^* = \max_{0\leq j\leq h-1}(\max_{1\leq i\leq p}|\lambda_{((t+h-j),i)}|)$; $\epsilon^* = \max_{1\leq j\leq h}||\mathcal{E}_{t+j}||_\infty$, where the $\mathcal{E}_{t+j}$ are matrices for $j = 1:k$ such that $\mathbf{K}_{t+j} = \mathbf{I} + \mathcal{E}_{t+j}$. Here, $||\cdot||_\infty$ denotes the l_∞ norm, defined as $||\mathbf{Q}||_\infty = \max_{1\leq i\leq p}\sum_{j=1}^{p}|q_{i,j}|$, with $q_{i,j}$ the ij-th element of a $p\times p$ matrix $\mathbf{Q}$.

Example 5.1 *Comparing forecast functions.* Consider a TVAR(p) such that, for all $j > t$, $||\mathcal{E}_{t+j}||_\infty \leq 10^{-4}$ and that all the characteristic reciprocal

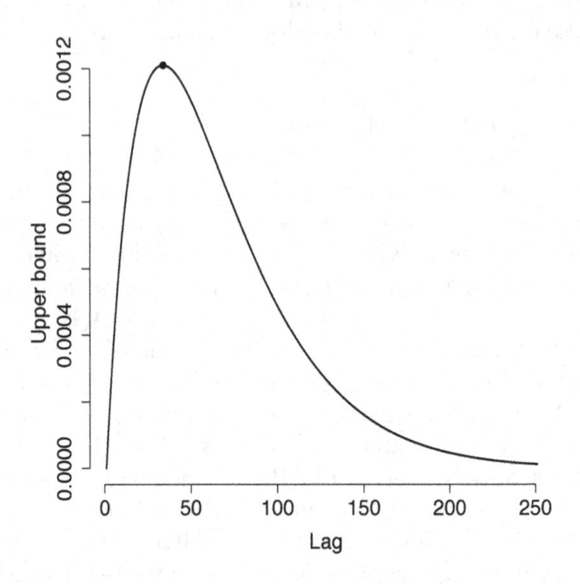

Figure 5.1 *Upper bound for the relative difference between the forecast functions of models $\mathcal{M}_1$ and $\mathcal{M}_2$.*

roots have moduli less than 0.97. Then, the above theory implies that

$$\frac{|f_t^{(1)}(h) - f_t^{(2)}(h)|}{\|\gamma_t\|_\infty} \leq 0.97^h \times \left[(1.0001)^h - 1\right]. \tag{5.9}$$

Figure 5.1 shows the upper bound in (5.9) for the relative differences between the forecast functions for h up to 250 steps ahead. The maximum value for the upper bound is obtained at 34 steps ahead. After $h = 34$ the difference decreases and tends to zero as h increases. The interpretation of this result is as follows. If $\mathcal{M}_2$ is used to approximate $\mathcal{M}_1$, the expected value of x_{t+h} in model $\mathcal{M}_2$ will differ from the expected value of x_{t+h} in model $\mathcal{M}_1$, relative to the magnitude of the state vector γ_t or, equivalently in this case, relative to the magnitude of the underlying signal x_t, in an amount that is at most 0.0012. This bound is negligible from a practical perspective, and so using $\mathcal{M}_2$ to approximate $\mathcal{M}_1$ will produce almost the same results in terms of predicting the future if $\mathbf{K}_t$ is very close to the identity, and if the TVAR(p) process is locally (instantaneously) stationary. Then, model $\mathcal{M}_2$, with a simpler latent structure than that of $\mathcal{M}_1$, can be used to interpret, at least approximately, attributes that characterize x_t such as the instantaneous amplitudes, frequencies, and moduli of its underlying components. Therefore, in this example the latent processes in the decomposition of x_t are basically TVAR(1) and TVARMA(2, 1) processes.

5.2 TVAR Model Specification and Posterior Inference

In order to complete the TVAR model specification given in (5.4), it is necessary to define the evolution structure of the TVAR coefficients $\boldsymbol{\phi}_t = (\phi_{t,1}, \ldots, \phi_{t,p}),'$ and that of v_t. These parameters describe the nonstationary behavior in the observed series over time.

In addition, prior distributions for $\boldsymbol{\phi}_t$ and v_t need to be provided. Sophisticated prior structures that can incorporate scientifically meaningful information have been developed in the past for standard AR models. Barnett, Kohn, and Sheather (1996, 1997) developed priors for AR and ARMA models on the partial regression coefficients, rather than on the ARMA coefficients. As discussed in Chapter 2, Huerta (1998) and Huerta and West (1999b) proposed a novel class of priors for parameters that characterize the number and structure of latent underlying components in AR processes. In the case of TVAR models, it is important to specify priors that define a reasonable evolution structure over time. Kitagawa and Gersch (1985, 1996a) introduce smoothness prior distributions for TVARs in two ways: by defining smoothness prior constraints on the AR coefficients and, alternatively, by imposing the same type of constraints on partial autocorrelation coefficients. A related approach is taken in Godsill, Doucet, and West (2004), where TVAR models are reparameterized in terms of partial correlation coefficients and normal priors, truncated to the $(-1, 1)$ interval for stationarity, and are used to describe the evolution of the partial correlation coefficients—and consequently, the evolution of the TVAR parameters—over time.

Here, random walk equations are used to describe the evolution of the model parameters, providing adaptation to the changing structure of the series through time without anticipating specific directions of changes (West and Harrison 1997, Chapter 3 and Sections 9.6 and 10.8). Specifically, we have

$$\boldsymbol{\phi}_t = \boldsymbol{\phi}_{t-1} + \boldsymbol{\xi}_t, \quad \boldsymbol{\xi}_t \sim N(\boldsymbol{\xi}_t \mid \mathbf{0}, \mathbf{U}_t), \tag{5.10}$$

and

$$v_t = \beta v_{t-1}/\eta_t, \quad \eta_t \sim Be(\eta_t \mid a_t, b_t), \tag{5.11}$$

where $\boldsymbol{\xi}_t$ and η_t are independent, mutually independent, and also independent of ϵ_t. Time variation in $\boldsymbol{\phi}_t$ is controlled by $\mathbf{U}_t$, whose specification is handled by the use of a single discount factor δ_ϕ. As explained in Chapter 4, a discount factor $\delta_\phi \in (0, 1]$ represents an increase of $100(1-\delta_\phi)/\delta_\phi\%$ in the evolution variance matrix from time $t-1$ to time t. A similar method determines the parameters (a_t, b_t) of the beta distribution defining the evolution of v_t. These parameters, defined as in West and Harrison (1997, Section 10.8; see also Chapter 4 of this book), are functions of δ_v, another discount factor in $(0,1]$ analogous to δ_ϕ.

Finally, model completion requires the specification of the prior distribution. A normal prior for $\boldsymbol{\phi}_0$ conditional on v_0, $(\boldsymbol{\phi}_0 \mid v_0, \mathcal{D}_0) \sim N(\mathbf{m}_0^*, v_0\mathbf{C}_0^*)$, and an inverse-gamma prior on v_0, $(v_0|\mathcal{D}_0) \sim IG(n_0, d_0)$, lead unconditionally to $(\boldsymbol{\phi}_0 \mid \mathcal{D}_0) \sim T_{n_0}(\mathbf{m}_0, \mathbf{C}_0)$, that is, a multivariate Student-t distribution with location $\mathbf{m}_0$, scale matrix $\mathbf{C}_0$, and n_0 degrees of freedom. Sequential updating and retrospective smoothing can then be applied to obtain posterior distributions of $\boldsymbol{\phi}_t$ and v_t based on observed information $\mathcal{D}_T$.

Choosing the values of the discount factors δ_ϕ and δ_v, and the model order p, is a relevant feature of the analysis. TVAR models of different orders can be compared using the last $T - p^*$ observations of the series, where p^* is the maximum model order that would be considered, and it is usually set prior to the analysis. Then, the discount factors and model order can be chosen to maximize joint log-likelihood functions defined as follows,

$$l(\delta_\phi, \delta_v, p) \equiv \log[p(x_{(p^*+1):T} \mid \mathcal{D}_{p^*})] = \sum_{t=p^*+1}^{T} \log[p(x_t \mid \mathcal{D}_{t-1})], \quad (5.12)$$

where each $p(x_t \mid \mathcal{D}_{t-1})$ is the one-step-ahead univariate Student-t density. Note that (5.12) depends on $\mathcal{D}_{p^*}$, the information up to time $t = p^*$. This includes the first p^* observations and the prior structure on v_{p^*} and $\boldsymbol{\phi}_{p^*}$, that also depends on the model order given that the dimension of $\boldsymbol{\phi}_t$ is determined by p.

Example 5.2 *Inferring latent structure in electroencephalograms (EEGs).* We show how TVAR models can be useful in analyzing nonstationary time series such as electroencephalograms. The top time series depicted in Figure 5.2 displays 3,600 EEG observations recorded on a patient who received electroconvulsive therapy (ECT). This series was recorded at channel *Cz*, a channel located at the center of the patient's scalp, and it is one of 19 EEG signals that were recorded at various locations. The original EEG recordings had more than 26,000 observations per channel. Approximately 2,000 observations were removed from the beginning and the end of the series, while the remaining central portion was subsampled every fifth observation, producing the 3,600 observations displayed in the figure. The original sampling rate was 256 Hz. For explanations about the clinical relevance of studying these series and further analyses see Prado (1998).

In order to choose optimal values of δ_ϕ, δ_v, and p, the function $l(\delta_\phi, \delta_v, p)$ in (5.12) was evaluated over a grid of δ_ϕ and δ_v values in $[0.9, 1] \times [0.9, 1]$ and p in $[4, 20]$. Then, the values of δ_ϕ, δ_v, and p that maximized $l(\delta_\phi, \delta_v, p)$ were chosen as the optimal values, which were found to be $\delta_\phi = 0.994$, $\delta_v = 0.95$, and $p = 12$ for this series. Various choices of the initial quantities that specify the normal/inverse-gamma priors for the model parameters were considered, leading to the same optimal values for δ_ϕ, δ_v, and p.

Posterior samples of the TVAR coefficients were obtained at selected time points. Posterior intervals for these parameters become wider with time (not shown), indicating that parameter uncertainty increases toward the end of the series. Standardized fitted and one-step-ahead forecast residuals were also computed. No patterns or extreme values were found on these residuals. The posterior samples of $\boldsymbol{\phi}_t$ and v_t led to posterior samples of the latent components that were computed via (5.3), using the TVAR DLM representation given by $\mathbf{F}_t = \mathbf{F} = (1, 0, \ldots, 0)'$, $\boldsymbol{\theta}_t = (x_t, \ldots, x_{t-p+1})'$ and the state matrices $\mathbf{G}_t$ in (5.5). The eigenstructure of $\mathbf{G}_t$ estimated from the EEG series and computed by fixing $\boldsymbol{\phi}_t$ at the posterior mean $E(\boldsymbol{\phi}_t \mid \mathcal{D}_T)$ at $t \leq T$ exhibits at least three complex components, each with modulus $r_{t,j}$ and argument $\omega_{t,j}$, both varying smoothly in time. All the components showed moduli less than the unity, and two of them had moduli consistently higher than 0.75, indicating that the series is locally stationary. At this point it is worth commenting on some complications that arise in going from the TVAR parameters to the set of characteristic roots $r_{t,j} \exp(\pm i\omega_{t,j})$. First, there is no inherent identification of the roots, and so no identification of the corresponding latent processes. A component that has the lowest frequency at a particular time point may have a higher frequency later. Similar comments apply to the moduli and amplitudes of the components. For posterior summary, identification must be enforced through an ordering. Usually, the characteristic roots at each time t are ordered in terms of their relative frequencies or moduli. Therefore, interpretation of posterior summaries must bear in mind that the components may switch from time to time as the data structure and the model's response evolve. The second closely related issue is that the number of real and complex pairs of eigenvalues may differ at different times. The decomposition (5.3) assumed fixed and constant numbers of real and complex components in time, but this is not generally the case in practice. The main reason for this is that collections of higher frequency components corresponding to complex roots will often have fairly low moduli and be apparent in the model decomposition as representations of high frequency noise; these components typically have low amplitude relative to more dominant components that have meaningful interpretations in applied contexts. With very high frequency ranges, relatively small changes in the $\boldsymbol{\phi}_t$ parameters can lead to one or more pair of such complex roots "disappearing," being substituted by two real roots with low values and correspondingly low amplitudes of the induced real components in the series. The reverse

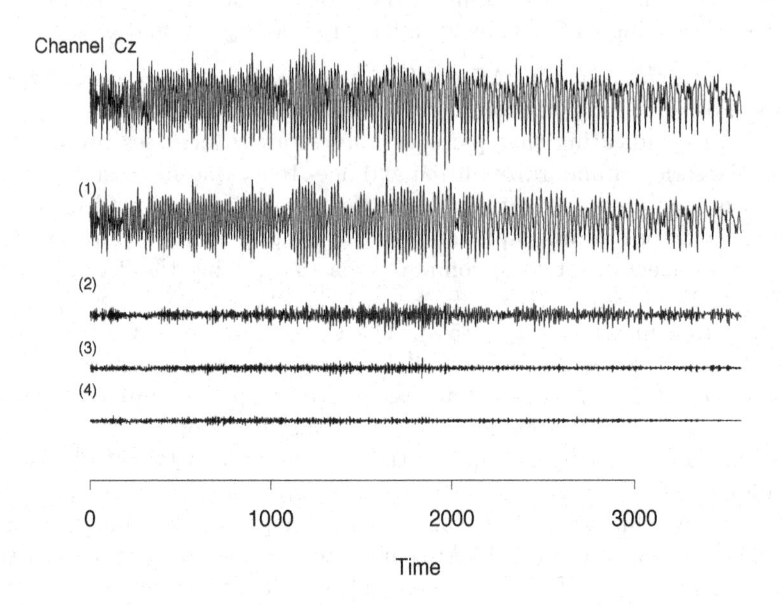

Figure 5.2 *Some estimated components in the decomposition of channel* Cz.

phenomenon, that is, complex roots substituting real roots, was also experienced.

Figure 5.2 displays estimates, based on $E(\phi_t \mid \mathcal{D}_T)$, of some of the latent components $x_{t,j}^{(1)}$ and $x_{t,j}^{(2)}$ in decomposition (5.3) for channel *Cz*. These components were computed using the software `tvar`, which is available and documented at `http://www2.stat.duke.edu/~mw/mwsoftware/TVAR/`. This plot shows the original time series followed by the first four estimated components. The components are ordered by increasing characteristic frequencies, and so component (1) has the lowest estimated frequency among those displayed. This component also dominates in amplitude basically driving the seizure for this episode. Component (2) is much lower in amplitude, but it is still significant. Higher frequency components also appear in the decomposition. Some of these components are relatively persistent during very short time periods but have much lower amplitudes than the lower frequency components.

For these data, the estimated $\mathbf{K}_t$ matrices were close to the identity for all t (in fact, the $\mathbf{K}_t$ matrices differed from the identity in the order of 10^{-5} element by element). Therefore, the latent processes $x_{t,j}^{(1)}$ and $x_{t,j}^{(2)}$ in (5.3)

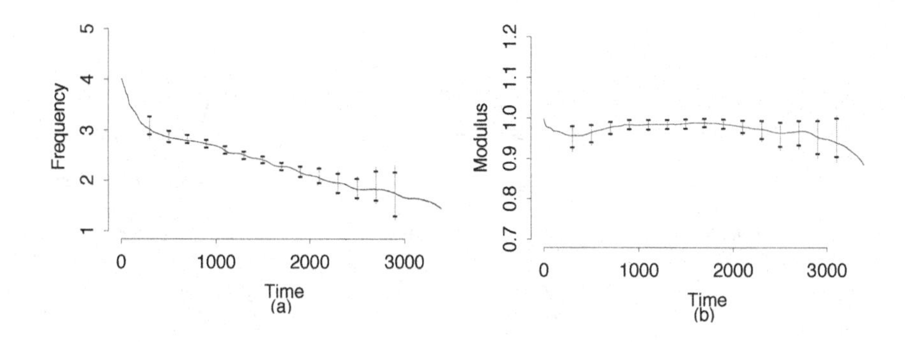

Figure 5.3 *(a) Trajectory and 95% posterior bands at selected time points of the lowest estimated frequency in channel Cz. (b) Trajectory and 95% posterior bands at selected time points of the modulus of the lowest frequency component in channel Cz.*

are approximately TVARMA(2,1) and TVAR(1) processes, respectively. In particular, components (1), (2), and (3) in Figure 5.2 can be thought of as realizations of approximate quasiperiodic TVARMA(2, 1) processes. Figure 5.3 shows the estimated trajectories over time of the frequency and modulus that characterize the latent process (1), as well as approximate 95% posterior intervals at selected time points. It is clear by looking at the modulus trajectory displayed in Figure 5.3(b) that component (1) is very persistent over the whole seizure course. There is a decay in the modulus at the end accompanied by a decrease in frequency, as well as an increase in the uncertainty of both the frequency and the modulus. This is consistent with the fact that the great intensity of the seizure in initial periods forces higher frequency oscillations in the early part of the seizure that gradually decrease and dissipate toward the end. The frequency lies in the so called "delta range" (0 to 4 Hz), which is the characteristic range of slow-waves manifested in the EEG during various periods, including middle and late phases of ECT seizures (Weiner and Krystal 1993). The frequency trajectories associated with components (2) and (3) were also computed. Component (2) is in the theta range (4–8 Hz) and subsidiary components at higher frequencies appear in the alpha band (8–13 Hz).

The decompositions presented here are time domain decompositions but they have a frequency domain interpretation. The frequencies $\omega_{t,j}$ represent peak frequencies of the correspondent component processes, while the trajectories of the frequencies over time represent movement in the peak of the corresponding component evolutionary spectra. The sharpness of the peak is an increasing function of $r_{t,j}$. Figure 5.4 shows the theoretical spec-

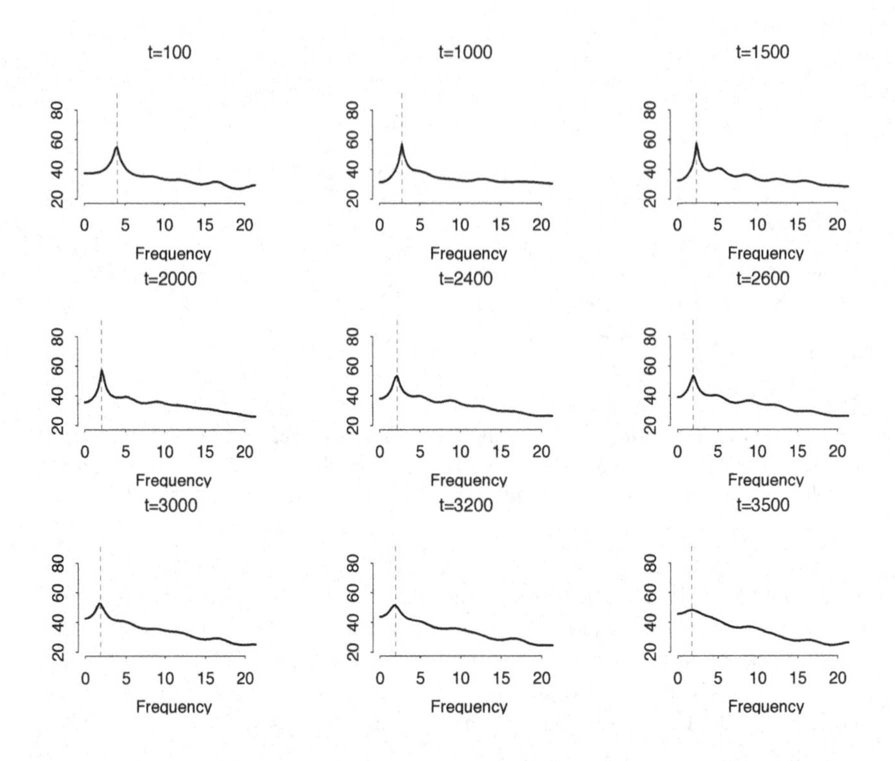

Figure 5.4 *Subset of instantaneous estimated AR spectra for channel* Cz.

tra of AR(12) models with parameters fixed at the posterior mean estimates of $\boldsymbol{\phi}_t$ and v at nine selected time points. The units of the instantaneous spectra are decibels (10 times logarithm to base 10 transformation). This figure is consistent with the results obtained via the time domain decomposition analysis; there is a dominant frequency, indicated by the vertical dotted line drawn at the sharpest peak of each spectra, that has a value around 4 Hz at $t = 100$ and eventually decreases to reach a value of approximately 2 Hz by the end of the seizure. The sharpness of the dominant frequency also diminishes toward the end. The estimated spectrum looks much smoother at time $t = 3,500$ than at the beginning of the seizure, due to the fact that there was a decrease in the modulus of this component towards the end of the seizure.

Example 5.3 *Oxygen isotope series.* We now revisit an example from geology that appears in Aguilar, Huerta, Prado, and West (1999) and emphasizes some interesting features of TVAR models and related decompositions.

Specifically, we fitted a TVAR(20) model to a detrended series that measures relative abundance of δ^{18}O to δ^{16}O on a time scale of 3,000 year (3 kyr) intervals, going back almost 2.5 Myr. Neither the original oxygen isotope series nor the detrended series are shown here. A graph of the original series and its estimated trend appears in Aguilar, Huerta, Prado, and West (1999). A detailed description of the data and the scientific implications of analyzing this type of data can be found in West (1997a, 1997b, 1997c).

The main interest lies on inferring the latent quasiperiodicities of the series. Previous analyses of the series included fitting a trend plus standard AR component model in which the AR parameters do not change over time. We fitted a TVAR(20) model to the detrended series setting the discount factor for the observational variance at one, i.e., $\delta_v = 1$, and determining the optimal discount factor that describes the evolution of the AR parameters over time, δ_ϕ, as explained previously in this chapter. The optimal discount factor in the interval $(0.99, 1]$ was found at $\delta_\phi = 0.994$. Figure 5.5 displays the trajectories of the two most dominant estimated frequencies (top graph) and their corresponding moduli (bottom graph). The solid lines in the top and bottom graphs correspond to the estimated period and modulus trajectories for the component with the highest period. The dotted lines correspond to the estimated period and modulus trajectories for the component with the second to largest period. We can see that the component with the largest period is the dominant component in terms of its modulus until about 1.1 million years ago. After this, the component with the second largest period becomes the dominant component. A similar behavior can be found if the components are ordered by amplitude instead of modulus (see Aguilar, Huerta, Prado, and West 1999). These components are associated with the main earth-orbital cycles. The component with the largest period, around 110 kyr, is associated with the *eccentricity* of the Earth's rotation about its axis, and that around 40 kyr corresponds to the *obliquity* of the Earth's orbit around the sun. The "switching" in terms of the moduli and amplitude of these two components is important in determining whether or not the increased significance of the second cycle was gradual, or, for instance, was the result of a significant structural climatic change (see Aguilar, Huerta, Prado, and West 1999 for further discussion and relevant scientific references).

A fully Bayesian analysis of the series would consider a model that simultaneously estimates the trend and quasiperiodicities in the data with a TVAR underlying component. In addition, we emphasize that, in spite of the fact that the TVAR analysis performed here does not formally allow us to model change points, it can capture interesting features present in the data that can be related to a change of regime—in this case related to which component has the highest modulus and amplitude at a given time. TVAR models have also been useful in detecting a switching of compo-

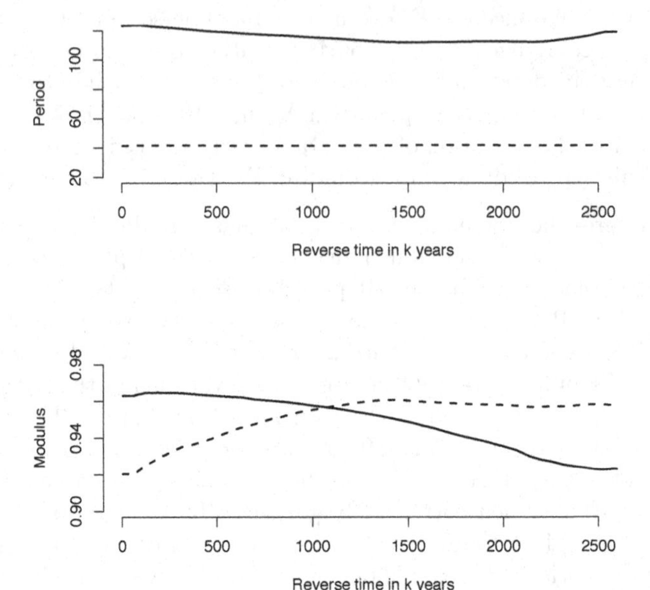

Figure 5.5 *Trajectories of the estimated periods (in thousands of years) of the two dominant quasiperiodic components in the detrended oxygen isotope series (top graph) and their corresponding moduli (bottom graph).*

nents that are scientifically meaningful in other data sets. For example, in the analyses of EEG data it is possible to see that some brain waves have changing patterns in terms of their amplitude and modulus over time.

5.3 Extensions

Extensions of TVAR models and related decompositions include models with time-varying orders. Prado and Huerta (2002) consider TVAR models with model order uncertainty. Specifically, the TVAR model order is assumed to evolve over time according to a discrete random walk. Prado and Huerta (2002) extend the decomposition results presented here to cases in which the number of latent components in the TVAR decomposition is not constant over time.

Godsill, Doucet, and West (2004) consider TVAR models in which the partial autocorrelation coefficients, and not the TVAR coefficients, vary over time according to a random walk. Parameterizing a TVAR model in terms of the partial autocorrelation coefficients is relevant in speech processing applications, because these coefficients can be interpreted as parameters of a linear acoustical tube model whose characteristics change over time. Such acoustical tube model approximates the characteristics of the vocal tract (again, see Godsill, Doucet, and West 2004 and references therein). Because of the implied nonlinear structure in the TVAR coefficients Godsill, Doucet, and West (2004) used sequential Monte Carlo methods for filtering and smoothing in this context. Such methods are discussed in Chapter 6. We now illustrate this parameterization below in the TVAR(2) case.

Example 5.4 *Partial autocorrelation function (PACF) parameterization of a TVAR(2).* Assume that x_t follows a TVAR(2), that is

$$x_t = \phi_{t,1} x_{t-1} + \phi_{t,2} x_{t-2} + \epsilon_t,$$

with $\epsilon_t \sim N(0, v)$.

Writing the Durbin-Levinson recursion at each time t (see Chapter 2) we have that $\phi^t(0,0) = 0$ and, for $h \geq 1$,

$$\phi^t(h,h) = \frac{\rho^t(h) - \sum_{l=1}^{h-1} \phi^t(h-1,l)\rho^t(h-l)}{1 - \sum_{l=1}^{h-1} \phi^t(h-1,l)\rho^t(l)},$$

where, for $h \geq 2$, $\phi^t(h,l) = \phi^t(h-1,l) - \phi^t(h,h)\phi^t(h-1,h-l)$, for $l = 1:(h-1)$. Then, in the TVAR(2) case we have

$$\begin{aligned} \phi^t(1,1) &= \rho^t(1) = \frac{\phi_{t,1}}{(1-\phi_{t,2})} \\ \phi^t(2,2) &= \frac{\rho^t(2) - [\rho^t(1)]^2}{1 - [\rho^t(1)]^2} = \phi_{t,2} \\ \phi^t(3,3) &= \frac{\rho^t(3) - \phi_{t,1}\rho^t(2) - \phi_{t,2}\rho^t(1)}{1 - \phi_{t,1}\rho^t(1) - \phi_{t,2}\rho^t(2)} = 0, \end{aligned}$$

and so the time-varying PACF coefficients are $\phi^t(1,1) = \frac{\phi_{t,1}}{(1-\phi_{t,2})}$ and $\phi^t(2,2) = \phi_{t,2}$.

The modeling approach of Godsill, Doucet, and West (2004) assumes the following evolution equation on the PACF coefficients:

$$p(\phi^t(h,h)|\phi^{t-1}(h,h), w) \propto \begin{cases} N(\beta\phi^{t-1}(h,h), w) & \text{if } \max\{|\phi^t(h,h)|\} < 1, \\ 0 & \text{otherwise,} \end{cases}$$

where $\beta < 1$.

Yang, Holan, and Wikle (2016) also proposed a dynamic linear model on the partial autocorrelation (PARCOR) domain that uses an efficient lattice filter approach for analyzing univariate non stationary time series. We briefly explain and illustrate this approach. We modify the notation above to make the modeling and inference clearer in this context.

Consider a TVAR(h) for the observed time series x_t. Denote $f_t^{(h)}$ and $b_t^{(h)}$ the forward and backward prediction errors associated to this model at time t, and $\phi_{t,j}^{f,h}$ and $\phi_{t,j}^{b,h}$ the forward and backward TVAR(h) coefficients, respectively, i.e.,

$$f_t^{(h)} = x_t - \sum_{j=1}^{h} \phi_{t,j}^{f,h} x_{t-j}, \quad b_t^{(h)} = x_t - \sum_{j=1}^{h} \phi_{t,j}^{b,h} x_{t+j}.$$

Let $\phi^{f,t}(h,j)$ and $\phi^{b,t}(h,j)$ denote the forward and backward PARCOR coefficients for $j = 1 : h$. At $j = h$ we have that the hth PARCOR coefficient is equal to the h TVAR(h) coefficient, i.e., $\phi^{f,t}(h,h) = \phi_{t,h}^{f,h}$. This was illustrated in the example above for a TVAR(2) model. Similarly, we have that $\phi^{b,t}(h,h) = \phi_{t,h}^{b,h}$. Yang, Holan, and Wikle (2016) proposes a modeling approach in the PARCOR domain with two sets of DLMs, one DLM related to the forward predictions and another one related to the backward predictions, each with observation equations given by

$$\begin{aligned} f_t^{(h-1)} &= \phi^{f,t}(h,h) b_{t-h}^{(h-1)} + f_t^{(h)}, \quad f_t^{(h)} \sim N(0, v_{t,f,h}), \\ b_t^{(h-1)} &= \phi^{b,t}(h,h) f_{t+h}^{(h-1)} + b_t^{(h)}, \quad b_t^{(h)} \sim N(0, v_{t,b,h}). \end{aligned}$$

In addition, the forward and backward PARCOR coefficients are assumed to evolve via random walk evolution equations, i.e.,

$$\begin{aligned} \phi^{f,t}(h,h) &= \phi^{f,t-1}(h,h) + \epsilon_{t,f,h}, \quad \epsilon_{t,f,h} \sim N(0, w_{t,f,h}), \\ \phi^{b,t}(h,h) &= \phi^{b,t-1}(h,h) + \epsilon_{t,b,h}, \quad \epsilon_{t,b,h} \sim N(0, w_{t,b,h}). \end{aligned}$$

Variances at the observational and system levels are specified via discount factors and posterior inference is obtained under conjugate normal inverse-gamma priors via the filtering and smoothing equations described in Chapter 4, Section 4.3.

Assuming a TVAR(p) for $x_{1:T}$, the posterior estimation algorithm of Yang, Holan, and Wikle (2016) requires running the filtering and smoothing equations for the forward and backward PARCOR time-varying coefficients for each stage $h = 1, \dots, p$ using the modeling structure detailed above. Note that at each stage h we have DLMs with univariate state parameters, which fully avoids matrix inversions required for the filtering and smoothing equations related to the TVAR(p) model described in Section 5.2. Once estimates or posterior samples of the forward and backward PARCOR parameters are available for each stage h, the forward and backward TVAR coefficients can be obtained as follows. For each $h = 1, \dots, p$, set $\phi_{t,h}^{f,h} = \phi^{f,t}(h,h)$ and $\phi_{t,h}^{b,h} = \phi^{b,t}(h,h)$, and

$$\begin{aligned} \phi_{t,j}^{f,h} &= \phi_{t,j}^{f,h-1} - \phi_{t,h}^{f,h} \phi_{t,h-j}^{b,h-1} \\ \phi_{t,j}^{b,h} &= \phi_{t,j}^{b,h-1} - \phi_{t,h}^{b,h} \phi_{t,h-j}^{f,h-1}, \end{aligned}$$

for $j = 1, \dots, h-1$.

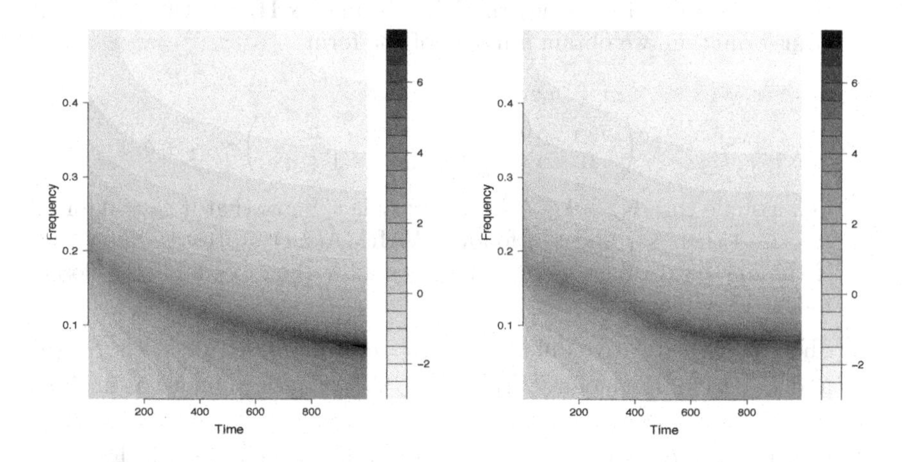

Figure 5.6 *Left plot: True spectral density of a TVAR(2) process. Right plot: Estimated spectral density.*

Example 5.5 *TVAR(2) PARCOR lattice filter.*

We simulated 1000 observations from a TVAR(2) process with a linearly increasing modulus from 0.9 to 0.97 and a linearly increasing period from 5 to 14 (or equivalently, linearly decreasing frequency from 0.2 to 0.07). A time-frequency representation of the true spectral density is shown in Figure 5.6 along with a time-frequency representation estimated from the lattice filter PARCOR approach of Yang, Holan, and Wikle (2016) summarized above. The model order was set to $p = 2$ so this required fitting DLMs to the forward and backward PARCOR coefficients with optimal discount factors at each stage set to maximize the likelihood for such stage. The time-varying PARCOR lattice filtering approach adequately estimates the TVAR model and its corresponding time-varying spectral representation.

5.4 Problems

1. Show that the processes $x_{t,j}^{(1)}$s that appear in the decomposition of an AR(p) are ARMA$(2, 1)$ processes with AR parameters $2r_j \cos(\omega_j)$ and $-r_j^2$. Note: this can be done by finding the linear canonical form associated to each of the j-th components in the decomposition.
2. Consider a TVAR(2) process x_t with two real reciprocal roots, $r_{t,1}$ and $r_{t,2}$. This model can be written in the general DLM form (5.1). Assume

that the model is reparameterized via the matrix $\mathbf{H}_t$ and that, after this transformation, we obtain a model of the form

$$\begin{aligned} x_t &= \gamma_{t,1} + \gamma_{t,2}, \\ \gamma_t &= \begin{pmatrix} r_{t,1} & 0 \\ 0 & r_{t,2} \end{pmatrix} \begin{pmatrix} 1+a_1 & a_2 \\ a_3 & 1+a_4 \end{pmatrix} \gamma_{t-1} + \boldsymbol{\delta}_t, \end{aligned}$$

i.e., assume that $\mathbf{K}_t = \mathbf{I} + \mathbf{A}$ for every time t. Show that if $a_k \neq 0$ for all $k = 1:4$ then $\gamma_{t,1}$ and $\gamma_{t,2}$ follow TVARMA(2,1) processes. Show that if the a_ks have values close to zero for all k then $\gamma_{t,1}$ and $\gamma_{t,2}$ closely follow TVAR(1) processes.

3. Show that the relative difference between the forecast functions $f_t^{(1)}(h)$ and $f_t^{(2)}(h)$ for the models $\mathcal{M}_1$ and $\mathcal{M}_2$ (5.7) can be bounded as in (5.8).

4. Consider the oxygen isotope series (`oxygen.dat`) analyzed in Example 5.3. Repeat the TVAR analysis described in such example with different discount factor values of $\delta_\phi \in (0.99, 1]$. Display the trajectories of the estimated periods of the two dominant quasiperiodic components and their corresponding moduli. Is the crossing point of the moduli trajectories (see Figure 5.5) very sensitive to the choice of the discount factor?

5. Simulate 100 observations from the model given by

$$\begin{aligned} y_t &= \phi_t y_{t-1} + \epsilon_t \\ \phi_t &= \frac{0.09}{99} \times t + \frac{89.01}{99} \end{aligned}$$

with $\epsilon_t \sim N(0,1)$. Fit a TVAR(1) model with conjugate priors to such data using a discount factor at the system level, $\delta_\phi \in (0,1]$ and constant but unknown variance v. Choose the optimal value of δ_ϕ based on your data $y_{2:100}$.

6. Repeat the TVAR analysis of the EEG series of channel *Cz* (`eegCz.dat`) shown in Example 5.2 for the data `eegFz.dat`, which corresponds to another EEG series recorded in the same individual but in a different scalp location (channel Fz). Compare your results to those presented in Example 5.2.

7. Consider again the EEG series of channel **Cz** in Example 5.2. Using the time-varying PARCOR lattice filtering approach of Yang, Holan, and Wikle (2016) (see Section 5.3) obtain estimates of a TVAR(12) model parameters, its time series decomposition and the corresponding time-frequency spectral representation. How do these results compare to those obtained from the TVAR analysis carried out in Problem 5?

8. Explore some univariate TVAR models for the monthly SOI index time series over the years up to 2017. Address the following specific questions,

with graphical and numerical summaries of your exploratory modeling analyses to support comments and discussions.

(a) What kinds of models seem to be suggested (in terms of discount factors and model order)?

(b) Does the model:data match support stationarity of the SOI series?

(c) On a chosen TVAR model (justify your choice of model):

   i. How do fitted/estimated values of the innovations appear? The model says they are normal–does that seem valid? If not, what else might be done?

   ii. What can you discover about underlying quasi-periodicities in the SOI series, and how stable are they over time?

   iii. Was the run of 60 months of consecutive negative SOI values back in 1995 a "rare event," or just business as usual for SOI? That is, do you think that was so unusual that, perhaps, we might link it to (human-induced) climate change, or is it just an excursion that—while unusual—is consistent with a model that does not include such an intervention effect but reflects "normal" evolution over the longer term.

(d) How might you revisit the above specific questions in the context of uncertainty about the values of discount factors and model order?

Chapter 6

General state-space models and sequential Monte Carlo methods

General state-space models, including non linear and non-Gaussian dynamic models, are described here. Posterior inference via MCMC algorithms that work well for these types of general models is challenging. Furthermore, MCMC schemes are typically not feasible for real-time (on-line) filtering and parameter learning in time series settings, since a new target posterior distribution—and therefore a new MCMC run—needs to be considered each time a new observation arrives. This chapter describes and illustrates sequential Monte Carlo approaches as an alternative and computationally feasible tool for online filtering and sequential posterior learning within the class of general state-space models.

6.1 General State-Space Models

Tools for posterior inference and forecasting in dynamic linear models (DLMs) and conditionally Gaussian dynamic linear models (CDLMs) were presented in Chapter 4. Filtering within the class of Normal DLMs (NDLMs) is available in closed form via the Kalman filtering recursions. Markov chain Monte Carlo (MCMC) algorithms can be used for filtering and prediction in conditionally Gaussian DLMs, as was also illustrated in Chapter 4. These classes of models are broad and flexible; however, more general state-space models that deal with nonlinear and often non-Gaussian structures at the state level and/or at the observational level are often needed in practice. MCMC algorithms can be customized to achieve posterior inference and prediction in general state-space models (e.g., Carlin, Polson, and Stoffer 1992). However, designing efficient algorithms when the models have strong nonlinearities can be very challenging.

In addition, in many practical scenarios filtering and parameter learning need to be performed on-line, so that MCMC approaches would not be computationally affordable. For instance, later in Chapter 7 we present a neuroscience application in which electroencephalograms (EEGs) of subjects performing a cognitive task continuously for an extensive period of time are recorded with the purpose of detecting fatigue. Regardless of the models being used, the objective is to determine if the subject is fatigued at any given time based on the available data, $\mathcal{D}_t = \{\mathbf{y}_{1:t}\}$, which are the EEG recordings at multiple scalp locations for each participant up to time t. Assuming that the models used to describe the brain dynamics are state-space models with states $\boldsymbol{\theta}_{1:t}$ and parameters $\boldsymbol{\phi}$, the interest lies in summarizing $p(\boldsymbol{\theta}_t, \boldsymbol{\phi}|\mathcal{D}_t)$ in real-time. MCMC-based algorithms would not be an option in this case since a new chain would need to be run each time a new observation is received. Other applications that require on-line filtering arise in the areas of wireless communications, tracking, and finance.

In this chapter we describe the class of general state-space models and discuss some recent methods for filtering, parameter learning, and smoothing within this class of models. More specifically, we summarize and illustrate some recent simulation-based methods generally referred to as Sequential Monte Carlo (SMC) algorithms. Most of the algorithms presented here can be used for on-line filtering in general state-space models, while some of them can also be used to simultaneously deal with parameter learning. Alternative algorithms for inference in general nonlinear dynamic models include the extended Kalman filter (EKF), the unscented Kalman filter (UKF), and the Gaussian quadrature Kalman filter (GKF) among others. These algorithms, as opposed to SMC methods, are not based on approximating the posterior distributions by a set of randomly generated particles. They have been applied successfully in practice; however, they do not work very well in cases where the noise at the observation or at the system level follows a distribution that is very different from the Gaussian distribution. We will not discuss these algorithms here, and so we refer the reader to Jazwinski (1970), Julier and Uhlmann (1997), van der Merwe, Doucet, de Freitas, and Wan (2000), Ito and Xiong (2000), and references therein for details, examples, and discussion.

A general Markovian state-space model is defined by

$$\mathbf{y}_t \sim p(\mathbf{y}_t|\boldsymbol{\theta}_t, \boldsymbol{\phi}) \quad \text{(observation density)}, \tag{6.1}$$

$$\boldsymbol{\theta}_t \sim p(\boldsymbol{\theta}_t|\boldsymbol{\theta}_{t-1}, \boldsymbol{\phi}) \quad \text{(state evolution density)}, \tag{6.2}$$

where $\boldsymbol{\theta}_t$ is the unobserved state vector and $\mathbf{y}_t$ is the vector of observations made at time t. Note that the model above and the filtering algorithms that will be described later in this chapter apply to the general case in which $\mathbf{y}_t$ is a vector instead of a scalar; however, the examples given below deal with univariate time series. Models for multivariate time series are discussed in Chapters 9 and 10.

The observation and state evolution densities in (6.1) and (6.2) may involve nonlinearities and/or non-Gaussian distributions. As was the case with the normal DLM, we are interested in obtaining the filtering density $p(\boldsymbol{\theta}_t, \boldsymbol{\phi}|\mathcal{D}_t)$, the predictive density $p(\mathbf{y}_t|\mathcal{D}_{t-1})$, and the smoothing density $p(\boldsymbol{\theta}_{0:T}, \boldsymbol{\phi}|\mathcal{D}_T)$.

Example 6.1 *Nonlinear time series model.* The equations below describe a simple nonlinear time series model that has been used extensively as a benchmark example for testing filtering techniques (see for instance Gordon, Salmond, and Smith 1993, Kitagawa 1996, West 1993b, and Cappé, Godsill, and Moulines 2007):

$$\begin{aligned} y_t &= a\theta_t^2 + \nu_t, \\ \theta_t &= b\theta_{t-1} + c\frac{\theta_{t-1}}{1+\theta_{t-1}^2} + d\cos(\omega t) + w_t, \end{aligned}$$

where ν_t and w_t are independent and mutually independent Gaussian random variables with $\nu_t \sim N(0, v)$ and $w_t \sim N(0, w)$. Note that the model is Gaussian but nonlinear at the observation and system levels. When the components of $\boldsymbol{\phi} = (a, b, c, d, \omega, v, w)$ are known, the interest lies in summarizing the filtering distribution $p(\theta_t|\mathcal{D}_t)$. In particular, the model with $a = 1/20$, $b = 1/2$, $c = 25$, $d = 8$, $\omega = 1.2$, and various fixed values of v and w has been widely studied. If some or all of the components of $\boldsymbol{\phi}$ are unknown, combined state and parameter estimation will be of interest.

Figure 6.1 shows some simulated data y_t and state θ_t, for $t = 1 : 200$, as well as plots of y_t versus θ_t, and θ_t versus θ_{t-1}, with y_t and θ_t obtained from a model with $a = 1/20$, $b = 1/2$, $c = 25$, $d = 8$, $\omega = 1.2$, $v = 10$, and $w = 1$.

Example 6.2 *Fat-tailed nonlinear state-space model.* The model given by

$$\begin{aligned} y_t &= \theta_t + \sqrt{\lambda_t}\nu_t, \\ \theta_t &= \beta\frac{\theta_{t-1}}{1+\theta_{t-1}^2} + w_t, \end{aligned}$$

with $\nu_t \sim N(0, v)$, $w_t \sim N(0, w)$ is a nonlinear model at the state level with non-Gaussian innovations at the observational level since $\lambda_t \sim IG(\nu/2, \nu/2)$ with ν known. The parameters are $\boldsymbol{\phi} = (\beta, v, w)$. Carvalho, Johannes, Lopes, and Polson (2010) use their particle learning algorithms (see Section 6.2) for simultaneous filtering and parameter learning and compare the performance of such algorithms with a MCMC algorithm based on the approach of Carlin, Polson, and Stoffer (1992) and the sequential Monte Carlo filter of Liu and West (2001) in simulated data sets.

Example 6.3 *AR*(1) *stochastic volatility model.* Univariate and multivariate stochastic volatility models will be discussed in detail and applied

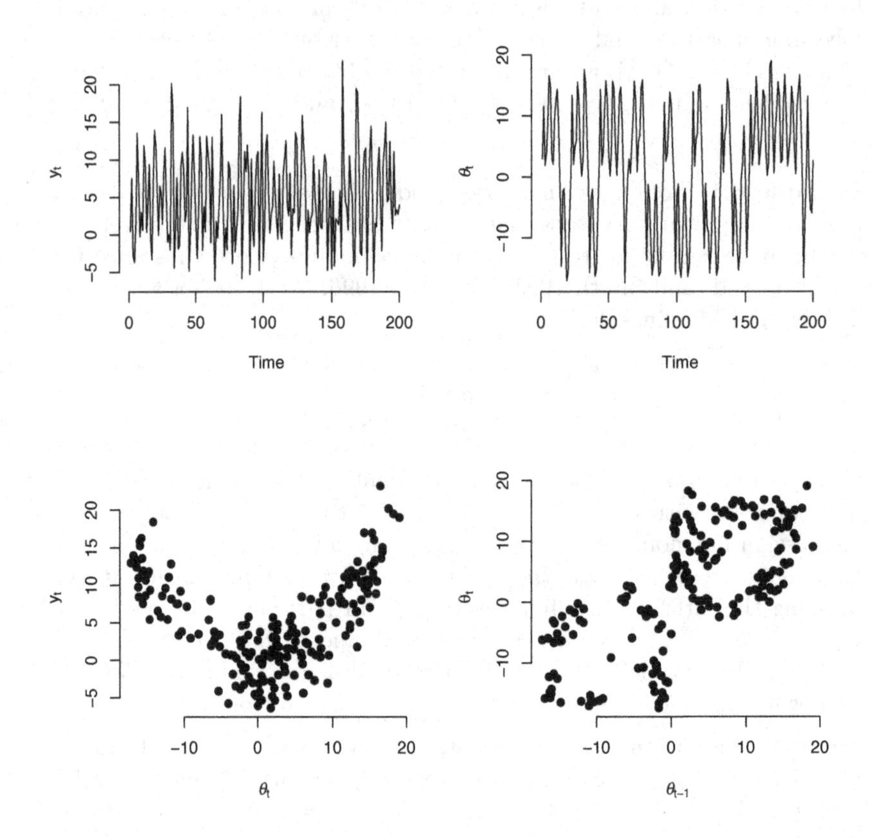

Figure 6.1 *Simulated data from the nonlinear time series model in Example 6.1.*

to real data in Chapters 7 and 10. Here we present the AR(1) (autoregressive) stochastic volatility (SV) model as another example of a nonlinear state-space model. The canonical AR(1) SV model (see for example Shephard 2005) is given by

$$\begin{aligned} y_t &= \exp(\theta_t/2)\nu_t, \\ \theta_t &= \mu + \phi\theta_{t-1} + \tau w_t, \end{aligned}$$

where ν_t and w_t are independent and mutually independent $N(0,1)$ innovations. This model is used in financial applications to describe the log-returns of an asset, i.e., $y_t = \log(P_t/P_{t-1})$, where P_t is a financial price series, such as a stock index or exchange rate. Kim, Shephard, and Chib (1998), Aguilar and West (2000), and Chib, Nadari, and Shephard (2002), among others, consider MCMC methods for posterior inference in this model. Details on a MCMC algorithm that rewrites the model above as a conditionally Gaussian state-space model and then uses the forward filtering backward sampling (FFBS) approach of Carter and Kohn (1994) and Frühwirth-Schnatter (1994) are given in Section 7.5. Sequential Monte Carlo algorithms have also been proposed in the literature (e.g., Pitt and Shephard 1999a).

Example 6.4 *Bearings-only target tracking model.* The equations below describe an angular time series model for target tracking that has been widely studied in the sequential Monte Carlo literature (e.g., Rubin 1988, Gordon, Salmond, and Smith 1993, Pitt and Shephard 1999a). More specifically, Pitt and Shephard (1999a) consider a model in which an observer is stationary at the origin of the $x-z$ plane, and a ship is assumed to gradually accelerate or decelerate randomly over time. If x_t and z_t represent the ship's horizontal and vertical positions at time t, and $v_t^{(x)}$, $v_t^{(z)}$ represent the corresponding velocities, a discretized version of the system is modeled via the following state equation:

$$\boldsymbol{\theta}_t = \begin{pmatrix} 1 & 1 & 0 & 0 \\ 0 & 1 & 0 & 0 \\ 0 & 0 & 1 & 1 \\ 0 & 0 & 0 & 1 \end{pmatrix} \boldsymbol{\theta}_{t-1} + \begin{pmatrix} \frac{1}{2} & 0 \\ 1 & 0 \\ 0 & \frac{1}{2} \\ 0 & 1 \end{pmatrix} w_t,$$

with $\boldsymbol{\theta}_t = (x_t, v_t^{(x)}, z_t, v_t^{(z)})'$ and $w_t \sim N(0, w)$. The initial state $\boldsymbol{\theta}_0 \sim N(\mathbf{m}_0, \mathbf{C}_0)$ describes the ship's initial position. The observation equation is based on a mean direction $\mu_t(\boldsymbol{\theta}_t) = \tan^{-1}(z_t/x_t)$, and the measured angle is assumed to be wrapped Cauchy with density

$$p(y_t|\boldsymbol{\theta}_t, \boldsymbol{\phi}) = \frac{1}{2\pi} \frac{1-\rho^2}{1+\rho^2 - 2\rho\cos(y_t - \mu_t(\boldsymbol{\theta}_t))},$$

with $y_t \in [0, 2\pi)$ and ρ, the mean resultant length, such that $0 \leq \rho \leq 1$. This is a nonlinear, non-Gaussian state-space model. Pitt and Shephard (1999a) consider an auxiliary variable particle filter algorithm (see Section 6.2) for obtaining the filtering distribution assuming that $\boldsymbol{\phi}$, with $\boldsymbol{\phi} = (\rho, w)$, is known.

Additional examples of nonlinear and non-Gaussian state-space models appear in the references listed above, in Cappé, Moulines, and Rydén (2005) and in Durbin and Koopman (2001), among others.

6.2 Posterior Simulation: Sequential Monte Carlo

We begin by describing algorithms for cases in which $\boldsymbol{\phi}$ is known, so we omit $\boldsymbol{\phi}$ from the notation until we revisit the filtering problem later for the general case in which $\boldsymbol{\phi}$ is unknown.

Note that, in order to update $p(\boldsymbol{\theta}_{t-1}|\mathcal{D}_{t-1})$ after $\mathbf{y}_t$ has been observed we need to compute the evolution and updating densities

$$p(\boldsymbol{\theta}_t|\mathcal{D}_{t-1}) = \int p(\boldsymbol{\theta}_t|\boldsymbol{\theta}_{t-1})p(\boldsymbol{\theta}_{t-1}|\mathcal{D}_{t-1})d\boldsymbol{\theta}_{t-1}$$

and

$$p(\boldsymbol{\theta}_t|\mathcal{D}_t) \propto p(\mathbf{y}_t|\boldsymbol{\theta}_t)p(\boldsymbol{\theta}_t|\mathcal{D}_{t-1}),$$

respectively, which are typically not available in closed form for general nonlinear and/or non-Gaussian state-space models.

We summarize some of the most commonly used sequential Monte Carlo algorithms for posterior inference, often referred to as *particle filters.* Such methods allow us to propagate rather general target distributions by using a combination of importance sampling, resampling, and MCMC steps. See Doucet, de Freitas, and Gordon (2001), Doucet, Godsill, and Andrieu (2000), Cappé, Godsill, and Moulines (2007), Del Moral, Jasra, and Doucet (2007), and references therein, for reviews of SMC methods for Bayesian computation. Migon, Gamerman, Lopes, and Ferreira (2005) review some SMC methods in the context of state and parameter estimation in dynamic models. In addition, several articles and technical reports on these types of methods can be found at `http://www-sigproc.eng.cam.ac.uk/smc/`.

Particle filters are simulation-based filters that approximate the distribution of $(\boldsymbol{\theta}_{t-1}|\mathcal{D}_{t-1})$ by particles $\boldsymbol{\theta}_{t-1}^{(1)}, \ldots, \boldsymbol{\theta}_{t-1}^{(M)}$, with corresponding weights $\omega_{t-1}^{(1)}, \ldots, \omega_{t-1}^{(M)}$ as follows:

$$p(\boldsymbol{\theta}_{t-1}|\mathcal{D}_{t-1}) \approx \hat{p}(\boldsymbol{\theta}_{t-1}|\mathcal{D}_{t-1}) = \sum_{m=1}^{M} \omega_{t-1}^{(m)} \delta_{\boldsymbol{\theta}_{t-1}^{(m)}}(\boldsymbol{\theta}_{t-1}),$$

where δ is the Dirac delta function. The particles at time $t-1$ are updated to obtain a new set of particles at time t via importance sampling and resampling. We discuss various algorithms for updating the particles below.

6.2.1 Sequential Importance Sampling and Resampling

Gordon, Salmond, and Smith (1993), a key early reference in the area of sequential Monte Carlo methods, developed an algorithm based on the sampling importance resampling (SIR) method (Rubin 1988) for implementing recursive Bayesian filters and applied it to data simulated from the bearings-only tracking model described in Example 6.4.

In the SIR algorithm, the goal is to draw a sample from a target density $p(\boldsymbol{\theta})$ based on draws from a so-called importance density, $g(\boldsymbol{\theta})$, as follows.

1. Sample $\tilde{\boldsymbol{\theta}}^{(m)}$ from $g(\boldsymbol{\theta})$, for $m = 1 : M$.
2. Compute the weights $\tilde{\omega}^{(m)} \propto p(\tilde{\boldsymbol{\theta}}^{(m)})/g(\tilde{\boldsymbol{\theta}}^{(m)})$.
3. Sample $\boldsymbol{\theta}^{(m)}$ from $\{(\tilde{\boldsymbol{\theta}}^{(j)}, \tilde{\omega}^{(j)});\ j = 1 : M\}$.

Finally, $\{(\boldsymbol{\theta}, \omega)^{(m)};\ m = 1 : M\}$, with $\omega^{(m)} = 1/M$ for all m approximates $p(\boldsymbol{\theta})$. The choice of $g(\cdot)$ is key for designing an algorithm that leads to a good particle approximation of $p(\boldsymbol{\theta})$. In settings where the target distribution is a posterior distribution, a natural but not necessarily good choice for $g(\boldsymbol{\theta})$ is the prior distribution.

The sequential importance sampling (SIS) algorithm described below uses importance sampling to obtain an approximation to $p(\boldsymbol{\theta}_{0:t}|\mathcal{D}_t)$. For details, discussion, and further references on the SIS algorithm see, for example, Liu and Chen (1998), Doucet, Godsill, and Andrieu (2000), and Cappé, Godsill, and Moulines (2007).

At time $t-1$, assume that $p(\boldsymbol{\theta}_{0:(t-1)}|\mathcal{D}_{t-1})$ is approximated by the weighted set of particles $\{(\boldsymbol{\theta}_{0:(t-1)}, \omega_{t-1})^{(m)}; m = 1 : M\}$. Then, for each m, with $m = 1 : M$:

1. Sample $\boldsymbol{\theta}_t^{(m)}$ from an importance density $g_t(\boldsymbol{\theta}_t|\boldsymbol{\theta}_{0:(t-1)}^{(m)}, \mathcal{D}_t)$ and set $\boldsymbol{\theta}_{0:t}^{(m)} = (\boldsymbol{\theta}_{0:(t-1)}^{(m)}, \boldsymbol{\theta}_t^{(m)})$.
2. Compute the importance weights $\omega_t^{(m)}$ as

$$\omega_t^{(m)} \propto \omega_{t-1}^{(m)} \frac{p(\mathbf{y}_t|\boldsymbol{\theta}_t^{(m)})p(\boldsymbol{\theta}_t^{(m)}|\boldsymbol{\theta}_{t-1}^{(m)})}{g_t(\boldsymbol{\theta}_t^{(m)}|\boldsymbol{\theta}_{0:(t-1)}^{(m)}, \mathcal{D}_t)}.$$

Finally, use $\{(\boldsymbol{\theta}_{0:t}, \omega_t)^{(m)}; m = 1 : M\}$ as a particle approximation to $p(\boldsymbol{\theta}_{0:t}|\mathcal{D}_t)$. As discussed in Doucet, Godsill, and Andrieu (2000), it can be shown (using a result found in Kong, Liu, and Wong 1994) that the variance of the importance weights increases over time, leading to particle degeneracy. That is, after some iterations of the algorithms all but one particle will have weights that are very close to zero. In order to limit degeneracy as much as possible, it is key to choose an importance density that minimizes the variance of the weights. Such density is the so-called optimal importance density. It can also be shown that the optimal importance density is given by

$$g_t(\boldsymbol{\theta}_t|\boldsymbol{\theta}_{0:(t-1)}^{(m)}, \mathcal{D}_t) = p(\boldsymbol{\theta}_t|\boldsymbol{\theta}_{t-1}^{(m)}, \mathbf{y}_t).$$

There are two requirements about the optimal importance density that make its use limited in practice. The first requirement is the ability to sample from $p(\boldsymbol{\theta}_t|\boldsymbol{\theta}_{t-1}^{(m)}, \mathbf{y}_t)$ and the second one is the ability to evaluate

$p(\mathbf{y}_t|\boldsymbol{\theta}_{t-1}^{(m)})$ in order to compute the weights in Step 2 of the SIS algorithm. Now, $p(\mathbf{y}_t|\boldsymbol{\theta}_{t-1}^{(m)}) = \int p(\mathbf{y}_t|\boldsymbol{\theta}_t)p(\boldsymbol{\theta}_t|\boldsymbol{\theta}_{t-1}^{(m)})d\boldsymbol{\theta}_t$ is generally not available in closed form for general dynamic models. For some models such as Gaussian DLMs and Gaussian dynamic models that are linear at the observational level and nonlinear at the system level, it is possible to evaluate $p(\mathbf{y}_t|\boldsymbol{\theta}_{t-1}^{(m)})$ analytically (see, e.g., Doucet, Godsill, and Andrieu 2000). For instance, in the case of NDLMs, with $(\boldsymbol{\theta}_t|\boldsymbol{\theta}_{t-1}) \sim N(\mathbf{G}_t\boldsymbol{\theta}_{t-1}, \mathbf{W}_t)$ and $(\mathbf{y}_t|\boldsymbol{\theta}_t) \sim N(\mathbf{F}_t'\boldsymbol{\theta}_t, \mathbf{V}_t)$, it can be shown that the optimal importance density is a Normal density with variance and mean given by

$$\begin{aligned}
V(\boldsymbol{\theta}_t|\boldsymbol{\theta}_{t-1}, \mathbf{y}_t) &= [\mathbf{W}_t^{-1} + \mathbf{F}_t\mathbf{V}_t^{-1}\mathbf{F}_t']^{-1} \\
E(\boldsymbol{\theta}_t|\boldsymbol{\theta}_{t-1}, \mathbf{y}_t) &= V(\boldsymbol{\theta}_t|\boldsymbol{\theta}_{t-1}, \mathbf{y}_t)[\mathbf{W}_t^{-1}\mathbf{G}_t\boldsymbol{\theta}_{t-1} + \mathbf{F}_t\mathbf{V}_t^{-1}\mathbf{y}_t].
\end{aligned}$$

The weights associated to this importance density are proportional to $p(\mathbf{y}_t|\boldsymbol{\theta}_{t-1}) = N(\mathbf{y}_t|\mathbf{F}_t'\mathbf{G}_t\boldsymbol{\theta}_{t-1}, \mathbf{V}_t + \mathbf{F}_t'\mathbf{W}_t\mathbf{F}_t)$.

Several approaches have been proposed in the literature to obtain importance densities that lead to importance weights with low variance. In the auxiliary variable particle filter of Pitt and Shephard (1999a), $p(\boldsymbol{\theta}_{0:t}|\mathcal{D}_t)$ is approximated by a mixture distribution which is considered to be the target distribution of the algorithm, and then importance distributions are proposed to efficiently sample from this mixture. Liu and West (2001) extended this algorithm to consider parameter learning in addition to filtering. More recently, Carvalho, Johannes, Lopes, and Polson (2010) extended and improved these methods by using a sufficient statistic structure for the parameters and/or the states in conditionally Gaussian dynamic linear models and nonlinear models at the state level. These three algorithms are described later in this chapter. Doucet, Godsill, and Andrieu (2000) proposed Gaussian importance densities whose parameters are evaluated via local linearizations to approximate the optimal importance density. Algorithms based on the "Rao-Blackwellization" method have also been proposed to reduce the variance of the importance weights (see, e.g., Liu and Chen 1998 and Doucet, Godsill, and Andrieu 2000).

In addition to the approaches described above, resampling methods have been used as a tool to reduce particle degeneracy by eliminating particles with small importance weights. The so-called *effective sample size* (see Liu 1996) given by

$$M_{t,\text{eff}} = \frac{1}{\sum_{m=1}^{M}(\omega_t^{(m)})^2}$$

can be used as a measure of degeneracy. Note that $M_{t,\text{eff}} = M$ (the total number of particles) when $\omega_t^{(m)} = 1/M$ for all the particles and $M_{t,\text{eff}} = 1$ when a single particle has weight equal to one, therefore indicating particle degeneracy.

A SIS algorithm with resampling (SISR) based on the effective number of particles can be considered, as follows. As usual, let $\{(\boldsymbol{\theta}_{0:(t-1)}, \omega_{t-1})^{(m)};\ m = 1 : M\}$ be a particle approximation to $p(\boldsymbol{\theta}_{0:(t-1)}|\mathcal{D}_{t-1})$.

1. Sample $\boldsymbol{\theta}_t^{(m)}$ from an importance density $g_t(\boldsymbol{\theta}_t|\boldsymbol{\theta}_{0:(t-1)}^{(m)}, \mathcal{D}_t)$ for $m = 1 : M$, and set $\boldsymbol{\theta}_{0:t}^{(m)} = (\boldsymbol{\theta}_{0:(t-1)}^{(m)}, \boldsymbol{\theta}_t^{(m)})$.
2. For $i = 1 : M$ compute the importance weights $\omega_t^{(m)}$ as
$$\omega_t^{(m)} \propto \omega_{t-1}^{(m)} \frac{p(\mathbf{y}_t|\boldsymbol{\theta}_t^{(m)})p(\boldsymbol{\theta}_t^{(m)}|\boldsymbol{\theta}_{t-1}^{(m)})}{g_t(\boldsymbol{\theta}_t^{(m)}|\boldsymbol{\theta}_{0:(t-1)}^{(m)}, \mathcal{D}_t)}.$$
3. Compute $M_{t,\text{eff}}$. If $M_{t,\text{eff}} < M_0$ for some prespecified minimum effective sample size M_0, then resample as follows:
   (a) For $m = 1 : M$ sample $\boldsymbol{\theta}_{0:t}^{(m)}$ with probability $\omega_t^{(m)}$ from the particle approximation obtained in Step 2.
   (b) Set $\omega_t^{(m)} = 1/M$ for all M.

Report $\{(\boldsymbol{\theta}_{0:t}, \omega_t)^{(m)};\ m = 1 : M\}$ as a particle approximation to $p(\boldsymbol{\theta}_{0:t}|\mathcal{D}_t)$.

The SIS and SISR schemes can be used to obtain a particle approximation to the filtering distribution $p(\boldsymbol{\theta}_t|\mathcal{D}_t)$ given by $\{(\boldsymbol{\theta}_t, \omega_t)^{(m)};\ m = 1 : M\}$. In other words, if the objective is to approximate $p(\boldsymbol{\theta}_t|\mathcal{D}_t)$ the first t components in each path $\boldsymbol{\theta}_{0:t}^{(m)}$ can be discarded as long as the calculation of weights depends only on $\boldsymbol{\theta}_t^{(m)}$ and $\boldsymbol{\theta}_{t-1}^{(m)}$. In principle, obtaining particle approximations to the smoothing distributions $p(\boldsymbol{\theta}_k|\mathcal{D}_t)$ for $k < t$ could also be done by marginalizing the particle approximation to $p(\boldsymbol{\theta}_{0:t}|\mathcal{D}_t)$; however, this would lead to poor approximations due to particle degeneracy. Alternative algorithms for obtaining particle approximations to the smoothing distributions are discussed later in this chapter.

We now describe the algorithms of Pitt and Shephard (1999a), Liu and West (2001), Storvik (2002), Polson, Stroud, and Müller (2008), and Carvalho, Johannes, Lopes, and Polson (2010) for obtaining particle approximations to $p(\boldsymbol{\theta}_t|\mathcal{D}_t)$ and $p(\boldsymbol{\theta}_t, \boldsymbol{\phi}|\mathcal{D}_t)$.

6.2.2 The Auxiliary Particle Filter

Pitt and Shephard (1999a) proposed the auxiliary variable particle filter method, now known as the auxiliary particle filter (APF). Here the goal is obtaining a sample from the joint density $p(\boldsymbol{\theta}_t, k|\mathcal{D}_t)$, where the *auxiliary variable* k is an index on the mixture

$$\hat{p}(\boldsymbol{\theta}_t|\mathcal{D}_{t-1}) = \sum_{m=1}^{M} p(\boldsymbol{\theta}_t|\boldsymbol{\theta}_{t-1}^{(m)})\omega_{t-1}^{(m)},$$

called the "empirical prediction density." More specifically, in defining

$$p(\boldsymbol{\theta}_t, k|\boldsymbol{\theta}_{t-1}, \mathcal{D}_t) \propto p(\mathbf{y}_t|\boldsymbol{\theta}_t)p(\boldsymbol{\theta}_t|\boldsymbol{\theta}_{t-1}^{(k)})\omega_{t-1}^{(k)}, \tag{6.3}$$

the idea is to draw samples from this joint density (or an approximate density) and discard the index to obtain a sample from the "empirical filtering density" $\hat{p}(\boldsymbol{\theta}_t|\mathcal{D}_t)$ given by

$$\hat{p}(\boldsymbol{\theta}_t|\mathcal{D}_t) \propto p(\mathbf{y}_t|\boldsymbol{\theta}_t) \sum_{m=1}^{M} p(\boldsymbol{\theta}_t|\boldsymbol{\theta}_{t-1}^{(m)})\omega_{t-1}^{(m)}.$$

Pitt and Shephard (1999a) approximate (6.3) by the importance density

$$g_t(\boldsymbol{\theta}_t, k|\boldsymbol{\theta}_{t-1}, \mathcal{D}_t) \propto p(\mathbf{y}_t|\boldsymbol{\mu}_t^{(k)})p(\boldsymbol{\theta}_t|\boldsymbol{\theta}_{t-1}^{(k)})\omega_{t-1}^{(k)},$$

where $\boldsymbol{\mu}_t^{(k)}$ is the mean, mode, a draw, or some other likely value associated with the density of $(\boldsymbol{\theta}_t|\boldsymbol{\theta}_{t-1}^{(k)})$ such that $Pr(k = i|\mathcal{D}_t) \propto p(\mathbf{y}_t|\boldsymbol{\mu}_t^{(i)})\omega_{t-1}^{(i)}$. Choosing $g_t(\cdot)$ is problem-specific and may be nontrivial (see examples in Pitt and Shephard 1999a).

The APF algorithm is summarized as follows. At time $t-1$, $p(\boldsymbol{\theta}_{t-1}|\mathcal{D}_{t-1})$ is approximated by $\{(\boldsymbol{\theta}_{t-1}, \omega_{t-1})^{(m)}; m = 1 : M\}$. Then, for $m = 1 : M$:

1. Sample an auxiliary variable $k^{(m)}$ from the set $\{1, \ldots, M\}$ with probabilities $Pr(k^{(m)} = k) \propto p(\mathbf{y}_t|\boldsymbol{\mu}_t^{(k)})\omega_{t-1}^{(k)}$.
2. Sample $\boldsymbol{\theta}_t^{(m)}$ from $p(\boldsymbol{\theta}_t|\boldsymbol{\theta}_{t-1}^{(k^{(m)})})$.
3. Compute the new weights $\omega_t^{(m)} \propto p(\mathbf{y}_t|\boldsymbol{\theta}_t^{(m)})/p(\mathbf{y}_t|\boldsymbol{\mu}_t^{(k^{(m)})})$.

The auxiliary variables in this algorithm are helpful in identifying those particles with larger predictive likelihoods and so this algorithm has lower computational cost and is more statistically efficient than other particle filters that had been proposed before. A resampling step based on the effective sample size $M_{t,\text{eff}}$ (similar to that in the SISR algorithm above) can be added.

6.2.3 SMC for Combined State and Parameter Estimation

We now discuss some SMC algorithms that allow us to achieve sequential filtering and parameter learning simultaneously in general state-space models. Particle filtering methods based on SIR (Gordon, Salmond, and Smith 1993, Liu and Chen 1995, Kitagawa 1996) and auxiliary particle filters can produce far from uniform particle weights and lead to particle degeneracy when outliers, model misspecification, or models with a large number of parameters are considered. In particular, particle degeneracy becomes an issue when sequential parameter learning is considered in conjunction with filtering in general state-space models (Andrieu, Doucet, and Tadić 2005). Several approaches have been proposed in recent years to deal with this problem. Berzuini, Best, Gilks, and Larizza (1997) considered MCMC moves within particle filters. Liu and West (2001) augmented the state vector to allow for artificial evolution on the static parameters, and combined APF with kernel density estimation techniques. Other algorithms based on off-line likelihood methods and recursive and batch maximum likelihood methods with stochastic gradients and expectation-maximization approaches, as well as algorithms that use MCMC sampling within a sequential MC framework appear in Hürzeler and Künsch (2001), Andrieu and Doucet (2003), Andrieu, Doucet, and Tadić (2005), and Del Moral, Doucet, and Jasra (2006). Polson, Stroud, and Müller (2008) developed a simulation-based approach that approximates the target posterior distribution by a mixture of fixed lag smoothing distributions. Fearnhead (2002), Storvik (2002), and, more recently, Carvalho, Johannes, Lopes, and Polson (2010) consider particle filters that use sufficient statistics for the parameters and/or states.

We begin by presenting the approach of Liu and West (2001) and then discuss some SMC methods that take advantage of a sufficient statistic structure on the static parameters and/or states. In particular, we focus on the algorithms of Storvik (2002), Polson, Stroud, and Müller (2008), and Carvalho, Johannes, Lopes, and Polson (2010).

6.2.3.1 Algorithm of Liu and West

The algorithm of Liu and West (2001) combines the kernel smoothing ideas presented in West (1993a,b) with the APF approach of Pitt and Shephard (1999a). Specifically, $p(\boldsymbol{\theta}_t, \boldsymbol{\phi}|\mathcal{D}_t)$ is written as

$$p(\boldsymbol{\theta}_t, \boldsymbol{\phi}|\mathcal{D}_t) \quad \propto \quad p(\mathbf{y}_t|\boldsymbol{\theta}_t, \boldsymbol{\phi})p(\boldsymbol{\theta}_t|\boldsymbol{\phi}, \mathcal{D}_{t-1})p(\boldsymbol{\phi}|\mathcal{D}_{t-1}),$$

and $p(\boldsymbol{\phi}|\mathcal{D}_{t-1})$ is then approximated using a smooth kernel density form. That is,

$$p(\boldsymbol{\phi}|\mathcal{D}_{t-1}) \approx \sum_{m=1}^{M} N(\boldsymbol{\phi}|\mathbf{m}_{t-1}^{(m)}, (1-a^2)\mathbf{V}_{t-1}), \tag{6.4}$$

where a is a smoothing parameter, $\mathbf{V}_{t-1} = V(\boldsymbol{\phi}|\mathcal{D}_{t-1})$, and the $\mathbf{m}_{t-1}^{(m)}$s are the locations of the mixture components. Liu and West (2001) use the idea of shrinkage of kernel locations and so

$$\mathbf{m}_{t-1}^{(m)} = a\boldsymbol{\phi}_{t-1}^{(m)} + (1-a)\bar{\boldsymbol{\phi}}_{t-1},$$

with $\{(\boldsymbol{\phi}_{t-1}, \omega_{t-1})^{(m)}; m = 1 : M\}$ a particle approximation to $p(\boldsymbol{\phi}|\mathcal{D}_{t-1})$, and $\bar{\boldsymbol{\phi}}_{t-1} = \sum_{m=1}^{M} \boldsymbol{\phi}_{t-1}^{(m)} \omega_{t-1}^{(m)}$. Furthermore, Liu and West (2001) set $a = (3\delta - 1)/2\delta$, where $\delta \in (0, 1]$ is a discount factor. In practice, relatively large discount factor values—with $\delta > 0.9$—are typically used. The algorithm is summarized below.

Assume that at time $t-1$ $\{(\boldsymbol{\theta}_{t-1}, \boldsymbol{\phi}_{t-1}, \omega_{t-1})^{(m)}; m = 1 : M\}$ approximates $p(\boldsymbol{\theta}_{t-1}, \boldsymbol{\phi}|\mathcal{D}_{t-1})$. Then, for each m perform the following steps:

1. Identify prior point estimates of $(\boldsymbol{\theta}_t, \boldsymbol{\phi})$ given by $(\boldsymbol{\mu}_t^{(m)}, \mathbf{m}_{t-1}^{(m)})$, where
$$\begin{aligned} \boldsymbol{\mu}_t^{(m)} &= E(\boldsymbol{\theta}_t|\boldsymbol{\theta}_{t-1}^{(m)}, \boldsymbol{\phi}_{t-1}^{(m)}), \\ \mathbf{m}_{t-1}^{(m)} &= a\boldsymbol{\phi}_{t-1}^{(m)} + (1-a)\bar{\boldsymbol{\phi}}_{t-1}. \end{aligned}$$
2. Sample an auxiliary integer variable $k^{(m)}$ from the set $\{1, \ldots, M\}$ with probabilities proportional to
$$Pr(k^{(m)} = k) \propto \omega_{t-1}^{(k)} p(\mathbf{y}_t|\boldsymbol{\mu}_t^{(k)}, \mathbf{m}_{t-1}^{(k)}).$$
3. Sample a new parameter vector $\boldsymbol{\phi}_t^{(m)}$ from
$$\boldsymbol{\phi}_t^{(m)} \sim N(\mathbf{m}_{t-1}^{(k^{(m)})}, (1-a^2)\mathbf{V}_{t-1}),$$
with $\mathbf{V}_{t-1} = \sum_{m=1}^{M} (\boldsymbol{\phi}_{t-1}^{(m)} - \bar{\boldsymbol{\phi}}_{t-1})(\boldsymbol{\phi}_{t-1}^{(m)} - \bar{\boldsymbol{\phi}}_{t-1})' \omega_{t-1}^{(m)}$.
4. Sample a value of the current state vector $\boldsymbol{\theta}_t^{(m)}$ from $p(\cdot|\boldsymbol{\theta}_{t-1}^{(k^{(m)})}, \boldsymbol{\phi}_t^{(m)})$.
5. Compute the weights
$$\omega_t^{(m)} \propto \frac{p(\mathbf{y}_t|\boldsymbol{\theta}_t^{(m)}, \boldsymbol{\phi}_t^{(m)})}{p(\mathbf{y}_t|\boldsymbol{\mu}_t^{(k^{(m)})}, \mathbf{m}_{t-1}^{(k^{(m)})})}.$$

Finally, $\{(\boldsymbol{\theta}_t, \boldsymbol{\phi}_t, \omega_t)^{(m)}; m = 1 : M\}$ approximates $p(\boldsymbol{\theta}_t, \boldsymbol{\phi}|\mathcal{D}_t)$. Note that in many practical scenarios some or all of the model parameters may need to be transformed so that the normal kernels in (6.4) are appropriate. For instance, if $\boldsymbol{\phi} = (p, v)$, where $p \in (0, 1)$ and v is a variance, then the algorithm should be applied to the transformed parameters $\boldsymbol{\phi}^* = (\log(p/(1-p)), \log(v))$. Alternatively, other kernels can be used, e.g., beta kernels can be used for parameters bounded in $(0, 1)$, while gamma kernels can be used for parameters that are always positive. This is illustrated in Chapter 5 of Petris, Petrone, and Campagnoli (2009).

Liu and West (2001) apply their algorithm to achieve state and parameter estimation in a dynamic latent factor model with multivariate stochastic volatility components. These authors also emphasize that SMC algorithms may need to be periodically combined with MCMC steps in order to minimize the accumulation of SMC approximation errors that tend to build up over time, as well as particle degeneracy.

6.2.3.2 Storvik's algorithm

The approaches of Fearnhead (2002) and Storvik (2002) use sufficient statistics, which offers advantages in terms of reducing the computational and memory requirements of the SMC algorithm—given that only the sufficient statistics need to be stored as opposed to the complete state trajectories—and provides a way to deal with the problem of particle impoverishment that typically occurs when sequential parameter learning and filtering are simultaneously considered. Below we summarize the algorithm of Storvik (2002).

Specifically, let $\mathbf{s}_t = \mathcal{S}(\boldsymbol{\theta}_{0:t}, \mathcal{D}_t)$ be a (typically low dimensional) sufficient statistic for $\boldsymbol{\phi}$. Assume that a particle approximation $\{(\boldsymbol{\theta}_{t-1}, \omega_{t-1})^{(m)}; m = 1:M\}$ of $p(\boldsymbol{\theta}_{t-1}|\mathcal{D}_{t-1})$ is available at time $t-1$, and that there is an updating rule $f(\cdot)$ for $\mathbf{s}_t$ such that $\mathbf{s}_t = f(\mathbf{s}_{t-1}, \boldsymbol{\theta}_t, \mathbf{y}_t)$. Then, for each $m = 1:M$ the importance sampling scheme of Storvik (2002) is as follows.

1. Sample $\boldsymbol{\phi} \sim g_{t,1}(\boldsymbol{\phi}|\boldsymbol{\theta}^{(m)}_{0:(t-1)}, \mathcal{D}_t)$.
2. Sample $\tilde{\boldsymbol{\theta}}^{(m)}_t \sim g_{t,2}(\boldsymbol{\theta}_t|\boldsymbol{\theta}^{(m)}_{0:(t-1)}, \mathbf{y}_t, \boldsymbol{\phi})$ and set $\tilde{\boldsymbol{\theta}}^{(m)}_{0:t} = (\boldsymbol{\theta}^{(m)}_{0:(t-1)}, \tilde{\boldsymbol{\theta}}^{(m)}_t)$.
3. Compute the importance sampling weights

$$\tilde{\omega}^{(m)}_t \propto \omega^{(m)}_{t-1} \frac{p(\boldsymbol{\phi}|\mathbf{s}^{(m)}_{t-1})p(\tilde{\boldsymbol{\theta}}^{(m)}_t|\boldsymbol{\theta}^{(m)}_{t-1}, \boldsymbol{\phi})p(\mathbf{y}_t|\tilde{\boldsymbol{\theta}}^{(m)}_t, \boldsymbol{\phi})}{g_{t,1}(\boldsymbol{\phi}|\boldsymbol{\theta}^{(m)}_{0:(t-1)}, \mathcal{D}_t)g_{t,2}(\tilde{\boldsymbol{\theta}}^{(m)}_t|\boldsymbol{\theta}^{(m)}_{0:(t-1)}, \mathbf{y}_t, \boldsymbol{\phi})}. \tag{6.5}$$

Resampling can also be performed for each $m = 1:M$ by sampling an index $k^{(m)}$ from $\{1, \ldots, M\}$ with probabilities $Pr(k^{(m)} = k) = \tilde{\omega}^{(k)}_t$, and then setting $\boldsymbol{\theta}^{(m)}_t = \tilde{\boldsymbol{\theta}}^{(k^{(m)})}_t$, $\mathbf{s}^{(m)}_t = f(\mathbf{s}^{(k^{(m)})}_{t-1}, \boldsymbol{\theta}^{(m)}_t, \mathbf{y}_t)$, and $\omega^{(m)}_t = 1/M$.

The distributions $g_{t,1}(\boldsymbol{\phi}|\boldsymbol{\theta}^{(m)}_{0:(t-1)}, \mathcal{D}_t)$ and $g_{t,2}(\boldsymbol{\theta}_t|\boldsymbol{\theta}^{(m)}_{0:(t-1)}, \mathbf{y}_t, \boldsymbol{\phi})$ are proposal distributions that are often set to $g_{t,1}(\boldsymbol{\phi}|\boldsymbol{\theta}^{(m)}_{0:(t-1)}, \mathcal{D}_t) = p(\boldsymbol{\phi}|\mathbf{s}^{(m)}_{t-1})$ and $g_{t,2}(\boldsymbol{\theta}_t|\boldsymbol{\theta}^{(m)}_{0:(t-1)}, \mathbf{y}_t, \boldsymbol{\phi}) = p(\boldsymbol{\theta}_t|\boldsymbol{\theta}^{(m)}_{t-1}, \boldsymbol{\phi})$. In this case the weights in (6.5) simplify to $\tilde{\omega}^{(m)}_t \propto \omega^{(m)}_{t-1}p(\mathbf{y}_t|\tilde{\boldsymbol{\theta}}^{(m)}_t, \boldsymbol{\phi})$. Storvik (2002) considers several Gaussian-based system models with general observation distributions, including linear and partial linear processes, as well as dynamic generalized linear models. In such models the sufficient statistics can be updated using Kalman-type filters. For cases where direct simulation from $p(\boldsymbol{\phi}|\mathbf{s}^{(m)}_t)$ cannot be achieved, samples from this distribution are approximately obtained by considering a few MCMC steps. This approach outperforms other approaches such as that of Liu and West (2001) in the simulation studies presented in Storvik (2002). An APF version of Storvik's algorithm can also be implemented (e.g., Polson, Stroud, and Müller 2008).

6.2.3.3 Practical filtering

Polson, Stroud, and Müller (2008) proposed an approach, referred to as the *practical filter*, based on a rolling window MCMC algorithm that approximates the target posterior distribution by a mixture of fixed lag smoothing distributions. As in Storvik (2002), the sufficient statistic structure is also exploited. This practical filter and parameter learning algorithm is particularly well-suited for models where efficient MCMC smoothing methods can be applied. In particular, Polson, Stroud, and Müller (2008) apply their algorithm to a benchmark autoregressive plus noise model with sequential parameter learning given by $y_t = \theta_t + \nu_t$, $\theta_t = \beta_0 + \beta_1\theta_{t-1} + \omega_t$, with $\nu_t \sim N(0, v)$ and $\omega_t \sim N(0, w)$. The authors show that schemes such as Storvik's SIR and APF algorithms lead to particle degeneracy when a change point is included at the state level, while this is not the case when the practical filter is considered. The algorithm is also applied to a higher dimensional spatio-temporal model. The practical filter summarized below requires three inputs: the number of independent state trajectories M, the number of MCMC iterations G, and the lag length of the rolling window h. Polson, Stroud, and Müller (2008) offer guidelines on how to choose these values.

The practical learning algorithm is as follows. Assume that, at time $t - 1$, M samples of $p(\boldsymbol{\theta}_{t-1}, \boldsymbol{\phi}|\mathcal{D}_{t-1})$ are available, i.e., assume that we have $\{(\boldsymbol{\theta}_{t-1}, \boldsymbol{\phi})^{(m)}; m = 1 : M\}$. During the initial warm-up period $t = 1 : h$, draws from the distribution $p(\boldsymbol{\theta}_{0:t}, \boldsymbol{\phi}|\mathcal{D}_t)$ are obtained using MCMC methods. Then, for every subsequent time t, with $t \geq (h + 1)$, and for each m perform the following steps:

1. Run G iterations of a MCMC algorithm with stationary distribution $p(\boldsymbol{\theta}_{(t-h+1):t}, \boldsymbol{\phi}|\mathbf{s}_{t-h}^{(m)}, \mathbf{y}_{(t-h+1):t})$.
2. Define $(\boldsymbol{\theta}_{(t-h+1):t}^{(m)}, \boldsymbol{\phi}^{(m)})$ as the last value of $(\boldsymbol{\theta}_{(t-h+1):t}, \boldsymbol{\phi})$ in the chain of Step 1.
3. Set $\mathbf{s}_{t-h+1}^{(m)} = f(\mathbf{s}_{t-h}^{(m)}, \boldsymbol{\theta}_{t-h+1}^{(m)}, \mathbf{y}_{t-h+1})$ and store $\mathbf{s}_{t-h+1}^{(m)}$ as a draw from $p(\mathbf{s}_{t-h+1}|\mathbf{y}_{1:t})$.

Finally, the new set of samples $\{(\boldsymbol{\theta}_t, \boldsymbol{\phi})^{(m)}; i = 1 : M\}$ are reported as samples from $p(\boldsymbol{\theta}_t, \boldsymbol{\phi}|\mathcal{D}_t)$.

6.2.3.4 Particle learning methods

Carvalho, Johannes, Lopes, and Polson (2010) describe particle learning (PL) methods for sequential filtering, parameter learning, and smoothing in rather general state-space models. More specifically, methods for filtering and parameter learning in conditionally Gaussian linear models, as well as methods for models that are nonlinear at the state level are provided, extending Liu and Chen's (2000) mixture Kalman filter (MKF) method by allowing parameter learning. Several simulation studies are shown in which the proposed PL methods outperform other particle methods such as that of Liu and West (2001), especially for large T. In these simulation studies it is also shown that the PL approach dominates MCMC with respect to computing time, and delivers similar accuracy when computing $p(\boldsymbol{\theta}_t|\mathcal{D}_t)$. Carvalho, Johannes, Lopes, and Polson (2010) also extend the results of Godsill, Doucet, and West (2004) for smoothing in state-space models—i.e., computing $p(\boldsymbol{\theta}_{0:T}, \boldsymbol{\phi}|\mathcal{D}_T)$—to the case in which $\boldsymbol{\phi}$ is unknown. We now discuss some features of this approach to sequential parameter learning and filtering. Smoothing methods are discussed later in this chapter.

The PL methods of Carvalho, Johannes, Lopes, and Polson (2010) have two main features. First, sufficient statistics $\mathbf{s}_t$ are used to represent the posterior distribution of $\boldsymbol{\phi}$, and sufficient statistics for the latent states, $\mathbf{s}_t^{\theta}$, are also exploited whenever the model structure allows it. This reduces the variance of the sampling weights, increasing the efficiency of the algorithm. Second, as opposed to other approaches that first propagate then resample the particles (e.g., those in Storvik 2002), the PL algorithm follows the resample and propagate framework of Johannes and Polson (2008). This avoids the decay in the particle approximation associated with the SIR type of methods. More specifically, the PL algorithm assumes that $p(\boldsymbol{\phi}|\boldsymbol{\theta}_{0:t}, \mathcal{D}_t) = p(\boldsymbol{\phi}|\mathbf{s}_t)$, with $\mathbf{s}_t = f(\mathbf{s}_{t-1}, \boldsymbol{\theta}_t, \mathbf{y}_t)$ where, as in Storvik (2002), $f(\cdot)$ is a deterministic updating rule, and then factorizes the posterior as

$$p(\boldsymbol{\phi}, \boldsymbol{\theta}_t, \mathbf{s}_t|\mathcal{D}_t) = p(\boldsymbol{\phi}|\mathbf{s}_t)p(\boldsymbol{\theta}_t, \mathbf{s}_t|\mathcal{D}_t),$$

and develops a particle approximation to $p(\boldsymbol{\theta}_t, \mathbf{s}_t|\mathcal{D}_t)$. Finally, parameter learning is performed by simulating from $p(\boldsymbol{\phi}|\mathbf{s}_t)$. If a particle approximation $\{(\boldsymbol{\theta}_{t-1}, \mathbf{s}_{t-1}, \boldsymbol{\phi})^{(m)}; m = 1 : M\}$ is available at time $t-1$, the PL method updates this approximation at time t, after observing $\mathbf{y}_t$, using the following rule:

$$\begin{aligned} p(\boldsymbol{\theta}_{t-1}, \mathbf{s}_{t-1}, \boldsymbol{\phi}|\mathcal{D}_t) &\propto p(\mathbf{y}_t|\boldsymbol{\theta}_{t-1}, \mathbf{s}_{t-1}, \boldsymbol{\phi})p(\boldsymbol{\theta}_{t-1}, \mathbf{s}_{t-1}, \boldsymbol{\phi}|\mathcal{D}_{t-1}) \quad (6.6)\\ p(\boldsymbol{\theta}_t, \mathbf{s}_t, \boldsymbol{\phi}|\mathcal{D}_t) &= \int p(\mathbf{s}_t|\boldsymbol{\theta}_t, \mathbf{s}_{t-1}, \mathbf{y}_t)p(\boldsymbol{\theta}_t|\boldsymbol{\theta}_{t-1}, \mathbf{s}_{t-1}, \boldsymbol{\phi}, \mathbf{y}_t) \times \\ &\quad p(\boldsymbol{\theta}_{t-1}, \mathbf{s}_{t-1}, \boldsymbol{\phi}|\mathcal{D}_t)d\boldsymbol{\theta}_{t-1}d\mathbf{s}_{t-1}. \quad (6.7)\end{aligned}$$

An approximation to $p(\boldsymbol{\theta}_{t-1}, \mathbf{s}_{t-1}, \boldsymbol{\phi}|\mathcal{D}_t)$ can be obtained by resampling the particles $\{(\boldsymbol{\theta}_{t-1}, \mathbf{s}_{t-1}, \boldsymbol{\phi})^{(m)}; m = 1 : M\}$ with weights proportional to $p(\mathbf{y}_t|\boldsymbol{\theta}_{t-1}, \mathbf{s}_{t-1}, \boldsymbol{\phi})$ (see Equation 6.6). Such an approximation can be used in (6.7) to generate and propagate samples from $p(\boldsymbol{\theta}_t|\boldsymbol{\theta}_{t-1}, \mathbf{s}_{t-1}, \boldsymbol{\phi}, \mathbf{y}_t)$, that are then used to update $\mathbf{s}_t$ via $\mathbf{s}_t = f(\mathbf{s}_{t-1}, \boldsymbol{\theta}_{t-1}, \mathbf{y}_t)$. Carvalho, Johannes,

Lopes, and Polson (2010) note that (6.7) is an abuse of notation, since $\mathbf{s}_t$ is updated deterministically from $\mathbf{s}_{t-1}, \boldsymbol{\theta}_{t-1}$, and $\mathbf{y}_t$. However, since $\mathbf{s}_{t-1}$ and $\boldsymbol{\theta}_{t-1}$ are random variables, the $\mathbf{s}_t$s are also random variables and so are treated as states in the filtering step.

Carvalho, Johannes, Lopes, and Polson (2010) develop PL and filtering algorithms for conditionally Gaussian dynamic linear models (CDLMs) of the form

$$\mathbf{y}_t = \mathbf{F}'_{\lambda_t}\boldsymbol{\theta}_t + \boldsymbol{\nu}_t, \quad \boldsymbol{\nu}_t \sim N(\mathbf{0}, \mathbf{V}_{\lambda_t}), \tag{6.8}$$

$$\boldsymbol{\theta}_t = \mathbf{G}_{\lambda_t}\boldsymbol{\theta}_{t-1} + \mathbf{w}_t, \quad \mathbf{w}_t \sim N(\mathbf{0}, \mathbf{W}_{\lambda_t}). \tag{6.9}$$

In such models the observation and system Equations (6.8) and (6.9) are linear and Gaussian conditional on the auxiliary (discrete or continuous) state $\boldsymbol{\lambda}_t$. PL methods are also developed for conditionally Gaussian nonlinear state spaces, or conditionally Gaussian dynamic models (CGDM), that have the same observation equation in (6.8), but for which a new system equation given by

$$\boldsymbol{\theta}_t = \mathbf{G}_{\lambda_t}\mathbf{Z}(\boldsymbol{\theta}_{t-1}) + \mathbf{w}_t, \quad \mathbf{w}_t \sim N(\mathbf{0}, \mathbf{W}_{\lambda_t}) \tag{6.10}$$

substitutes (6.9), where $\mathbf{Z}(\cdot)$ is a nonlinear function. PL methods for models defined by (6.8) and (6.9) update sufficient statistics $\mathbf{s}_t$ and $\mathbf{s}_t^{\theta}$ for the parameters and the states $\boldsymbol{\phi}$ and $\boldsymbol{\theta}_t$, respectively. Given the nonlinearity in (6.10), PL methods for CGDMs work exclusively with the sufficient statistic $\mathbf{s}_t$. In spite of this, it is possible to evaluate the predictive density to perform the resampling step in CGDMs, given that the linear and Gaussian structure is preserved at the observational level. Below we summarize the PL algorithm for CDLMs and that for CGDMs with discrete auxiliary states $\boldsymbol{\lambda}_t$. Details on PL algorithms for CDLMs and CGDMs with continuous auxiliary states appear in Carvalho, Johannes, Lopes, and Polson (2010).

Assume a model structure described by (6.8) and (6.9). Begin with a particle set $\{(\boldsymbol{\theta}_{t-1}, \boldsymbol{\phi}, \boldsymbol{\lambda}_{t-1}, \mathbf{s}_{t-1}^{\theta}, \mathbf{s}_{t-1})^{(m)}; m = 1 : M\}$, at time $t-1$. Then, for each m perform the following steps:

1. Sample an index $k(m)$ from $\{1, \ldots, M\}$ with

$$Pr(k(m) = k) \propto p(\mathbf{y}_t | (\boldsymbol{\lambda}_{t-1}, \mathbf{s}_{t-1}^{\theta}, \boldsymbol{\phi})^{(k)}).$$

2. Propagate states via

$$\boldsymbol{\lambda}_t^{(m)} \sim p(\boldsymbol{\lambda}_t | (\boldsymbol{\lambda}_{t-1}, \boldsymbol{\phi})^{k(m)}, \mathbf{y}_t),$$

$$\boldsymbol{\theta}_t^{(m)} \sim p(\boldsymbol{\theta}_t | (\boldsymbol{\theta}_{t-1}, \boldsymbol{\phi})^{k(m)}, \boldsymbol{\lambda}_t^{(m)}, \mathbf{y}_t).$$

3. Propagate sufficient statistics for parameters and states using

$$\begin{aligned}\mathbf{s}_t^{\theta,(m)} &= \mathcal{K}(\mathbf{s}_{t-1}^{\theta,k(m)}, \boldsymbol{\phi}^{k(m)}, \boldsymbol{\lambda}_t^{(m)}, \mathbf{y}_t),\\ \mathbf{s}_t^{(m)} &= f(\mathbf{s}_{t-1}^{k(m)}, \boldsymbol{\theta}_t^{(m)}, \boldsymbol{\lambda}_t^{(m)}, \mathbf{y}_t),\end{aligned}$$

where $\mathcal{K}(\cdot)$ denotes the Kalman filter recursion. Finally, $\boldsymbol{\phi}^{(m)}$ can then be sampled from $p(\boldsymbol{\phi}^{(m)}|\mathbf{s}_t^{(m)})$, and the particle set $\{(\boldsymbol{\theta}_t, \boldsymbol{\phi}, \boldsymbol{\lambda}_t, \mathbf{s}_t^{\theta}, \mathbf{s}_t)^{(m)}; m = 1 : M\}$ is reported at time t.

For the Kalman filter recursion $\mathcal{K}(\cdot)$ above, define $\mathbf{s}_{t-1}^{\theta} = (\mathbf{m}_{t-1}, \mathbf{C}_{t-1})$ as the Kalman filter first and second moments at time $t-1$. Then, $\mathbf{s}_t^{\theta} = (\mathbf{m}_t, \mathbf{C}_t)$ are obtained via

$$\begin{aligned}\mathbf{m}_t &= \mathbf{G}_{\lambda_t}\mathbf{m}_{t-1} + \mathbf{A}_t(\mathbf{y}_t - \mathbf{F}'_{\lambda_t}\mathbf{G}_{\lambda_t}\mathbf{m}_{t-1}),\\ \mathbf{C}_t^{-1} &= \mathbf{R}_t^{-1} + \mathbf{F}_{\lambda_t}\mathbf{V}_{\lambda_t}^{-1}\mathbf{F}'_{\lambda_t},\end{aligned}$$

where $\mathbf{A}_t = \mathbf{R}_t\mathbf{F}_{\lambda_t}\mathbf{Q}_t^{-1}$ and $\mathbf{Q}_t = \mathbf{F}'_{\lambda_t}\mathbf{R}_t\mathbf{F}_{\lambda_t} + \mathbf{V}_{\lambda_t}$. Also, the predictive distribution in Step 1 of the algorithm, assuming that $\boldsymbol{\lambda}_t$ is discrete, is computed via

$$p(\mathbf{y}_t|(\boldsymbol{\lambda}_{t-1}, \mathbf{s}_{t-1}^{\theta}, \boldsymbol{\phi})^{(m)}) = \sum_{\boldsymbol{\lambda}_t} p(\mathbf{y}_t|\boldsymbol{\lambda}_t, (\mathbf{s}_{t-1}^{\theta}, \boldsymbol{\phi})^{(m)})p(\boldsymbol{\lambda}_t|(\boldsymbol{\lambda}_{t-1}, \boldsymbol{\phi})^{(m)}),$$

with

$$p(\mathbf{y}_t|\boldsymbol{\lambda}_t, \mathbf{s}_{t-1}^{\theta}, \boldsymbol{\phi}) = \int p(\mathbf{y}_t|\boldsymbol{\theta}_t, \boldsymbol{\lambda}_t, \boldsymbol{\phi})p(\boldsymbol{\theta}_t|\mathbf{s}_{t-1}^{\theta}, \boldsymbol{\phi})d\boldsymbol{\theta}_t.$$

When the auxiliary variable $\boldsymbol{\lambda}_{t-1}$ is continuous, integrating out $\boldsymbol{\lambda}_t$ in Step 1 is not always possible. In such cases the above algorithm can be modified to propagate $\boldsymbol{\lambda}_t$ via $\boldsymbol{\lambda}_t \sim p(\boldsymbol{\lambda}_t|(\boldsymbol{\lambda}_{t-1}, \boldsymbol{\phi})^{(m)})$, and then resample the particle $(\boldsymbol{\theta}_{t-1}, \boldsymbol{\lambda}_t, \mathbf{s}_{t-1})^{(m)}$ with the appropriate predictive $p(\mathbf{y}_t|(\boldsymbol{\theta}_{t-1}, \boldsymbol{\lambda}_t, \boldsymbol{\phi})^{(m)})$ (see details in Carvalho, Johannes, Lopes, and Polson 2010).

Similarly, the PL algorithm for CGDMs with discrete auxiliary states is as follows. Assume that the particle set $\{(\boldsymbol{\theta}_{t-1}, \boldsymbol{\phi}, \boldsymbol{\lambda}_{t-1}, \mathbf{s}_{t-1})^{(m)}; m = 1 : M\}$ is available at time $t-1$. Then, for each m follow the steps below.

1. Sample an index $k(m)$ from $\{1, \ldots, M\}$, with

$$Pr(k(m) = k) \propto p(\mathbf{y}_t|(\boldsymbol{\theta}_{t-1}, \boldsymbol{\lambda}_{t-1}, \boldsymbol{\phi})^{(k)}).$$

2. Propagate states via

$$\begin{aligned}\boldsymbol{\lambda}_t^{(m)} &\sim p(\boldsymbol{\lambda}_t|(\boldsymbol{\lambda}_{t-1}, \boldsymbol{\phi})^{k(m)}, \mathbf{y}_t),\\ \boldsymbol{\theta}_t^{(m)} &\sim p(\boldsymbol{\theta}_t|(\boldsymbol{\theta}_{t-1}, \boldsymbol{\phi})^{k(m)}, \boldsymbol{\lambda}_t^{(m)}, \mathbf{y}_t).\end{aligned}$$

3. Propagate the parameter sufficient statistics by computing

$$\mathbf{s}_t^{(m)} = f(\mathbf{s}_{t-1}^{k(m)}, \boldsymbol{\theta}_t^{(m)}, \boldsymbol{\lambda}_t^{(m)}, \mathbf{y}_t).$$

Once again, details on algorithms for CGDMs with continuous auxiliary states are given in Carvalho, Johannes, Lopes, and Polson (2010). Note that this algorithm does not use the state sufficient statistics $\mathbf{s}_t^\theta$ due to the nonlinear structure in the evolution equation.

6.2.4 Smoothing

Godsill, Doucet, and West (2004) proposed an approach for performing smoothing computations in general state-space models for which the parameters $\boldsymbol{\phi}$ are known. This approach assumes that filtering has already been performed using any particle filtering scheme so that an approximate representation of $p(\boldsymbol{\theta}_t|\mathcal{D}_t)$ is available at each $t = 1 : T$ via the weighted set of particles $\{(\boldsymbol{\theta}_t, \omega_t)^{(m)}; m = 1 : M\}$. In order to obtain sample representations of $p(\boldsymbol{\theta}_{0:T}|\mathcal{D}_T)$, the following factorization is used:

$$p(\boldsymbol{\theta}_{0:T}|\mathcal{D}_T) = p(\boldsymbol{\theta}_T|\mathcal{D}_T) \prod_{t=0}^{T-1} p(\boldsymbol{\theta}_t|\boldsymbol{\theta}_{(t+1):T}, \mathcal{D}_T), \tag{6.11}$$

where

$$\begin{aligned} p(\boldsymbol{\theta}_t|\boldsymbol{\theta}_{(t+1):T}, \mathcal{D}_T) &= p(\boldsymbol{\theta}_t|\boldsymbol{\theta}_{t+1}, \mathcal{D}_t) \\ &= \frac{p(\boldsymbol{\theta}_t|\mathcal{D}_t)p(\boldsymbol{\theta}_{t+1}|\boldsymbol{\theta}_t)}{p(\boldsymbol{\theta}_{t+1}|\mathcal{D}_t)} \\ &\propto p(\boldsymbol{\theta}_t|\mathcal{D}_t)p(\boldsymbol{\theta}_{t+1}|\boldsymbol{\theta}_t). \end{aligned} \tag{6.12}$$

Then, it is possible to obtain the modified particle approximation

$$p(\boldsymbol{\theta}_t|\boldsymbol{\theta}_{t+1}, \mathcal{D}_T) \approx \sum_{m=1}^{M} \omega_{t|t+1}^{(m)} \delta_{\boldsymbol{\theta}_t^{(m)}}(\boldsymbol{\theta}_t),$$

with

$$\omega_{t|t+1}^{(m)} = \frac{\omega_t^{(m)} p(\boldsymbol{\theta}_{t+1}|\boldsymbol{\theta}_t^{(m)})}{\sum_{i=1}^{M} \omega_t^{(i)} p(\boldsymbol{\theta}_{t+1}|\boldsymbol{\theta}_t^{(i)})},$$

leading to the following particle smoothing algorithm:

1. At time T choose $\tilde{\boldsymbol{\theta}}_T = \boldsymbol{\theta}_T^{(m)}$ with probability $\omega_T^{(m)}$.
2. For $t = (T-1) : 0$,
   (a) calculate $\omega_{t|t+1}^{(m)} \propto \omega_t^{(m)} p(\tilde{\boldsymbol{\theta}}_{t+1}|\boldsymbol{\theta}_t^{(m)})$ for $m = 1 : M$;
   (b) choose $\tilde{\boldsymbol{\theta}}_t = \boldsymbol{\theta}_t^{(m)}$ with probability $\omega_{t|t+1}^{(m)}$.
3. Report $\tilde{\boldsymbol{\theta}}_{0:T} = (\tilde{\boldsymbol{\theta}}_0, \ldots, \tilde{\boldsymbol{\theta}}_T)$ as an approximate realization from the distribution $p(\boldsymbol{\theta}_{0:T}|\mathcal{D}_T)$.

Steps 1 to 2 can be repeated several times to obtain further independent approximate realizations of $p(\boldsymbol{\theta}_{0:T}|\mathcal{D}_T)$. Godsill, Doucet, and West (2004) showed convergence of the smoothed trajectories in the mean squared error sense and tested this method on a speech signal processing application represented by time-varying autoregressive models that are parameterized in terms of time-varying partial correlation coefficients.

Carvalho, Johannes, Lopes, and Polson (2010) extend the algorithm summarized above to consider smoothing in models for which the fixed parameters $\boldsymbol{\phi}$ are unknown. Basically, these authors make explicit the dependence on $\boldsymbol{\phi}$ in (6.11) and (6.12), and use the sufficient statistics structure to obtain the following algorithm:

1. At time T, randomly choose $(\tilde{\boldsymbol{\theta}}_T, \tilde{\mathbf{s}}_T)$ from a particle approximation to $p(\boldsymbol{\theta}_T, \mathbf{s}_T|\mathcal{D}_T)$, such as that obtained using the PL algorithm summarized in the previous section. Then, sample $\tilde{\boldsymbol{\phi}} \sim p(\boldsymbol{\phi}|\tilde{\mathbf{s}}_T)$.
2. For $t = (T-1) : 0$ choose $\tilde{\boldsymbol{\theta}}_t = \boldsymbol{\theta}_t^{(m)}$ from the filtered weighted particles $\{(\boldsymbol{\theta}_t, \omega_{t|t+1})^{(m)}; m = 1 : M\}$ with weights
$$\omega_{t|t+1}^{(m)} \propto \omega_t^{(m)} p(\tilde{\boldsymbol{\theta}}_{t+1}|\boldsymbol{\theta}_t^{(m)}, \tilde{\boldsymbol{\phi}}).$$
3. Report $\tilde{\boldsymbol{\theta}}_{0:T} = (\tilde{\boldsymbol{\theta}}_0, \ldots, \tilde{\boldsymbol{\theta}}_T)$ as an approximate realization from the distribution $p(\boldsymbol{\theta}_{0:T}|\mathcal{D}_T)$.

Note that the computations in Steps 2 and 3 above can also be performed if the filtering was done using algorithms that do not assume a sufficient statistics structure such as that of Liu and West (2001).

6.2.5 Examples

Example 6.5 *Nonlinear time series model.* We now illustrate how the algorithm of Liu and West (2001) can be used for filtering and parameter learning when applied to the 200 observations simulated from the nonlinear time series model of Example 6.1 displayed in Figure 6.1. The R code for this example was provided by Hedibert Lopes. The parameters a and ω were considered known and set at $a = 1/20$ and $\omega = 1.2$, and so the algorithm of Liu and West (2001) was applied for filtering and learning of the parameters b, c, d, v, and w. Recall from Example 6.1 that the data were simulated using $v = 10$, $w = 1$, $b = 0.5$, $c = 25$, and $d = 8$. Figure 6.2 shows the results of applying the algorithm of Liu and West (2001) to the simulated data. A graph of the effective sample size $M_{t,\text{eff}}$ at each time t and the means and quartiles (2.5% and 97.5%) of the particle approximations to the posteriors of b, c, d, v, and w are shown. The algorithm was run with $\delta = 0.75$. The results show that the particle approximation to the posterior distribution is not very good for some of the parameters (e.g., see plot (f)). The approximation can be improved by increasing the number of particles, adding resampling steps when the effective sample size is relatively small, considering different discount factor values or kernels that are non-Gaussian, and/or by using an importance sampling transition density different from the prior distribution. The next example illustrates how particle approximations can be improved by increasing the number of particles. Petris, Petrone, and Campagnoli (2009) consider a modification of Liu and West's algorithm to allow for nonnormal mixtures in the approximation of the posterior distribution. In particular, algorithms that use beta kernels when the model parameters are probabilities and gamma kernels when the model parameters are variances are considered.

Example 6.6 *AR*(1) *plus noise and sequential parameter learning.* Consider the model given by

$$\begin{aligned} y_t &= \theta_t + \nu_t, \quad \nu_t \sim N(0, v), && (6.13) \\ \theta_t &= \phi\theta_{t-1} + w_t, \quad w_t \sim N(0, w), && (6.14) \end{aligned}$$

with $\boldsymbol{\phi} = (\phi, v, w)$ the vector of fixed unknown parameters and the following prior distributions: $\theta_0 \sim N(\mu_{\theta,0}, C_{\theta,0})$, $v \sim IG(\alpha_{v,0}, \beta_{v,0})$, $w \sim IG(\alpha_{w,0}, \beta_{w,0})$, and $\phi \sim N(\mu_{\phi,0}, C_{\phi,0})$.

For this model the algorithm of Liu and West (2001) can be applied to obtain a particle approximation to $p(\theta_t, \boldsymbol{\phi}^*|\mathcal{D}_t)$, with $\boldsymbol{\phi}^* = (\phi_1, \phi_2, \phi_3)' = (\phi, \log(v), \log(w))'$. More specifically, at time $t = 0$ a set of equally weighted particles are simulated from the prior, leading to $\{(\theta_0, \boldsymbol{\phi}_0^*, \omega_0)^{(m)}; m = 1 : M\}$. Then, the following steps are repeated each time a new observation y_t is received, for $m = 1 : M$.

1. Compute $\mu_t^{(m)} = \boldsymbol{\phi}_{t-1,1}^{*,(m)}\theta_{t-1}^{(m)}$ and $\mathbf{m}_{t-1}^{*,(m)} = a\boldsymbol{\phi}_{t-1}^{*,(m)} + (1-a)\bar{\boldsymbol{\phi}}_{t-1}^*$, where $\bar{\boldsymbol{\phi}}_{t-1}^* = \sum_{m=1}^M \omega_{t-1}^{(m)}\boldsymbol{\phi}_{t-1}^{*,(m)}$.
2. Sample $k^{(m)}$ from the set $\{1, \dots, M\}$ with probabilities
$$Pr(k^{(m)} = k) \propto \omega_{t-1}^{(k)} N(y_t|\mu_t^{(k)}, \exp(m_{t-1,2}^{*,(k)})).$$
3. Sample a new parameter vector $\boldsymbol{\phi}_t^{*,(m)}$ from
$$\boldsymbol{\phi}_t^{*,(m)} \sim N(\mathbf{m}_{t-1}^{*,k(m)}, (1-a^2)\mathbf{V}_{t-1}^*),$$
with $\mathbf{V}_{t-1}^* = \sum_{m=1}^M \omega_{t-1}^{(m)}(\boldsymbol{\phi}_{t-1}^{*,(m)} - \bar{\boldsymbol{\phi}}_{t-1}^*)(\boldsymbol{\phi}_{t-1}^{*,(m)} - \bar{\boldsymbol{\phi}}_{t-1}^*)'$.
4. Sample $\theta_t^{(m)}$ from $N(\theta_t|\phi_{t,1}^{*,(m)}\theta_{t-1}^{k(m)}, \exp(\phi_{t,3}^{*,(m)}))$.
5. Compute the weights
$$\omega_t^{(m)} \propto \frac{N(y_t|\theta_t^{(m)}, \exp(\phi_{t,2}^{*,(m)}))}{N(y_t|\mu_t^{k(m)}, \exp(m_{t-1,2}^{*,k(m)}))}.$$

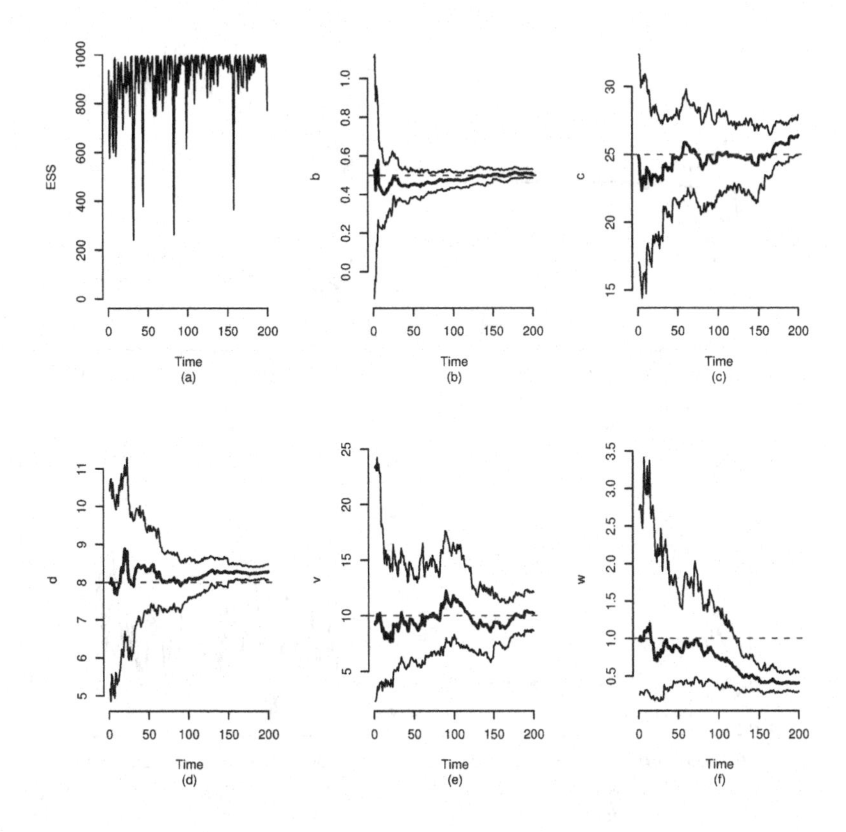

Figure 6.2 *Results of applying the algorithm of Liu and West (2001) to 200 observations from the nonlinear time series model of Example 6.1. (a) Effective sample sizes at each time t. (b)–(f) Trajectories of the mean and quartiles (2.5% and 97.5%) of the posteriors of b, c, d, v, and w, respectively, based on 1,000 particles. The dotted lines in plots (b)–(f) display the true parameter values.*

Each weighted particle set $\{(\theta_t, \boldsymbol{\phi}_t^*, \omega_t)^{(m)}; m = 1 : M\}$ approximates $p(\boldsymbol{\theta}_t, \boldsymbol{\phi}^* | \mathcal{D}_t)$.

In order to illustrate the performance of this algorithm, $T = 300$ values of θ_t and y_t were sampled from the model in (6.13) and (6.14) with $\phi = 0.9$, $v = 4$, $w = 1$, and $\theta_0 = 0$. Figure 6.3 depicts plots of the time trajectories of the mean, 2.5% and 97.5% quantiles of the approximate posterior distributions of ϕ, v, and w (plots (a), (b) and (c), respectively), as well as the true and estimated values of θ_t obtained from applying the algorithm of Liu and West (2001) with $M = 500$ particles and $\delta = 0.95$ (which leads to

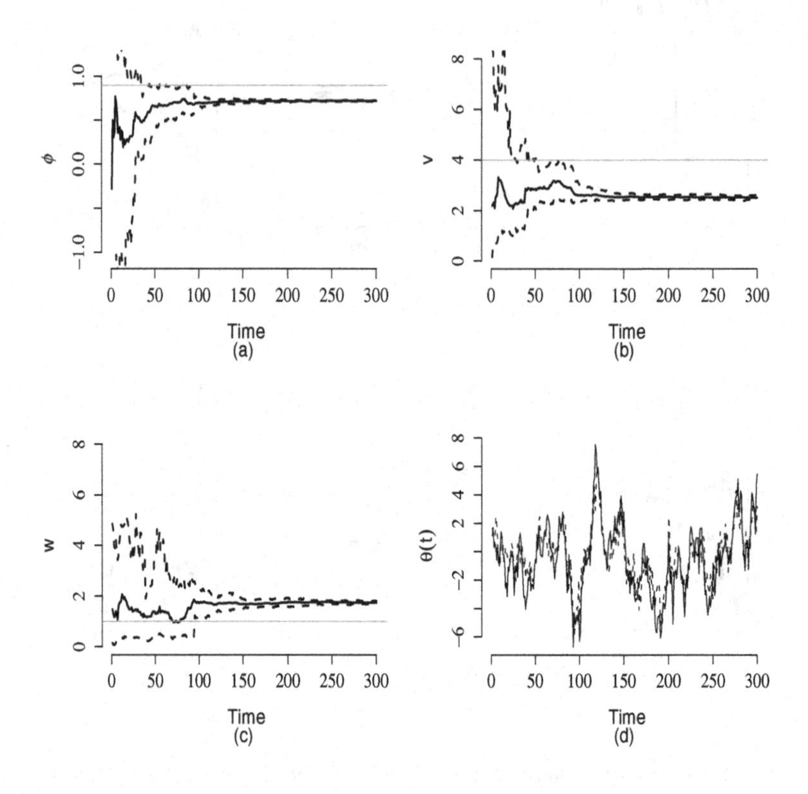

Figure 6.3 *Results of applying the algorithm of Liu and West (2001) with $M = 500$ particles in the $AR(1)$ plus noise model. (a) Time trajectories of the mean (solid line) and quantiles (2.5% and 97.5%) of the posterior distribution of the AR coefficient ϕ. (b) Time trajectories of the mean (solid line) and quantiles (2.5% and 97.5%) of the posterior distribution of the observational variance v. (c) Time trajectories of the mean (solid line) and quantiles (2.5% and 97.5%) of the posterior distribution of w. (d) Time traces of the true state θ_t (solid line) and posterior mean of the particle approximation to the distribution of θ_t.*

$a = 0.9737$). The algorithm produces a good approximation to θ_t (see plot (d)); however, particle degeneracy is evident.

Figure 6.4 shows the results of running Liu and West's algorithm with 5,000 particles instead of 500. Increasing the number of particles considerably

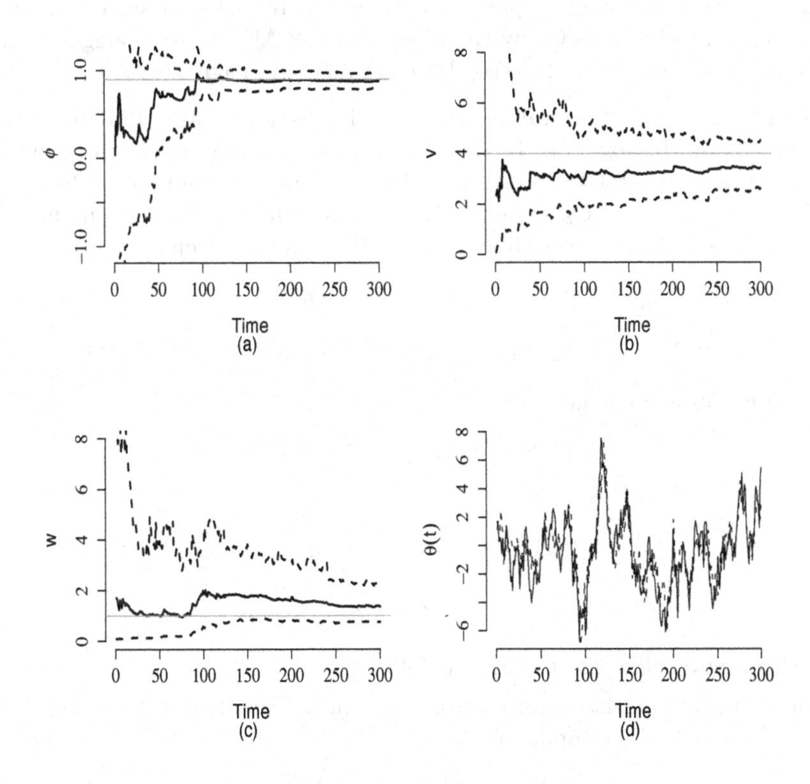

Figure 6.4 *Results of applying the algorithm of Liu and West (2001) with $M =$ 5,000 particles in the AR(1) plus noise model. (a) Time trajectories of the mean (solid line) and quantiles (2.5% and 97.5%) of the posterior distribution of the AR coefficient ϕ. (b) Time trajectories of the mean (solid line) and quantiles (2.5% and 97.5%) of the posterior distribution of the observational variance v. (c) Time trajectories of the mean (solid line) and quantiles (2.5% and 97.5%) of the posterior distribution of w. (d) Time traces of the true state θ_t (solid line) and posterior mean of the particle approximation to the distribution of θ_t.*

improves the approximation in this example. We should also mention at this point that the quality of the approximation will depend on the number of particles and how this number relates to T. Eventually, for T large, the algorithm will lead to particle degeneracy. In order to avoid this the algorithm can be periodically combined with MCMC steps as suggested in Liu and West (2001) and Petris, Petrone, and Campagnoli (2009).

The PL approach of Carvalho, Johannes, Lopes, and Polson (2010) is also illustrated in this example. In the case of the state-space AR(1) model with sequential parameter learning, Steps 1 to 3 in the algorithm are as follows. Define $\mathbf{s}_{t-1}^{\theta} = (m_{t-1}, C_{t-1})$ as the Kalman filter first and second moments at time $t-1$. Then, for each m perform the following steps:

1. Sample an index $k^{(m)}$ from $\{1, \ldots, M\}$ with

$$Pr(k^{(m)} = k) \propto N(y_t|\phi^{(k)} m_{t-1}^{(k)}, (\phi^{(k)})^2 C_{t-1}^{(k)} + w^{(k)} + v^{(k)}).$$

2. Propagate states via

$$\theta_t^{(m)} \sim N(\theta_t | d_t^{k^{(m)}}, D_t^{k^{(m)}}),$$

   with

$$\begin{aligned} D_t^{-1} &= (w^{-1} + v^{-1}), \\ d_t &= D_t \left(\frac{\phi \theta_{t-1}}{w} + \frac{y_t}{v} \right). \end{aligned}$$

3. Propagate sufficient statistics as follows:
   (a) Sufficient statistics for states. The sufficient statistics $\mathbf{s}_t^{\theta} = (m_t, C_t)$ for $t = 1 : T$ are updated via

$$\begin{aligned} C_t^{-1} &= (\phi^2 C_{t-1} + w)^{-1} + v^{-1} \\ m_t &= \phi m_{t-1} + A_t(y_t - \phi m_{t-1}), \end{aligned}$$

   with $A_t = (\phi^2 C_{t-1} + w)/q_t$, and $q_t = \phi^2 C_{t-1} + w + v$.
   (b) Sufficient statistics for the parameters. $p(\phi, v, w|\mathbf{s}_t)$ is decomposed in the following way.
      i. $p(\phi|w, \mathbf{s}_t) \sim N(\mu_{\phi,t}, C_{\phi,t})$ with

$$\begin{aligned} C_{\phi,t}^{-1} &= C_{\phi,t-1}^{-1} + \frac{\theta_{t-1}^2}{w}, \\ \mu_{\phi,t} &= C_{\phi,t}(C_{\phi,t-1}^{-1}\mu_{\phi,t-1} + \theta_t\theta_{t-1}/w). \end{aligned}$$

      ii. $p(v|\mathbf{s}_t) \sim IG(\alpha_{v,t}, \beta_{v,t})$ with

$$\begin{aligned} \beta_{v,t} &= \beta_{v,t-1} + (y_t - \theta_t)^2/2, \\ \alpha_{v,t} &= \alpha_{v,t-1} + 1/2. \end{aligned}$$

iii. $p(w|\phi, \mathbf{s}_t) \sim IG(\alpha_{w,t}, \beta_{w,t})$ with

$$\begin{aligned}\beta_{w,t} &= \beta_{w,t-1} + (\theta_t - \phi\theta_{t-1})^2/2,\\ \alpha_{w,t} &= \alpha_{w,t-1} + 1/2.\end{aligned}$$

Then, $\mathbf{s}_t = (\mu_{\phi,t}, C_{\phi,t}, \alpha_{v,t}, \beta_{v,t}, \alpha_{w,t}, \beta_{w,t})$.

Figures 6.5 and 6.6 show the performance of the PL algorithm of Carvalho, Johannes, Lopes, and Polson (2010) when it was applied to the same $T = 300$ values of y_t simulated from the AR(1) state plus noise model that were used to illustrate the approach of Liu and West (2001). More specifically, Figures 6.5 and 6.6 depict plots of the time trajectories of the mean, 2.5% and 97.5% quantiles of the approximate posterior distributions of ϕ, v, and w (plots (a), (b), and (c), respectively), as well as the true and estimated values of θ_t obtained from applying the PL algorithm described above with 500 and 5,000 particles, respectively. It is clear from these plots that in this example the PL algorithm is much more robust to particle degeneracy than the algorithm of Liu and West (2001). This is also the case when other models are considered, as discussed in Carvalho, Johannes, Lopes, and Polson (2010).

6.3 Problems

1. Consider the AR(1) state plus noise model discussed in Example 6.6. Implement the algorithm of Storvik (2002) for filtering with parameter estimation and compare it to other SMC algorithms such as that of Liu and West (2001) and the PL algorithm of Carvalho, Johannes, Lopes, and Polson (2010) discussed in Example 6.6.
2. Consider the PACF TVAR(2) parameterization in Example 5.4. Simulate data $x_{1:T}$ from this model.
   (a) Sketch and implement a SMC algorithm for filtering and smoothing assuming that β, v, and w are known.
   (b) Sketch and implement a SMC algorithm for filtering and smoothing assuming that β, v, and w are unknown.
3. Consider the fat-tailed nonlinear state-space model studied in Carvalho, Johannes, Lopes, and Polson (2010) and given by

$$\begin{aligned}y_t &= \theta_t + v\sqrt{\lambda_t}\nu_t,\\ \theta_t &= \beta\frac{\theta_{t-1}}{(1+\theta_{t-1}^2)} + w\omega_t,\end{aligned}$$

   where $\nu_t \sim N(0,1)$, $\omega_t \sim N(0,1)$ and $\lambda_t \sim IG(\nu/2, \nu/2)$.
   (a) Assume that ν, v, w, and β are known. Propose, implement, and compare SMC approaches for filtering without parameter learning.

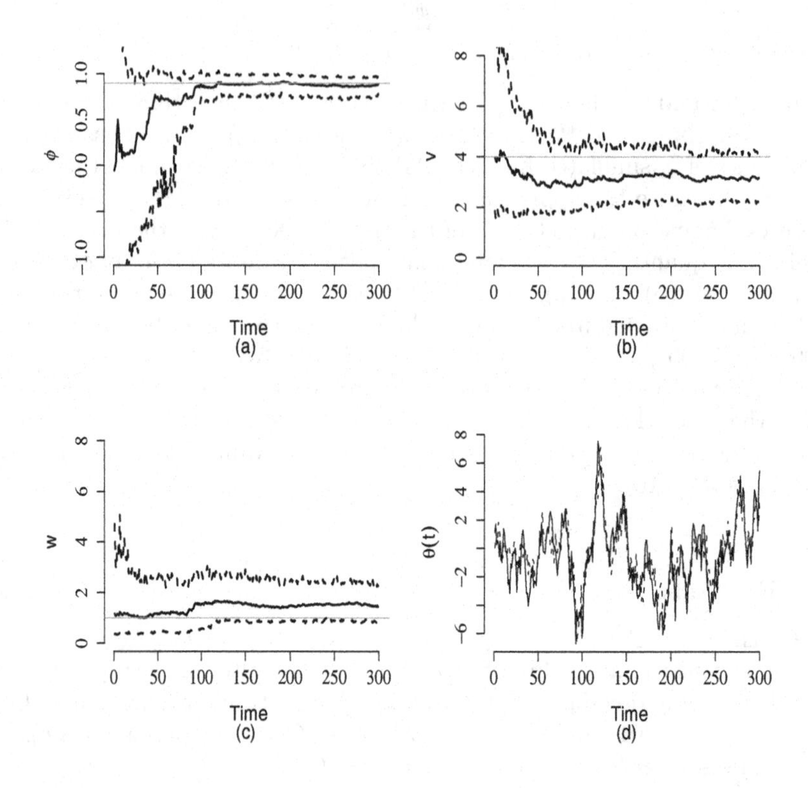

Figure 6.5 *Results of applying the PL algorithm of Carvalho, Johannes, Lopes, and Polson (2008) with* $M = 500$ *particles in the* $AR(1)$ *plus noise model. (a) Time trajectories of the mean (solid line) and quantiles (2.5% and 97.5%) of the posterior distribution of the AR coefficient* ϕ*. (b) Time trajectories of the mean (solid line) and quantiles (2.5% and 97.5%) of the posterior distribution of the observational variance* v*. (c) Time trajectories of the mean (solid line) and quantiles (2.5% and 97.5%) of the posterior distribution of* w*. (d) Time traces of the true state* θ_t *(solid line) and posterior mean of the particle approximation to the distribution of* θ_t*.*

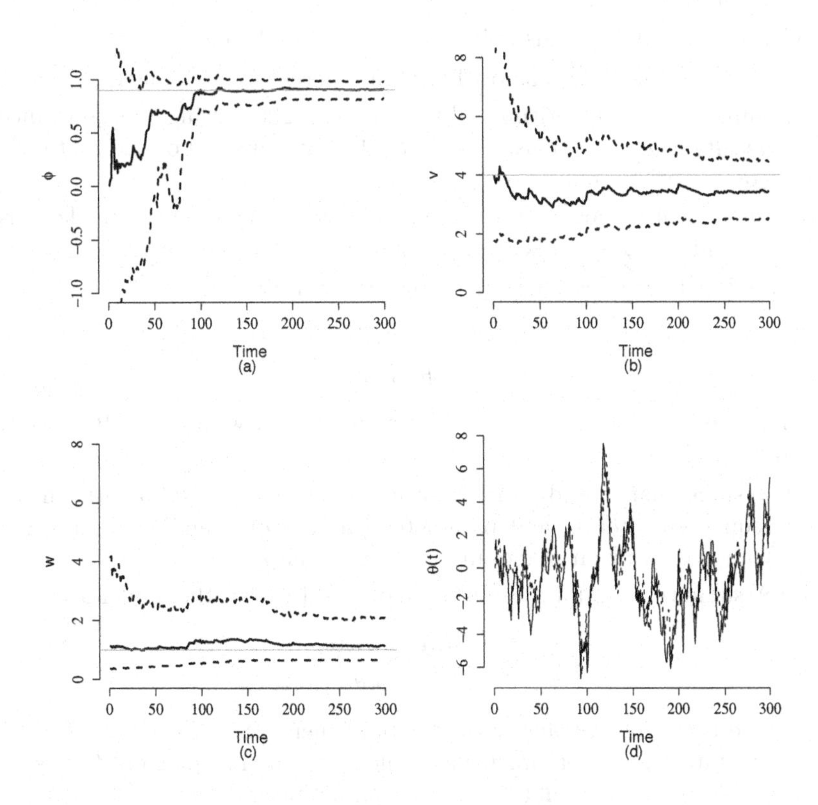

Figure 6.6 *Results of applying the PL algorithm of Carvalho, Johannes, Lopes, and Polson (2008) with $M = 5{,}000$ particles in the $AR(1)$ plus noise model. (a) Time trajectories of the mean (solid line) and quantiles (2.5% and 97.5%) of the posterior distribution of the AR coefficient ϕ. (b) Time trajectories of the mean (solid line) and quantiles (2.5% and 97.5%) of the posterior distribution of the observational variance v. (c) Time trajectories of the mean (solid line) and quantiles (2.5% and 97.5%) of the posterior distribution of w. (d) Time traces of the true state θ_t (solid line) and posterior mean of the particle approximation to the distribution of θ_t.*

(b) Assume that ν is known but v, w, and β are unknown. Propose, implement, and compare SMC approaches for filtering and parameter learning.

4. Consider the AR(1) plus noise model in Example 6.6.

(a) Run Liu and West's algorithm with different discount factors.

(b) Implement the algorithm of Liu and West (2001) using an importance density that conditions on $\mathbf{y}_t$, $p(\boldsymbol{\theta}_t|\boldsymbol{\theta}_{t-1}, \mathbf{y}_t)$, instead of one that does not condition on $\mathbf{y}_t$.

(c) Implement the algorithm of Liu and West (2001) using gamma kernels instead of Gaussian kernels to sample the variance parameters.

5. Consider the logistic DLM given by (see Storvik 2002)

$$\begin{aligned} y_t &\sim Bin(r, \text{logit}(\alpha + \beta\theta_t)), \\ \theta_t &\sim N(\phi\theta_{t-1}, w). \end{aligned}$$

Simulate $T = 300$ observations from this model with $\phi = 0.9$, $w = 1$, and $\alpha = \beta = 0.5$.

(a) Assume that α and β are known. Implement a SMC algorithm for sequential filtering and parameter learning of ϕ and w assuming a Gaussian prior on ϕ and an inverse-gamma prior on w.

(b) Assume that α and β are also unknown. Rewrite the model as

$$\begin{aligned} y_t &\sim Bin(r, \text{logit}(\theta_t^*)), \\ \theta_t^* &\sim N(\alpha + \phi(\theta_{t-1}^* - \alpha), w^*), \end{aligned}$$

where $\theta_t^* = \alpha + \beta\theta_t$ and $w^* = \beta w$. Implement a SMC algorithm for online filtering and parameter learning assuming independent Gaussian priors on ϕ and α and an inverse-gamma prior on w^*.

Chapter 7

Mixture models in time series

This chapter concerns mixture models in time series analysis. Examples include Markov switching models, mixtures of dynamic linear models (DLMs) also known as multi-process models, and examples of nonlinear and/or non-Gaussian models that can be approximated with normal mixtures of some form. Bayesian analysis using such models can employ computational strategies that exploit the conditionally normal, linear model structures implied by expansion of the parameter space to include the inherent latent "mixing" variables. These models and inference methods are developed and exemplified. Two detailed illustrations involve an analysis of electroencephalogram signals via mixtures of autoregressive processes, and an analysis of daily returns on exchange rates via stochastic volatility models.

7.1 Markov Switching Models

The use of discrete and continuous mixtures of distributions of tractable forms, such as mixtures of normals, is pervasive in time series analysis and forecasting, as in many other areas. Mixtures arise broadly as components of practical models, and mixture structure is often usefully exploited in model analysis and fitting. This chapter discusses a selection of mixture models that we have found useful in specific applied settings, pointing the reader to several key references. Extensive development of several mixture modeling approaches in time series and forecasting-of relevance more broadly in statistical analysis-are introduced and covered in several chapters of West and Harrison (1997), with references. The more recent book on finite mixtures and Markov switching models, Frühwirth-Schnatter (2006), is a key reference that deals with theory, inference, and application of finite mixture models for temporal and nontemporal data. In particular, Bayesian inference for Markov switching models and switching state-space models as well as related theory and examples are discussed in Chapters 10 to 13 of Frühwirth-Schnatter (2006) and further references therein.

Example 7.1 *Markov mixture of Gaussian distributions.* This model (see Frühwirth-Schnatter 2006, and references therein) assumes that

$$p(y_t|S_t = k) = N(m_k, C_k),$$

for $k = 1 : K$, where S_t is a hidden (unobserved) discrete Markov process with $K \times K$ transition probability matrix $\boldsymbol{\xi}$ whose (i, j)-th element is the transition probability of going from state i to state j, that is, $\xi_{i,j} = Pr(S_t = j|S_{t-1} = i)$.

Example 7.2 *Markov switching regression.* In a Markov switching regression model the regression parameters depend on a hidden state S_t, that is,

$$y_t = \mathbf{f}_t' \boldsymbol{\beta}_{S_t} + \epsilon_t, \quad \epsilon_t \sim N(0, v_{S_t}). \tag{7.1}$$

Again, S_t is assumed to be a hidden Markov process with transition probability matrix $\boldsymbol{\xi}$ and $\mathbf{f}_t$ is a p-dimensional vector of explanatory variables.

Example 7.3 *Markov switching autoregression.* This is a particular case of the regression model in (7.1) with $\mathbf{f}_t' = (y_{t-1}, \ldots, y_{t-p})$ and $\boldsymbol{\beta}_{S_t} = (\phi_1^{(S_t)}, \ldots, \phi_p^{(S_t)})'$. Then

$$y_t = \phi_1^{(S_t)} y_{t-1} + \ldots + \phi_p^{(S_t)} y_{t-p} + \epsilon_t, \tag{7.2}$$

where S_t is a hidden Markov process with K states and $\epsilon_t \sim N(0, v_{S_t})$.

As pointed out in Frühwirth-Schnatter (2006), the mixture autoregressive (MAR) models discussed in Juang and Rabiner (1985) and Wong and Li (2000) are special cases of (7.2) in which the process S_t is an i.i.d. process instead of a Markov process. Wong and Li (2001) extended MAR models models to consider mixture autoregressive conditional heteroscedastic models (MAR-ARCH). MAR-ARCH models describe the mean of the observed time series as an AR process and its conditional variance as an autoregressive conditional heteroscedastic process (for a definition of ARCH processes see Engle 1982).

Threshold autoregressive processes (or TAR; see Tong 1983, 1990) whose AR parameters switch according to the value of y_{t-d} for some integer d, are special cases of the Markov switching autoregressive models in (7.2). An example is given below.

Example 7.4 *Threshold autoregressive model of order one.* Consider the TAR(1) model

$$y_t = \phi^{(S_t)} y_{t-1} + \epsilon_t,$$

where $\epsilon_t \sim N(0, v_{S_t})$ and

$$S_t = \begin{cases} 1 & \text{if} \quad \theta + y_{t-d} > 0 \\ 2 & \text{if} \quad \theta + y_{t-d} \leq 0. \end{cases}$$

7.1.1 Parameter Estimation

Frühwirth-Schnatter (2006) discusses various algorithms for inference in Markov switching models, including the following: algorithms for estimation of the hidden states $\mathbf{S} = \{S_{1:T}\}$ given known transition parameters $\boldsymbol{\xi}$ and known state-specific parameters $\boldsymbol{\theta}$; algorithms for estimation of the transition parameters and the state-specific parameters $\boldsymbol{\xi}$ and $\boldsymbol{\theta}$, given the states $\mathbf{S}$; and algorithms for simultaneous Bayesian estimation of $\mathbf{S}, \boldsymbol{\xi}$, and $\boldsymbol{\theta}$ via Markov chain Monte Carlo (MCMC) algorithms.

Here we briefly outline the MCMC algorithm presented in Frühwirth-Schnatter (2006) for Bayesian estimation in Markov switching models. This algorithm is based on sampling $\boldsymbol{\xi}, \boldsymbol{\theta}$ conditional on the states $\mathbf{S}$, and then sampling $\mathbf{S}$ conditional on $\boldsymbol{\xi}$ and $\boldsymbol{\theta}$. More specifically, the algorithm starts at some initial state process $\mathbf{S}^{(0)}$ and then, Steps 1 to 3 below are repeated for iterations $i = 1 : I$ (until MCMC convergence is reached).

1. Sample $\boldsymbol{\xi}^{(i)}$ from $p(\boldsymbol{\xi}|\mathbf{S}^{(i-1)}, \mathcal{D}_T)$. Assuming that the rows of the transition probability matrix $\boldsymbol{\xi}$, denoted by $\boldsymbol{\xi}_{j\cdot}$, are independent a priori, each with a Dirichlet distribution
$$\boldsymbol{\xi}_{j\cdot} \sim Dir(\boldsymbol{\alpha}_j),$$
with $\boldsymbol{\alpha}_j = (\alpha_{j,1}, \ldots, \alpha_{j,K})'$, it is possible to draw each $\boldsymbol{\xi}_{j\cdot}$ for $j = 1 : K$ using a Gibbs step. That is, for $j = 1 : K$, $\boldsymbol{\xi}_{j\cdot}$ is sampled from a Dirichlet distribution
$$\boldsymbol{\xi}_{j\cdot} \sim Dir(\alpha_{j,1} + N_{j,1}(\mathbf{S}), \ldots, \alpha_{j,K} + N_{j,K}(\mathbf{S})),$$
with $N_{j,k}(\mathbf{S})$ the number of transitions from S_j to S_k in $\mathbf{S}$.
2. Sample $\boldsymbol{\theta}^{(i)}$ from $p(\boldsymbol{\theta}_{1:K}|\mathbf{S}^{(i-1)}, \mathcal{D}_T)$. The form of this conditional posterior depends on which distributions are chosen for each component in the mixture, requiring custom made MCMC steps that do not have a general form.
3. Sample $\mathbf{S}^{(i)}$ from $p(\mathbf{S}|\boldsymbol{\xi}^{(i)}, \boldsymbol{\theta}^{(i)}, \mathcal{D}_T)$. This step is done using forward filtering backward sampling (FFBS, see Chapter 4, Section 4.5) and is implemented as follows:
   (a) for $t = 1 : T$, compute $Pr(S_t = k|\boldsymbol{\xi}^{(i)}, \boldsymbol{\theta}^{(i)}, \mathcal{D}_t)$, with
$$Pr(S_t = k|\boldsymbol{\xi}^{(i)}, \boldsymbol{\theta}^{(i)}, \mathcal{D}_t) \propto p(y_t|S_t = k, \mathcal{D}_{t-1}, \boldsymbol{\xi}^{(i)}, \boldsymbol{\theta}^{(i)}) \times Pr(S_t = k|\mathcal{D}_{t-1}, \boldsymbol{\xi}^{(i)}, \boldsymbol{\theta}^{(i)}),$$

and sample $S_T^{(i)}$ from $Pr(S_T = k|\boldsymbol{\xi}^{(i)}, \boldsymbol{\theta}^{(i)}, \mathcal{D}_T)$;

(b) for $t = (T-1) : 1$, sample the hidden state at time t, $S_t^{(i)}$, from $Pr(S_t = j|S_{t+1}^{(i)} = k, \boldsymbol{\xi}^{(i)}, \boldsymbol{\theta}^{(i)}, \mathcal{D}_T)$, where

$$Pr(S_t = j|S_{t+1} = k, \boldsymbol{\xi}, \boldsymbol{\theta}, \mathcal{D}_T) \propto Pr(S_t = j|S_{t+1} = k, \boldsymbol{\xi}, \boldsymbol{\theta}, \mathcal{D}_t)$$

and

$$\begin{aligned} Pr(S_t = j|S_{t+1} = k, \boldsymbol{\xi}, \boldsymbol{\theta}, \mathcal{D}_t) \;\; &\propto \;\; Pr(S_{t+1} = k|S_t = j, \boldsymbol{\xi}, \boldsymbol{\theta}, \mathcal{D}_t) \\ & \quad\; \times Pr(S_t = j|\boldsymbol{\xi}, \boldsymbol{\theta}, \mathcal{D}_t). \end{aligned}$$

Finally, set $\mathbf{S}^{(i)} = \{S_{1:T}^{(i)}\}$.

Frühwirth-Schnatter (2006) discusses variations in the algorithm described above to obtain efficient samplers that explore the full Markov mixture posterior distribution. For instance, a modified algorithm based on the random permutation MCMC sampling scheme of Frühwirth-Schnatter (2001) is outlined.

7.1.2 Other Models

Huerta, Jiang, and Tanner (2003) consider a class of mixture models for time series analysis based on the hierarchical mixtures-of-experts (HME) approach of Jordan and Jacobs (1994). Specifically, if $p(y_t|\mathbf{X}, \boldsymbol{\theta}, \mathcal{D}_{t-1})$ is the probability density function (pdf) of y_t given all the past information $\mathcal{D}_{t-1}$ and the external information $\mathbf{X}$ (e.g., covariates), then it is assumed that

$$p(y_t|\mathbf{X}, \boldsymbol{\theta}, \mathcal{D}_{t-1}) = \sum_{o=1}^{O}\sum_{m=1}^{M} g_t(o, m|\mathcal{D}_{t-1}, \mathbf{X}, \boldsymbol{\gamma}) p(y_t|\mathcal{D}_{t-1}, \boldsymbol{\eta}, o, m), \quad (7.3)$$

where $\boldsymbol{\theta} = (\boldsymbol{\gamma}', \boldsymbol{\eta}')'$ with $\boldsymbol{\eta}$ the parameters of the sampling distribution and $\boldsymbol{\gamma}$ the parameters that define $g_t(o, m|\cdot)$. In this representation an "expert," indexed by (o, m), corresponds to an overlay index o, with $o = 1 : O$, and a model-type index m, with $m = 1 : M$. The same type of model m may assume different parameter values at each possible overlay. The functions $g_t(o, m|\cdot)$ in (7.3) are the mixture weights, often referred to as "gating functions." They may depend on exogenous information or exclusively on time. It is also assumed that these mixture weights have a particular parametric form given by

$$g_t(o, m|\mathbf{X}, \boldsymbol{\gamma}, \mathcal{D}_{t-1}) \;\; = \;\; \left[\frac{e^{v_o + \mathbf{u}_o'\mathbf{X}_t}}{\sum_{s=1}^{O} e^{v_s + \mathbf{u}_s'\mathbf{X}_s}}\right] \times \left[\frac{e^{v_{m|o} + \mathbf{u}_{m|o}'\mathbf{X}_t}}{\sum_{l=1}^{M} e^{v_{l|o} + \mathbf{u}_{l|o}'\mathbf{X}_t}}\right].$$

The parameters $\mathbf{v}$ and $\mathbf{u}$ above are the components of $\boldsymbol{\gamma}$. If the interest lies on assessing the weight of various models across different time periods, and no other covariates are available, $\mathbf{x}_t$ is a scalar taken as $x_t = t/T$. In such case $\boldsymbol{\gamma}$ includes the components $v_{1:(O-1)}$, $u_{1:(O-1)}$, $v_{1:(M-1)|o}$, and $u_{1:(M-1)|o}$ for $o = 1 : O$. For identifiability $v_O = u_O = v_{M|o} = u_{M|o} = 0$ for all $o = 1 : O$.

These HME models allow us to consider time-varying weights for the different mixture components (experts) via the gating functions. Huerta, Jiang, and Tanner (2001) proposed an expectation-maximization (EM) algorithm for parameter estimation. Huerta, Jiang, and Tanner (2003) considered a Bayesian modeling approach and outlined a MCMC algorithm for posterior inference. These authors, following Kim, Shephard, and Chib (1998) and Elerian, Chib, and Shephard (2001), also proposed the use of one-step-ahead predictive distributions for model checking and model diagnostics.

Example 7.5 *Difference stationary vs. trend stationary model.* Huerta, Jiang, and Tanner (2003) considered a HME model to describe the US industrial production index as reported by the Federal Reserve Statistical Release G.17 (see McCulloch and Tsay 1994). Such time series corresponds to seasonally adjusted monthly data from January 1947 to December 1993. One of the main objectives in the analysis of these data is determining if the observed trend has a stochastic or a deterministic behavior. In order to assess this, Huerta, Jiang, and Tanner (2003) fit a HME model with two overlays and two models, namely, a difference stationary model given by

$$y_t = \phi_0 + y_{t-1} + \phi_{1,1}(y_{t-1} - y_{t-2}) + \phi_{1,2}(y_{t-2} - y_{t-3}) + \epsilon_{1,t},$$

and a trend stationary model given by

$$y_t = \beta_0 + \beta_1 t/T + \phi_{2,1} y_{t-1} + \phi_{2,2} y_{t-2} + \epsilon_{2,t},$$

with $\epsilon_{m,t} \sim N(0, v_m)$ for $m = 1 : 2$. Since each model m has two overlays, we have that the parameters for the difference stationary model are $(\phi_0^{(o)}, \phi_{1,1}^{(o)}, \phi_{1,2}^{(o)})'$ for $o = 1 : 2$ and v_1, while the parameters for the trend stationary model are $(\beta_0^{(o)}, \beta_1^{(o)}, \phi_{2,1}^{(o)}, \phi_{2,2}^{(o)})'$ for $o = 1 : 2$ and v_2. This implies that

$$p(y_t|\mathcal{D}_{t-1}, o, m = 1) = \frac{1}{\sqrt{2\pi v_1}} e^{-\left[y_t - \phi_0^{(o)} - y_{t-1}(1 + \phi_{1,1}^{(o)} - \phi_{1,2}^{(o)}) - y_{t-2}(\phi_{1,2}^{(o)} - \phi_{1,1}^{(o)})\right]^2 / 2v_1},$$

and

$$p(y_t|\mathcal{D}_{t-1}, o, m = 2) = \frac{1}{\sqrt{2\pi v_2}} e^{-\left[y_t - \beta_0^{(o)} - \beta_1^{(o)} t/T - \phi_{2,1}^{(o)} y_{t-1} - \phi_{2,2}^{(o)} y_{t-2}\right]^2 / 2v_2}.$$

No covariates were considered in the analysis and so

$$g_t(o, m|\mathcal{D}_{t-1}, \boldsymbol{\gamma}) = \left[\frac{e^{v_o + u_o t/T}}{\sum_{s=1}^{2} e^{v_s + u_s t/T}}\right] \times \left[\frac{e^{v_{m|o} + u_{m|o} t/T}}{\sum_{l=1}^{2} e^{v_{l|o} + u_{l|o} t/T}}\right],$$

and $\boldsymbol{\gamma} = (v_1, u_1, v_{1|1}, u_{1|1}, v_{1|2}, u_{1|2})'$.

Huerta, Jiang, and Tanner (2003) use a MCMC algorithm to achieve posterior inference in the above model assuming independent Gaussian priors for $\boldsymbol{\phi}_1^{(o)} = (\phi_0^{(o)}, \phi_{1,1}^{(o)}, \phi_{1,2}^{(o)})'$ and $\boldsymbol{\phi}_2^{(o)} = (\beta_0^{(o)}, \beta_1^{(o)}, \phi_{2,1}^{(o)}, \phi_{2,2}^{(o)})'$, namely $\boldsymbol{\phi}_m^{(o)} \sim N(\mathbf{m}_m, \mathbf{C}_m)$, for $m = 1:2$, and $o = 1:2$. In addition, inverse-gamma priors are assumed on v_m, and uniform priors are assumed on the components of $\boldsymbol{\gamma}$. In their analysis of the US Industrial Production Index Huerta, Jiang, and Tanner (2003) found that, for approximately the first half of the observations, the difference stationary model was favored, while for more recent data points the trend stationary model was favored.

More recently, Villagrán and Huerta (2006) showed how including more than one covariate in the gating functions may lead to substantial changes in the estimates of some of the HME model parameters. In particular, these authors considered ME models ($O = 1$) for stochastic volatility in a time series of returns of the Mexican stock market where time and the Dow Jones index were included as covariates. HME models were also considered to describe multivariate time series in Prado, Molina, and Huerta (2006). In this approach the experts were assumed to follow a vector autoregressive (VAR) structure. The order of each vector autoregressive component in the mixture was assumed to depend on the model m, while the VAR coefficients and the variance-covariance matrices for each expert were assumed to depend on both the model m and the overlay o. HME-VAR models are briefly described in Chapter 9.

7.2 Multiprocess Models

West and Harrison (1997) dedicate a full chapter (Chapter 12) to describe a class of models consisting of mixtures of DLMs. Such models, referred to as *multiprocess* models, were originally introduced into the statistics literature by Harrison and Stevens (1971, 1976). These authors distinguish two classes of multiprocess models, namely, class I and class II. We summarize the main features of these two classes of DLM mixtures and revisit the methods for posterior inference detailed in West and Harrison (1997). Later in this chapter we present an application of mixture models to the analysis of EEG series recorded during a cognitive fatigue experiment.

7.2.1 Definitions and Examples

Let $\boldsymbol{\alpha}$ be a set of uncertain quantities that define a particular DLM at time t, denoted as $\mathcal{M}_t = \mathcal{M}_t(\boldsymbol{\alpha})$, for $\boldsymbol{\alpha} \in \mathcal{A}$, with $\mathcal{A}$ the possible set of values for $\boldsymbol{\alpha}$. The set $\mathcal{A}$ can be discrete or continuous, finite, countable, or uncountable. In addition, the initial prior in the model may depend on $\boldsymbol{\alpha}$. Following West and Harrison (1997), two classes of models can be considered:

1. *Class I multiprocesses.* This class of multiprocesses assumes that a single model $\mathcal{M}_t(\boldsymbol{\alpha})$ holds for all t for some specific value of $\boldsymbol{\alpha} \in \mathcal{A}$. Therefore, a single DLM is appropriate for all t, but there is uncertainty about which value of $\boldsymbol{\alpha}$ defines such model.
2. *Class II multiprocesses.* In this case it is assumed that at time t, $\boldsymbol{\alpha}$ takes a value in $\mathcal{A}$, so that $\mathcal{M}_t(\boldsymbol{\alpha}_t)$ holds at time t. This implies that the sequence $\boldsymbol{\alpha}_{1:T}$ defines the DLMs for each $t = 1 : T$ and, as opposed to the class I models, no single DLM is appropriate at all times.

Example 7.6 *Mixtures of first order DLMs with different discount factors.* Let $\boldsymbol{\alpha} = \delta$ and let $\mathcal{A} = \{\delta_1, \ldots, \delta_K\}$ be a discrete set of discount factors, with $\delta_k \in (0, 1]$, for all k. Define $\mathcal{M}_t(\delta)$ as

$$\begin{aligned} y_t &= \theta_t + \nu_t, \;\; \nu_t \sim N(0, v), \\ \theta_t &= \theta_{t-1} + w_t, \;\; w_t \sim N(0, w(\delta)), \end{aligned}$$

indicating that the evolution variance $w(\delta)$ is fully specified by the discount factor $\delta \in (0, 1]$. If a class I multiprocess is used, it is assumed that a single discount factor $\delta^* \in \mathcal{A}$ is appropriate for all t. A more realistic scenario in practical settings is that described by a class II multiprocess in which the model at time t is represented by $\mathcal{M}_t(\delta_t)$, with $\delta_t \in \mathcal{A}$.

Example 7.7 *Mixtures of autoregressions.* Let $\mathcal{M}_t(\phi^{(k)})$ be defined as

$$y_t = \phi^{(k)} y_{t-1} + \nu_t, \;\; \nu_t \sim N(0, v),$$

where $k = 1 : 2$. Furthermore, assume that the parameters of the two models have different priors, e.g., $\phi^{(1)} \sim U(0, 1)$, and $\phi^{(2)} \sim U(-1, 0)$, and that the sequence $\boldsymbol{\alpha}_{1:T}$ defines the models at each time t, with $\boldsymbol{\alpha}_t = \phi^{(k)}$, for $k = 1$ or $k = 2$. Then, for $\boldsymbol{\alpha}_t = \phi^{(k)}$, the DLM is $\{\phi^{(k)}, 0, v, 0\}$, which is in fact a static autoregression with coefficient $\phi^{(k)}$.

7.2.2 Posterior Inference

7.2.2.1 Posterior inference in class I models

This class of multiprocess models assumes that a single DLM, $\mathcal{M}_t(\boldsymbol{\alpha})$, holds for all t, and so inference is summarized as follows.

1. Posterior $p(\boldsymbol{\alpha}|\mathcal{D}_t)$. Starting with a prior density, $p(\boldsymbol{\alpha}|\mathcal{D}_0)$, $p(\boldsymbol{\alpha}|\mathcal{D}_t)$ is sequentially updated via
$$p(\boldsymbol{\alpha}|\mathcal{D}_t) \propto p(\boldsymbol{\alpha}|\mathcal{D}_{t-1})p(y_t|\boldsymbol{\alpha}, \mathcal{D}_{t-1}). \tag{7.4}$$
2. Posterior $p(\boldsymbol{\theta}_t|\mathcal{D}_t)$. This is given by
$$p(\boldsymbol{\theta}_t|\mathcal{D}_t) = \int_{\mathcal{A}} p(\boldsymbol{\theta}_t|\boldsymbol{\alpha}, \mathcal{D}_t)p(\boldsymbol{\alpha}|\mathcal{D}_t)d\boldsymbol{\alpha}. \tag{7.5}$$
3. One-step-ahead forecast $p(y_{t+1}|\mathcal{D}_t)$. This density is obtained via
$$p(y_{t+1}|\mathcal{D}_t) = \int_{\mathcal{A}} p(y_{t+1}|\boldsymbol{\alpha}, \mathcal{D}_t)p(\boldsymbol{\alpha}|\mathcal{D}_t)d\boldsymbol{\alpha}. \tag{7.6}$$

West and Harrison (1997) discuss in detail posterior inference when $\mathcal{A}$ is a discrete set, $\mathcal{A} = \{\boldsymbol{\alpha}_1, \ldots, \boldsymbol{\alpha}_K\}$, for some integer $K \geq 1$. In such case, the density (7.4) is a probability mass function with $p_t(k) = Pr(\boldsymbol{\alpha} = \boldsymbol{\alpha}_k|\mathcal{D}_t)$, for $k = 1 : K$. Then, denoting $l_t(k) = p(y_t|\boldsymbol{\alpha}_k, \mathcal{D}_{t-1})$, (7.4) is updated via $p_t(k) = c_t p_{t-1}(k)l_t(k)$, with $c_t = (\sum_{k=1}^{K} p_{t-1}(k)l_t(k))^{-1}$. Similarly, (7.5) and (7.6) are given by
$$p(\boldsymbol{\theta}_t|\mathcal{D}_t) = \sum_{k=1}^{K} p(\boldsymbol{\theta}_t|\boldsymbol{\alpha}_k, \mathcal{D}_t)p_t(k)$$
and
$$p(y_{t+1}|\mathcal{D}_t) = \sum_{k=1}^{K} p(y_{t+1}|\boldsymbol{\alpha}_k, \mathcal{D}_t)p_t(k).$$
When each $\mathcal{M}_t(\boldsymbol{\alpha}_k)$ has a normal DLM (NDLM) form, the densities above are discrete mixtures of normal or Student-t distributions. West and Harrison (1997) also discuss and illustrate the use of class I multiprocess models in automatic model identification (see Section 12.2 of West and Harrison 1997 for details and examples). It is also possible to consider $\boldsymbol{\alpha}$ as an unknown parameter and use Gibbs sampling to achieve posterior inference within this class of models.

7.2.2.2 Posterior inference in class II models

Assuming that $\mathcal{A}$ is a discrete set, $\mathcal{A} = \{\boldsymbol{\alpha}_1, \ldots, \boldsymbol{\alpha}_K\}$, a multiprocess model of class II, assumes that no single DLM is appropriate to describe the behavior of the observed time series. Instead, a collection of models indexed by $\boldsymbol{\alpha}_{1:T}$ are used to describe such series. West and Harrison (1997) discuss posterior inference within this class of models in settings where NDLMs are used as model components in the mixture, and in cases where the prior distribution does not depend on $\boldsymbol{\alpha}$. We now enumerate and briefly describe the steps needed to perform posterior inference in such settings. Later in this chapter we discuss and illustrate extensions to non-Gaussian mixture models and applications. Posterior inference in such cases can be done using approximations or simulation-based algorithms such as particle filters. Posterior inference via approximations in general multiprocesses follows the same general steps described below for Gaussian multiprocesses.

In order to simplify the notation, we follow West and Harrison (1997), setting $\mathcal{A} = \{1, \dots, K\}$ and referring to model $\mathcal{M}_t(k)$ as model k at time t. Let $\pi_t(k) = Pr(\mathcal{M}_t(k)|\mathcal{D}_{t-1})$ be the prior probability of selecting model $\mathcal{M}_t(k)$ before observing y_t. Assume that the probability of choosing a given model at time t depends on which model was chosen at time $t-1$, but not on which models were selected prior to $t-1$. Denote $\pi_t(k|i) = Pr(\mathcal{M}_t(k)|\mathcal{M}_{t-1}(i), \mathcal{D}_{t-1})$, the first order Markov transition probability at time t. Then, the prior probability of model $\mathcal{M}_t(k)$ at time t is given by

$$\pi_t(k) = \sum_{i=1}^{K} \pi_t(k|i) p_{t-1}(i),$$

where $p_{t-1}(i) = Pr(\mathcal{M}_{t-1}(i)|\mathcal{D}_{t-1})$ is the posterior probability of model $\mathcal{M}_{t-1}(i)$ at time $t-1$. Higher order Markov transition probabilities can also be considered. In many applications it is also assumed that the prior selection probabilities and the transition probabilities are fixed in time, i.e., $\pi_t(k) = \pi(k)$ and $\pi_t(k|i) = \pi(k|i)$ for all t. In addition, for each integer $0 \leq h < t$, define the posterior probability of the path of models indexed by $k_{t-h}, \dots, k_t$ as

$$p_t(k_t, k_{t-1}, \dots, k_{t-h}) = Pr(\mathcal{M}_t(k_t), \mathcal{M}_{t-1}(k_{t-1}), \dots, \mathcal{M}_{t-h}(k_{t-h})|\mathcal{D}_t).$$

At time t we are interested in obtaining the posterior density $p(\boldsymbol{\theta}_t|\mathcal{D}_t)$ which, conditioning on the K models that can be selected at time t, can be written as a mixture of K components

$$p(\boldsymbol{\theta}_t|\mathcal{D}_t) = \sum_{k_t=1}^{K} p(\boldsymbol{\theta}_t|\mathcal{M}_t(k_t), \mathcal{D}_t) p_t(k_t). \tag{7.7}$$

However, in order to proceed with standard DLM analyses it is necessary to condition on the models selected at times $t-1, t-2, \dots, 1$, since only the posteriors $p(\boldsymbol{\theta}_t|\mathcal{M}_t(k_t), \mathcal{M}_{t-1}(k_{t-1}), \dots, \mathcal{M}_1(k_1), \mathcal{D}_t)$ have a standard DLM form. Then, at time t, (7.7) is written as a mixture with K^t components. That is,

$$\begin{aligned} p(\boldsymbol{\theta}_t|\mathcal{D}_t) &= \sum_{k_t=1}^{K} \cdots \sum_{k_1=1}^{K} p(\boldsymbol{\theta}_t|\mathcal{M}_t(k_t), \dots, \mathcal{M}_1(k_1), \mathcal{D}_t) \\ &\quad \times p_t(k_t, \dots, k_1). \end{aligned} \tag{7.8}$$

In practice, obtaining these posterior densities at each time t is computationally demanding, particularly for t large. West and Harrison (1997) propose reducing the number of components in these mixtures by approximations. More specifically, it can be assumed that the conditional posterior will depend only on which models were chosen in the last h-steps back in time, where $h \geq 1$ but not too large (typically $h = 1$ or $h = 2$ are adequate in practical settings). In other words, it is assumed that the dependence on early models becomes negligible as time passes. That is

$$\begin{aligned} & p(\boldsymbol{\theta}_t | \mathcal{M}_t(k_t), \mathcal{M}_{t-1}(k_{t-1}), \ldots, \mathcal{M}_1(k_1), \mathcal{D}_t) \approx \\ & p(\boldsymbol{\theta}_t | \mathcal{M}_t(k_t), \mathcal{M}_{t-1}(k_{t-1}), \ldots, \mathcal{M}_h(k_{t-h}), \mathcal{D}_t), \end{aligned} \tag{7.9}$$

and so the mixture with K^t components in (7.8) is approximated by the following mixture with at most K^{h+1} components:

$$\sum_{k_t=1}^{K} \sum_{k_{t-1}=1}^{K} \cdots \sum_{k_{t-h}=1}^{K} p(\boldsymbol{\theta}_t | \mathcal{M}_t(k_t), \mathcal{M}_{t-1}(k_{t-1}), \ldots, \mathcal{M}_h(k_{t-h}), \mathcal{D}_t) \times p_t(k_t, k_{t-1}, \ldots, k_{t-h}). \tag{7.10}$$

Further approximations to (7.10) can be considered to reduce the number of components, such as ignoring those components with very small probabilities, combining components that are roughly equal into a single component, and replacing a collection of components by a single distribution. West and Harrison (1997) discuss techniques for collapsing components using a method based on the Kullback-Leibler divergence. The idea behind this method is to approximate any given mixture with a certain number of components by a single "optimal" distribution that somehow summarizes the contribution of all the components in the original mixture. The Kullback-Leibler divergence is simply used to measure the distance between a particular candidate distribution and the mixture, and the "optimal" distribution, chosen from a collection of candidate approximating distributions, is the one that minimizes the Kullback-Leibler divergence. For instance, it can be shown that if the original distribution is a mixture of K Gaussian distributions, each with corresponding weight $p(k)$, mean $\mathbf{m}(k)$, and variance $\mathbf{C}(k)$, the optimal approximating distribution that minimizes the Kullback-Leibler divergence is a normal distribution with mean $\mathbf{m}$ and variance-covariance matrix $\mathbf{C}$ given by

$$\mathbf{m} = \sum_{k=1}^{K} \mathbf{m}(k) p(k) \quad \text{and} \quad \mathbf{C} = \sum_{k=1}^{K} [\mathbf{C}(k) + (\mathbf{m} - \mathbf{m}(k))(\mathbf{m} - \mathbf{m}(k))'] p(k).$$

West and Harrison (1997) discuss posterior estimation in multiprocesses with $K = 4$ possible models at each time t and with $h = 1$ in (7.9). In this setting, each $\mathcal{M}_t(k_t)$ for $k_t = 1:4$ was a DLM defined by the quadruple $\{\mathbf{F}, \mathbf{G}, v_t v(k_t), \mathbf{W}_t(k_t)\}$, with $v(k_t)$ known for all k_t. Two scenarios are presented: one in which v_t is assumed known for all t and a prior $(\boldsymbol{\theta}_0|\mathcal{D}_0) \sim N(\mathbf{m}_0, \mathbf{C}_0)$ is considered, and another scenario where $v_t = v = \phi^{-1}$ for all t, with ϕ unknown, and where priors of the form $(\boldsymbol{\theta}_0|v, \mathcal{D}_0) \sim N(\mathbf{m}_0, \mathbf{C}_0 v/s_0)$ and $(\phi|\mathcal{D}_0) \sim G[n_0/2, d_0/2]$ are considered. Below we summarize the steps for posterior estimation in this last case.

- Posterior densities at time $t-1$. Given $\mathcal{M}_{t-1}(k_{t-1})$, we have that

$$\begin{aligned} (\boldsymbol{\theta}_{t-1}|\mathcal{M}_{t-1}(k_{t-1}), \mathcal{D}_{t-1}) &\sim T_{n_{t-1}}(\mathbf{m}_{t-1}(k_{t-1}), \mathbf{C}_{t-1}(k_{t-1})), \\ (\phi|\mathcal{M}_{t-1}(k_{t-1}), \mathcal{D}_{t-1}) &\sim G(n_{t-1}/2, d_{t-1}(k_{t-1})/2), \end{aligned}$$

  for some values $\mathbf{m}_{t-1}(k_{t-1})$, $\mathbf{C}_{t-1}(k_{t-1})$, n_{t-1} and $d_{t-1}(k_{t-1})$.
- Prior densities at time t. Conditional on $\mathcal{M}_t(k_t)$ and $\mathcal{M}_{t-1}(k_{t-1})$, we have that

$$\begin{aligned} (\boldsymbol{\theta}_t|\mathcal{M}_t(k_t), \mathcal{M}_{t-1}(k_{t-1}), \mathcal{D}_{t-1}) &\sim T_{n_{t-1}}(\mathbf{a}_t(k_{t-1}), \mathbf{R}_t(k_t, k_{t-1})), \\ (\phi|\mathcal{M}_t(k_t), \mathcal{M}_{t-1}(k_{t-1}), \mathcal{D}_{t-1}) &\sim G(n_{t-1}/2, d_{t-1}(k_{t-1})/2), \end{aligned}$$

  where $\mathbf{a}_t(k_{t-1}) = \mathbf{G}\mathbf{m}_{t-1}(k_{t-1})$ and $\mathbf{R}_t(k_t, k_{t-1}) = \mathbf{G}\mathbf{C}_{t-1}(k_{t-1})\mathbf{G}' + \mathbf{W}_t(k_t)$.
- One-step-ahead forecast. Conditioning on $\mathcal{M}_t(k_t)$ and $\mathcal{M}_{t-1}(k_{t-1})$ we have

$$(y_t|\mathcal{M}_t(k_t), \mathcal{M}_{t-1}(k_{t-1}), \mathcal{D}_{t-1}) \sim T_{n_{t-1}}(f_t(k_{t-1}), q_t(k_t, k_{t-1})),$$

  where $f_t(k_{t-1}) = \mathbf{F}'\mathbf{a}_t(k_{t-1})$, and

$$q_t(k_t, k_{t-1}) = \mathbf{F}'\mathbf{R}_t(k_t, k_{t-1})\mathbf{F} + s_{t-1}(k_{t-1})v(k_t).$$

  Then, the unconditional density of $(y_t|\mathcal{D}_{t-1})$ is a mixture of Student-t components

$$\begin{aligned} p(y_t|\mathcal{D}_{t-1}) &= \sum_{k_t=1}^{K} \sum_{k_{t-1}=1}^{K} [p(y_t|\mathcal{M}_t(k_t), \mathcal{M}_{t-1}(k_{t-1}), \mathcal{D}_{t-1}) \\ &\quad \times \pi(k_t) p_{t-1}(k_{t-1})]. \end{aligned}$$

- Posterior densities and posterior model probabilities at time t. The posterior densities of $\boldsymbol{\theta}_t$ and ϕ given $\mathcal{M}_t(k_t)$ and $\mathcal{M}_{t-1}(k_{t-1})$ are

$$\begin{aligned} (\boldsymbol{\theta}_t|\mathcal{M}_t(k_t), \mathcal{M}_{t-1}(k_{t-1}), \mathcal{D}_t) &\sim T_{n_t}(\mathbf{m}_t(k_t, k_{t-1}), \mathbf{C}_t(k_t, k_{t-1})), \\ (\phi|\mathcal{M}_t(k_t), \mathcal{M}_{t-1}(k_{t-1}), \mathcal{D}_t) &\sim G(n_t/2, d_t(k_t, k_{t-1})/2), \end{aligned}$$

where

$$\begin{aligned}
\mathbf{m}_t(k_t, k_{t-1}) &= \mathbf{a}_t(k_{t-1}) + \mathbf{A}_t(k_t, k_{t-1})e_t(k_{t-1}), \\
\mathbf{C}_t(k_t, k_{t-1}) &= [s_t(k_t, k_{t-1})/s_{t-1}(k_{t-1})] \times \\
&\quad [\mathbf{R}_t(k_t, k_{t-1}) - \mathbf{A}_t(k_t, k_{t-1})\mathbf{A}_t'(k_t, k_{t-1})q_t(k_t, k_{t-1})], \\
e_t(k_{t-1}) &= y_t - f_t(k_{t-1}), \\
\mathbf{A}_t(k_t, k_{t-1}) &= \mathbf{R}_t(k_t, k_{t-1})\mathbf{F}/q_t(k_t, k_{t-1}), \\
d_t(k_t, k_{t-1}) &= d_{t-1}(k_{t-1}) + s_{t-1}(k_{t-1})e_t(k_{t-1})^2/q_t(k_t, k_{t-1}).
\end{aligned}$$

and $n_t = n_{t-1} + 1$. In addition, the posterior model probabilities are given by

$$p_t(k_t, k_{t-1}) \propto \frac{\pi(k_t)p_{t-1}(k_{t-1})}{q_t(k_t, k_{t-1})^{1/2}[n_{t-1} + e_t(k_{t-1})^2/q_t(k_t, k_{t-1})]^{n_t/2}}.$$

The unconditional posterior density for $\boldsymbol{\theta}_t$ is then based on the following mixture with K^2 components:

$$p(\boldsymbol{\theta}_t|\mathcal{D}_t) = \sum_{k_t=1}^{K} \sum_{k_{t-1}=1}^{K} p(\boldsymbol{\theta}_t|\mathcal{M}_t(k_t), \mathcal{M}_{t-1}(k_{t-1}), \mathcal{D}_t)p_t(k_t, k_{t-1}). \quad (7.11)$$

Similarly, the distribution of $(\phi|\mathcal{D}_t)$ is a mixture of gamma distributions. Prior to evolving to time $t+1$, these mixtures are collapsed over all possible models at time $t-1$. As mentioned before, in order to find optimal approximating distributions to mixtures, the Kullback-Leibler divergence is used as a distance measure (for details see Section 12.3 of West and Harrison 1997). More specifically, for each k_t, the mixture posteriors $p(\boldsymbol{\theta}_t|\phi, \mathcal{M}_t(k_t), \mathcal{D}_t)$ and $p(\phi|\mathcal{M}_t(k_t), \mathcal{D}_t)$ are approximated by single Gaussian/gamma posteriors. That is,

$$\begin{aligned}
(\boldsymbol{\theta}_t|\mathcal{M}_t(k_t), \mathcal{D}_t) &\approx T_{n_t}(\mathbf{m}_t(k_t), \mathbf{C}_t(k_t)), && (7.12) \\
(\phi|\mathcal{M}_t(k_t), \mathcal{D}_t) &\approx G(n_t/2, d_t(k_t)/2), && (7.13)
\end{aligned}$$

where $d_t(k_t) = n_t s_t(k_t)$, $\mathbf{m}_t(k_t) = \sum_{k_{t-1}=1}^{K} \mathbf{m}_t(k_t, k_{t-1})p_t^*(k_t, k_{t-1})$, and

$$\begin{aligned}
\mathbf{C}_t(k_t) &= \sum_{k_{t-1}=1}^{K} [\mathbf{C}_t(k_t, k_{t-1}) + (\mathbf{m}_t(k_t) - \mathbf{m}_t(k_t, k_{t-1})) \\
&\quad \times(\mathbf{m}_t(k_t) - \mathbf{m}_t(k_t, k_{t-1}))']\, p_t^*(k_t, k_{t-1}),
\end{aligned}$$

with

$$\begin{aligned}
s_t(k_t)^{-1} &= \sum_{k_{t-1}=1}^{K} s_t(k_t, k_{t-1})^{-1}p_t(k_t, k_{t-1})/p_t(k_t), \\
p_t^*(k_t, k_{t-1}) &= s_t(k_t)s_t(k_t, k_{t-1})^{-1}p_t(k_t, k_{t-1})/p_t(k_t).
\end{aligned}$$

The distributions in (7.12) and (7.13) approximate the components in the mixture posteriors $p(\boldsymbol{\theta}_t|\phi, \mathcal{M}_t(k_t), \mathcal{D}_t)$ and $p(\phi|\mathcal{M}_t(k_t), \mathcal{D}_t)$, respectively, and so the mixture with K^2 components in (7.11) has been collapsed into K normal/gamma components. In addition, we can write

- $p_t(k_t) = Pr(\mathcal{M}_t(k_t)|\mathcal{D}_t) = \sum_{k_{t-1}=1}^{K} p_t(k_t, k_{t-1})$,
- $Pr(\mathcal{M}_{t-1}(k_{t-1})|\mathcal{D}_t) = \sum_{k_t=1}^{K} p_t(k_t, k_{t-1})$,
- $Pr(\mathcal{M}_{t-1}(k_{t-1})|\mathcal{M}_t(k_t), \mathcal{D}_t) = p_t(k_t, k_{t-1})/p_t(k_t)$.

West and Harrison (1997) illustrate class II multiprocess models in the context of describing time series that behave in general as a second order polynomial, but that are occasionally subject to outliers and changes in level or growth. Such a model was introduced in Harrison and Stevens (1971, 1976). Extensions and applications also appear in Green and Harrison (1973), Smith and West (1983) and Ameen and Harrison (1985a). We refer the reader to these references for extensions and applications of multiprocess models. Later in this chapter we revisit the topic to discuss and illustrate extensions of mixture models for which the mixture components are not normal DLMs.

7.3 Mixtures of General State-Space Models

Mixtures of linear and Gaussian state-space models can be written as

$$y_t = \mathbf{F}'_{S_t}\boldsymbol{\theta}_t + \nu_t, \quad (7.14)$$

$$\boldsymbol{\theta}_t = \mathbf{G}_{S_t}\boldsymbol{\theta}_{t-1} + \mathbf{w}_t, \quad (7.15)$$

with S_t a hidden state, $\nu_t \sim N(0, v_{S_t})$, and $\mathbf{w}_t \sim N(\mathbf{0}, \mathbf{W}_{S_t})$. Conditional on the hidden state S_t, the model above is a DLM defined by the quadruple $\{\mathbf{F}_{S_t}, \mathbf{G}_{S_t}, v_{S_t}, \mathbf{W}_{S_t}\}$. These models are also referred to as conditionally Gaussian dynamic linear models, or CDLMs.

Example 7.8 *Markov switching dynamic factor model.* Carvalho and Lopes (2007) and Carvalho, Johannes, Lopes, and Polson (2010) consider a Markov switching dynamic factor model to describe a bivariate time series $\mathbf{y}_t = (y_{t,1}, y_{t,2})'$. The model is given by

$$\mathbf{y}_t = \mathbf{F}'_{S_t}\theta_t + \nu_t,$$

$$\theta_t = \theta_{t-1} + w_t,$$

where $\mathbf{F}'_{S_t} = (1, \beta_{S_t})'$, $\nu_t \sim N(0, v\mathbf{I})$, and $w_t \sim N(0, w)$. S_t is assumed to follow a two-state Markov switching process with transition probabilities p and q. That is, $Pr(S_t = i|S_{t-1} = i) = p$ for $i = 1:2$ and $Pr(S_t = i|S_{t-1} = j) = q$ for $i, j = 1:2$, and $i \neq j$.

Example 7.9 *DLMs with outliers.* Outliers in time series can be modeled via

$$\begin{aligned} y_t &= \mathbf{F}_t'\boldsymbol{\theta}_t + \nu_t, \\ \boldsymbol{\theta}_t &= \mathbf{G}'\boldsymbol{\theta}_{t+1} + \mathbf{w}_t, \end{aligned}$$

with $\nu_t \sim N(0, v_{S_t})$ and

$$v_{S_t} = \begin{cases} v & \text{if} \quad S_t = 1 \\ \kappa^2 v & \text{if} \quad S_t = 2. \end{cases}$$

This model adopts a normal mixture distribution for the measurement error, i.e., $\nu_t \sim (1-\pi)N(\nu_t|0,v) + \pi N(\nu_t|0,\kappa^2 v)$, with $\pi = Pr(S_t = 2)$, π typically small, and $\kappa > 1$ (see Chapter 4 of this book and Chapter 12 of West and Harrison 1997). In particular, West (1997c) considers an AR(p) latent model with a mixture distribution for the measurement error to analyze single deep ocean core oxygen isotope time series. Specifically, the model of West (1997c) has the form

$$\begin{aligned} y_t &= \mu_t + x_t + \nu_t, \\ \mu_t &= \mu_{t-1} + w_{t,1} \\ x_t &= \phi_1 x_{t-1} + \ldots + \phi_p x_{t-p} + w_{t,2}, \end{aligned}$$

with a normal mixture distribution on ν_t, and with the indicator variables S_t assumed independent so that the resulting model is a switching state-space model. West (1997c) uses a MCMC algorithm based on the FFBS approach of Carter and Kohn (1994) and Frühwirth-Schnatter (1994) to achieve posterior inference. A more general model would assume that S_t follows a hidden Markov chain instead of an i.i.d. process.

A description of various off-line methods (e.g., MCMC-based methods) for inference within the class of mixtures of linear and Gaussian state-space models can be found in Frühwirth-Schnatter (2006). Methods for sequential (on-line) filtering, parameter learning, and smoothing in mixtures of conditionally Gaussian DLMs, and mixtures of conditional dynamic models-referred to as PL methods-have been proposed in Carvalho, Johannes, Lopes, and Polson (2010). These PL algorithms outlined and illustrated in Chapter 6.1 can be easily applied to the models described above. As pointed out in Chapter 6.1, Carvalho, Johannes, Lopes, and Polson (2010) also developed PL algorithms for conditionally Gaussian nonlinear state-space models (CGDMs) in which the linear system Equation (7.15) is substituted by

$$\boldsymbol{\theta}_{S_t} = \mathbf{G}_{S_t}\mathbf{Z}(\boldsymbol{\theta}_{t-1}) + \mathbf{w}_t, \quad \mathbf{w}_t \sim N(\mathbf{0}, \mathbf{W}_{S_t}),$$

where $\mathbf{Z}(\cdot)$ is a nonlinear function. Such algorithms for inference within mixtures of Gaussian nonlinear models are also outlined in Section 6.2.3.4.

Note that these algorithms also work when the latent variable S_t is continuous. Examples of models with continuous latent variables are given below.

Example 7.10 *A conditionally Gaussian nonlinear model.* Carvalho, Johannes, Lopes, and Polson (2010) (see Problem 3 in Chapter 6.1) consider a conditionally Gaussian nonlinear dynamic model with heavy tails at the observational level. The model is given by

$$\begin{aligned} y_t &= \theta_t + v\sqrt{\lambda_t}\nu_t, \\ \theta_t &= \frac{\theta_{t-1}}{1+\theta_{t-1}^2} + w_t, \end{aligned}$$

where $\lambda_t \sim IG(\nu/2, \nu/2)$, ν_t and w_t are independent, $\nu_t \sim N(0,1)$, $w_t \sim N(0,w)$, and ν is known. At each time t the latent variable λ_t is a continuous variable. Carvalho, Johannes, Lopes, and Polson (2010) apply their PL approach for CGDMs to simultaneously obtain on-line filtering and parameter learning for this model. The PL algorithm for CGDMs is outlined in Chapter 6.1.

Example 7.11 *Robust latent autoregressive moving average model.* Chapter 13 of Frühwirth-Schnatter (2006) discusses a model used in Godsill (1997) for enhancing speech and audio signals that can be described via autoregressive moving average, ARMA(p,q), processes. More specifically, assuming that $p > q$, the model of Godsill (1997) has the form

$$\begin{aligned} y_t &= \mathbf{F}'\boldsymbol{\theta}_t + \nu_t, \\ \boldsymbol{\theta}_t &= \mathbf{G}(\boldsymbol{\phi})\boldsymbol{\theta}_{t-1} + \mathbf{w}_t, \end{aligned}$$

with $\mathbf{F} = (1, -\theta_1, \ldots, -\theta_q, 0, \ldots, 0)'$ and $\boldsymbol{\theta}_t = (x_t, x_{t-1}, \ldots, x_{t-p+1})'$ vectors of dimension p, $\mathbf{w}_t = w\epsilon_t$, and

$$\mathbf{G}(\boldsymbol{\phi}) = \begin{pmatrix} \phi_1 & \phi_2 & \cdots & \phi_{p-1} & \phi_p \\ 1 & 0 & \cdots & 0 & 0 \\ 0 & 1 & \cdots & 0 & 0 \\ \vdots & \ddots & \cdots & \vdots & \vdots \\ 0 & 0 & \cdots & 1 & 0 \end{pmatrix}.$$

In addition, it was assumed that the noise sources at the observational and system levels followed Gaussian Markov switching models. That is, $\nu_t \sim N(0, v_{S_t^{(1)}})$ and $\epsilon_t \sim N(0, w_{S_t^{(2)}})$, with

$$v_{S_t^{(1)}} = \begin{cases} v & \text{if } S_t^{(1)} = 1 \\ v\lambda_t^{(1)} & \text{if } S_t^{(1)} = 2, \end{cases}$$

and

$$w_{S_t^{(2)}} = \begin{cases} w & \text{if} \quad S_t^{(2)} = 1 \\ w\lambda_t^{(2)} & \text{if} \quad S_t^{(2)} = 2. \end{cases}$$

Here $S_t^{(i)}$, for $i = 1 : 2$, are hidden Markov processes with transition probabilities $\boldsymbol{\xi}^{(i)}$, while $\lambda_t^{(i)}$ are continuous with $\lambda_t^{(1)} \sim IG(\alpha_1, \beta_1)$ and $\lambda_t^{(2)} \sim IG(\alpha_2, \beta_2)$. Godsill (1997) implements a MCMC algorithm for posterior inference in this setting.

Example 7.12 *Stochastic volatility models.* As developed in detail in Section 7.5, canonical stochastic volatility (hereafter SV) models used in financial time series lead to DLMs with nonlinear observation error terms that have been widely analyzed using normal mixture approximations. A basic mathematical form for components of such models, such as in Shephard (1994) and Jacquier, Polson, and Rossi (1994), yields to the observation equation

$$y_t = \mu + x_t + \nu_t \tag{7.16}$$

where μ is a constant level, x_t follows an autoregressive process (typically of order one, though the framework can of course admit more general models), and the independent errors ν_t are a random sample from a specific nonnormal distribution. In particular, $\nu_t = \log(\kappa_t)/2$ where $\kappa_t \sim \chi_1^2$. Shephard (1994) approximates the density of ν_t by a discrete mixture of univariate normal components, viz.,

$$p(\nu_t) \approx \sum_{j=1}^{J} q_j N(\nu_t | b_j, w_j). \tag{7.17}$$

Very good approximations can be obtained by choosing J as low as five and with appropriate choices of the component weights, means, and variances. This strategy converts the model into a conditionally Gaussian DLM; introducing the inherent, latent normal mixture component indicators $\gamma_t \in \{1, \ldots, J\}$ at each t, we have

$$(\nu_t | \gamma_t) \sim N(\nu_t | b_{\gamma_t}, w_{\gamma_t}) \text{ and } Pr(\gamma_t = j) = q_j \tag{7.18}$$

independently over t. Extending the analysis to include inference on the sequence of γ_t opens up the model fitting strategy of conditional Gaussian models, so that posterior inference can be performed via standard MCMC methods; see Shephard (1994), Kim, Shephard, and Chib (1998), Aguilar and West (2000) and Chib, Nadari, and Shephard (2002), among others. A core part of the analysis involves iterative simulation of the latent process x_t over the time period of observations, and this is nowadays generally performed using the efficient FFBS methods of Carter and Kohn (1994) and Frühwirth-Schnatter (1994) described in Section 4.5. We develop the analysis of this framework in detail in Section 7.5 below.

The idea of using a mixture of normals to approximate the distribution of non-Gaussian observation errors has also been applied to long memory time series models and long memory stochastic volatility models (see Petris 1997 and Chan and Petris 2000). The model for long range dependence proposed in Petris (1997) and briefly discussed in Chapter 3 is now described within the mixture models context.

Example 7.13 *Frequency domain models for long range dependence.* A stationary process z_t observed at $t = 1 : T$ is said to be a long memory process of index d, with $d \in (0, 1)$, if its spectral density $f(\omega)$ behaves like ω^{-d} as ω goes to zero. In particular, Petris (1997) assumed that $f(\omega) = \omega^{-d} \exp(g(\omega))$, with $g(\omega)$ a continuous, bounded function on $[0, \pi)$. By discretizing this model and using asymptotic theory, Petris (1997) considered a regression model of the form

$$y_k = -d \log(\omega_k) + g_k + \nu_k,$$

with $y_k = \log(I_T(\omega_k))$ and $g_k = g(\omega_k)$, where $I_T(\cdot)$ is the periodogram and $\omega_k = 2\pi k/T$, for $k = 1 : K$ with $K = (\lfloor T/2 \rfloor - 1)$, are the Fourier frequencies (see Chapter 3). The ν_k terms are assumed independently drawn from a $\log(\chi_2^2/2)$ distribution that is approximated by a discrete mixture of univariate normal components. In other words, the density of ν_k is approximated by Equation (7.17) (now with the time index t replaced by the frequency index k). In addition, the g_ks were modeled as $g_k = \tilde{g}_k + m$, with $\tilde{g}$ a Gaussian process with an autoregressive structure of the form $(1 - \rho B)^p \tilde{g}_k = u_k$, where $u_k \sim N(0, w)$ and $0 < \rho < 1$. Again, exploiting the conditional Gaussian structure obtained by introducing latent normal component indicator variables $\gamma_{1:K}$ as in Equation (7.18), Petris (1997) develops a MCMC algorithm to obtain samples from the joint posterior distribution of $(g_{1:K}, m, w, d, \gamma_{1:K} | y_{1:T})$, using technical steps very similar to those underlying the analysis of the SV model of Example 7.12.

7.4 Case Study: Detecting Fatigue from EEGs

Prado (2010) presents analyses of multichannel electroencephalogram signals (EEGs) via mixtures of structured autoregressive models, referred to as multi-AR processes. Here we briefly describe the structure of such multi-AR processes, as well as the steps needed to perform posterior inference when structured prior distributions are placed on the characteristic roots of each of the autoregressive components in the mixture. We then illustrate how these models can be used to study some of the EEG traces previously analyzed in Prado (2010).

We begin by describing the motivating application that led to the use of multi-AR process. The EEG data considered here is part of a much larger data set, recorded at the NASA Ames Research Center by L. Trejo and collaborators, during an experiment designed to study and characterize cognitive fatigue. For a detailed description of the experiment and various analyses of these data see Trejo, Kochavi, Kubitz, Montgomery, Rosipal, and Matthews (2006) and Trejo, Knuth, Prado, Rosipal, Kubitz, Kochavi, Matthews, and Zhang (2007). In particular, the EEG signal analyzed here was recorded in one of 16 subjects in a study where participants were asked to solve simple summations-involving only four randomly generated single digits, three operators, and a target sum (e.g., $1 + 3 - 2 = 3$)-continuously for up to 3 hours. Each participant sat in front of a computer screen and, whenever an equation appeared on the screen, he/she had to decide if the summation on the left was less than, equal to, or greater than the target sum on the right by pressing the appropriate button on a key pad. After an answer was received the monitor was blank for 1 second before a new summation appeared. The subjects were asked to solve the summations as quickly as possible without sacrificing accuracy. EEGs were recorded at 30 channel locations in each participant for 180 minutes or less if the participant quit the experiment from exhaustion before 3 hours had elapsed. Artifacts were removed manually and automatically from the signals (see Trejo, Kochavi, Kubitz, Montgomery, Rosipal, and Matthews 2006 for details). The EEGs were also "epoched" around the times of the stimuli. That is, each EEG trace consists of a collection of consecutive epochs, where each epoch is a time series with 1,664 observations taken from 5 seconds prior to the stimulus (appearance of a given equation on the screen), to 8 seconds after such stimulus. The sampling rate in these data is 128 Hz.

One of the goals of this EEG study was the development of automatic methods for on-line (real-time) detection of cognitive fatigue. Trejo, Knuth, Prado, Rosipal, Kubitz, Kochavi, Matthews, and Zhang (2007), and Prado (2009, 2010) use approaches based on AR models and related time series decompositions to discover spectral features of the EEG signals that may be associated with cognitive fatigue. Prado (2009) shows differences in the estimated spectral features of EEG epochs recorded in the first 15 minutes of the experiment-when participants were supposed to be rested-and those of EEG epochs recorded in the last 15 minutes-when participants were expected to show signs of exhaustion. Such spectral features are estimated using autoregressive models. Motivated by such analyses, Prado (2010) considered mixtures of structured autoregressive models to describe epochs that are assumed to arise from two or more possible mental states of alertness. The main assumption behind this approach is that no single AR model is adequate to capture the structure of the EEG signal over time. Instead, a collection of K models, $\mathcal{M}_q(1), \ldots, \mathcal{M}_q(K)$, is used to represent K different mental states. Each of these mental states or models is assumed to be characterized in terms of the prior structure on their corresponding AR parameters. Here q indexes the epoch, and so it is also assumed that a single model is appropriate to describe the data of an entire epoch. In other words, a collection of models indexed by $k_1, \ldots, k_q$ is used to describe the data from q consecutive epochs; however, all the observations within an epoch are assumed to arise from a single model or mental state.

More specifically, Prado (2010) assumes that each component $\mathcal{M}_q(k)$ in the mixture has the following form,

$$\mathcal{M}_q(k): \ \mathbf{y}_q = \mathbf{F}'_q \boldsymbol{\theta}_q^{(k)} + \boldsymbol{\epsilon}_q^{(k)}, \ \ \boldsymbol{\epsilon}_q^{(k)} \sim N(\mathbf{0}, \phi^{-1}\mathbf{I}), \tag{7.19}$$

where T =1,664, $\mathbf{y}_q = (y_{p_k+1,q}, \ldots, y_{T,q})'$ is a vector of dimension $(T - p_k)$, with $y_{t,q}$ the t-th observation from epoch q, $\boldsymbol{\theta}_q^{(k)}$ is a p_k-dimensional parameter vector, and $\boldsymbol{\epsilon}_q^{(k)}$ is a $(T - p_k)$-dimensional vector of innovations. If a standard structure were assumed to model the evolution of $\boldsymbol{\theta}_q^{(k)}$, that is, if $\boldsymbol{\theta}_q^{(k)} = \mathbf{G}_q^{(k)} \boldsymbol{\theta}_{q-1}^{(k)} + \boldsymbol{\nu}_q^{(k)}$, with $\boldsymbol{\nu}_q^{(k)} \sim N(\mathbf{0}, \mathbf{W}_q^{(k)})$, and if the prior distributions on the state parameters and the precision parameter were conjugate Gaussian/gamma priors, then we would have been in the context of the class II multiprocess models described earlier in this chapter. However, in Prado (2010), the prior distributions are specific to each component in the mixture, and do not have conjugate normal/gamma forms. It is also assumed that $\mathbf{G}_q^{(k)} = \mathbf{0}$ and $\mathbf{W}_q^{(k)} = \mathbf{0}$ for all q and k, implying that the model components are not dynamic but static, and so $\boldsymbol{\theta}_q^{(k)} = \boldsymbol{\theta}^{(k)}$ for all k. In addition, $\mathbf{F}'_q$ is given by

$$\mathbf{F}'_q = \begin{pmatrix} y_{p_k,q} & \cdots & y_{1,q} \\ y_{p_k+1,q} & \cdots & y_{2,q} \\ \vdots & \vdots & \vdots \\ y_{T-1,q} & \cdots & y_{T-p_k,q} \end{pmatrix},$$

so that, under model $\mathcal{M}_q(k)$,

$$y_{t,q} = \theta_1^{(k)} y_{t-1,q} + \ldots + \theta_{p_k}^{(k)} y_{t-p_k,q} + \epsilon_{t,q}^{(k)}.$$

In the EEG application context it is also assumed that $p_k = p$ for all k, to indicate that the different mental states do not differ in the number of spectral components, but rather in the ranges that define the values of such components. The prior distributions on the model components are chosen following an approach similar to that proposed in Huerta and West (1999b) in the context of standard AR models (see Chapter 2). Such prior structure is summarized below.

7.4.1 Structured Priors in Multi-AR Models

Let $\gamma^{(k)} = (\gamma_1^{(k)}, \ldots, \gamma_p^{(k)})$ denote the reciprocal roots of the AR(p) model with characteristic polynomial

$$\Theta^{(k)}(u) = (1 - \theta_1^{(k)} u - \theta_2^{(k)} u^2 - \cdots - \theta_p^{(k)} u^p),$$

with u a complex number. Some of these roots will be real and the remaining roots will appear in complex conjugate pairs. Assume that all the AR mixture components have the same number of real and complex reciprocal roots, denoted by n_r and n_c, respectively. Each pair of complex reciprocal roots can be represented by their modulus $r_j^{(k)}$ and wavelength $\lambda_j^{(k)}$. Then, $\gamma_{2j-1}^{(k)} = r_j^{(k)} \exp(-2\pi i/\lambda_j^{(k)})$ and $\gamma_{2j}^{(k)} = r_j \exp(+2\pi i/\lambda_j^{(k)})$, for $j = 1 : n_c$. Similarly, for the n_r real roots, $\gamma_j^{(k)} = r_j^{(k)}$, for $j = (2n_c + 1) : p$, with $p = 2n_c + n_r$.

The prior structure of Prado (2010) is represented in terms of the moduli and periods of the characteristic roots for each of the K models as follows. For the complex reciprocal roots, it is assumed that

$$(r_j^{(k)} | \mathcal{D}_0) \sim f_{j,k}(r_j^{(k)}) \text{ and } (\lambda_j^{(k)} | \mathcal{D}_0) \sim g_{j,k}(\lambda_j^{(k)}), \quad j = 1 : n_c, \tag{7.20}$$

with $f_{j,k}(\cdot)$ a continuous distribution on the interval $(l_{j,k}^c(1), u_{j,k}^c(1))$, with $0 < l_{j,k}^c(1) < u_{j,k}^c(1) < 1$, and $g_{j,k}^c(\cdot)$ a continuous distribution on the interval $(l_{j,k}^c(2), u_{j,k}^c(2))$, with $2 \leq l_{j,k}^c(2) < u_{j,k}^c(2) \leq \lambda_{j,k}^*$, for some fixed value $\lambda_{j,k}^* \geq 2$. Similarly, for the real roots it is assumed that

$$(r_j^{(k)} | \mathcal{D}_0) \sim h_{j,k}(r_j^{(k)}), \quad j = (2n_c + 1) : p, \tag{7.21}$$

with $h_{j,k}(\cdot)$ a continuous distribution on $(l_{j,k}^r, u_{j,k}^r)$, with $-1 < l_{j,k}^r < u_{j,k}^r < 1$. Note that this prior structure implies that each mixture component is modeled a stationary AR process, since $|\gamma_j^{(k)}| < 1$ for all j and k. In addition, if the precision parameter ϕ in (7.19) is unknown, a gamma prior distribution $(\phi | \mathcal{D}_0) \sim G(n_0/2, d_0/2)$ is considered.

Example 7.14 *Truncated normal priors in quasiperiodic AR(2) processes.* In this example we illustrate how truncated normal distributions can be used to specify the prior distribution in a model with two mixture components, each of them with a quasiperiodic AR(2) structure. We also show how truncated normal priors on the AR coefficients can be used to approximate the prior structure assumed on the characteristic roots.

Suppose that we have a mixture model with $K = 2$ AR(2) components, and that each of such components is quasiperiodic, implying that $n_c = 1$ and $n_r = 0$. Assume also that these two components have different periods and that both processes are rather persistent, and so relatively large associated moduli are expected a priori. Such prior structure would be useful, for example, in modeling time series that are known to switch between two quasicyclical states, as may be the case with brain signals that are recorded during different cognitive states.

In order to illustrate some features of these priors, consider the following prior structure on the moduli and wavelengths of the two AR(2) mixture components:

$$(r^{(1)}|\mathcal{D}_0) \sim TN(0.8, 0.01, \mathcal{R}_1^{(1)}), \quad (\lambda^{(1)}|\mathcal{D}_0) \sim TN(5, 1, \mathcal{R}_2^{(1)}), \quad (7.22)$$

$$(r^{(2)}|\mathcal{D}_0) \sim TN(0.8, 0.01, \mathcal{R}_1^{(2)}), \quad (\lambda^{(2)}|\mathcal{D}_0) \sim TN(14, 1, \mathcal{R}_2^{(2)}), \quad (7.23)$$

where $TN(a, b, \mathcal{R})$ denotes a truncated normal with parameters a and b, and truncation region $\mathcal{R}$. Assume that the truncation regions above are given by $\mathcal{R}_1^{(k)} = (0.7, 1)$, for $k = 1:2$, $\mathcal{R}_2^{(1)} = (3, 8)$, and $\mathcal{R}_2^{(2)} = (11, 17)$. Figure 7.1 shows histograms of 2,000 samples from the prior distributions of $r^{(1)}$, $\lambda^{(1)}$, $r^{(2)}$, and $\lambda^{(2)}$ (plots (a), (b), (d), and (e), respectively), as well as the implied prior distributions on $\phi_1^{(1)}, \phi_2^{(1)}, \phi_1^{(2)}$, and $\phi_2^{(2)}$ (plots (c) and (f)).

In some cases, the implied priors on the AR coefficients that result from using the structure (7.20) and (7.21) on the moduli and wavelength of the characteristic roots can be well approximated by using truncated normal priors directly on the AR coefficients. For instance, the prior on $\phi_1^{(2)}$ and $\phi_2^{(2)}$ obtained from assuming the prior (7.23) on the associated characteristic modulus and wavelength, shown in Figure 7.2 (a), can be well approximated by a truncated normal prior on $\boldsymbol{\phi}^{(2)} = (\phi_1^{(2)}, \phi_2^{(2)})'$, $\boldsymbol{\phi}^{(2)} \sim TN(\mathbf{m}, \mathbf{C}, \mathcal{R})$, with

$$\mathbf{m} = (1.6\cos(2\pi/14), -0.8^2)', \quad \mathbf{C} = \begin{pmatrix} 1 & -0.9997 \\ -0.9997 & 1 \end{pmatrix} \quad (7.24)$$

and $\mathcal{R} = (1.4\cos(2\pi/11), 2\cos(2\pi/17)) \times (-1^2, -0.7^2)$. Samples from this truncated normal prior on $\boldsymbol{\phi}^{(2)}$ are displayed in Figure 7.2 (b).

7.4.2 Posterior Inference

Following the multiprocess notation, denote $\pi_q(k) = Pr(\mathcal{M}_q(k)|\mathcal{D}_{q-1})$, the prior probability given to model k in epoch q before observing the data for such epoch, $\mathbf{y}_q$. Similarly, let $p_q(k) = Pr(\mathcal{M}_q(k)|\mathcal{D}_q)$ be the probability of model k for epoch q after observing $\mathbf{y}_q$. Assuming that the transition probabilities are known and do not depend on the epoch, that is, $\pi(k|i) = Pr(\mathcal{M}_q(k)|\mathcal{M}_{q-1}(i), \mathcal{D}_{q-1}) = Pr(\mathcal{M}_q(k)|\mathcal{M}_{q-1}(i), \mathcal{D}_0)$ for all q and for all $k, i = 1:K$, we have that $\pi_q(k) = \sum_{i=1}^{K} \pi(k|i)p_{q-1}(i)$. The parameters of the mixture model with K autoregressive components are the vectors of AR coefficients $\boldsymbol{\theta}^{(1)}, \ldots, \boldsymbol{\theta}^{(K)}$, or $\boldsymbol{\theta}^{(1:K)}$, and the precision ϕ. On-line posterior

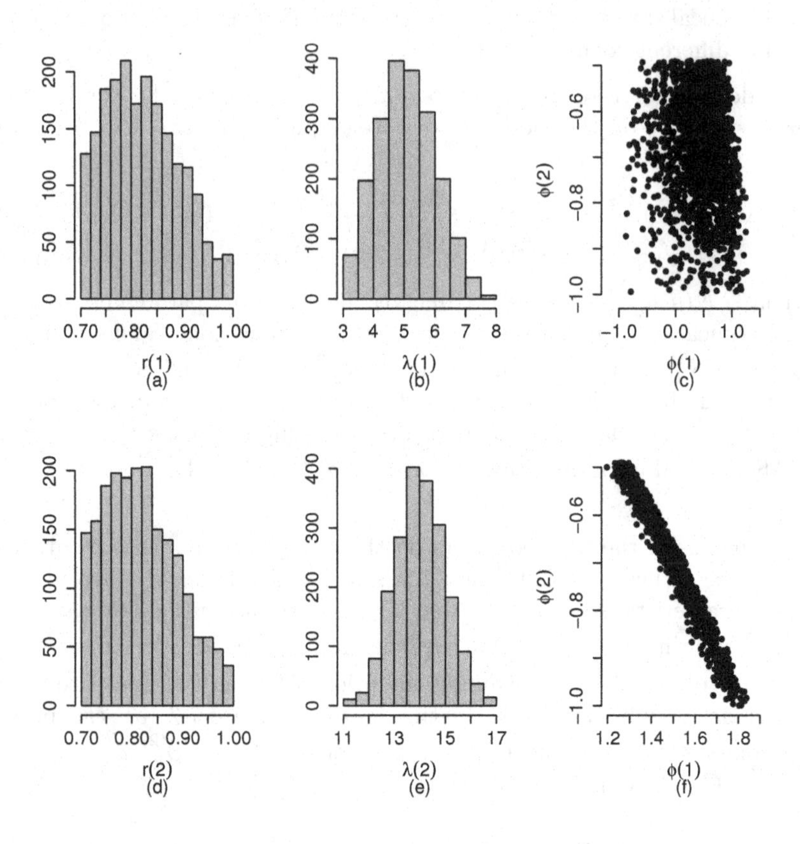

Figure 7.1 *Histograms of 2,000 samples from the priors of* $r^{(1)}$, $\lambda^{(1)}$, $r^{(2)}$, *and* $\lambda^{(2)}$*-graphs (a), (b), (d), and (e), respectively-and implied priors on* $(\phi_1^{(1)}, \phi_2^{(1)})$ *and* $(\phi_1^{(2)}, \phi_2^{(2)})$*-graphs (c) and (f).*

inference requires updating $p_q(k_q)$ and $p(\boldsymbol{\theta}^{(1:K)}, \phi|\mathcal{D}_q)$ sequentially as the epoched EEG data arrive. Starting with the first epoch $q = 1$ we have that

$$p(\boldsymbol{\theta}^{(1:K)}, \phi|\mathcal{D}_1) = \sum_{k_1=1}^{K} p(\boldsymbol{\theta}^{(1:K)}, \phi|\mathcal{M}_1(k_1), \mathcal{D}_1) \times p_1(k_1), \tag{7.25}$$

where

$$p(\boldsymbol{\theta}^{(1:K)}, \phi|\mathcal{M}_1(k_1), \mathcal{D}_1) = \frac{p(y_1|\boldsymbol{\theta}^{(1:K)}, \phi, \mathcal{M}_1(k_1))p(\boldsymbol{\theta}^{(1:K)}, \phi|\mathcal{D}_0)}{p(y_1|\mathcal{M}_1(k_1), \mathcal{D}_0)} \tag{7.26}$$

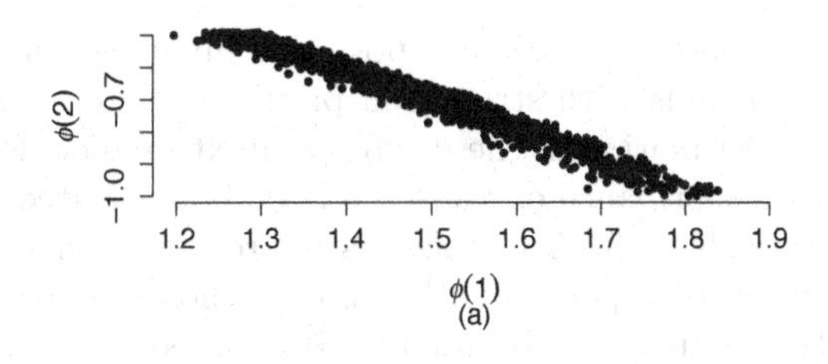

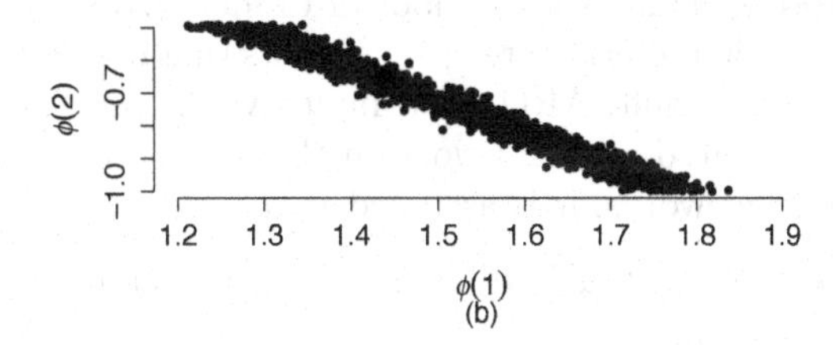

Figure 7.2 *Plot (a): samples from the implied prior distribution on* $\phi_1^{(2)}$ *and* $\phi_2^{(2)}$ *obtained from considering the prior on* $r^{(2)}$ *and* $\lambda^{(2)}$ *given in (7.23). Plot (b): samples from a* $TN(\boldsymbol{\phi}^{(2)}|\mathbf{m}, \mathbf{C}, \mathcal{R})$*, with* $\mathbf{m}$ *and* $\mathbf{C}$ *given by (7.24) and* $\mathcal{R} = (1.4\cos(2\pi/12), 2\cos(2\pi/20)) \times (-1, -.7^2)$.

and

$$p(y_1|\mathcal{M}_1(k_1), \mathcal{D}_0) = \int \cdots \int \left\{ \frac{1}{(2\pi|\phi^{-1}\mathbf{I}|)^{1/2}} \times \exp\left[-\frac{\phi(\mathbf{y}_1 - \mathbf{F}_1'\boldsymbol{\theta}^{(k_1)})'(\mathbf{y}_1 - \mathbf{F}_1'\boldsymbol{\theta}^{(k_1)})}{2} \right] \times p(\boldsymbol{\theta}^{(1:K)}, \phi|\mathcal{D}_0) \right\} d\boldsymbol{\theta}^{(1:K)} d\phi. \quad (7.27)$$

In addition,

$$p_1(k_1) = Pr(\mathcal{M}_1(k_1)|\mathcal{D}_1) \propto p(y_1|\mathcal{M}_1(k_1), \mathcal{D}_0) Pr(\mathcal{M}_1(k_1)|\mathcal{D}_0), \quad (7.28)$$

with $p_1(k_1)$ normalized such that $\sum_{k_1=1}^{K} p_1(k_1) = 1$. Computing (7.25) is nontrivial given that the prior structure on the characteristic roots described above does not lead to closed form expressions of the posteriors. Furthermore, (7.27) is typically not available in closed form. Implementing a MCMC algorithm similar to that proposed in Huerta and West (1999b) would not be helpful in this setting since it would not allow us to directly compute (7.27). More importantly, a MCMC scheme would not lead to on-line posterior inference, particularly in cases where K and q are relatively large, since the posterior is a mixture of K^q components.

Prado (2010) developed approximate posterior inference for multi-AR(1) and multi-AR(2) models with structured priors. The steps to achieve approximate on-line estimation in these types of mixture models are summarized below. The performance of these methods is illustrated in simulated and real EEG data. If multi-AR processes of orders higher than two need to be considered, other types of methods, e.g., those based on sequential Monte Carlo algorithms, could be used for on-line posterior inference.

The first assumption behind the methods of Prado (2010) is that the structured priors on the characteristic reciprocal roots that define the AR(1) and AR(2) components in multi-AR(1) and multi-AR(2) processes can be approximated by truncated normal priors on the implied coefficients of such components. In other words, it is assumed that

$$p(\boldsymbol{\theta}^{(k)}|\phi, \mathcal{D}_0) \approx TN(\boldsymbol{\theta}^{(k)}|\mathbf{m}_0(k), \phi^{-1}\mathbf{C}_0^*(k), \mathcal{R}^{(k)}),$$

and so the joint prior has the form

$$p(\boldsymbol{\theta}^{(1:K)}, \phi|\mathcal{D}_0) \approx \prod_{k=1}^{K} TN(\boldsymbol{\theta}^{(k)}|\mathbf{m}_0(k), \phi^{-1}\mathbf{C}_0^*(k), \mathcal{R}^{(k)}) \times G(\phi|n_0/2, d_0/2).$$

Then, under this prior structure, we have that at $q = 1$, i.e., after observing $\mathbf{y}_1$,

$$\begin{aligned}(\boldsymbol{\theta}^{(k)}|\mathcal{M}_1(k_1), \mathcal{D}_1, \phi) &= TN(\boldsymbol{\theta}^{(k)}|\mathbf{m}_1^{(k)}(k_1), \phi^{-1}C_1^{*,(k)}(k_1), \mathcal{R}^{(k)}),\\ (\phi|\mathcal{M}_1(k_1), \mathcal{D}_1) &\approx G(n_1/2, d_1(k_1)/2),\end{aligned}$$

with $n_1 = n_0 + n$, where n is the dimension of $\mathbf{y}_q$,

$$\mathbf{C}_1^{*,(k)}(k_1) = \begin{cases} \mathbf{C}_0^*(k) & \text{if } k \neq k_1, \\ (\mathbf{C}_0^*(k)^{-1} + \mathbf{F}_1\mathbf{F}_1')^{-1} & \text{if } k = k_1, \end{cases}$$

$$\mathbf{m}_1^{(k)}(k_1) = \begin{cases} \mathbf{m}_0(k) & \text{if } k \neq k_1, \\ \mathbf{C}_1^{*,(k_1)}(\mathbf{C}_0^*(k_1)^{-1}\mathbf{m}_0(k_1) + \mathbf{F}_1\mathbf{y}_1) & \text{if } k = k_1, \end{cases}$$

and

$$d_1(k_1) = d_0 + (\mathbf{y}_1 - \mathbf{F}_1'\mathbf{m}_0(k_1))'[\mathbf{Q}_1^*(k_1)]^{-1}(\mathbf{y}_1 - \mathbf{F}_1'\mathbf{m}_0(k_1))$$

where $\mathbf{Q}_1^*(k_1) = (\mathbf{F}_1'\mathbf{C}_0^*(k_1)\mathbf{F}_1 + \mathbf{I})$. It can also be shown that

$$p(k_1) \propto \frac{\kappa_1(k_1)\pi(k_1)}{\kappa_0^*(k_1)} \times \frac{|\mathbf{C}_1^{(k)}(k_1)|^{1/2} d_0^{(n_0/2)} \gamma(n_1/2)}{|\mathbf{C}_0(k_1)|^{1/2}\Gamma(n_0/2)[d_1(k_1)]^{n_1/2}},$$

where $\mathbf{C}_1^{(k)}(k_1) = \mathbf{C}_1^{*,(k)}(k_1)S_1(k_1)$, $S_1(k_1) = d_1(k_1)/n_1$, and $\kappa_0^*(\cdot)$ and

$\kappa_1(\cdot)$ are given by

$$\begin{aligned}\kappa_0^*(k_1) &= \int TT_{n_0}(\boldsymbol{\theta}^{(k_1)}|\mathbf{m}_0(k_1), \mathbf{C}_0(k_1), \mathcal{R}^{(k_1)})d\boldsymbol{\theta}^{(k_1)}, \\ \kappa_1(k_1) &= \int TT_{n_1}(\boldsymbol{\theta}^{(k_1)}|\mathbf{m}_1^{(k_1)}(k_1), \mathbf{C}_1^{(k_1)}(k_1), \mathcal{R}^{(k_1)})d\boldsymbol{\theta}^{(k_1)}.\end{aligned}$$

In these equations $TT_\nu(\mathbf{m}, \mathbf{C}, \mathcal{R})$ denotes a truncated Student-t distribution with ν degrees of freedom, location $\mathbf{m}$, scale matrix $\mathbf{C}$, and truncation region $\mathcal{R}$.

Similarly, at $q > 1$, we can obtain the following approximations.

- $(\phi|\mathcal{M}_q(k_q), \mathcal{D}_q) \approx G(n_q/2, d_q(k_q)/2)$, with $n_q = n_{q-1} + n$, $d_q(k_q) = n_q S_q(k_q)$ and

$$S_q^{-1}(k_q) = \sum_{k_{q-1}=1}^{K} S_q^{-1}(k_q, k_{q-1})p_q(k_q, k_{q-1})/p_q(k_q).$$

In this equation, $p_q(k_q) = \sum_{k_{q-1}=1}^{K} p_q(k_q, k_{q-1})$, and

$$\begin{aligned}p_q(k_q, k_{q-1}) \;\propto\; & \frac{\pi(k_q|k_{q-1})\kappa_q(k_q, k_{q-1})}{\kappa_{q-1}^*(k_q, k_{q-1})} \times \\ & \frac{|\mathbf{C}_q^{(k_q)}(k_q, k_{q-1})|^{1/2}(d_{q-1}(k_{q-1}))^{n_{q-1}/2}\Gamma(n_q/2)}{|\mathbf{C}_{q-1}^{(k_q)}(k_{q-1})|^{1/2}(d_q(k_q, k_{q-1}))^{n_q/2}\Gamma(n_{q-1}/2)},\end{aligned}$$

with

$$\begin{aligned}\kappa_{q-1}^*(k_q, k_{q-1}) &= \int TT_{n_{q-1}}\left(\boldsymbol{\theta}^{(k_q)}\middle|\, \mathbf{m}_{q-1}^{(k_q)}(k_{q-1}),\right. \\ &\qquad \left.\mathbf{C}_{q-1}^{*,(k_q)}(k_{q-1})\frac{d_{q-1}(k_{q-1})}{n_{q-1}}, \mathcal{R}^{(k_q)}\right) d\boldsymbol{\theta}^{(k_q)},\end{aligned}$$

$$\begin{aligned}\kappa_q(k_q, k_{q-1}) &= \int TT_{n_q}\left(\boldsymbol{\theta}^{(k_q)} \mid \mathbf{m}_q^{(k_q)}(k_q, k_{q-1})\right. \\ &\qquad \left.\mathbf{C}_{q-1}^{*,(k_q)}(k_q, k_{q-1})\frac{d_q(k_q)}{n_q}\mathcal{R}^{(k_q)}\right) d\boldsymbol{\theta}^{(k_q)},\end{aligned}$$

$$\mathbf{C}_q^{*,(k)}(k_q, k_{q-1}) = \begin{cases} \mathbf{C}_{q-1}^{*,(k)}(k_{q-1}) & k \neq k_q, \\ ([\mathbf{C}_q^{*,(k)}(k_q, k_{q-1})]^{-1} + \mathbf{F}_q\mathbf{F}'_q)^{-1} & k = k_q, \end{cases}$$

$$\mathbf{m}_q^{(k)}(k_q, k_{q-1}) = \begin{cases} \mathbf{m}_{q-1}^{(k)}(k_{q-1}) & k \neq k_q, \\ \mathbf{C}_q^{*,(k)}(k_q, k_{q-1})\times & k = k_q, \\ ([\mathbf{C}_{q-1}^{*,(k)}(k_{q-1})]^{-1}\mathbf{m}_q^{(k)}(k_{q-1}) + \mathbf{F}_q\mathbf{y}_q) & \end{cases}$$

$$\begin{aligned} d_q(k_q, k_{q-1}) &= d_{q-1}(k_{q-1}) + \Big\{ (\mathbf{y}_q - \mathbf{F}'_q \mathbf{m}_{q-1}^{(k_q)}(k_{q-1}))' \times \\ &\quad [\mathbf{Q}_q^*(k_q, k_{q-1})]^{-1} (\mathbf{y}_q - \mathbf{F}'_q \mathbf{m}_{q-1}^{(k_q)}(k_{q-1})) \Big\}, \end{aligned}$$

$S_q(k_q, k_{q-1}) = d_q(k_q, k_{q-1})/n_q$, and $\mathbf{Q}_q^*(k_q, k_{q-1}) = (\mathbf{F}'_q \mathbf{C}_{q-1}^{(k_q)}(k_{q-1}) \mathbf{F}_q + \mathbf{I})$.

- $(\boldsymbol{\theta}^{(k)} | \mathcal{M}_q(k_q), \mathcal{D}_q, \phi^{-1}) \approx TN(\boldsymbol{\theta}^{(k)} | \mathbf{m}_q^{(k)}(k_q), \mathbf{C}_q^{*,(k)}(k_q)/\phi, \mathcal{R}^{(k)})$, with

$$\mathbf{C}_q^{*,(k)}(k_q) = \frac{\mathbf{C}_q^{(k)}(k_q)}{S_q(k_q)}.$$

The values of $\mathbf{m}_q^{(k)}(k_q)$ and $\mathbf{C}_q^{(k)}(k_q)$ are computed as follows:

$$\begin{aligned} \mathbf{m}_q^{(k)}(k_q) &= \sum_{k_{q-1}=1}^{K} \mathbf{m}_q^{(k)}(k_q, k_{q-1}) p_q^*(k_q, k_{q-1}), \\ \mathbf{C}_q^{(k)}(k_q) &= \sum_{k_{q-1}=1}^{K} \Big\{ \Big[\mathbf{C}_q^{(k)}(k_q, k_{q-1}) + (\mathbf{m}_q^{(k)}(k_q) - \mathbf{m}_q^{(k)}(k_q, k_{q-1}))' \\ &\quad \times (\mathbf{m}_q^{(k)}(k_q) - \mathbf{m}_q^{(k)}(k_q, k_{q-1})) \Big] p_q^*(k_q, k_{q-1}) \Big\}, \end{aligned}$$

where $p_q^*(k_q, k_{q-1}) = S_q(k_q) S_q^{-1}(k_q, k_{q-1}) p_q(k_q, k_{q-1})/p_q(k_q)$, normalized such that, for all k_q, $\sum_{k_{q-1}=1}^{K} p_q^*(k_q, k_{q-1}) = 1$.

These approximations use the fact that $p(\boldsymbol{\theta}^{(k)} | \mathcal{M}_q(k_q), \mathcal{D}_q, \phi)$ can be written as a mixture of K components and that these can be collapsed into a single component using Kullback-Leibler divergence arguments similar to those discussed in Section 7.2 of this Chapter and also found in Chapter 12 of West and Harrison (1997). The calculations above also used approximation (7.9) with $h = 1$ (see Prado 2010 for details).

Example 7.15 *Simulated multi-AR(2) data.* A time series with 10,000 data points was simulated, in batches of 100 observations, from two AR(2) processes as follows. The first 20 batches (epochs) were simulated from a quasiperiodic autoregression with modulus $r_1 = 0.95$ and wavelength $\lambda_1 = 6$. The following 30 epochs were simulated from a quasiperiodic autoregression with modulus $r_2 = 0.99$ and wavelength $\lambda_2 = 16$. Then, epochs 51–80 were simulated from the AR(2) process with characteristic reciprocal roots $r_1 e^{\pm 2\pi i/\lambda_1}$, while epochs 81–100 were simulated from the AR(2) process with characteristic reciprocal roots $r_2 e^{\pm 2\pi i/\lambda_2}$. The innovations for both types of epochs were assumed to follow independent Gaussian distributions centered at zero with variance $v = 1/\phi = 100$. The top panel in Figure 7.3 shows the simulated data. Vertical dotted lines are set every 100 data points to indicate the epochs.

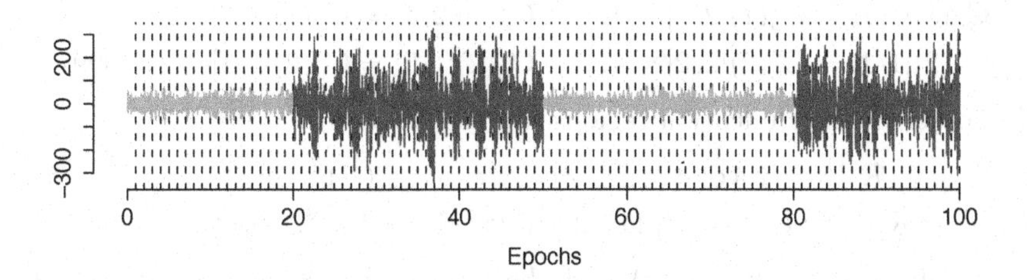

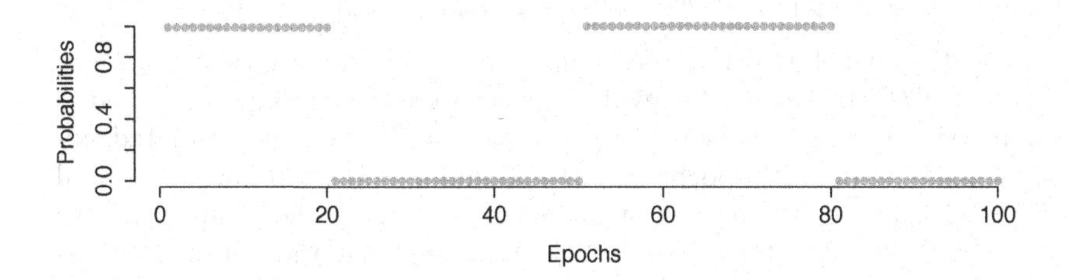

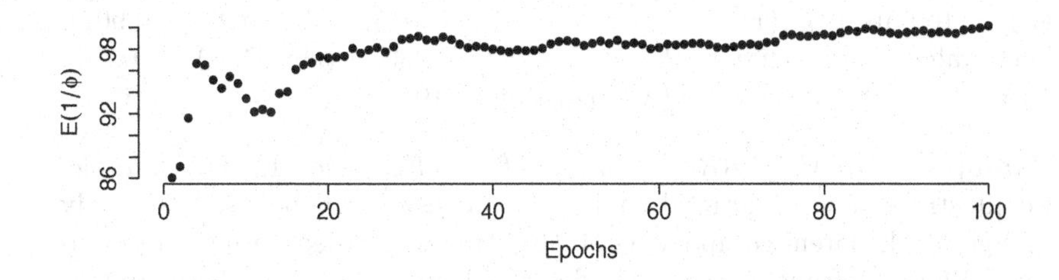

Figure 7.3 *Top: data simulated from two AR(2) processes; data in epochs 1–20 and 51–80 were simulated according to a quasiperiodic AR(2) process with modulus 0.95 and wavelength 6; data in epochs 21–50 and 81–100 were simulated from an AR(2) process with modulus 0.99 and wavelength 16. Middle: estimated values of* $p_q(1) = Pr(\mathcal{M}_q(1)|\mathcal{D}_q)$, *for* $q = 1:100$. *Bottom: estimated values of* $E(1/\phi|\mathcal{D}_q)$ *for* $q = 1:100$.

We fitted a multi-AR(2) model with $K = 2$ and a prior structure of the form

$$\begin{aligned} p(\boldsymbol{\theta}^{(k)}) &= TN(\boldsymbol{\theta}^{(k)}|\mathbf{m}_0(k), \phi^{-1}\mathbf{C}_0^*(k), \mathcal{R}^{(k)}), \\ p(\phi|\mathcal{D}_0) &= G(1/2, 1/2), \end{aligned}$$

where $\mathbf{m}_0(1) = (1.6\cos(2\pi/5), -0.8^2)$, $\mathbf{m}_0(2) = (1.6\cos(2\pi/14), -0.8^2)$, and the $\mathbf{C}_0^*(k)$ matrices for $k = 1:2$ are given by:

$$\mathbf{C}_0^*(1)[1,1] = \mathbf{C}_0^*(1)[2,2] = 1, \qquad \mathbf{C}_0^*(1)[2,1] = \mathbf{C}_0^*(1)[1,2] = 0.4,$$
$$\mathbf{C}_0^*(2)[1,1] = \mathbf{C}_0^*(2)[2,2] = 1, \qquad \mathbf{C}_0^*(2)[1,2] = \mathbf{C}_0^*(2)[2,1] = -0.9997.$$

In addition, the truncation regions were set to

$$\mathcal{R}^{(1)} = (1.4\cos(2\pi/3), 2\cos(2\pi/8)) \times (-1^2, -0.7^2),$$
$$\mathcal{R}^{(2)} = (1.4\cos(2\pi/11), 2\cos(2\pi/17)) \times (-1^2, -0.7^2).$$

The values of $\mathbf{m}_0(k)$, $\mathbf{C}_0^*(k)$ and the truncation regions were chosen to approximate the prior structure on $r^{(k)}$, $\lambda^{(k)}$ discussed in Example 7.14. Finally, we set $\pi(1|1) = \pi(2|2) = 0.9$, and $\pi_0(1) = 0.5$.

The middle panel in Figure 7.3 shows approximate values of $p_q(1) = Pr(\mathcal{M}_q(1)|\mathcal{D}_q)$ for $q = 1:100$, which correctly capture the structure used to simulate the data, i.e., epochs $q = 1:20$ and $q = 51:80$ have estimated values of $p_q(1) = 1$, while epochs $q = 21:50$ and $q = 81:100$ have estimated values of $p_q(1) = 0$. The bottom panel in the figure shows approximate values of $E(\phi^{-1}|\mathcal{D}_q)$ for $q = 1:100$, indicating that the approximations discussed above also work well in terms of the posterior inference for ϕ. In addition, we find that the location parameters of the truncated normals at $q = 100$ are $\mathbf{m}_{100}(1) = (1.827, -0.976)$ and $\mathbf{m}_{100}(2) = (0.956, -0.907)$. Such values correspond to $E(r^{(1)}|\mathcal{D}_{100}) \approx 0.956$, $E(r^{(2)}|\mathcal{D}_{100}) \approx 0.988$, $E(\lambda^{(1)}|\mathcal{D}_{100}) \approx 6.015$, and $E(\lambda^{(2)}|\mathcal{D}_{100}) \approx 16.104$.

Example 7.16 *Multi-AR*(1) *analysis of an EEG series.* In this example, we illustrate how structured multi-AR processes can be used to study changes in the latent components of EEG traces recorded during the cognitive fatigue experiment previously described. In order to use these models, some preprocessing of the EEG data was performed (a brief description appears below). The analysis presented here uses data recorded at channels Pz and P_4, both located in the parietal region, in one of the participants. The EEG data for these channels consist of a series of 864 epochs of 1,664 observations, i.e., a total of 1,437,696 observations per channel.

First, for each channel, AR(10) models were fitted to the 864 epochs, and the AR reciprocal roots estimates, as well as their corresponding time series decompositions (see Chapter 2), were obtained by fixing the AR coefficients at their posterior means. Then, the estimated latent processes in these decompositions were extracted, and those processes associated to the real reciprocal roots with the largest moduli were considered to be our data set. In other words, in this analysis the data $y_{P_z,1:864}$ and $y_{P_4,1:864}$ consist of 864 univariate time series (epochs) with 1,664 observations each obtained from extracting the latent component associated to the real reciprocal root with the largest modulus in the AR-based decomposition of each epoch. Due to the structure of the estimated latent processes in the AR decompositions,

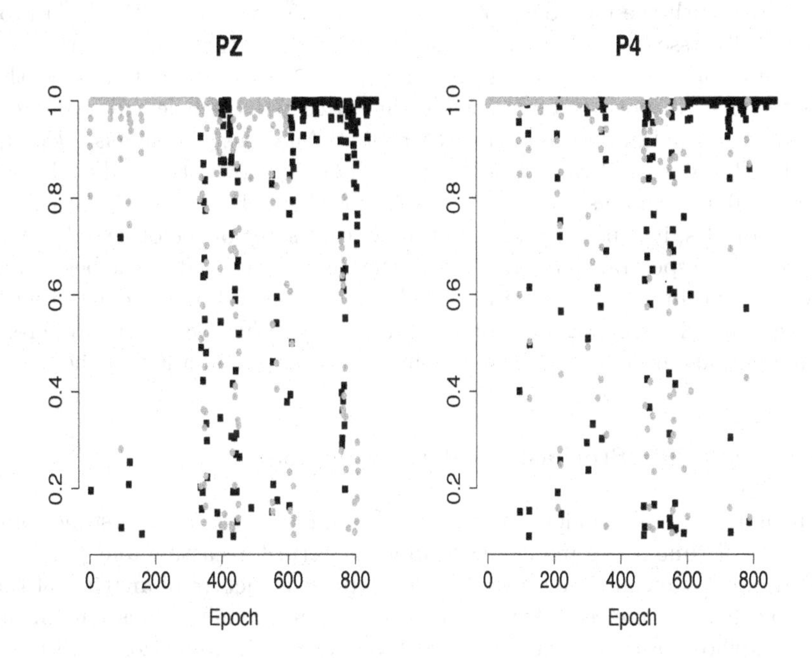

Figure 7.4 *Estimated values of $p_q(1)$ (light dots) and $p_q(2)$ (dark squares) for the latent processes with the highest moduli in channels P_z and P_4. These values are based on a multi-AR(1) analysis.*

each of these 864 time series should have, at least approximately, an AR(1) structure.

Then, we assume that $y_{P_z,1:864}$ and $y_{P_4,1:864}$ can be modeled via multi-AR(1) processes with two states, i.e., it is assumed that $K = 2$. Furthermore, it is also assumed that both states are rather persistent, one of them with a modulus in the $(0.9, 0.975)$ range and the other one with a modulus in the $(0.975, 1.0)$ range. More specifically, we assume a mixture model with two AR(1) components and the prior structure

$$p(\theta^{(k)}|\phi, \mathcal{D}_0)) \sim TN(\theta^{(k)}|m_0(k), \phi^{-1}C_0^*(k), \mathcal{R}^{(k)}),$$

with $m_0(1) = 0.94$, $m_0(2) = 0.98$, $C_0^*(k) = 1$ for $k = 1:2$, $\mathcal{R}^{(1)} = (0.9, 0.975)$, and $\mathcal{R}^{(2)} = (0.975, 1)$. In addition, we set $\pi(1|1) = \pi(2|2) = 0.999$ and $\pi_0(1) = 0.5$.

Figure 7.4 shows the estimated values of $p_q(1) = Pr(\mathcal{M}_q(1)|\mathcal{D}_q)$ (light dots) and $p_q(2) = Pr(\mathcal{M}_q(2)|\mathcal{D}_q)$ (dark squares) for both channels. Only estimated probabilities above 0.05 are shown. These estimated values were obtained with the methods for approximate inference in multi-AR(1) processes discussed earlier in this chapter. Based on this analysis, the AR mixture component with coefficient in the $(0.9, 0.975)$ range dominates the first epochs in both channels, while that with AR coefficient above 0.975 dominates the last epochs of the experiment. Analyses of channels CP_4, P_3, and CP_z appear in Prado (2010). These channels show estimated $p_q(1)$ values similar to those shown in Figure 7.4 for channels P_z and P_4. Based on these results, it is possible to hypothesize that some of the observed differences in the spectral characteristics of the signals recorded at the beginning of the experiment and those recorded toward the end are associated with cognitive fatigue. It was also found that spectral differences across different time periods are not evident in all the channels, just in a few of them.

7.5 Univariate Stochastic Volatility models

The univariate SV model introduced in Example 7.12 is a cornerstone component of time series models used in short term forecasting and portfolio studies in financial time series. We develop the structure of analysis of the simple model here as a key example of a conditionally Gaussian, linear model whose use has been amplified based on the introduction of discrete normal mixture model approximations coupled with Monte Carlo methods for model fitting. The following discusses and exemplifies the model development and use of a standard, Gibbs sampling MCMC approach for model fitting.

7.5.1 Zero-Mean AR(1) SV Model

Consider a financial price series P_t, such as a stock index or exchange rate, observed at equally spaced time points t. Model the per-period *returns* $r_t = P_t/P_{t-1} - 1$ as a zero mean but time-varying volatility process

$$\begin{aligned} r_t &\sim N(r_t|0, \sigma_t^2), \\ \sigma_t &= \exp(\mu + x_t), \\ x_t &= \phi x_{t-1} + \epsilon_t, \quad \epsilon_t \sim N(\epsilon_t|0, v), \end{aligned} \tag{7.29}$$

with the ϵ_t independent over time. In reality, such models are typically components of more useful models in which the mean of the returns series is nonzero and is modeled via regression on economic and financial predictors. The parameter μ defines baseline log-volatility; the AR(1) parameter ϕ defines persistence in deviations in volatility from the baseline, and the innovations variance v "drives" the levels of activity in the volatility process. Typically ϕ is close to one, and in any case the AR(1) process is assumed stationary, so that the marginal distribution for the missing initial value is $N(x_0|0, v/(1-\phi^2))$.

Based on data $r_{1:T}$ over a time interval of T periods, the uncertain quantities to infer are $(x_{0:n}, \mu, \phi, v)$. The major complication arises due to the nonnormality inherent in the observation equation. One approach to dealing with this transforms the data to $y_t = \log(r_t^2)/2$ so that

$$y_t = \mu + x_t + \nu_t, \quad \nu_t = \log(\kappa_t)/2, \quad \kappa_t \sim \chi_1^2. \tag{7.30}$$

Here the volatility process x_t is assumed independent of the ν_t, though more general models may allow for potential correlations. This is the model exhibited in Example 7.12, now being specific about its origin; it is a DLM but with nonnormal observation errors ν_t.

7.5.2 Normal Mixture Approximation

The basic idea is to approximate $p(\nu_t)$ by a discrete normal mixture of the form

$$p(\nu_t) \approx \sum_{j=1}^{J} q_j N(\nu_t|b_j, w_j)$$

with specific values of $\{J, q_{1:J}, b_{1:J}, w_{1:J}\}$. The exact $p(\nu_t)$ is continuous, unimodal, and heavier tailed on negative values. We can approximate a defined continuous pdf as accurately as desired using a discrete mixture of normals, and a number of mixtures of a small number ($J = 5-9$, say) have been used in the literature, fitted typically using nonlinear optimization based on some measure of difference. An example from Kim, Shephard, and Chib (1998) with $J = 7$ has parameters as follows:

q_j :	0.0073	0.0000	0.1056	0.2575	0.3400	0.2457	0.0440
b_j :	−5.7002	−4.9186	−2.6216	−1.1793	−0.3255	0.2624	0.7537
w_j :	1.4490	1.2949	0.6534	0.3157	0.1600	0.0851	0.0418

Mixtures with more components can refine the approximation; the theory and MCMC analysis are structurally the same.

Introduce latent *indicator variables* $\gamma_t \in \{1 : J\}$ where, independently over t and of all other random quantities, $Pr(\gamma_t = j) = q_j$. Then the mixture of normals can be constructed from the conditionals

$$(\nu_t|\gamma_t = j) \sim N(\nu_t|b_{\gamma_t}, w_{\gamma_t}) \equiv N(\nu_t|b_j, w_j), \quad j = 1 : J.$$

The implied marginal distribution of ν_t, averaging with respect to the discrete density $p(\gamma_t)$, is the mixture of normals above. This data augmentation trick induces conditionally normal DLMs that are amenable to analysis, and MCMC allows us to embed them within an overall computational strategy.

7.5.3 Centered Parameterization

We reexpress the model with the baseline volatility moved from the observation equation to that of the state, i.e., volatility process. That is, we use the equivalent model representation

$$\begin{aligned} y_t &= z_t + \nu_t, \\ z_t &= \mu + \phi(z_{t-1} - \mu) + \epsilon_t, \end{aligned} \tag{7.31}$$

where $z_t = \mu + x_t$ is the AR(1) volatility process now centered around the baseline level μ. We do this for the following reason.

MCMC approaches that iteratively resample from conditional posteriors for $(\mu|x_{0:T}, -)$ and $(x_{0:T}|\mu, -)$ suffer from slow convergence due to the inherent, negative correlations *always* evident in posteriors between μ and each of the x_t. The y_t data provide direct observations on the sum $z_t = \mu + x_t$, so even with informative priors the implied negative posterior correlation is clear: $z_t = (\mu + c) + (x_t - c)$ for any c. Gibbs sampling-based MCMC applied to the original parameterization then tend to have very poor convergence properties, generating samples such that the implied $z_t = \mu + x_t$ are consistent with the data, but with values of μ wandering far from regions consistent with the prior-and, in a financial times series analysis, with economic reality-coupled with corresponding inverse drifts in the simulated x_t. Monitoring the MCMC will show clear evidence of the negative dependence as well as high positive autocorrelations within the MCMC iterates. This is resolved by developing the posterior simulation on the posterior for $z_{0:T}, \mu$, and other quantities, i.e., using the centered volatility process z_t rather than x_t.

This is a common problem in hierarchical models; other examples appear in, for example, Ferreira, Gamerman, and Migon (1997) and Gamerman and Lopes (2006) (Sections 5.3.4 and 6.4.4). The practicality is one of appropriate choice of parameterization to define a well-behaved MCMC algorithm, with a representation that weakens structural posterior dependencies. This is often clarified by viewing the directed graphical model representations of the model. For the model in the initial parameterization now augmented by the latent indicators of normal mixture components, $\gamma_{1:T}$, the full joint density over all quantities over $t = 0 : T$ is

$$\begin{aligned} &p(y_{1:T}, x_{0:T}, \gamma_{1:T}, \mu, \phi, v) \propto \\ &\qquad p(\mu, \phi, v) p(x_0|\phi, v) \prod_{t=1}^{T} p(y_t|\mu, x_t, \gamma_t) p(\gamma_t) p(x_t|x_{t-1}, \phi, v) \end{aligned} \tag{7.32}$$

based on any specified prior (μ, ϕ, v).

The conditional independence structure is exhibited in the graphical model

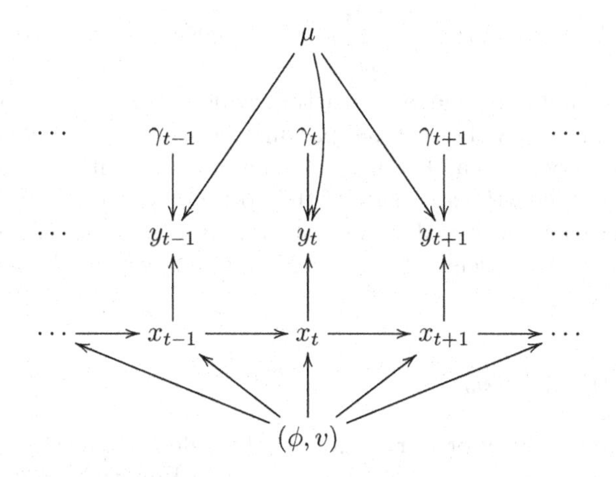

(a) Directed graph using initial volatility parameterization.

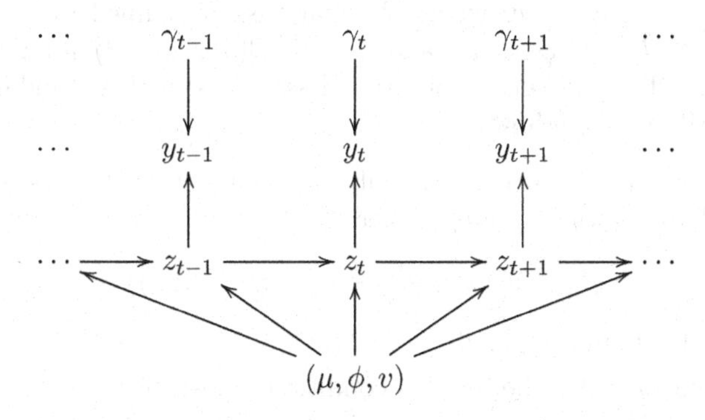

(b) Directed graph using centered volatility parameterization.

Figure 7.5 *Directed graphical representation of the joint distribution of data, states, and parameters in the AR(1) stochastic volatility model: (a) using the original parameterization with zero-mean volatility process* x_t*, and (b) in terms of the centered volatility process* $z_t = \mu + x_t$.

in which directed arrows represent conditional dependencies in the set of conditional distributions comprising the model; see Figure 7.5(a). In moving to the centered parameterization of Equation (7.31) with the volatility process z_t having mean μ, the joint density is

$$p(y_{1:T}, z_{0:T}, \gamma_{1:T}, \mu, \phi, v) \propto \\ p(\mu, \phi, v)p(z_0|\mu, \phi, v) \prod_{t=1}^{T} p(y_t|z_t, \gamma_t)p(\gamma_t)p(z_t|z_{t-1}, \mu, \phi, v) \tag{7.33}$$

with conditional independence structure evident in Figure 7.5(b). It is clear from the figures how the centered parameterization completely decouples μ from the observations and results in a simplified conditional independence structure. Further, the structure of the posterior is such that the resulting dependencies are weakened, as we now have direct (unbiased) observations y_t on each z_t. We discuss the direct MCMC analysis of the model in this parameterization.

7.5.4 MCMC Analysis

All conditional posteriors for MCMC analysis derive directly from the posterior $p(z_{0:T}, \gamma_{1:T}, \mu, \phi, v|y_{1:T})$ proportional to Equation (7.33). This specific algorithm combines direct sampling from complete conditionals with Metropolis-Hastings steps. For illustration, and consistent with standard practice in applications, we take independent priors on the parameters with $p(\mu, \phi, v) = p(\mu)p(\phi)p(v)$ where the margins are defined by $\mu \sim N(g, G)$, $\phi \sim N(c, C)I(0 < \phi < 1)$, and $v^{-1} \sim G(a/2, av_0/2)$ for given hyperparameters. The full set of conditional samplers in this standard, convergent MCMC are as follows.

1. Resample $\gamma_{1:T}$ from conditionally independent posteriors for each γ_t, with $Pr(\gamma_t = j|y_t, z_t) = q^*_{t,j}$ where
$$q^*_{t,j} \propto q_j \exp\{-(y_t - b_j - z_t)^2/(2w_j)\}/w_j^{1/2}$$
for $j = 1 : J$, and $\sum_{j=1}^{J} q^*_{t,j} = 1$.
2. Resample ϕ under the implied conditional posterior
$$p(\phi|z_{0:T}, \mu, v) \propto a(\phi)p^*(\phi)I(0 < \phi < 1)$$
where $p^*(\phi) = N(\phi|c^*, C^*)$ is the normal density given by
$$p^*(\phi) \propto N(\phi|c, C) \prod_{t=1}^{T} N(z_t|\mu + \phi(z_{t-1} - \mu), v)$$
and
$$a(\phi) \propto p(z_0|\mu, \phi, v) = (1 - \phi^2)^{1/2} \exp(\phi^2(z_0 - \mu)^2/(2v)).$$

The truncated normal $p^*(\phi)I(0 < \phi < 1)$ part of the conditional posterior is generally dominant, being "worth" T observations relative to the $a(\phi)$ term based on only one, and this underlies an efficient Metropolis component: use the truncated normal as a proposal distribution in a Metropolis step that accepts/rejects candidates using ratios of the $a(\phi)$ terms.

3. Resample μ from the implied conditional normal posterior proportional to $p(\mu)N(z_0|\mu, v/(1-\phi^2))\prod_{t=1}^T N(z_t|\mu + \phi(z_{t-1}-\mu), v)$.
4. Resample v from the implied conditional inverse-gamma posterior proportional to $p(v)N(z_0|\mu, v/(1-\phi^2))\prod_{t=1}^T N(z_t|\mu + \phi(z_{t-1}-\mu), v)$.
5. Resample from the conditional posterior for the full volatility sequence $z_{0:T}$. Given the parameters and data, this is the posterior in the conditional, hidden AR(1) DLM of Equation (7.31) but now with the observation equation modified to that conditional on γ_t. That is,

$$\begin{aligned} y_t &= z_t + b_{\gamma_t} + \nu_t^*, \quad \nu_t^* \sim N(\nu_t^*|0, w_{\gamma_t}), \\ z_t &= \mu + \phi(z_{t-1} - \mu) + \epsilon_t, \quad \epsilon_t \sim N(\epsilon_t|0, v). \end{aligned} \tag{7.34}$$

So the posterior for $z_{0:T}$ is multivariate normal and most efficiently sampled using the forward filtering backward sampling (FFBS) algorithm, discussed in Chapter 4, Section 4.5, as follows.

- *Forward Filtering:*
  Set $t = 0$ and define $m_0 = \mu$ and $M_0 = v/(1-\phi^2)$. Filter forward over $t = 1 : T$ to sequentially compute and update the on-line posteriors

$$(z_t|y_{1:t}, \gamma_{1:t}, \mu, \phi, v) \sim N(z_t|m_t, M_t)$$

  using the standard DLM updating equations applied to the model of Equation (7.34).
- *Backward Sampling:*
  Sample from $N(z_T|m_T, M_T)$. Then, for each $t = (T-1) : 0$, sample from the implied sequence of normals

$$p(z_t|z_{t+1}, y_{1:t}, \gamma_{1:t}, \mu, \phi, v) \propto N(z_t|m_t, M_t)N(z_{t+1}|\mu + \phi(z_t - \mu), v).$$

This generates-in reverse order-the volatility trajectory $z_{0:T}$.

Iterating through these steps provides one iterate of the overall MCMC.

Example 7.17 *Daily exchange rates.* Frames (a) in Figures 7.6 and 7.7 show zero-centered, daily returns on the £UK:$USA and ¥Japan:$USA exchange rates, respectively, over a period of 1,000 business days beginning in fall 1992 and ending in early August 1996. The corresponding frames (b) in each figure show the absolute returns, evidencing changes in volatility over the four year period. In particular, there are clearly periods of increased volatility common to each currency, while the volatility is substantially higher than normal for the £UK series near the start of this time window. This corresponds to economic circumstances that were associated with the withdrawal of the United Kingdom from the European Monetary System (EMS) in September 1992, resulting in increased uncertainties and financial volatilities, in particular. Additional spurts of increased volatility in the ¥Japan series were associated with imposed target currency bands for several EU currencies in late 1993, events that played a key role in breaking the EMS. There is some, lesser reflection of this in the £UK series. See further discussion of the economic circumstances related to some of these periods of higher volatility, as well as some aspects of additional volatility modeling, in Quintana and Putnam (1996), and also in Aguilar, Huerta, Prado, and West (1999).

Practicalities of understanding and specifying prior hyper-parameters distributions are key. The SV model is one of "signal-in-noise" in which the z_t signal is weak compared to the noise by ν_t. Careful judgments about relative scales of variation, as well as of persistence of the SV processes, are necessary to define priors that appropriately reflect the practical context and constrain posteriors. We use the same prior specification for separate analyses of each of the two series, based on the following specifications. First, on a daily basis changes in volatility are always quite sustained, suggesting values of ϕ close to one. We use a truncated normal prior $N(\phi|c, C)I(0 < \phi < 1)$ with $c = 1, C = 0.01$, that has an increasing density function relatively concentrated on $0.8 - 1$. It turns out that the posteriors in each of the two analyses are very heavily concentrated on higher values, and this specific prior is really diffuse relative to the posterior; compared to other hyper-parameters, analysis is relatively robust to this component of the prior.

The model implies a marginal variance of $V(y_t) = V(z_t) + V(\nu_t)$ where $V(z_t) = v/(1 - \phi^2)$ and $V(\nu_t)$ is easily computed under the normal mixture error model; for the specific values in the table above, and used here, $V(\nu_t) = 1.234$. Exploration of returns data for a few hundred days prior to the start of the time series indicate sample variances for $y_t = \log(r_t^2)/2$ around 1.3 or so. Note first that this clearly indicates that the noise substantially dominates the signal in these kinds of series. Second, we can use this to guide prior specification, noting that it suggests relevant ranges of values of (ϕ, v) will be roughly consistent with $v = 0.066(1-\phi^2)$. For $\phi = 0.98$, for example, corresponding to a highly persistent AR process, this indicates $v = 0.0026$. With this guideline, we specify the prior $G(v^{-1}|a/2, av_0/2)$, having mean $v_0 = 0.002$ and prior degree of freedom $a = 1{,}000$ so that the prior has the same weight as the data in defining the posterior.

The MCMC analysis was initialized with a zero volatility process, $\phi = 0.95$ and other parameters set at prior means, and repeat analyses confirm rapid convergence and fast mixing for all model parameters. Posterior summaries use a final 5,000 posterior draws after discarding 500 burn-in iterates. Posterior summaries include MCMC-based posterior means (and standard deviations) for ϕ of 0.995 (0.003), and for $v^{1/2}$ of 0.050 (0.001), for the £UK:$USA series; the corresponding figures for the ¥Japan:$USA model are 0.988 (0.005) for ϕ and 0.049 (0.001) for $v^{1/2}$. These are similar, though evidencing somewhat higher levels of persistence in volatility changes for the £UK series. Frames (c) of Figures 7.6 and 7.7 plot the MCMC-based posterior mean volatility trajectories over time for each series. These are accompanied by additional trajectories based on a random selection of 50 posterior samples for $z_{0:T}$ in each case; as a collective, the latter provide some indication of posterior uncertainties and potential levels of volatility excursion around the posterior mean. They also help to convey the point that the volatility process for the £UK:$USA series is substantially more uncertain as well as more variable than that for the ¥Japan:$USA series. Note that, as with frames (a) and (b), the volatility plots in frames (c) maintain the same axis scales across the two analyses to aid the comparison.

7.5.5 Further Comments

It is of interest to note that, though the discrete mixture of normals is introduced as an approximation to the exact error density $p(\nu_t)$, the latter is itself based on the initial assumption of conditional normality of the returns r_t in Equation (7.29). The mixture form itself may be viewed as a direct SV model specification, as an alternative to the initial model. We must also be open to questioning the adequacy of the assumed form in applied work, of course. Exercise 4 in the problems of Section 7.6 is relevant to this general question, and we also note that the overall adequacy of finite mixture approximations, especially in the tails of the error distribution, can be problematic. In SV models, this is currently an area for further research. For example, Niemi and West (2010) have recently introduced a novel computational strategy for model fitting based on adaptive mixture model approximations in state-space models. Applied to the SV model context, this new approach can be used to incorporate the exact form of the error density $p(\nu_t)$ in overall Metropolis-Hastings analysis, and so provide information on the adequacy of assumed approximations as well as improve the overall MCMC. This and other developments will in future extend the computational machinery for SV models, all beginning from and aiming to improve upon the standard and widely used Gibbs sampling approach discussed above. We also note that there is an extensive literature on Monte Carlo methods in SV models that, in addition to MCMC methods as referenced above, includes increasing roles for sequential and adaptive Monte Carlo methods; see references in the discussion of particle filtering and learning in Chapter 6.1.

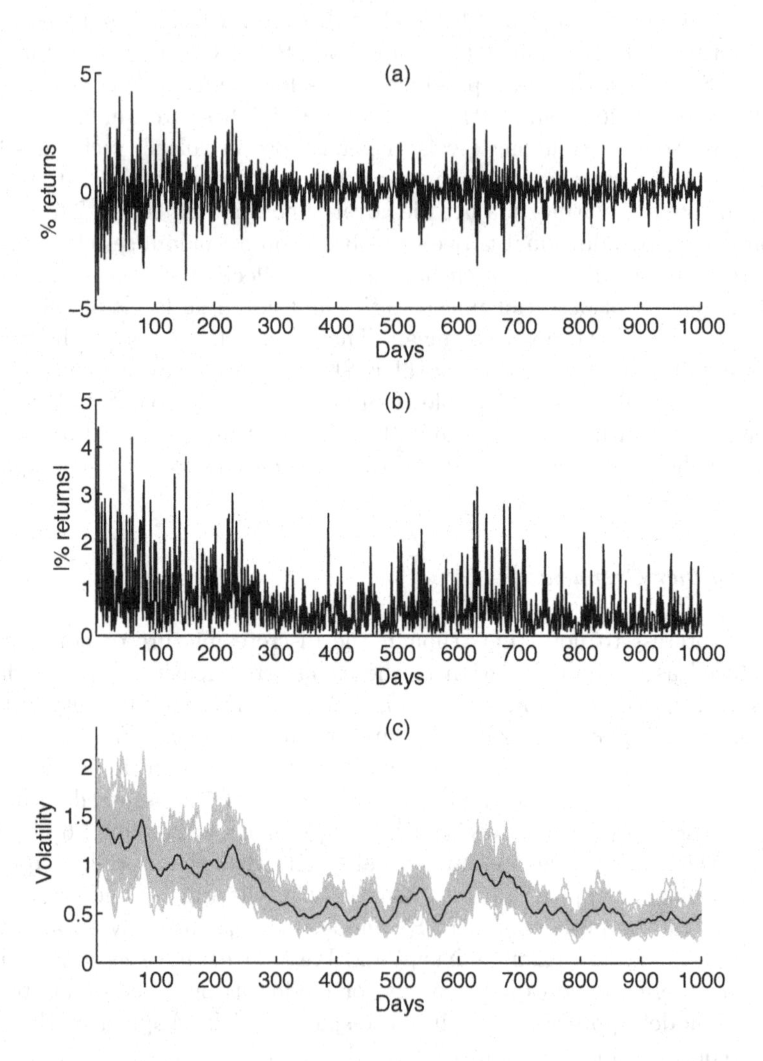

Figure 7.6 *(a) Zero-centered returns on daily £UK:\$USA exchange rate over a period of 1,000 business days ending on August 9, 1996. (b) Absolute returns. (c) Estimated volatility process in the standard univariate SV model. The full line indicates the posterior mean of* $\exp(z_t)$*, plotted over days* $t = 0 : T$*, from the MCMC analysis; the grey shading is 50 similar time plots representing 50 randomly selected trajectories from the posterior.*

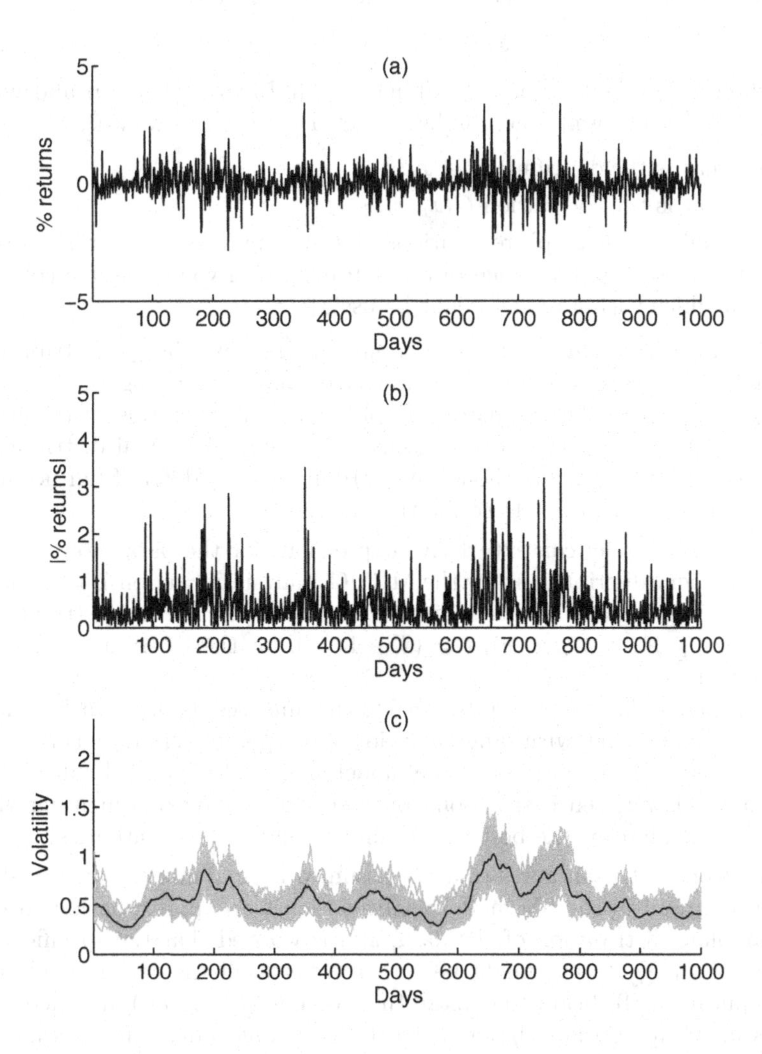

Figure 7.7 *(a) Zero-centered returns on daily ¥Japan:\$USA exchange rate over a period of 1,000 business days ending on August 9, 1996. (b) Absolute returns. (c) Estimated volatility process in the standard univariate SV model. The full line indicates the posterior mean of* $\exp(z_t)$, *plotted over days* $t = 0 : T$, *from the MCMC analysis; the grey shading is 50 similar time plots representing 50 randomly selected trajectories from the posterior.*

7.6 Problems

1. Suppose that $x_{t-1} \sim N(x_{t-1}|0, s)$ and x_t is generated by

$$x_t = z_t x_{t-1} + (1 - z_t)\omega_t$$

where $\omega_t \sim N(0, s)$ and z_t is binary with $Pr(z_t = 1) = a$, and with x_{t-1}, z_t, ω_t being mutually independent. Here s, a are known.

   (a) What is the distribution of x_t?
   (b) What is the correlation $C(x_t, x_{t-1})$?
   (c) Simulate and graph realizations of such a process $x_{1:T}$ for T =1,000 with $s = 1$ and for each of $a = 0.9, 0.99$. Can you suggest a context in which such a model might be useful?

2. Generate a reasonably large random sample from the χ_1^2 distribution and look at histograms of the implied values of $\nu_t = \log(\kappa_t)/2$ where $\kappa_t \sim \chi_1^2$ as in the SV model. Explore the shape of this distribution using quantile-quantile plots against the standard normal distribution to generate insight into how the distribution of ν_t differs from normal, especially in the tails. Describe the differences.

3. In the SV model context, derive expressions for the parameters of the component distributions for the MCMC analysis of Section 7.5.4 including the expressions for the sequences of conditional normal distributions $p(z_t|z_{t+1}, y_{1:t}, \gamma_{1:t}, \mu, \phi, v)$, $t = (T-1) : 0$, for the FFBS algorithm to simulate $z_{0:T}$.
   Implement the MCMC and replicate the analyses reported in Example 7.17. Experiment with different prior hyper-parameters to generate insights into the sensitivity of the model analysis to specified values; the analysis is very sensitive to some of these values, and experimenting with repeat analyses is the best way to understand these sensitivities.

4. An econometrician notes that the assumed normal mixture error distribution used in the SV model analysis is just an approximation to the sampling distribution of the data, and is worried that the specific values of the (q_j, b_j, w_j) might be varied to better fit the data in a specific application. To begin to explore this, assume the (b_j, w_j) are fixed and assumed appropriate, but now treat the mixing probabilities as uncertain. That is, assume the more general model

$$(y_t|z_t, \eta_{1:J}) \sim \sum_{j=1}^{J} \eta_j N(b_j + z_t, w_j)$$

where the J probabilities $\eta_{1:J}$ are now parameters to be estimated too. The original $q_{1:J}$ are viewed as "good first guesses" and used as prior means for a Dirichlet prior

$$p(\eta_{1:J}) \propto \prod_{j=1}^{J} \eta_j^{\alpha q_j - 1}$$

over $0 < \eta_j < 1$ and subject to $\sum_{j=1}^{J} \eta_j = 1$. Taking a reasonably large value of the total *precision* α, such as $\alpha = 500$, indicates that each η_j is likely to be close to its prior mean q_j under this Dirichlet prior, but still allows for variation in the underlying sampling distribution.

(a) Conditional on $\eta_{1:J}$, the MCMC analysis of the SV model remains precisely the same, but with each q_j replaced by η_j. We now need to add one further conditional posterior distribution to generate, at each Gibbs step, a new value $\eta_{1:J}$. What is the appropriate conditional distribution?

(b) Implement the SV MCMC analysis now extended to incorporate this uncertainty about the normal mixture model form, resampling this distribution for $\eta_{1:J}$ at each step. Explore how the analysis results are impacted in examples.

(c) Describe how this general idea might be extended to incorporate some uncertainty about the chosen values of the (b_j, w_j) as well as the q_j.

5. In the univariate SV model assuming the normal mixture form of $p(\nu_t)$, what is the stationary *marginal* distribution $p(y_t|\mu, \phi, v)$ for any t implied by this model? Investment analysis focuses, in part, on questions of just how large or small a per-period return might be, and this marginal distribution of a future return is relevant in that context.

6. A stationary, first order Markov process is generated from $p(x_t|x_{t-1})$ given by

$$x_t \sim \begin{cases} N(x_t|\phi x_{t-1}, v), & \text{with probability } \pi \\ N(x_t|0, s), & \text{otherwise,} \end{cases}$$

where $|\phi| < 1$, $s = v/(1-\phi^2)$ and for some probability π; in the application of interest, π is relatively large, such as 0.9.

(a) What is the distribution of $(x_t|x_{t-1})$?

(b) What is the conditional mean $E(x_t|x_{t-1})$ of this state transition distribution?

(c) Show that the conditional variance $V(x_t|x_{t-1})$ depends quadratically on x_{t-1}. Interpret this.

(d) The process is stationary. What is the marginal distribution $p(x_t)$ for all t?

(e) Comment on the forms of trajectories generated by such a model, and speculate on possible applications. Use simulation examples to help generate insights. Discuss how it differs from the basic, normal AR(1) model (the special case $\pi = 1$).

(f) Is the process reversible? Either prove or disprove.

7. Simulate data from the following model:

$$\begin{aligned} y_t &= \phi^{(1)} y_{t-1} + \nu_t, \quad t = 1:200, \\ y_t &= \phi^{(2)} y_{t-1} + \nu_t, \quad t = 201:400, \end{aligned}$$

where $\phi^{(1)} = 0.9$, $\phi^{(2)} = -0.9$, and $\nu_t \sim N(0, v)$ with $v = 1$. Assuming priors of the form $\phi^{(1)} \sim TN(0.5, 1, \mathcal{R}_1)$ and $\phi^{(2)} \sim TN(-0.5, 1, \mathcal{R}_2)$, with $\mathcal{R}_1 = (0, 1)$ and $\mathcal{R}_2 = (-1, 0)$ on the AR coefficients and an inverse-gamma prior on v, compute $Pr(\mathcal{M}_t(1)|\mathcal{D}_t)$ with $\mathcal{M}(1)$ the AR(1) model with coefficient $\phi^{(1)}$.

Chapter 8

Topics and examples in multiple time series

Multivariate time series analysis develops models and methodology for simultaneous description and forecasting of multiple time series that are inherently interconnected and often have common underlying structures. This chapter shows how univariate analyses of such time series can be useful in discovering some of the common latent structure that may be underlying the relationships that are foundational to a vector time series. Beyond the immediate interest in specific applications, a detailed understanding of these specific examples of linked univariate models aids in motivating classes of multivariate models in following chapters.

8.1 Multichannel Modeling of EEG Data

Prado and West (1997), Prado (1998), and Prado, West, and Krystal (2001) consider various analyses of electroencephalogram (EEG) series recorded during a seizure induced by electroconvulsive therapy (ECT) in 19 locations over a patient's scalp. In this chapter, we revisit, discuss, and extend such analyses.

8.1.1 Multiple Univariate TVAR Models

We begin by showing how common patterns across multiple EEG series can be detected via repeated TVAR univariate analyses. A TVAR analysis of one of 19 EEG series recorded at different scalp locations on a patient who received electroconvulsive therapy—the series recorded at channel *Cz* located at the center of the scalp—was presented and discussed in detail in the

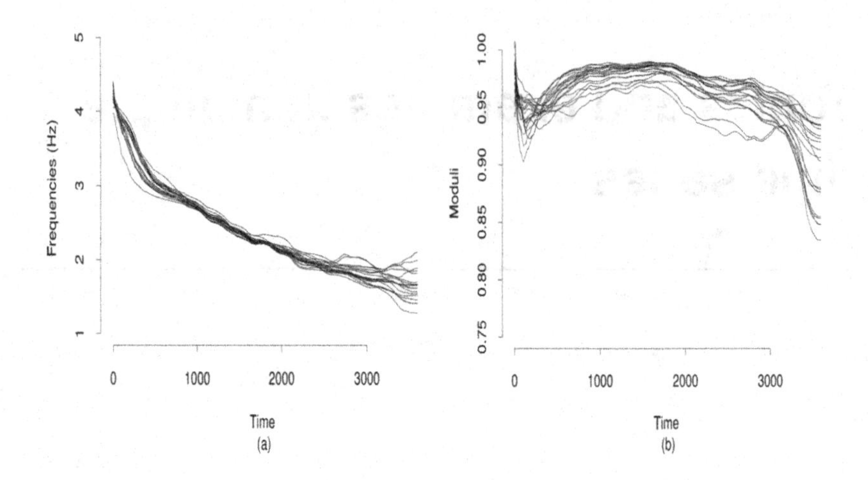

Figure 8.1 *(a) Estimated trajectories of the frequencies of the dominant quasiperiodic components of the 19 channels. (b) Estimated trajectories of the moduli of the dominant quasiperiodic components of the 19 channels.*

previous chapter. Here we show the results obtained by fitting TVAR(12) models to each of these 19 series (for a schematic representation of the locations of the EEG channels over the scalp see Prado, West, and Krystal 2001). For consistency across models, we used the same system discount factor for all the series, $\delta_\phi = 0.994$, and no discounting at the observational level, i.e., $\delta_v = 1.0$, implying that the observational variance was assumed constant over time. We computed estimates of the TVAR decompositions, based on the estimated posterior means of the model parameters, and extracted the trajectories of the frequencies, moduli, and amplitudes of the latent processes in the decompositions over time for all the channels. The latent components underlying the 19 EEG series exhibited and maintained at least four pairs of complex conjugate roots over the entire seizure course. Two of such components, those with the highest moduli and amplitude, had estimated frequencies lying in the delta (0–4 Hz) and theta (4–8 Hz) bands with relatively high estimated moduli (typically above 0.9).

Figure 8.1 (a) shows the estimated trajectories of the frequencies of the dominant seizure waveforms as a function of time for each of the 19 channels. Similarly, Figure 8.1 (b) displays the estimated moduli trajectories of the dominant latent components for all channels. From these graphs we can infer that there is a common pattern of frequency content across series. The trajectories of dominant frequencies over time for the 19 channels lie consistently in the 1–5 cycles per second band and exhibit a decay toward the end of the seizure. Similarly, the corresponding moduli are relatively large and stable during the beginning and middle parts of the seizure, while gradually decaying toward the end of the seizure.

8.1.2 A Simple Factor Model

Factor models are formally discussed and illustrated later in Chapter 11. Here we use the simplest possible factor model—one with a single factor—to infer underlying structure in multiple nonstationary signals. Such a model can also be viewed as a regression with an m-dimensional response at time t, $\mathbf{y}_t = (y_{t,1}, \ldots, y_{t,m})'$, and whose single explanatory variable is an unknown latent process x_t with a given underlying structure. In this context, we consider models in which x_t is a TVAR with known model order p, but unknown parameters $\boldsymbol{\phi}_t = (\phi_{t,1}, \ldots, \phi_{t,p})'$ that vary over time according to a random walk as discussed previously in Chapter 5. Specifically, we consider the following model:

$$\begin{aligned} y_{t,i} &= \beta_i x_t + \nu_{t,i}, \\ x_t &= \textstyle\sum_{j=1}^{p} \phi_{t,j} x_{t-j} + w_t, \\ \boldsymbol{\phi}_t &= \boldsymbol{\phi}_{t-1} + \boldsymbol{\xi}_t, \end{aligned} \tag{8.1}$$

for $i = 1 : m$ and $t = 1 : T$. The β_i's are the factor weights (or regression coefficients); $\boldsymbol{\phi}_t = (\phi_{t,1} \ldots, \phi_{t,p})'$ is the vector of TVAR coefficients at time t; $\nu_{t,i}$, w_t, and $\boldsymbol{\xi}_t$ are independent and mutually independent zero-mean innovations with distributions $N(\nu_{i,t} \mid 0, v_i)$, $N(w_t \mid 0, w)$, and $N(\boldsymbol{\xi}_t \mid 0, \mathbf{U}_t)$ for some observational variance v, AR structural variance w, and system variance-covariance matrices $\mathbf{U}_t$. Here $v_i = v$, with v and w assumed unknown and the $\mathbf{U}_t$s can be specified by means of a discount factor δ_ϕ.

The general idea behind factor analysis and factor models is to reduce the dimensionality of a multivariate set of observations in order to extract their main features. In the particular case of observations that are time series, the goal is to explain the observed variability in the m series in terms of a much smaller number of factors, say $k \ll m$. Some references regarding factor analysis and factor models in context of time series analysis include Peña and Box (1987), who developed a method to identify common hidden factors in a vector of time series assuming an underlying ARMA structure on the common factors; Tiao and Tsay (1989), who proposed a method to model vector autoregressive moving average (VARMA) processes in a parsimonious way via scalar components models; Molenaar, de Gooijer, and Schmitz (1992), where a dynamic factor model with a linear trend model component was developed for analyzing nonstationary multivariate time series. Additional recent references that consider fairly sophisticated factor models, related methodology for posterior inference and model selection, as well as applications in areas such as financial time series and general spatio-temporal models include Aguilar and West (2000), Lopes (2000), Lopes and West (2004), and Lopes, Salazar, and Gamerman (2008). Some of the models developed in these references and their applications will be discussed in Chapter 11. Further additional references to factor models in the time domain are also provided in Chapters 10 and 11. Factor analysis in the frequency domain has been considered by Priestley, Subba-Rao, and Tong (1974), Geweke and Singleton (1981), and Stoffer (1999), among others.

More general factor models can be considered to extend (8.1) to include more than one latent factor, i.e., $\mathbf{x}_t = (x_{1,t}, \ldots, x_{k,t})'$, and possibly time-varying factor weights. In such cases, the first equation in (8.1) would have the following general form,

$$\mathbf{y}_t = \mathbf{B}_t \mathbf{x}_t + \boldsymbol{\nu}_t, \tag{8.2}$$

where $\mathbf{B}_t$ are $m \times k$ matrices of time-varying factor weights, $\mathbf{B}_t = \boldsymbol{\beta}_{1:k,t}$ with $\boldsymbol{\beta}_{j,t} = (\beta_{(1,j,t)}, \ldots \beta_{(m,j,t)})'$, and each $\boldsymbol{\nu}_t$ is an m-dimensional vector usually assumed to follow a zero-mean Gaussian distribution. The latent process vector $\mathbf{x}_t$ can be modeled with a general dynamic linear model (DLM) structure. One important class of models is that based on lagged latent factors. Suppose the first factor $x_{1,t} = x_t$ is a TVAR process and that additional factors are lagged values of this factor, $x_{2,t} = x_{t-1}, \ldots, x_{k,t} = x_{t-k+1}$. Then, it follows that

$$\phi_t(B) y_{i,t} = \sum_{j=1}^{k} \beta_{(i,j,t)} w_{t-j+1} + \phi_t(B) \nu_{i,t}, \tag{8.3}$$

and so, $y_{i,t}$ is a TVARMA(p,q) with $q = max(p,k)$. When $\boldsymbol{\phi}_t = \boldsymbol{\phi}$ and $\mathbf{B}_t = \mathbf{B}$ for all t, we are in the context of the models of Peña and Box (1987).

Note that model (8.1) is not identifiable. Taking $\beta_i^* = \beta_i / c$ for some nonzero constant c, the first two equations in (8.1) can be written in terms of β_i^*, x_t^* and w_t^* as

$$\begin{aligned} y_{i,t} &= \beta_i^* x_t^* + \nu_{i,t}, \\ x_t^* &= \textstyle\sum_{j=1}^{p} \phi_{t,j} x_{t-j}^* + w_t^*, \end{aligned}$$

with $x_t^* = c x_t$ and $w_t^* = c w_t$. One way to deal with this identifiability issue is to impose restrictions on the factor weights, such as setting $\beta_i = 1$ for some i. Other restrictions can be used. In particular, for the general factor model, Peña and Box (1987) take $\mathbf{B}'\mathbf{B} = \mathbf{I}_k$. For the model in (8.1), such restriction implies that $\sum_{i=1}^{m} {\beta_i}^2 = 1$. Another identifiability problem arises when the observational and system variances v and w are both unknown (the variances and x_t can be rescaled, obtaining again the same model representation given in Equation 8.1). To deal with this it can be assumed that the signal-to-noise ratio, $r = w/v$, is a known quantity.

Posterior inference in model (8.1) can be achieved following a Gibbs sampling scheme. Specifically, denote $\beta_{1:m}$ the set of factor weights; $x_{1:T}$ the set of latent values; $\boldsymbol{\phi}_{1:T}$ the set of TVAR coefficients; and $\mathbf{U}_{1:T}$ the collection of system variance-covariance matrices. Then, if the matrices in $\mathbf{U}_{1:T}$ are assumed known, or specified by means of a discount factor δ_ϕ, posterior simulation methods based on the basic DLM theory and the Gibbs sampler for state-space models summarized in Chapter 4 can be applied to obtain samples from the joint posterior by iteratively sampling from the conditional densities as follows.

1. Sample $x_{1:T}$ from $p(x_{1:T}|\mathbf{y}_{1:T}, \boldsymbol{\beta}_{1:m}, \boldsymbol{\phi}_{1:T}, v)$. Sampling from this distribution is done via the forward filtering backward sampling (FFBS) algorithm.
2. Sample $\boldsymbol{\phi}_{1:T}$ from $p(\boldsymbol{\phi}_{1:T}|x_{1:T}, \mathbf{U}_{1:T}, v)$. Sampling from this distribution is done via the FFBS algorithm.
3. Sample $\boldsymbol{\beta}_{1:m}$ from $p(\boldsymbol{\beta}_{1:m}|\mathbf{y}_{1:T}, x_{1:T}, v)$.
4. Sample v from $p(v|\mathbf{y}_{1:T}, \boldsymbol{\beta}_{1:m}, x_{1:T})$.

Repeating these steps iteratively in a Gibbs sampling form will allow us to obtain samples from the posterior $p(x_{1:T}, \boldsymbol{\phi}_{1:T}, \boldsymbol{\beta}_{1:m}, v|\mathbf{y}_{1:T})$ once the Markov chain Monte Carlo (MCMC) algorithm has converged. Details related to the posterior sampling algorithm are left to the reader (see Problem 1 in this Chapter).

Example 8.1 *An analysis of the Ictal-19 data set.* The model in (8.1) was fitted to the 19 EEG series analyzed above. We set the factor weight of channel *Cz* to one for identifiability, i.e., $\beta_{\mathbf{Cz}} = 1.0$. Reference priors were adopted on β_i and v, while relatively diffuse normal priors were chosen for $\boldsymbol{\phi}_0$. A discount factor $\delta_\phi \in (0.9, 1]$ was used to specify the structure of the matrices $\mathbf{U}_t$ that control the evolution of the AR coefficients over time. The signal-to-noise ratio $r = s/v$ was set at a fixed value. Choices of r in the $(5, 15)$ interval produced reasonable results in terms of the x_t process. Estimates of x_t obtained with these values of r had amplitudes similar to the amplitude of the process obtained by adding the first two to four components in the TVAR decompositions of the *Cz* EEG series. TVAR model orders between four and eight were considered to represent the fact that typically about two to four quasiperiodic components are needed to describe most of the variability in the EEG series based on the TVAR analyses presented in Chapter 5.

Figures 8.2 and 8.3 summarize the results obtained for a model with $p = 6$, $\delta_\phi = 0.994$, and $r = w/v = 10$, based on a posterior sample of 3,000 draws

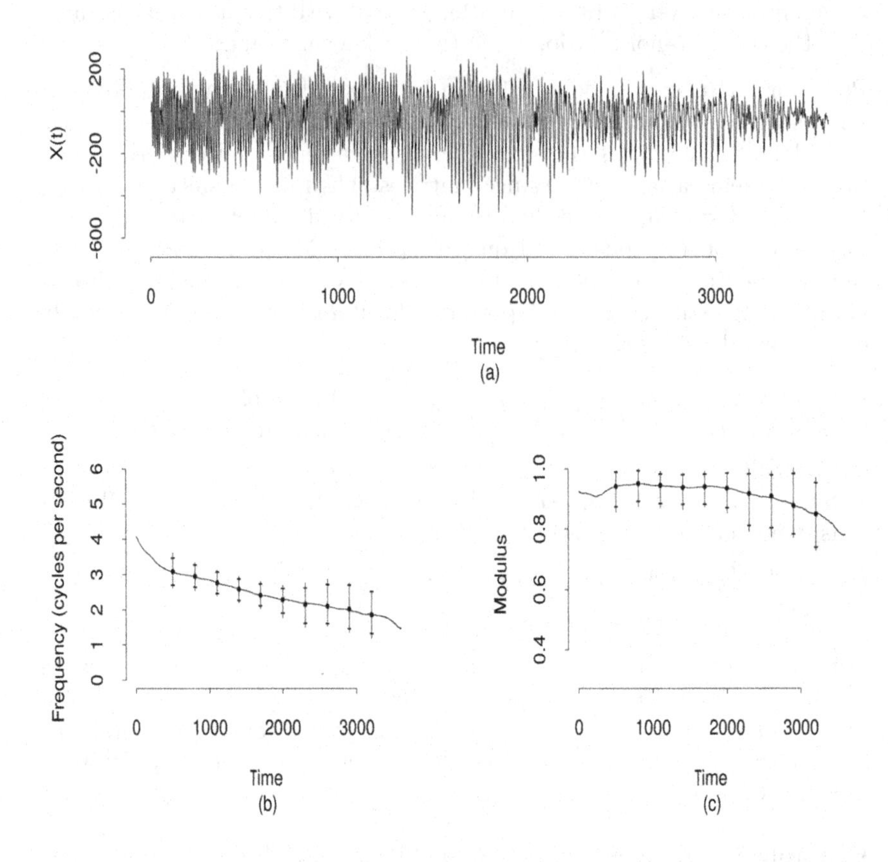

Figure 8.2 *(a) Estimated latent factor process. (b) Frequency trajectory of the estimated dominant component for the latent process and approximate 95% posterior intervals at selected points. (c) Modulus trajectory of the estimated dominant component for the latent process and approximate 95% posterior intervals at selected points.*

taken after the MCMC chain had achieved convergence. Figure 8.2 displays the estimated posterior mean of the latent process x_t as well as time trajectories for the frequency and modulus of the dominant quasiperiodic component of x_t. The TVAR(6) structure assumed on x_t consistently exhibited two pairs of complex roots: one whose frequency and modulus trajectories are shown in Figure 8.2, graphs (b) and (c), and a second component with a frequency usually lying on the alpha and theta ranges (4–13 Hz) that switches during some very brief periods of time to a much higher frequency (usually > 15 Hz) and very low modulus. The instantaneous characteristic frequencies and moduli were computed by solving the characteristic equa-

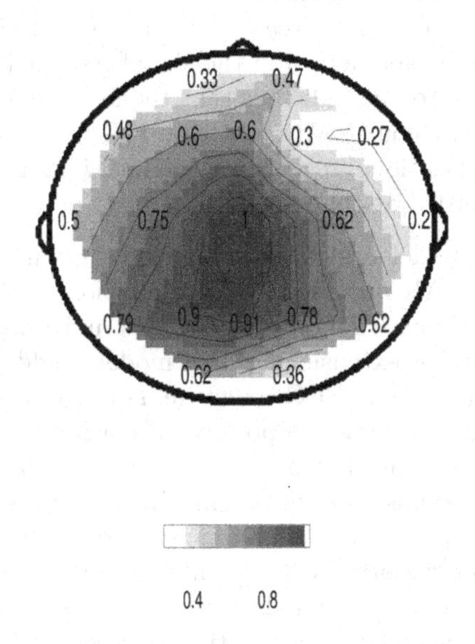

Figure 8.3 *Image and contour plot of the posterior means of the factor weights for the 19 EEG channels.*

tions evaluated at the estimated posterior means of the TVAR parameters. Figures 8.2(b) and 8.2(c) also show approximate 95% posterior intervals for the instantaneous frequency and moduli. These graphs show the same type of behavior found in single channel analyses via univariate TVAR models: the dominant component of x_t is quasiperiodic with decreasing frequency in the 0–4 Hz range, and rather persistent modulus.

Figure 8.3 shows a spatial display of the estimated posterior means of β_i for $i = 1:19$. The picture was drawn by interpolating the posterior means of the β_is onto a grid defined by the approximate electrode locations on the scalp. Dark intensities correspond to high values of β_i while light intensities are related to low values. Given that channel *Cz*, located at the very center of the scalp, was set to have a constant factor weight $\beta_{\mathbf{Cz}} = 1$, the EEG signal recorded at this channel is being modeled as the latent process x_t plus noise, while signals recorded at other locations are represented in terms of the underlying process x_t weighted by a factor of β_i plus noise. Figure 8.3 exhibits a strong pattern of spatial relationship across channels. Those channels located closer to channel *Cz* have higher factor weights. It can also be seen that channels located at right fronto-temporal sites have smaller weights than those located at left fronto-temporal sites with respect to *Cz*. Approximate 95% intervals for all the β_is were also computed (not shown).

Such intervals showed that all the factor weights were smaller than unity in magnitude. Each β_i can be seen as a measure of similarity between the signal recorded at channel i and the signal recorded at channel Cz. A value of β_i very close to one indicates that the two signals are very similar throughout the seizure course. Alternatively, if β_i is smaller than one then, at least on average, the signal recorded at channel i has a smaller amplitude level than that recorded at channel Cz.

This simple factor model basically reveals the same underlying structure suggested by the TVAR analyses, but, in addition, it shows spatial patterns across the estimated factor weights that univariate TVAR models cannot capture. Further extensions of this model would give us additional insight into the multichannel EEG structure. For instance, the residuals for the different channels are not temporally or spatially independent, showing quasiperiodic patterns that are left unexplained by the model. Such structure in the residuals may be arising from many sources. First, the assumption of a single latent common factor is quite simplistic, as it is to assume constant factor weights throughout the seizure course. A simple eye inspection of a couple of EEG signals recorded at sites located relatively far away on the scalp shows that the relation between their amplitudes does not remain constant in time: while both series display roughly the same amplitude levels at the beginning of the seizure, the EEG signal recorded at one of the locations shows a much larger amplitude than that of the signal recorded at the other location toward the end of the seizure. Also, including lags and/or leads of x_t into the model might be key to explore temporal relationships across the multiple series. Finally, models that formally take into account the spatial dependence across the signals recorded at the different channel locations could be used. We further investigate some of these issues later in this Chapter, after introducing some theory and notation related to multivariate spectral analysis.

8.2 Some Spectral Theory

We now summarize some aspects of multivariate spectral theory. For a more detailed theoretical development and illustrations see, for example, Brockwell and Davis (1991).

Let $y_{t,1}$ and $y_{t,2}$ be two stationary time series processes with means μ_1 and μ_2, respectively. In Chapter 1, we defined the cross-covariance function as

$$\gamma_{1,2}(h) = E\{(y_{t+h,1} - \mu_1)(y_{t,2} - \mu_2)\}.$$

Assuming that this function is absolutely summable, i.e., if $\sum_h |\gamma_{1,2}(h)| <$

∞, we have the following spectral representation,

$$\gamma_{1,2}(h) = \int_{-\pi}^{\pi} e^{i\omega h} f_{1,2}(\omega) d\omega, \tag{8.4}$$

for $h = 0, \pm 1, \pm 2, \ldots$, where $f_{1,2}(\omega)$ is referred to as the *cross-spectrum*, or *cross-spectral density* of $y_{t,1}$ and $y_{t,2}$. We also have that

$$f_{1,2}(\omega) = \frac{1}{2\pi} \sum_{h=-\infty}^{\infty} \gamma_{1,2}(h) e^{-i\omega h}, \tag{8.5}$$

for $\omega \in [-\pi, \pi]$. The *spectral density matrix* or *spectrum* of the two-dimensional process $\mathbf{y}_t = (y_{t,1}, y_{t,2})'$ is then defined as

$$\mathbf{f}(\omega) = \begin{pmatrix} f_{1,1}(\omega) & f_{1,2}(\omega) \\ f_{2,1}(\omega) & f_{2,2}(\omega) \end{pmatrix} = \frac{1}{2\pi} \sum_{h=-\infty}^{\infty} \Gamma(h) e^{-ih\omega},$$

where $\Gamma(h)$ is the autocovariance matrix given by

$$\Gamma(h) = \begin{pmatrix} \gamma_{1,1}(h) & \gamma_{1,2}(h) \\ \gamma_{2,1}(h) & \gamma_{2,2}(h) \end{pmatrix}.$$

The *coherence function* is defined as

$$\mathcal{C}_{1,2}(\omega) = \frac{|f_{1,2}(\omega)|}{[f_{1,1}(\omega) f_{2,2}(\omega)]^{1/2}}, \tag{8.6}$$

which satisfies $0 \leq |\mathcal{C}_{1,2}(\omega)|^2 \leq 1$, for $\omega \in [-\pi, \pi]$. A value of $|\mathcal{C}_{1,2}(\omega)|^2$ close to one indicates a strong linear relation between the spectral density of $y_{t,1}$ and that of $y_{t,2}$. We can also define the *cospectrum*, $c_{1,2}(\omega)$, and the *quadrature* spectrum, $q_{1,2}(\omega)$, of $y_{t,1}$ and $y_{t,2}$ as

$$c_{1,2}(\omega) = Re\{f_{1,2}(\omega)\} \quad \text{and} \quad q_{1,2}(\omega) = -Im\{f_{1,2}(\omega)\}, \tag{8.7}$$

respectively.

Equations (8.4) and (8.5) summarize the spectral representation for a two-dimensional time series process. This can be extended for an m-dimensional process, $\mathbf{y}_t = (y_{t,1}, \ldots, y_{t,m})'$, with absolutely summable autocovariance matrix $\Gamma(h)$. In other words, $\Gamma(h)$ has the representation

$$\Gamma(h) = \int_{-\pi}^{\pi} e^{ih\omega} \mathbf{f}(\omega) d\omega, \tag{8.8}$$

for $h = 0, \pm 1, \pm 2, \ldots$, if $\sum_h |\gamma_{i,j}(h)| < \infty$ for all $i, j = 1 : m$. Here $\mathbf{f}(\omega)$ is the spectral density matrix, whose ij-th element is the cross-spectrum between the components i and j of $\mathbf{y}_t$. This matrix has the representation

$$\mathbf{f}(\omega) = \frac{1}{2\pi} \sum_{h=-\infty}^{\infty} \Gamma(h) e^{-ih\omega}, \tag{8.9}$$

for $\omega \in [-\pi, \pi]$.

The cross-spectrum of $y_{t,1}$ and $y_{t,2}$ can also be written in polar coordinates, i.e.,

$$f_{1,2}(\omega) = \alpha_{1,2}(\omega)\exp(i\phi_{1,2}(\omega)),$$

where $\alpha_{1,2}(\omega) = c_{1,2}^2(\omega) + q_{1,2}^2(\omega)$ is the *amplitude spectrum* and $\phi_{1,2}(\omega) = arg(c_{1,2}(\omega) - iq_{1,2}(\omega))$ is the *phase spectrum.*

Example 8.2 *Cross-spectrum of a bivariate time series.* Let $\mathbf{y}_t$ be a bivariate time series process defined as follows:

$$\begin{aligned} y_{t,1} &= \phi y_{t-1,1} + \epsilon_t, \\ y_{t,2} &= \beta y_{t+2,1}, \end{aligned} \tag{8.10}$$

where $\phi \in (-1, 1)$, $\beta > 0$, and $\epsilon_t \sim N(0, 1)$. Then,

$$\Gamma(h) = \begin{pmatrix} \frac{\phi^{|h|}}{1-\phi^2} & \frac{\beta\phi^{|h-2|}}{1-\phi^2} \\ \frac{\beta\phi^{|h+2|}}{1-\phi^2} & \frac{\beta^2\phi^{|h|}}{1-\phi^2} \end{pmatrix}$$

for $h = 0, \pm1, \pm2, \ldots$, and so,

$$\mathbf{f}(\omega) = \begin{pmatrix} \frac{1}{2\pi(1+\phi^2-2\phi\cos(\omega))} & \frac{\beta e^{-i2\omega}}{2\pi(1+\phi^2-2\phi\cos(\omega))} \\ \frac{\beta e^{i2\omega}}{2\pi(1+\phi^2-2\phi\cos(\omega))} & \frac{\beta^2}{2\pi(1+\phi^2-2\phi\cos(\omega))} \end{pmatrix}. \tag{8.11}$$

This implies that the cospectrum and quadrature spectrum of $y_{t,1}$ and $y_{t,2}$ are, respectively,

$$c_{1,2}(\omega) = \frac{\beta\cos(2\omega)}{2\pi(1 - 2\phi\cos(\omega) + \phi^2)} \quad \text{and} \quad q_{1,2}(\omega) = \frac{\beta\sin(2\omega)}{2\pi(1 - 2\phi\cos(\omega) + \phi^2)}.$$

It is easy to see that the coherency is one, indicating a perfect linear relation between the spectral densities of $y_{t,1}$ and $y_{t,2}$.

8.2.1 The Cross-Spectrum and Cross-Periodogram

Let $\mathbf{y}_t$ be bivariate time series vectors for $t = 1 : T$. The *cross-periodogram* at Fourier frequencies $\omega_j = 2\pi j/T$ for $j = 1 : \lfloor T/2 \rfloor$ is given by

$$I_{1,2}(\omega_j) = \frac{2}{T}\left(\sum_{t=1}^{T} y_{t,1}e^{-i\omega_j t}\right)\left(\sum_{t=1}^{T} y_{t,2}e^{i\omega_j t}\right).$$

In addition, if $\mathbf{y}_t$ is stationary with mean $\boldsymbol{\mu}$ and covariance matrices $\Gamma(h)$ with absolutely summable components, it can be shown that $E(I_{1,2}(0)) - T\mu_1\mu_2$ goes to $2\pi f_{1,2}(0)$, and $E(I_{1,2}(\omega))$ goes to $2\pi f_{1,2}(\omega)$, for $\omega \neq 0$, when T goes to infinity.

Example 8.3 *Cross-spectrum of a bivariate time series (continued).* We simulated 200 observations from the process (8.10) with $\phi = 0.9$ and $\beta = 0.8$. Figure 8.4 shows the theoretical spectral densities of $y_{t,1}$ (solid line) and $y_{t,2}$ (dotted line), as well as the estimated spectral densities obtained from R (top left and right graphs, respectively). The figure also shows the estimated squared coherency (bottom left graph). From this graph, we see an estimated coherency of almost one at all frequencies. The estimated phase spectrum appears at the bottom right. Note that this graph is piecewise linear. For a process like the one described by (8.10), the theoretical phase spectrum consists of parallel lines with slope -2, indicating that the series $y_{t,2}$ leads the series $y_{t,1}$ by two time units. The estimated phase spectrum is consistent with this behavior.

These spectral estimates were obtained using R's `spec.pgram` function. Confidence intervals around the estimated spectra, coherency, and phase spectrum can also be plotted (not shown).

Example 8.4 *Estimated spectra, coherency, and phase in EEG data.* We now illustrate the use of traditional multivariate spectral density estimation methods in the analysis of three channels of EEG data recorded on a single patient during ECT therapy. The data analyzed here were recorded at channels O_1, located in the left occipital region; C_z, located at the center of the scalp; and Fp_2, located in the parietal frontal region. Figure 8.3 shows a schematic representation of the locations of these three channels: O_1 is the channel at the back of the scalp with estimated factor weight of 0.62, C_z has a factor weight of 1.0, and Fp_2 is the frontal channel with estimated factor weight of 0.47.

We considered the analysis of these series during two time periods: a period consisting of 100 observations from the beginning of the recording, i.e., $t = 1 : 100$, and a period of the middle portion of the seizure course, also consisting of 100 observations in the period $t = 1101 : 1200$. The top left graph in Figure 8.5 shows the estimated spectra of the EEG series recorded at O_1 (solid line) and that of the EEG series recorded at channel Fp_2 (dotted line) during the initial period of the seizure. Both estimated spectra show a peak around 4.3 Hz, characteristic of dominant frequencies at initial seizure stages. The bottom left graph shows the estimated spectra later in the seizure course and the peak in both estimated spectra is below 4.3 Hz, indicating that the dominant frequency decreases toward the end of the seizure, as was already shown in multichannel TVAR analyses of the same data. A similar behavior is found in the estimated spectra of the series recorded at channel C_z (see top and bottom left graphs in Figure 8.6). The top and bottom middle graphs in Figures 8.5 and 8.6 display the estimated squared coherency at different time periods. From these graphs we see that the estimated coherency values are very large at the dominant frequencies,

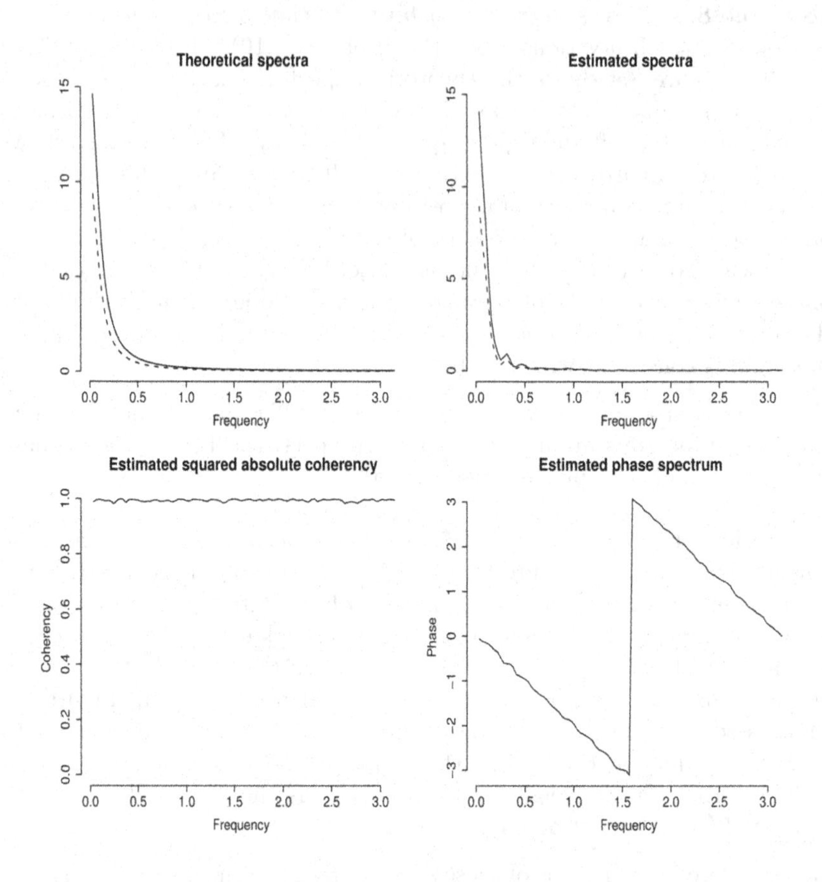

Figure 8.4 *Top left: theoretical spectra of* $y_{t,1}$ *(solid line) and* $y_{t,2}$ *(dotted line). Top right: estimated spectra of* $y_{t,1}$ *(solid line) and* $y_{t,2}$ *(dotted line). Bottom left: estimated squared coherency. Bottom right: estimated phase spectrum.*

indicating that at these frequencies the series are correlated. Note that the highest estimated coherency values are achieved at the dominant frequency between the series recorded O_1 and the series recorded at C_z during middle seizure portions. Finally, the right top and bottom graphs in Figures 8.5 and 8.6 display the estimated phase spectra. The estimated phase spectrum for the series recorded at O_1 and Fp_2 during the middle portion of the seizure is piecewise, with approximate slope of -0.52 at low frequencies and approximate slope of -1.24 at high frequencies, providing evidence that the series recorded at Fp_2 may lead the series recorded at O_1 by 0.5 to 1.2 time units during $t = 1101 : 1200$. Similarly, the estimated phase spectrum for the series recorded at O_1 and C_z, again during middle

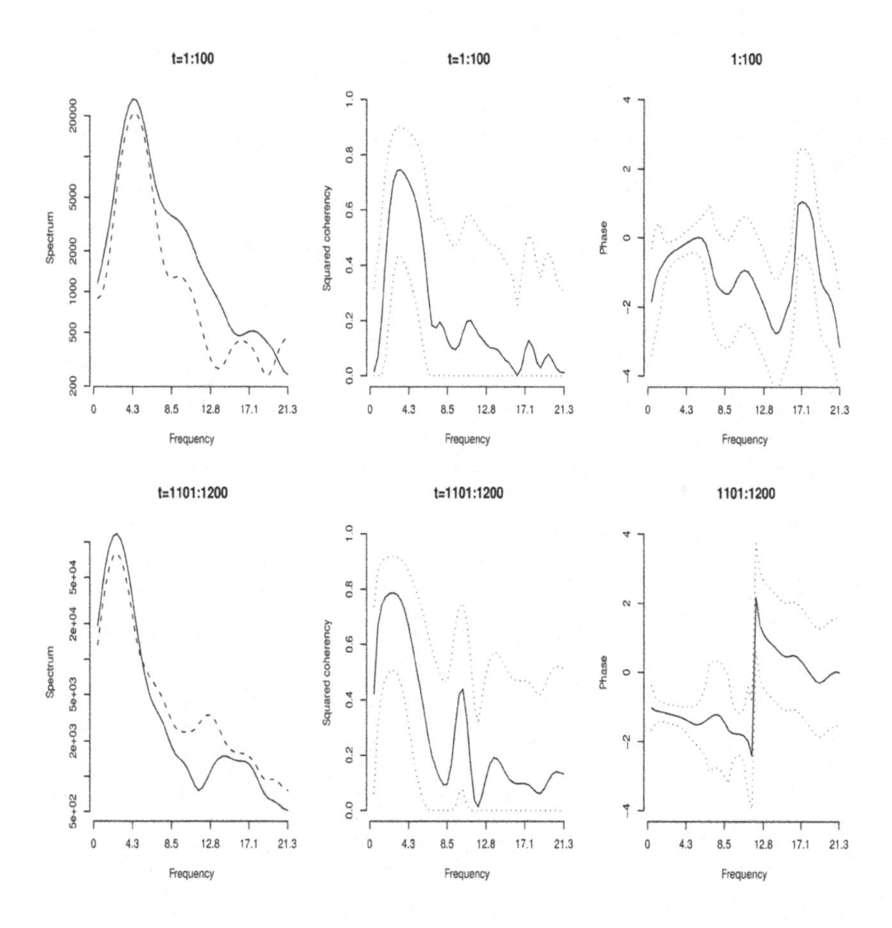

Figure 8.5 *Left top and bottom graphs: estimated spectra of the series for channel* O_1 *(solid line) and channel* Fp_2 *(dotted line). Middle and right top graphs: estimated square coherency and estimated phase with corresponding confidence bands for the initial seizure period* $t = 1 : 100$. *Middle and right bottom graphs: same graphs displayed at the top but now for the middle period of the seizure* $t = 1101 : 1200$.

portions of the seizure, is piecewise with approximate slope of -0.99 at low frequencies and approximate slope of -0.66 at high frequencies, providing evidence that the series recorded at C_z may lead the series recorded at O_1 by 0.6 to 1 time units during $t = 1101 : 1200$.

These results motivate the models discussed next in the analysis of the multichannel EEG series.

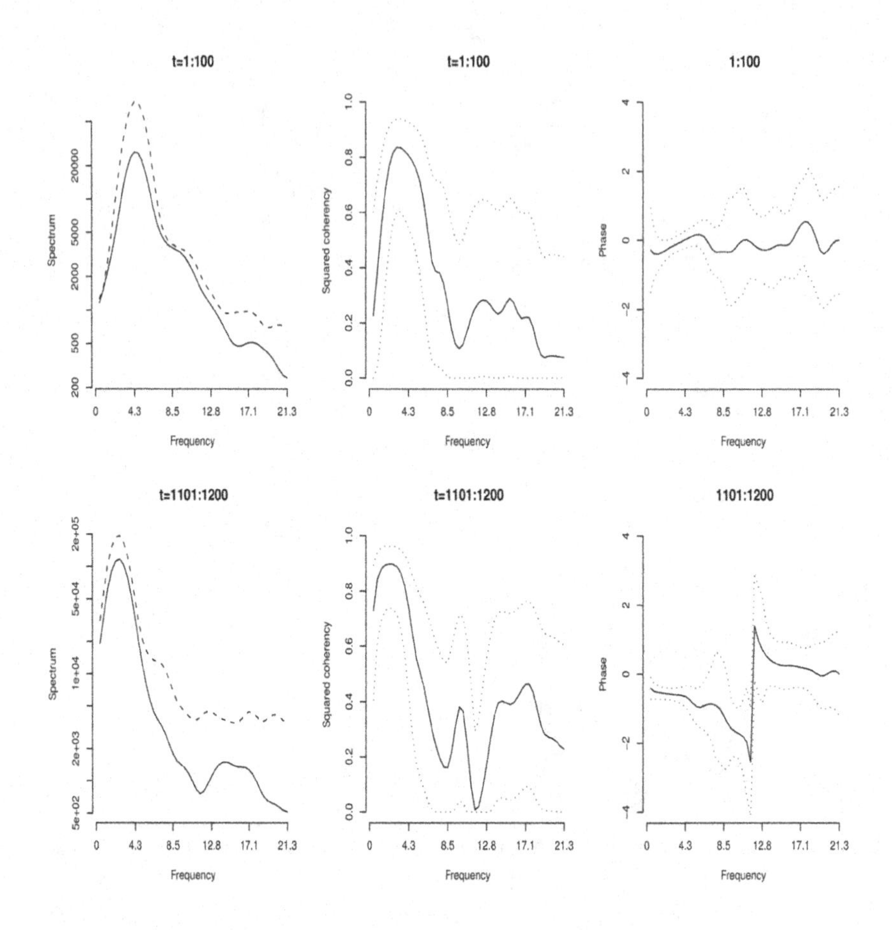

Figure 8.6 *Left top and bottom graphs: estimated spectra of the series for channel O_1 (solid line) and channel Cz (dotted line). Middle and right top graphs: estimated square coherency and estimated phase with corresponding confidence bands for the initial seizure period $t = 1 : 100$. Middle and right bottom graphs: same graphs displayed at the top but now for the middle period of the seizure $t = 1101 : 1200$.*

8.3 Dynamic Lag/Lead Models

The spectral characteristics of the EEG series recorded at channels O_1, Fp_2, and C_z inferred in the previous section suggest that there is a frontal-to-occipital relation between channels. Channels located at the occipital region of the scalp exhibit signals that are delayed with respect to those recorded from channels located at the front of the scalp during some periods of seizure course. This is related to the concept of driving or causality studied in Kitagawa and Gersch (1996a, Section 12.5) in epileptic human signals. Specifically, Kitagawa and Gersch (1996a) fit time-varying vector AR models to multiple series with smoothness priors on the partial autocorrelation coefficients. Their models are time domain models, but their definition of causality is a frequency domain concept that explores instantaneous power spectral densities, spectral coherences, and partial spectral coherences to determine which channels drive the others at a given time.

We now illustrate how the univariate models proposed in Prado, West, and Krystal (2001) can be used to explore dynamic lag/lead structures across EEG series recorded at multiple channel locations. The main idea behind these models is to regress each series on a given underlying process, or lagged/leaded values of such process, at each time t. In particular, Prado, West, and Krystal (2001) consider the following model,

$$\begin{aligned} y_{i,t} &= \beta_{(i,t)} x_{t-l_{i,t}} + \nu_{i,t}, \\ \beta_{(i,t)} &= \beta_{(i,t-1)} + \xi_{i,t}, \end{aligned} \tag{8.12}$$

with $y_{i,t}$ the observation recorded at time t and site i on the scalp for $i = 1:19$, and the following specifications:

- x_t is an underlying process assumed known at each time t, such as the EEG signal recorded at a particular location on the scalp.
- $l_{i,t}$ is the lag/lead that $y_{i,t}$ exhibits with respect to x_t, where $l_{i,t} \in \{-k_0, \ldots, -1, 0, 1, \ldots, k_1\}$. The evolution on $l_{i,t}$ is specified by the transition probabilities $Pr(l_{i,t} = k | l_{i,t-1} = m)$, with $-k_0 \leq k, m \leq k_1$, and $-k_0$, k_1 some suitable bounds chosen a priori.
- $\beta_{(i,t)}$ is the dynamic regression coefficient that weights the influence of x_t, or its lagged/leaded values, on the series recorded at channel i at time t.
- $\nu_{i,t}$ are independent zero-mean Gaussian innovations with variance v_i; $\xi_{i,t}$ are independent zero-mean system innovations assumed normally distributed with variance $w_{i,t}$ at each time t.

Based on these assumptions we have that

$$E(y_{i,t}) = E(\beta_{(i,t)}) \Big[\textstyle\sum_{k=-k_0}^{k=k_1} x_{t-k} \times Pr(l_{i,t} = k) \Big].$$

That is, $E(y_{i,t})$ is a weighted average of the processes $x_{t+k_0}, x_{t+k_0-1}, \ldots, x_t, \ldots, x_{t-k_1}$ with weights $Pr(l_{i,t} = -k_0), \ldots, Pr(l_{i,t} = k_1)$ amplified or reduced by $E(\beta_{(i,t)})$. Given that x_t is the same fixed underlying process for all channels, it is possible to make comparisons between any two channels by comparing each channel with respect to x_t via $\beta_{(i,t)}$ and $l_{i,t}$.

Prado, West, and Krystal (2001) set $x_t = y_{t,C_z}$ so that x_t is the actual signal recorded at channel Cz. Any other signal recorded at any of the 19 available channels could have been chosen as x_t and the results, though not the same in terms of parameter values, would preserve the same relationships across channels. Additional model components that need to be specified are priors on $\beta_{i,0}$, v_i, and the transition probabilities. Relatively diffuse normal/inverse-gamma priors were placed on the regression coefficients at time $t = 0$ and the observational variances for each i. The system innovation variances $w_{i,t}$ were specified via discount factors. The transition probabilities were fixed for all t as follows:

$$Pr(l_{i,t} = k | l_{i,t-1} = m) = \begin{cases} 0.9999 & \text{if } k = m, \\ 0.0001 & \text{if } k = -1 \text{ and } m = -2 \text{ or} \\ & \text{if } k = 1 \text{ and } m = 2, \\ 0.0005 & \text{if } |k - m| = 1 \text{ and} \\ & k \text{ and } m \text{ are neither} \\ & (-1, -2) \text{ nor } (1, 2), \\ 0 & \text{otherwise.} \end{cases}$$

Models with $k_0 = k_1 = 3$ were also considered leading to similar results. Prior distributions $Pr(l_{i,0} = k \mid D_0) = p_{i,k}$ were taken so that all the lags/leads considered in the model had the same weight a priori. Discount factors close to unity were chosen to control the variability of the regression coefficients. Posterior inference was achieved via Gibbs sampling using an FFBS scheme (see Chapter 4).

The posterior summaries displayed here are based on 5,000 draws from the posterior distributions of the model parameters for each channel. A discount factor of $\delta_v = 0.996$ was used for all the channels. Figure 8.7 shows some of the time-varying structure exhibited by the regression coefficients across channels. The image plots at selected time points were built by interpolating the values of the estimated posterior means of the regression coefficients onto a grid defined by the approximate location of the electrodes over the scalp. Dark intensities correspond to higher values of the regression coefficients, while light intensities are related to lower values. Note the asymmetry, more evident toward the end of the seizure, between the estimated coefficients of channels located at the frontal right side and those located at the frontal left side of the scalp. In addition, channels located at the occipital region exhibit, particularly toward the end of the seizure, higher regression coefficients than those located in frontal regions.

Figure 8.8 shows a spatial display of the estimated lag/lead structure across channels based on the posterior means at different time points during the seizure. Circles correspond to leads with respect to channel Cz, while squares represent lags. Channels with estimated lags/leads of zero have no squares or circles. The size of the circle/square is proportional to the ab-

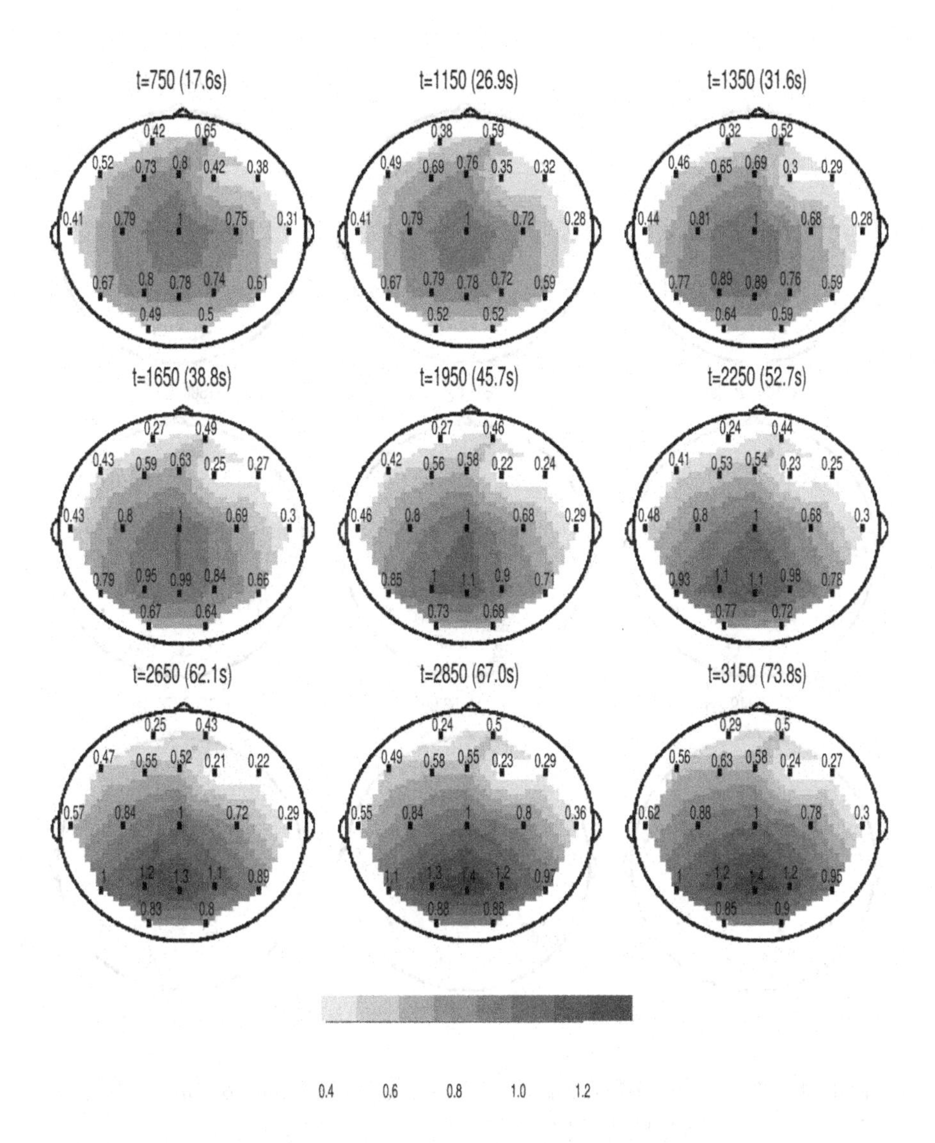

Figure 8.7 *Image plots of estimated posterior means of the dynamic factor weights for all channels at selected points during the seizure course.*

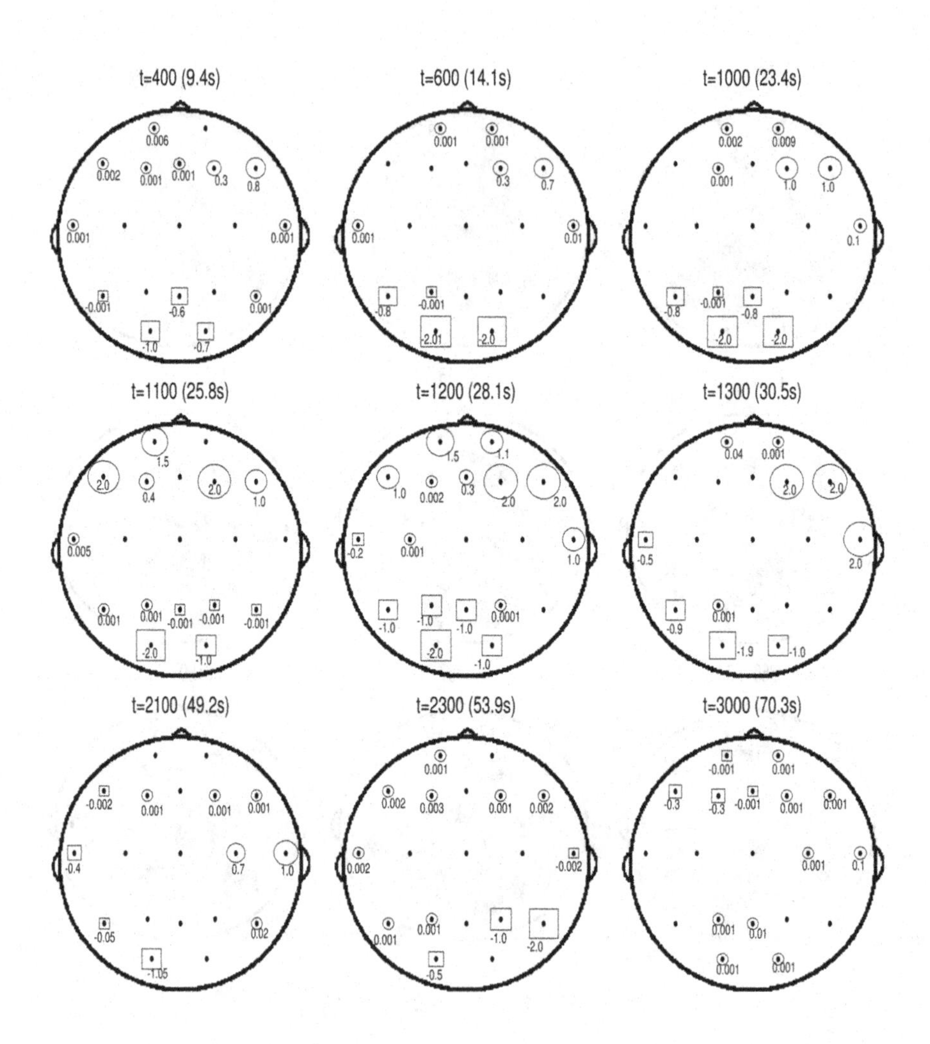

Figure 8.8 *Image plot displaying the dynamic lag/lead structure on the Ictal-19 data set, based on posterior mean estimates.*

solute value of the estimated lag/lead. The plots at times t =1,100, 1,200, and 1,300 display intense "lag/lead activity" with respect to the signal recorded at Cz. Such activity is characterized by lags in the occipital region and leads in the frontal and prefrontal regions; the level of activity decreases considerably toward the end of the seizure. These results illustrate that the Ictal-19 data set shows a complex structure across channels over time.

8.4 Other Approaches

Chapters 9 and 10 consider several time domain approaches for analyzing multivariate time series. Different frequency domain approaches for analysis and classification of multivariate stationary and nonstationary time series have also been considered in recent years. For a review of frequency domain methods for analysis—including principal components, canonical, and factor analysis—discrimination and clustering of multivariate *stationary* time series see Shumway and Stoffer (2017). Also, more recently, Rosen and Stoffer (2007) proposed a Bayesian approach that uses MCMC to fit smoothing splines to each component (real and imaginary) of the Cholesky decomposition of the periodogram matrix.

Frequency domain approaches for analysis, discrimination, and clustering of nonstationary multivariate time series are also available. In particular, Ombao, Raz, von Sachs, and Malow (2001) introduced a method for analyzing bivariate nonstationary time series based on the smooth localized complex exponential (SLEX) transform. Such a method automatically segments the time series into blocks that are approximately stationary and segments the span to be used in obtaining smoothed estimates of the time-varying spectra and coherence. The smoothed periodograms obtained via SLEX are shown to be consistent estimators. Ombao, Raz, von Sachs, and Malow (2001) used their methods to analyze electroencephalograms recorded in two channels during an epileptic seizure. Later, Huan, Ombao, and Stoffer (2004) use the SLEX model for discrimination and classification of nonstationary time series. More specifically, these authors propose a discriminant method for nonstationary time series that has two parts, namely, a feature extraction part and a classification part. The feature extraction step consists of automatically selecting a SLEX basis that captures the key differences among the signals in the different training groups. Then, the SLEX periodograms of the time series that need to be classified are computed and each time series is assigned to a specific class using a criterion based on the Kullback-Leibler divergence. Ombao, von Sachs, and Guo (2005) generalized the approach of Ombao, Raz, von Sachs, and Malow (2001) to deal with multivariate time series by taking into account the coherence between all the components of such time series, as opposed to just looking at the coherence of pairs of components.

8.5 Problems

1. Work out the details of the Gibbs sampling algorithm sketched in Section 8.1 for obtaining samples from the posterior distribution of the parameters in model (8.1). In particular, show that sampling from the full conditional distributions in Steps 1 and 2 can be achieved via FFBS.
2. Show that the cospectrum and the quadrature spectrum defined in (8.7) can be written as

$$\begin{aligned} c_{1,2}(\omega) &= \frac{1}{2\pi}\sum_{h=-\infty}^{\infty} \gamma_{1,2}(h)\cos(h\omega), \\ q_{1,2}(\omega) &= \frac{1}{2\pi}\sum_{h=-\infty}^{\infty} \gamma_{1,2}(h)\sin(h\omega). \end{aligned}$$

   In addition, show that $c_{1,2}(\omega) = c_{2,1}(\omega)$ and $q_{1,2}(\omega) = -q_{2,1}(\omega)$.
3. Show that the coherency can be written in terms of the amplitude and phase spectra, $\alpha_{1,2}(\omega)$ and $\phi_{1,2}(\omega)$, as

$$C_{1,2}(\omega) = \frac{\alpha_{1,2}(\omega)}{[f_{1,1}(\omega)f_{2,2}(\omega)]^{1/2}}\exp\{i\phi_{1,2}(\omega)\}.$$

4. Show that the bivariate process defined by (8.10) has the spectrum given in (8.11) and therefore the implied coherency of the process is one.
5. Consider the two-dimensional process $\mathbf{y}_t = (y_{t,1}, y_{t,2})'$ defined as

$$\begin{aligned} y_{t,1} &= \epsilon_{t,1}, \\ y_{t,2} &= \beta y_{t+d,1} + \epsilon_{t,2}, \end{aligned}$$

   where $\epsilon_{t,i}$ are independent, mutually independent, zero-mean processes with $\epsilon_{t,i} \sim N(0,1)$. Find the cross-spectrum, the amplitude, and phase spectra, as well as the squared coherency of $y_{t,1}$ and $y_{t,2}$.
6. Show that if $y_{t,1} \sim N(0,1)$ for all t and $y_{t,2}$ is given by

$$y_{t,2} = \frac{1}{3}(y_{t-1,1} + y_{t,1} + y_{t+1,1}),$$

   the coherency of the process $\mathbf{y}_t = (y_{t,1}, y_{t,2})'$ is zero.
7. Find the phase spectrum for the process in (8.10).
8. Sketch the MCMC algorithm for obtaining samples from the joint posterior distribution in the dynamical lag/lead model presented in Section 8.3.

Chapter 9

Vector AR and ARMA models

This chapter overviews aspects of the class of vector autoregressive (VAR) and vector autoregressive moving average (VARMA) models. We discuss their structure, some properties, and methods for parameter estimation, and provide discussion and links to relevant literature. VAR models, in particular, are workhorses of applied time series analysis and forecasting in many areas, especially based on Bayesian analysis. The basic theory of VAR model fitting and analysis extends to classes of time-varying parameter extensions, or TV-VAR models. The latter are particularly prominent in areas such as macroeconomic modeling and forecasting, as well as engineering signal processing. VAR and TV-VAR models are also special cases of more general multivariate DLMs that follow in Chapter 10.

9.1 Vector Autoregressive Models

A multivariate k-dimensional process $\mathbf{y}_t$ follows a vector autoregressive model of order p, denoted as $\text{VAR}_k(p)$, if $\mathbf{y}_t$ can be written in terms of its p most recent past values. This is

$$\mathbf{y}_t = \mathbf{\Phi}_1 \mathbf{y}_{t-1} + \mathbf{\Phi}_2 \mathbf{y}_{t-2} + \cdots + \mathbf{\Phi}_p \mathbf{y}_{t-p} + \boldsymbol{\epsilon}_t, \tag{9.1}$$

where the $\mathbf{\Phi}_j$s are $k \times k$ matrices of VAR coefficients for $j = 1:p$, and $\boldsymbol{\epsilon}_t$ is a k-dimensional zero-mean vector with variance-covariance matrix Σ. In addition, it is assumed that $\boldsymbol{\epsilon}_q$ and $\boldsymbol{\epsilon}_s$ are independent for any times q and s such that $q \neq s$.

9.1.1 State-Space Representation of a VAR Process

A $\text{VAR}_k(p)$ process is *stable* (e.g., Lütkepohl 2005), if the polynomial

$$\mathbf{\Phi}(u) = \det(\mathbf{I}_k - \mathbf{\Phi}_1 u - \cdots - \mathbf{\Phi}_p u^p), \tag{9.2}$$

with $\mathbf{I}_k$ the $k \times k$ identity matrix, has no roots within or on the complex unit circle. In other words, a $\text{VAR}_k(p)$ process is stable if $\mathbf{\Phi}(u) \neq 0$ for $|u| \leq 1$.

A VAR process is *stationary* if its mean and covariance functions are time invariant. If the process is stable then it is stationary, and so stability is a sufficient stationarity condition for a VAR. However, an unstable process is not necessarily nonstationary.

The $\text{VAR}_k(p)$ in (9.1) can be written in state-space or dynamic linear model (DLM) form, also called $\text{VAR}_{kp}(1)$ form, as follows:

$$\begin{aligned} \mathbf{y}_t &= \mathbf{F}'\boldsymbol{\theta}_t + \boldsymbol{\nu}_t \\ \boldsymbol{\theta}_t &= \mathbf{G}\boldsymbol{\theta}_{t-1} + \mathbf{w}_t, \end{aligned}$$

with $\boldsymbol{\nu}_t = \mathbf{0}$, and $\mathbf{F}'$, $\boldsymbol{\theta}_t$, and $\mathbf{w}_t$ given by

$$\mathbf{F}' = \begin{pmatrix} \mathbf{I}_k & \mathbf{0}_k & \cdots & \mathbf{0}_k \end{pmatrix}; \quad \boldsymbol{\theta}_t = \begin{pmatrix} \mathbf{y}_t \\ \mathbf{y}_{t-1} \\ \vdots \\ \mathbf{y}_{t-p+1} \end{pmatrix}; \quad \mathbf{w}_t = \begin{pmatrix} \boldsymbol{\epsilon}_t \\ 0 \\ \vdots \\ 0 \end{pmatrix}.$$

$\mathbf{I}_k$ and $\mathbf{0}_k$ above denote, respectively, the $k \times k$ identity matrix and the $k \times k$ matrix of zeros. Finally, $\mathbf{G}$ is the $(kp) \times (kp)$ state evolution matrix given by

$$\mathbf{G} = \begin{pmatrix} \mathbf{\Phi}_1 & \mathbf{\Phi}_2 & \cdots & \mathbf{\Phi}_{p-1} & \mathbf{\Phi}_p \\ \mathbf{I}_k & \mathbf{0}_k & \cdots & \mathbf{0}_k & \mathbf{0}_k \\ \mathbf{0}_k & \mathbf{I}_k & \cdots & \mathbf{0}_k & \mathbf{0}_k \\ \vdots & & \ddots & & \vdots \\ \mathbf{0}_k & \mathbf{0}_k & \cdots & \mathbf{I}_k & \mathbf{0}_k \end{pmatrix}. \tag{9.3}$$

The eigenvalues of $\mathbf{G}$ satisfy the equation

$$\det(\mathbf{I}_k \lambda^p - \mathbf{\Phi}_1 \lambda^{p-1} - \mathbf{\Phi}_2 \lambda^{p-2} - \cdots - \mathbf{\Phi}_p) = 0.$$

Therefore, the eigenvalues of $\mathbf{G}$ are the reciprocal roots of the characteristic polynomial $\mathbf{\Phi}(u)$, and the process is stable if all the eigenvalues of $\mathbf{G}$ have moduli less than one.

9.1.2 The Moving Average Representation of a VAR Process

When a k-dimensional $\text{VAR}_k(p)$ process is stable, it is possible to write the following vector moving average (VMA) representation

$$\mathbf{y}_t = \mathbf{\Psi}(B)\epsilon_t = \left(\sum_{j=0}^{\infty} \mathbf{\Psi}_j B^j\right) \epsilon_t. \tag{9.4}$$

The matrices $\mathbf{\Psi}_0, \mathbf{\Psi}_1, \dots,$ are obtained by inverting the VAR matrix characteristic polynomial. That is, $\mathbf{\Psi}(B)$ is such that $\mathbf{\Phi}(B)\mathbf{\Psi}(B) = \mathbf{I}$, which leads to

$$\mathbf{\Psi}_k = \sum_{i=1}^{p} \mathbf{\Phi}_i \mathbf{\Psi}_{k-i},$$

with $\mathbf{\Psi}_0 = \mathbf{I}$ and $\mathbf{\Psi}_j = \mathbf{0}$ for $j < 0$.

9.1.3 VAR Time Series Decompositions

Prado (1998) uses the DLM representation of a VAR process to derive decomposition results analogous to those given in Chapter 5 for univariate AR and TVAR processes. Specifically, for each scalar component $y_{i,t}$ of $\mathbf{y}_t$, the DLM representation given by $\{\mathbf{F}_i', \mathbf{G}, 0, \mathbf{W}\}$ is considered, with $\mathbf{F}_i' = (\mathbf{e}_i', 0, \dots, 0)$, $\mathbf{e}_i$ is a vector whose components are all zero except for the i-th component which is equal to one, and $\mathbf{W} = \text{blockdiag}(\Sigma, \mathbf{0}_k, \dots, \mathbf{0}_k)$. Then, assuming that $\mathbf{G}$ has exactly kp distinct eigenvalues with n_c pairs of complex conjugate eigenvalues denoted by $\lambda_{2j-1} = r_j \exp(-i\omega_j)$ and $\lambda_{2j} = r_j \exp(+i\omega_j)$ for $j = 1 : n_c$, and n_r real eigenvalues $\lambda_j = r_j$, for $j = (2n_c + 1) : (2n_c + n_r)$, $y_{i,t}$ can be written as

$$\begin{aligned} y_{i,t} &= \mathbf{1}'\boldsymbol{\gamma}_{i,t}, \\ \boldsymbol{\gamma}_{i,t} &= \mathbf{A}\boldsymbol{\gamma}_{i,t-1} + \boldsymbol{\delta}_{i,t}. \end{aligned}$$

In this representation, $\mathbf{1}' = (1, \dots, 1)$ and $\mathbf{A}$ is the $kp \times kp$ diagonal matrix of eigenvalues of $\mathbf{G}$. In addition, $\boldsymbol{\gamma}_{i,t} = \mathbf{H}_i\boldsymbol{\theta}_t$ and $\boldsymbol{\delta}_{i,t} = \mathbf{H}_i\mathbf{w}_t$, with $\mathbf{H}_i = \text{diag}(\mathbf{B}'\mathbf{F}_i)\mathbf{B}^{-1}$, where $\mathbf{B}$ the matrix of eigenvectors of $\mathbf{G}$ ordered according to $\mathbf{A}$. This implies that $y_{i,t}$ for $i = 1 : k$ can be expressed as a sum of $n_c + n_r$ components, i.e.,

$$y_{i,t} = \sum_{l=1}^{n_c} z_{i,l,t} + \sum_{l=1}^{n_r} x_{i,l,t}, \tag{9.5}$$

where each $z_{i,l,t}$ is a quasiperiodic ARMA(2,1) process with characteristic modulus and frequency given by (r_l, ω_l) for $l = 1 : n_c$ and each $x_{i,l,t}$ has an AR(1) structure with AR parameter r_l for $l = (2n_c + 1) : (2n_c +$

n_r). Thus, each univariate element of $\mathbf{y}_t$ can be decomposed into processes whose latent ARMA(2,1) and AR(1) components are characterized by the same frequencies and moduli across i, but whose phases and amplitudes are specific to each scalar component $y_{i,t}$.

Example 9.1 *Decomposition of a VAR process.* Two time series with 200 data points each were simulated independently as follows. The first series, $x_{1,t}$, was simulated from an AR(2) with a single pair of complex characteristic roots with modulus $r_1 = 0.95$ and frequency $\omega_1 = 2\pi/18$. The second series, $x_{2,t}$, was simulated from an AR(2) with modulus $r_2 = 0.95$ and frequency $\omega_2 = 2\pi/7$. That is,

$$\begin{aligned} x_{1,t} &= 2 \times 0.95 \times \cos(2\pi/18)x_{1,t-1} - 0.95^2 x_{1,t-2} + \eta_{1,t}, \\ x_{2,t} &= 2 \times 0.95 \times \cos(2\pi/7)x_{2,t-1} - 0.95^2 x_{2,t-2} + \eta_{2,t}, \end{aligned}$$

with $\boldsymbol{\eta}_t = (\eta_{1,t}, \eta_{2,t})'$ such that $\boldsymbol{\eta}_t \sim N(\mathbf{0}, \mathbf{I}_2)$. We then used $\mathbf{x}_t = (x_{1,t}, x_{2,t})'$ to obtain 200 data points $\mathbf{y}_{1:200}$, with $\mathbf{y}_t = (y_{1,t}, y_{2,t}, y_{3,t})'$, as follows,

$$\mathbf{y}_t = \begin{pmatrix} 1 & 1 \\ 1 & 0 \\ 0 & 0.9 \end{pmatrix} \mathbf{x}_t + \boldsymbol{\epsilon}_t, \tag{9.6}$$

with $\boldsymbol{\epsilon}_t \sim N(\mathbf{0}, 4\mathbf{I}_3)$. The simulated series $y_{1,t}, y_{t,2}$, and $y_{t,3}$ are displayed at the top in Figure 9.1.

A VAR$_3$(4) was fitted to the three-dimensional series $\mathbf{y}_t$ using the `ar` function in `R`. Details about the estimation procedure are discussed later in this chapter. Based on the parameter estimates, we computed estimates of the VAR decomposition as follows. First, the matrix $\mathbf{G}$ was built as in (9.3) using the estimated values of the AR coefficients. It was found that the estimated $\mathbf{G}$ matrix had exactly five pairs of distinct complex eigenvalues and two real eigenvalues. The first two of these eigenvalues, taken in order of decreasing wavelengths, had moduli and periods given by $(0.91, 15.43)$ and $(0.89, 7.16)$. Figure 9.1 shows the corresponding estimated components for these two eigenvalues in the decomposition of the three series. The top series in each plot display the data followed by the estimated latent components in order of decreasing period. In these multivariate decompositions, each univariate component of the observed time series vector is broken up into latent components that have exactly the same characteristic frequencies and moduli. And so, the components labeled as $(j, 1)$ for $j = 1:3$ in Figure 9.1 are quasiperiodic ARMA(2,1) with randomly varying amplitudes and phases that are different for each series, but with common estimated wavelength and modulus of 15.43 and 0.91, respectively. Similarly, the components labeled as $(j, 2)$ for $j = 1:3$ are quasiperiodic ARMA(2,1) with common estimated wavelength 7.16 and modulus 0.89. The decomposition of series $y_{1,t}$ shows that the latent processes $(1, 1)$ and $(1, 2)$-with estimated

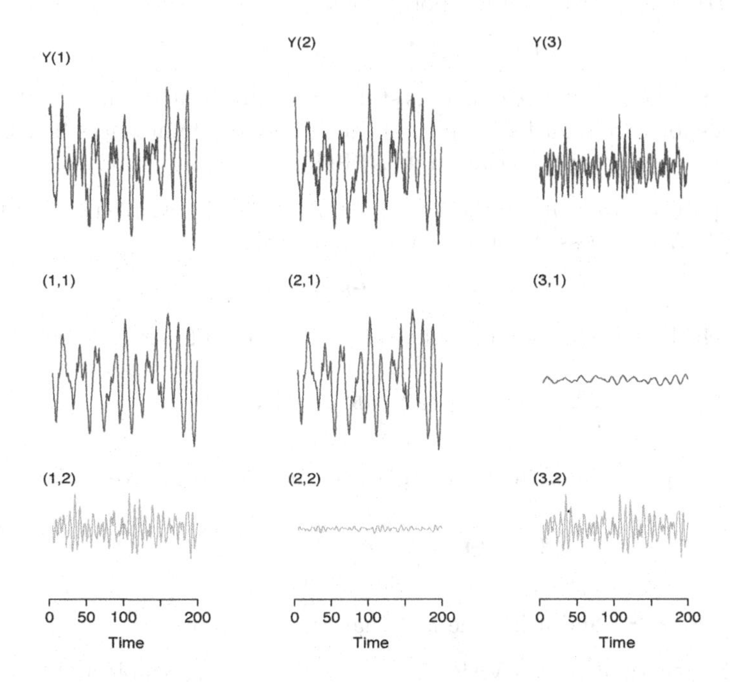

Figure 9.1 *Simulated quasiperiodic processes and estimated VAR decompositions.*

wavelengths 15.43 and 7.16, respectively-contribute significantly to this series. In contrast, the amplitude of the estimated component $(2, 2)$ in the decomposition of $y_{2,t}$ is very close to zero, indicating that such component is not present in this series. Similarly, the amplitude of the estimated component $(3, 1)$ is close to zero. These results are consistent with the structure in (9.6).

9.2 Vector ARMA Models

We say that $\mathbf{y}_t$ follows a vector autoregressive moving average model of orders p and q, denoted as VARMA$_k(p, q)$, if $\mathbf{y}_t$ can be written as

$$\mathbf{y}_t = \sum_{i=1}^{p} \mathbf{\Phi}_i \mathbf{y}_{t-i} + \sum_{j=1}^{q} \mathbf{\Theta}_j \boldsymbol{\epsilon}_{t-j} + \boldsymbol{\epsilon}_t, \tag{9.7}$$

with $\boldsymbol{\epsilon}_t \sim N(\mathbf{0}, \Sigma)$. The process $\mathbf{y}_t$ is stable and stationary if the polynomial $\mathbf{\Phi}(u)$ has no roots within or on the complex unit circle. The process $\mathbf{y}_t$ is *invertible* if the roots of the polynomial

$$\boldsymbol{\Theta}(u) = \det(\mathbf{I}_k + \boldsymbol{\Theta}_1 u + \cdots + \boldsymbol{\Theta}_q u^q)$$

lie outside the unit circle. If $\mathbf{y}_t$ is stable, it can be written as a purely VMA process of infinite order as in (9.4). In such case, the matrices $\boldsymbol{\Psi}_0, \boldsymbol{\Psi}_1, \ldots,$ are obtained from the relation $\boldsymbol{\Phi}(B)\boldsymbol{\Psi}(B) = \boldsymbol{\Theta}(B)$.

If a purely VMA process $\mathbf{y}_t$ is invertible, it can be written as an infinite order VAR process. In other words, assume that

$$\mathbf{y}_t = \boldsymbol{\epsilon}_t + \boldsymbol{\Theta}_1\boldsymbol{\epsilon}_{t-1} + \boldsymbol{\Theta}_2\boldsymbol{\epsilon}_{t-2} + \cdots + \boldsymbol{\Theta}_q\boldsymbol{\epsilon}_{t-q}$$

is such that $\boldsymbol{\Theta}(u)$ has no roots inside or on the unit circle. Then,

$$\mathbf{y}_t = \sum_{i=1}^{\infty} \boldsymbol{\Phi}_i^* \mathbf{y}_{t-i} + \boldsymbol{\epsilon}_t.$$

The computation of the $\boldsymbol{\Phi}_i$ matrices can be done recursively via

$$\boldsymbol{\Phi}_i^* = \boldsymbol{\Theta}_i - \sum_{j=1}^{i-1} \boldsymbol{\Phi}_{i-j}^* \boldsymbol{\Theta}_j, \quad i \geq 2,$$

with $\boldsymbol{\Phi}_1^* = \boldsymbol{\Theta}_1$ and using the fact that $\boldsymbol{\Theta}_j = \mathbf{0}$ for $j > q$.

In the general case of a VARMA process that is both stable and invertible, we have that such process can be written as a pure VMA process or, alternatively, as a pure VAR process. In both cases, the resulting processes are of infinite order. More specifically, if a k-dimensional process represented by (9.7) is stable, we have

$$\mathbf{y}_t = \sum_{i=1}^{\infty} \boldsymbol{\Psi}_i^* \boldsymbol{\epsilon}_{t-i} + \boldsymbol{\epsilon}_t,$$

where $\boldsymbol{\Psi}_i^* = \boldsymbol{\Theta}_i + \sum_{j=1}^{i} \boldsymbol{\Phi}_j \boldsymbol{\Psi}_{i-j}^*$, for $i \geq 1$ and $\boldsymbol{\Psi}_0^* = \mathbf{I}_k$. Alternatively, if the process is also invertible, we have that

$$\mathbf{y}_t = \sum_{i=1}^{\infty} \boldsymbol{\Phi}_i^* \mathbf{y}_{t-i} + \boldsymbol{\epsilon}_t,$$

where

$$\boldsymbol{\Phi}_i^* = \boldsymbol{\Phi}_i + \boldsymbol{\Theta}_i - \sum_{j=1}^{i-1} \boldsymbol{\Theta}_{i-j} \boldsymbol{\Phi}_j^*, \quad \text{for } i \geq 1. \tag{9.8}$$

9.2.1 *Autocovariances and Cross-covariances*

It is possible to show that the lag h covariance matrix for a VARMA$_k(p,q)$ process is given by

$$\Gamma(h) = E(\mathbf{y}_t\mathbf{y}'_{t-h}) = \begin{cases} \sum_{j=1}^{r} \mathbf{\Phi}_j\Gamma(h-j) + \sum_{j=0}^{r-h} \mathbf{\Theta}_{j+h}\Sigma\mathbf{\Psi}_j, & h \leq r \\ \sum_{j=1}^{r} \mathbf{\Phi}_j\Gamma(h-j) & h > r, \end{cases}$$

where $\mathbf{\Psi}_0 = \mathbf{I}_k$, $\mathbf{\Theta}_0 = \mathbf{I}_k$, $r = \max(p,q)$, and $\mathbf{\Phi}_{p+1} = \cdots = \mathbf{\Phi}_r = \mathbf{0}$, when $p < q$, while $\mathbf{\Theta}_{q+1} = \cdots = \mathbf{\Theta}_r = \mathbf{0}$ when $q < p$. Note that $\Gamma(-h) = \Gamma'(h)$.

In the case of VMA$_k(q)$, i.e., when $p = 0$, we have that

$$\Gamma(h) = \begin{cases} \sum_{j=0}^{q-h} \mathbf{\Theta}_j\Sigma\mathbf{\Theta}'_{j+h}, & h \leq q \\ \mathbf{0} & h > q, \end{cases} \tag{9.9}$$

and so, for VMA$_k(q)$ processes, $\Gamma(h) = 0$ for all $h > q$. Alternatively, if $\mathbf{y}_t$ is a VAR process — i.e., if $q = 0$ — the autocorrelations and cross-correlations will decay to zero as h increases.

9.2.2 *Partial Autoregression Matrix Function*

The *partial autoregression* matrix function $\mathbf{P}(h)$ (e.g., see Tiao 2001b; Tiao and Box 1981) is analogous to the partial autocorrelation function in the univariate case and is defined as

$$\mathbf{P}'(h) = \begin{cases} \Gamma^{-1}(0)\Gamma(1) & h = 1 \\ \left[\Gamma(0) - \mathbf{a}_2(h)\mathbf{A}^{-1}(h)\mathbf{a}_1(h)\right]^{-1} \times & \\ \left[\Gamma(0) - \mathbf{a}_2(h)\mathbf{A}^{-1}(h)\mathbf{a}_3(h)\right] & h > 1, \end{cases}$$

where

$$\mathbf{A}(h) = \begin{pmatrix} \Gamma(0) & \Gamma(-1) & \cdots & \Gamma(-h+3) & \Gamma(-h+2) \\ \Gamma(1) & \Gamma(0) & \cdots & \Gamma(-h) & \Gamma(-h+1) \\ \vdots & \vdots & \ddots & \vdots & \vdots \\ \Gamma(h-3) & \Gamma(h-4) & \cdots & \Gamma(0) & \Gamma(-1) \\ \Gamma(h-2) & \Gamma(h-3) & \cdots & \Gamma(1) & \Gamma(0) \end{pmatrix},$$

$\mathbf{a}_1(h) = (\Gamma(-h+1), \ldots, \Gamma(-1))'$, $\mathbf{a}_2(h) = (\Gamma'(h-1), \ldots, \Gamma'(1))$ and $\mathbf{a}_3(h) = (\Gamma(1), \ldots, \Gamma(h-1))'$.

In the case of a VAR$_k(p)$ $q = 0$, $\mathbf{P}(h) = \mathbf{\Phi}_h$ if $h = p$, and $\mathbf{P}(h) = \mathbf{0}$ for $h > p$. Note that unlike the PACF in the univariate case, the partial autoregression matrix defined does not lead to proper correlation entries. Heyse and Wei (1985) define the so-called lag partial autocorrelation matrix which has the desired property of being a matrix of zeros after lag p if the process is a VAR$_k(p)$, while also being a proper correlation matrix.

9.2.3 VAR(1) and DLM Representations

A k-dimensional VARMA process of the form (9.7) can be written as

$$\mathbf{y}_t^* = \mathbf{G}\mathbf{y}_{t-1}^* + \mathbf{e}_t^*, \tag{9.10}$$

with

$$\mathbf{y}_t^* = \begin{pmatrix} \mathbf{y}_t \\ \vdots \\ \mathbf{y}_{t-p+1} \\ \boldsymbol{\epsilon}_t \\ \vdots \\ \boldsymbol{\epsilon}_{t-q+1} \end{pmatrix}, \quad \mathbf{G} = \begin{pmatrix} \mathbf{G}_{11} & \mathbf{G}_{12} \\ \mathbf{0}_{kq\times kp} & \mathbf{G}_{22} \end{pmatrix}, \text{ and } \mathbf{e}_t^* = \begin{pmatrix} \mathbf{e}_{t,1}^* \\ \mathbf{e}_{t,2}^* \end{pmatrix},$$

where $\mathbf{e}_{t,1}^*$ is a kp-dimensional vector given by $\mathbf{e}_{t,1}^* = (\boldsymbol{\epsilon}_t', \mathbf{0}', \dots, \mathbf{0}')$ and $\mathbf{e}_{t,2}^*$ is a kq-dimensional vector also of the form $\mathbf{e}_{t,2}^* = (\boldsymbol{\epsilon}_t', \mathbf{0}', \dots, \mathbf{0}')'$. Finally, $\mathbf{G}_{11}$ is a $kp \times kp$ matrix, $\mathbf{G}_{12}$ is a $kp \times kq$ matrix, and $\mathbf{G}_{22}$ is a $kq \times kq$ matrix. These matrices are given by

$$\mathbf{G}_{11} = \begin{pmatrix} \boldsymbol{\Phi}_1 & \cdots & \boldsymbol{\Phi}_{p-1} & \boldsymbol{\Phi}_p \\ \mathbf{I}_k & & \mathbf{0} & \mathbf{0} \\ \vdots & \ddots & & \vdots \\ \mathbf{0} & \cdots & \mathbf{I}_k & \mathbf{0} \end{pmatrix}, \quad \mathbf{G}_{12} = \begin{pmatrix} \boldsymbol{\Theta}_1 & \cdots & \boldsymbol{\Theta}_{q-1} & \boldsymbol{\Theta}_q \\ \mathbf{0} & \cdots & \mathbf{0} & \mathbf{0} \\ \vdots & & \vdots & \vdots \\ \mathbf{0} & \cdots & \mathbf{0} & \mathbf{0} \end{pmatrix},$$

and

$$\mathbf{G}_{22} = \begin{pmatrix} \mathbf{0} & \cdots & \mathbf{0} & \mathbf{0} \\ \mathbf{I}_k & & \mathbf{0} & \mathbf{0} \\ & \ddots & & \vdots \\ \mathbf{0} & \cdots & \mathbf{I}_k & \mathbf{0} \end{pmatrix}.$$

The process $\mathbf{y}_t^*$ is stable if and only if $\mathbf{y}_t$ is stable. If $\mathbf{y}_t^*$ is stable, it has a vector MA representation of infinite order, i.e., $\mathbf{y}_t^* = \sum_{i=0}^{\infty} \mathbf{G}^i \mathbf{e}_{t-i}^*$. This implies that $\mathbf{y}_t = \sum_{i=0}^{\infty} \boldsymbol{\Psi}_i \boldsymbol{\epsilon}_{t-i}$, where $\boldsymbol{\Psi}_i = \mathbf{J}\mathbf{G}^i\mathbf{H}$, with $\mathbf{J}$ the $k \times k(p+q)$ matrix $\mathbf{J} = [\mathbf{I}_k\ \mathbf{0}\ \cdots\ \mathbf{0}]$, and $\mathbf{H}$ a $(k(p+q)) \times k$ with $\mathbf{H}' = [\mathbf{H}_1', \mathbf{H}_2']$, with $\mathbf{H}_1$ a $kp \times k$ matrix given by $\mathbf{H}_1' = [\mathbf{I}_k\ \mathbf{0}\ \cdots\ \mathbf{0}]$ and $\mathbf{H}_2$ a $kq \times k$ matrix with $\mathbf{H}_2' = [\mathbf{I}_k\ \mathbf{0}\ \cdots\ \mathbf{0}]$.

The DLM representation of a VARMA$_k(p,q)$ process $\mathbf{y}_t$ is given by

$$\begin{aligned} \mathbf{y}_t &= \mathbf{F}'\boldsymbol{\theta}_t + \boldsymbol{\nu}_t, \\ \boldsymbol{\theta}_t &= \mathbf{G}\boldsymbol{\theta}_{t-1} + \mathbf{w}_t, \end{aligned}$$

with $\boldsymbol{\theta}_t$ a $k \times m$-dimensional state vector, where $m = \max(p, q+1)$, $\boldsymbol{\nu}_t = \mathbf{0}$,

$\mathbf{F}' = (\mathbf{I}_k\ \mathbf{0}_k\ \ldots\ \mathbf{0}_k)$, and $\mathbf{w}_t = \mathbf{U}\boldsymbol{\epsilon}_t$ with

$$\mathbf{U} = \begin{pmatrix} \mathbf{I}_k \\ \boldsymbol{\Theta}_1 \\ \boldsymbol{\Theta}_2 \\ \vdots \\ \boldsymbol{\Theta}_{m-1} \end{pmatrix} \quad \text{and} \quad \mathbf{G} = \begin{pmatrix} \boldsymbol{\Phi}_1 & \mathbf{I}_k & \mathbf{0} & \cdots & \mathbf{0} \\ \boldsymbol{\Phi}_2 & \mathbf{0} & \mathbf{I}_k & \cdots & \mathbf{0} \\ \vdots & \vdots & & \ddots & \vdots \\ \boldsymbol{\Phi}_{m-1} & \mathbf{0} & \mathbf{0} & \cdots & \mathbf{I}_k \\ \boldsymbol{\Phi}_m & \mathbf{0} & \mathbf{0} & \cdots & \mathbf{0} \end{pmatrix}.$$

Here, as in the univariate case, we take $\boldsymbol{\Phi}_j = \mathbf{0}$ for $j > p$ and $\boldsymbol{\Theta}_j = \mathbf{0}$ for $j > q$.

9.3 Estimation in VARMA

9.3.1 Identifiability

In the case of a stable and invertible VARMA$_k(p, q)$ with $p, q > 0$, some restrictions need to be imposed on $\boldsymbol{\Phi}(B)$ and $\boldsymbol{\Theta}(B)$ for parameter identifiability. Detailed discussions on this issue appear, for example, in Lütkepohl (2005) and Reinsel (1993). Specifically, it is possible that two model representations, say, $\boldsymbol{\Phi}(B)\mathbf{y}_t = \boldsymbol{\Theta}(B)\boldsymbol{\epsilon}_t$ and $\boldsymbol{\Phi}^*(B)\mathbf{y}_t = \boldsymbol{\Theta}^*(B)\boldsymbol{\epsilon}_t$, lead to an infinite order VMA representation of the process, namely

$$\mathbf{y}_t = \sum_{i=0}^{\infty} \boldsymbol{\Psi}_j \boldsymbol{\epsilon}_{t-j}$$

such that $\boldsymbol{\Psi}(B) = (\boldsymbol{\Phi}(B))^{-1}\boldsymbol{\Theta}(B) = (\boldsymbol{\Phi}^*(B))^{-1}\boldsymbol{\Theta}^*(B)$, with both representations associated to the same covariance structure. So, a VARMA$_k(p, q)$ is *identifiable* if the parameter matrices $\boldsymbol{\Phi}_j$ and $\boldsymbol{\Theta}_j$ are uniquely determined by the matrices $\boldsymbol{\Psi}_j$ in the unique, infinite order, VMA representation of $\mathbf{y}_t$.

The following conditions are sufficient for identifiability of a VARMA process:

1. $\boldsymbol{\Phi}(B)$ and $\boldsymbol{\Theta}(B)$ are *left coprime*. This is, if $\boldsymbol{\Phi}(B) = \mathbf{U}(B)\boldsymbol{\Phi}^*(B)$ and $\boldsymbol{\Theta}(B) = \mathbf{U}(B)\boldsymbol{\Theta}^*(B)$, then $\mathbf{U}(B)$ must be such that $\det(\mathbf{U}(B)) = c$ with $c \neq 0$.
2. q and p are as small as possible so that the matrix $(\boldsymbol{\Phi}_p, \boldsymbol{\Theta}_q)$ is of full rank k.

We now discuss estimation procedures for VARMA processes that are identifiable.

9.3.2 Least Squares Estimation

Assume we observe a k-dimensional VAR process of order p, for $t = 1 : T$. Then we can write

$$\mathbf{Y} = \mathbf{\Phi X} + \mathbf{E} \tag{9.11}$$

where $\mathbf{Y} = (\mathbf{y}_{p+1}, \ldots, \mathbf{y}_T)$, is a $k \times (T-p)$ matrix; $\mathbf{\Phi} = (\mathbf{\Phi}_1, \ldots, \mathbf{\Phi}_p)$ is a $k \times kp$ matrix; and $\mathbf{E} = (\boldsymbol{\epsilon}_{p+1}, \ldots, \boldsymbol{\epsilon}_T)$ is a $k \times (T-p)$ matrix. In addition, $\mathbf{X}$ is a $kp \times (T-p)$ matrix given by $\mathbf{X} = (\mathbf{y}_p^p, \ldots, \mathbf{y}_{T-1}^p)$, with $\mathbf{y}_t^p$ kp-dimensional vectors defined as $\mathbf{y}_t^p = (\mathbf{y}'_t, \ldots, \mathbf{y}'_{t-p+1})'$.

Then the LS estimator of $\mathbf{\Phi}$ is given by

$$\hat{\mathbf{\Phi}} = \mathbf{YX}'(\mathbf{XX}')^{-1}. \tag{9.12}$$

In addition, if $\mathbf{y}_t$ is a stationary process and $\boldsymbol{\epsilon}_t \sim N(\mathbf{0}, \Sigma)$ for all t, it can be shown that, when T is large, $\sqrt{(T-p)}(\text{vec}(\hat{\mathbf{\Phi}}) - \text{vec}(\mathbf{\Phi}))$ converges in distribution to a multivariate Gaussian with mean $\mathbf{0}$ and covariance matrix $\Sigma_{\hat{\mathbf{\Phi}}}$, where $\Sigma_{\hat{\mathbf{\Phi}}} = \Gamma_Y(0)^{-1} \bigotimes \Sigma$ and $\Gamma_Y(0) = E[\mathbf{y}_t^p(\mathbf{y}_t^p)']$.

The estimator in (9.12) can also be derived from the Yule-Walker equations, namely

$$\Gamma(h) = \mathbf{\Phi}_1\Gamma(h-1) + \cdots + \mathbf{\Phi}_p\Gamma(h-p), \quad h > 0, \tag{9.13}$$

$$\Gamma(0) = \mathbf{\Phi}_1\Gamma(1)' + \cdots + \mathbf{\Phi}_p\Gamma(p)' + \Sigma. \tag{9.14}$$

From these equations we can write

$$\begin{aligned} [\Gamma(1), \ldots, \Gamma(p)] &= [\mathbf{\Phi}_1, \ldots, \mathbf{\Phi}_p] \times \begin{bmatrix} \Gamma(0) & \ldots & \Gamma(p-1) \\ \vdots & \ldots & \vdots \\ \Gamma(-p+1) & \cdots & \Gamma(0) \end{bmatrix} \\ &= \mathbf{\Phi}\Gamma^*(0), \end{aligned}$$

and so, $\mathbf{\Phi} = (\Gamma(1), \ldots, \Gamma(p))(\Gamma^*(0))^{-1}$. Then, if we estimate $\Gamma(0)$ by $\mathbf{XX}'/(T-p)$ and $(\Gamma(1), \ldots, \Gamma(p))$ by $\mathbf{YX}'/(T-p)$, the resulting estimator for $\mathbf{\Phi}$ obtained from the Yule-Walker equations is the LS estimator.

9.3.3 Maximum Likelihood Estimation

As is the case in the univariate framework, it is possible to consider the exact likelihood in multivariate ARMA processes, as well as an approximation to the exact likelihood by means of a conditional likelihood.

9.3.3.1 Conditional likelihood

Conditioning on the first p values of $\mathbf{y}_t$, $\mathbf{y}_{1:p}$, and on the values of $\boldsymbol{\epsilon}_{(p-q+1):p}$ being equal to zero, the conditional likelihood function can be written as

$$p^*(\mathbf{y}_{(p+1):T}|\mathbf{y}_{1:p}, \mathbf{\Phi}, \mathbf{\Theta}, \Sigma) \propto |\Sigma|^{-(T-p)/2} \exp\left\{-\frac{1}{2}\text{tr}[\Sigma^{-1}S(\mathbf{\Phi}, \mathbf{\Theta})]\right\} \quad (9.15)$$

where $S(\mathbf{\Phi}, \mathbf{\Theta}) = \sum_{t=p+1}^{T} \boldsymbol{\epsilon}_t \boldsymbol{\epsilon}_t'$.

The conditional likelihood in (9.15) can be used to approximate the exact likelihood in (9.16). However, as pointed out by Tiao (2001b), Hillmer and Tiao (1979) showed that such approximation can be inappropriate when T is not sufficiently large, or when at least one of the roots of $\mathbf{\Theta}(u)$ lies on or is close to the unit circle, since the estimates of the MA parameters can be very biased.

9.3.3.2 Exact likelihood

The exact likelihood for VARMA processes takes the form

$$p(\mathbf{y}_{1:T}|\mathbf{\Phi}, \mathbf{\Theta}, \Sigma) \propto p^*(\mathbf{y}_{(p+1):T}|\mathbf{y}_{1:p}, \mathbf{\Phi}, \mathbf{\Theta}, \Sigma)p^{**}(\mathbf{y}_{1:T}|\mathbf{\Phi}, \mathbf{\Theta}, \Sigma) \quad (9.16)$$

where $p^{**}(\cdot)$ depends only on $\mathbf{y}_{1:p}$ if $q = 0$ and on $\mathbf{y}_{1:T}$ if $q \neq 0$. A close approximation to this likelihood can be obtained by considering the transformation $\mathbf{w}_t = \mathbf{\Phi}(B)\mathbf{y}_t$, so that $\mathbf{w}_t = \mathbf{\Theta}(B)\boldsymbol{\epsilon}_t$, and then applying the exact likelihood results of Hillmer and Tiao (1979) for $\text{VMA}_k(q)$ processes to $\mathbf{w}_t$, with $t = (p+1) : T$.

An alternative representation of an approximation to (9.16) is given by (e.g., see Lütkepohl 2005)

$$p^{***}(\mathbf{y}_{1:T}|\mathbf{\Phi}, \mathbf{\Theta}, \Sigma) \propto |\Sigma|^{-T/2} \exp\left(-\frac{1}{2}\sum_{t=1}^{T} \tilde{\boldsymbol{\epsilon}}_t'\Sigma^{-1}\tilde{\boldsymbol{\epsilon}}_t\right), \quad (9.17)$$

where

$$\tilde{\boldsymbol{\epsilon}}_t = \mathbf{y}_t - \sum_{i=1}^{t-1} \mathbf{\Phi}_i^* \mathbf{y}_{t-i}, \quad (9.18)$$

and $\mathbf{\Phi}_i^*$ is given by (9.8). Here it is also assumed that $\mathbf{y}_{-p+1} = \cdots = \mathbf{y}_0 = \mathbf{0}$ and $\boldsymbol{\epsilon}_{-q+1} = \cdots = \boldsymbol{\epsilon}_0 = \mathbf{0}$.

As in the univariate case, maximizing the likelihood in a general VARMA setting is a nonlinear maximization problem. Therefore, algorithms such as Newton-Raphson and those based on scoring methods are used for parameter estimation. Details of these algorithms appear, for example, in Lütkepohl (2005).

9.4 Bayesian VAR, TV-VAR, and DDNMs

The representation of a $\text{VAR}_k(p)$ model in the multivariate linear regression form of eqn. (9.11) underlies the traditional Bayesian analysis that has become popular in many fields. Conditional on p initial data values, the resulting likelihood function for $(\mathbf{\Phi}, \mathbf{\Sigma})$ leads immediately to conjugate Bayesian analysis under multivariate normal-inverse Wishart priors. With series dimension k and/or VAR lag p at all large, the challenges raised by resulting large numbers of VAR parameters in practical models can be addressed via informative priors. This recognition is partly responsible for the popularity of Bayesian VAR models, and the use of smoothness and shrinkage priors to regularize the analysis. This can engender shrinkage of VAR coefficients toward zero when data:model matches suggests that is appropriate.

Thinking of VAR models as regression models in which the explanatory variables are lagged values of the response variable $\mathbf{y}_t$ allows for consideration of the same type of priors used in variable selection settings. Each (i, j) element of the matrix $\mathbf{\Phi}_l$, can be set to zero if the corresponding variable, $y_{t-l,j}$, is not included as a predictor for $y_{t,i}$. A set of indicator variables, say γ, can be used to determine which components of $\mathbf{\Phi}$ are included in the model. E.g., George, Sun, and Ni (2008) and Jochmann, Koop, and Strachan (2010) consider a stochastic search variable selection (SSVS) approach that assumes a mixture of two normal distributions on the VAR coefficients conditional on the indicator variables. In other words, using the VAR representation in (9.11) and stacking the VAR coefficients in $\boldsymbol{\beta} = \text{vec}(\mathbf{\Phi})$, it is assumed that

$$\beta_m \sim (1 - \gamma_m)N(0, \tau_{0,m}^2) + \gamma_m N(0, \tau_{1,m}^2),$$

with γ_m a 0-1 variable such that $\gamma_m \sim \textit{Bernoulli}(\theta_m)$ with $\theta_m = 0.5$ (a prior distribution can also be imposed on θ_m instead of using a constant value). The same type of prior structure is assumed on $\Sigma^{-1} = \mathbf{\Psi}'\mathbf{\Psi}$, with $\mathbf{\Psi}$ the lower triangular Cholesky root of Σ^{-1}. Discussion on how to model $\tau_{0,m}^2$ and $\tau_{1,m}^2$ and details of the posterior sampling MCMC scheme are also provided in George, Sun, and Ni (2008). More recently, Kastner and Huber (2020) develop a model VAR model with a parsimonious factor stochastic volatility structure, global-local shrinkage priors on the VAR coefficients, and fast sampling tools for efficient posterior computation. Alternative approaches for dealing with overparameterization in VAR processes include dimension reduction methods such as factor models (see Chapter 11 and references therein) and approaches based on data compression such as those proposed in Koop, Korobilis, and Pettenuzzo (2019) that extend the Bayesian compression ideas of Guhaniyogi and Dunson (2015) to the VAR settings.

The extension of the constant parameter model to time-varying VAR models, or TV-VAR models, is immediate. Allowing for both the AR coefficient matrices and the innovations variance matrix to be potentially time-varying opens up major advances in applicability. Here we note that TV-VAR models are in fact special cases of the broader classes of multivariate time series developed in the following chapter; see Section 10.2. While TV-VAR models are widely used in science and engineering applications (Kitagawa and Gersch 1996b), one of the main areas of application for some decades now has been in macroeconomic studies. In this area, Bayesian TV-VAR modeling is a mainstay of applied time series analysis in monetary policy and related studies. The papers by Cogley and Sargent (2005) and Primiceri (2005), in particular, defined the applied opportunities for some of the core methodology that has become standard in Bayesian TV-VAR analyses in macroeconomics. Some additional links to this very large and fast-growing literature can be found in, for example, Benati (2008), Benati and Surico (2008), Koop, Leon-Gonzalez, and Strachan (2009), Koop and Korobilis (2010) and Nakajima, Kasuya, and Watanabe (2011). In addition to extending methodology, examples in these papers involve studies that aim to assess dynamic relationships between monetary policy and economic variables, typically focusing on changes in the exercise of monetary policy and the resulting effect on the rest of the economy. Structural shocks hitting the economy and simultaneous interactions between macroeconomic variables are identified by TV-VAR models.

One important and increasingly exploited extension of TV-VAR modeling is to use so-called recursive or ordered models. This has been extended and generalized under the name of *dynamic dependence network models (DDNMs)* by Zhao, Xie, and West (2016). These model developments nucleated the concept of *decouple/recouple* that has since been more broadly developed; see West (2020), and references therein as well as further discussion on more general and recent advances in the following chapter, Section 10.6. DDNMs define coherent multivariate dynamic models via coupling of sets of customized univariate DLMs, and extend TV-VAR models as well as *multiregression dynamic models* (Queen and Smith 1993, Queen 1994, Queen, Wright, and Albers 2008 and Anacleto, Queen, and Albers 2013). These models incorporate directed graphical model structure into a multivariate time series, allowing contemporaneous values of some univariate series to appear as predictors of other series. Originally introduced to preserve certain conditional independence structures related to causality over time (Queen and Smith 1993), the approach has been developed and applied to multivariate time series in areas such as forecasting of brand sales and traffic flows (e.g., Queen 1994, Anacleto, Queen, and Albers 2013), and as empirical models of dynamic network structures generating inter-related time series in areas such as neuroscience, engineering signal processing, and financial econometrics (e.g., Costa, Smith, Nichols, Cussens, Duff, and Makin 2015, Nakajima and West 2013a).

Extension to the richer class of DDNMs defines more general methodology allowing TV-VAR and dynamic regression components as well as a flexible approach to multivariate stochastic volatility modeling; they create coherent multivariate dynamic models via coupling of sets of customized univariate DLMs. The basic idea links to traditional recursive systems of structural (and/or simultaneous) equation models in econometrics (e.g., Bodkin, Klein, and Marwah 1991 and references therein). At one level, DDNMs extend this traditional thinking to time-varying parameter/state-space models within the Bayesian framework, while also defining a richer class of TV-VAR and dynamic regression structures. In connection with multivariate volatility, DDNMs are able capture and quantify time-variations in patterns of conditional independence structures, linking to the increasingly popular Cholesky-style approach to modeling multivariate stochastic volatility (e.g., Pinheiro and Bates 1996, Smith and Kohn 2002, Primiceri 2005, Lopes, McCulloch, and Tsay 2018).

More recent developments that exploit the DDNM structure have led to structured Bayesian analysis based on the concept of dynamic latent thresholding, defining extensions referred to as *latent threshold models (LTMs)*. LTMs allow TV-VAR parameters, regression parameters, and parameters defining elements of multivariate volatility matrices to be completely shrunk to zero over periods of time, while being non zero and time-varying otherwise. The utility of this in higher-dimensional models is key, and the methodology builds critically on the use of DDNM structures. Selected examples of these methodological extensions as well as detailed applications in finance, econometrics, neuroscience, and other areas, can be found in Nakajima and West (2013a), Nakajima and West (2013b), Zhou, Nakajima, and West (2014), Nakajima and West (2015), Nakajima and West (2017), and Irie and West (2019).

Alternative approaches are also available for dealing with inference and forecasting in large-dimensional TV-VAR settings. Koop and Korobilis (2013) use dynamic model averaging and discounting ideas to obtain approximate posterior inference that avoids MCMC. Huber, Koop, and Onorante (2020) use global-local priors for inducing shrinkage on the model parameters and then impose additional sparsity extending the sparsification strategy of Hahn and Carvalho (2015). Other Bayesian frameworks and corresponding `R` packages that can be used for obtaining inference on VAR and TV-VAR models with certain prior structures include Krueger (2015) and Knaus, Bitto-Nemling, Cadonna, and Frühwirth-Schnatter (2019).

9.5 Mixtures of VAR Processes

Krolzig (1997) considers maximum likelihood and Bayesian inference in Markov switching vector autoregressive models, or MSVAR. A k-dimensional MSVAR of order p can be written as

$$\mathbf{y}_t = \mathbf{\Phi}_{S_t}^{(1)} \mathbf{y}_{t-1} + \cdots + \mathbf{\Phi}_{S_t}^{(p)} \mathbf{y}_{t-p} + \boldsymbol{\epsilon}_t,$$

with $\boldsymbol{\epsilon}_t \sim N(\mathbf{0}, \Sigma_{S_t})$. As in Chapter 7, S_t denotes a hidden Markov process.

Prado, Molina, and Huerta (2006) consider hierarchical mixture-of-experts (HME) models (see Section 7.1.2) in which the experts are vector autoregressions. Estimation within this modeling framework is achieved via the EM algorithm. These models are used to analyze multichannel electroencephalogram data. Following the notation introduced in Section 7.1.2, a VAR-HME model with O overlays and M experts is given by

$$p(\mathbf{y}_t | \mathcal{D}_{t-1}, \mathbf{X}, \boldsymbol{\theta}) = \sum_{o=1}^{O} \sum_{m=1}^{M} g_t(o, m | \mathcal{D}_{t-1}, \mathbf{X}, \boldsymbol{\gamma}) p(\mathbf{y}_t | \mathcal{D}_{t-1}, \boldsymbol{\eta}, o, m),$$

where $\boldsymbol{\theta} = (\boldsymbol{\gamma}', \boldsymbol{\eta}')'$ and

$$p(\mathbf{y}_t | \mathcal{D}_{t-1}, \boldsymbol{\eta}, o, m) = N(\mathbf{y}_t | \sum_{j=1}^{p_m} \mathbf{\Phi}_j^{o,m} \mathbf{y}_{t-j}, \Sigma_{o,m}).$$

Each $\mathbf{\Phi}_j^{o,m}$ is a $k \times k$ matrix that contains the jth lag coefficient matrix of the VAR process indexed by o and m, and $\Sigma_{o,m}$ is the corresponding variance covariance matrix.

More recently, Fox, Sudderth, Jordan, and Willsky (2009, 2011) consider nonparametric Bayesian switching VAR processes in which a hierarchical Dirichlet process prior is used to learn about the unknown number of persistent smooth dynamical modes that may describe a given process, as well as the VAR model orders. These models are applied to different real data sets, including sequences of honey bee dances that need to be appropriately segmented.

9.6 PARCOR Representations and Spectral Analysis

Zhao and Prado (2020) extends the PARCOR model representation of Yang, Holan, and Wikle (2016) for univariate TVAR processes illustrated in Section 5.3 to the multivariate TV-VAR$_k(p)$ case by considering multivariate dynamic linear models (see Chapter 10) on the k-dimensional forward and backward predictors of the TV-VAR. Discount factors are to specify the system covariance matrices in the multivariate PARCOR DLM representation. Covariance matrices at the observational level are assumed to be constant and estimated using the approximations of Triantafyllopoulos (2007). As it was the case in the univariate setting, the time-varying PARCOR multivariate representation offers advantages in terms of flexibility and computational efficiency when compared to TV-VAR representations.

Regarding the frequency domain, the spectral theory developed in Chapter 8 for bivariate time series also applies to k-dimensional time series processes $\mathbf{y}_t$. In particular, if $\mathbf{\Gamma}(h)$ denotes the autocovariance function of a stationary k-dimensional process, and $\mathbf{f}(\omega)$ denotes the $k \times k$ spectral density matrix, we have that, if each (i, j) element of $\mathbf{\Gamma}(h)$ is absolutely summable, we can write

$$\mathbf{\Gamma}(h) = \int_{-\pi}^{\pi} e^{i\omega h}\mathbf{f}(\omega)d\omega,$$

and

$$\mathbf{f}(\omega) = \frac{1}{2\pi}\sum_{h=-\infty}^{\infty} \mathbf{\Gamma}(h)e^{-i\omega h}.$$

The diagonal elements of $\mathbf{f}(\omega)$, $f_{i,i}(\omega)$ correspond to the spectral density of the ith component of the k-dimensional time series process $\mathbf{y}_t$, while the off-diagonal elements $f_{i,j}(\omega)$ correspond to the cross-spectrum or the cross-spectral density of $y_{t,i}$ and $y_{t,j}$. Note that $\mathbf{f}(\omega)$ is positive semi-definite, i.e., $\mathbf{d}'\mathbf{f}(\omega)\mathbf{d} \geq 0$ for any k-dimensional complex-valued vector $\mathbf{d}$, and also Hermitian, i.e., $\mathbf{f}^*(\omega) = \mathbf{f}(\omega)$, where $\mathbf{f}^*$ denotes the transpose conjugate of $\mathbf{f}$.

The *squared coherence* between $y_{t,i}$ and $y_{t,j}$ is given by

$$C^2_{i,j}(\omega) = \frac{|f_{i,j}(\omega)|^2}{f_{i,i}(\omega)f_{j,j}(\omega)},$$

which satisfies $0 \leq C^2_{i,j}(\omega) \leq 1$. When $C^2_{i,j}(\omega) = 1$ for all ω, the frequency components of series i and j are linearly related, while having $C^2_{i,j}(\omega) = 0$ for all ω implies no linear relationship in the frequency domain.

Several approaches are available for estimating the spectral density matrix of a stationary process. In the Bayesian setting, Rosen and Stoffer (2007) presents an approach based on Whittle's likelihood approximation that fits smoothing splines to the real and imaginary components of the Cholesky decomposition of the periodogram matrix. The periodogram matrix is of $\mathbf{y}_t$ is computed using the discrete Fourier transform with

$$I_{i,j}(\omega_l) = \frac{2}{T}\left(\sum_{t=1}^{T} y_{t,i}e^{i\omega_l t}\right)\left(\sum_{t=1}^{T} y_{t,j}e^{i\omega_l t}\right),$$

being the (i, j) component of such matrix. Zhang (2016) and Li and Krafty

(2019) extend the approach of Rosen and Stoffer (2007) to the non stationary case. For theoretical results and further discussion on multivariate spectral analysis and related topics see Brockwell and Davis (1991) and Wei (2019).

9.6.1 Spectral Matrix of a VAR and VARMA processes

If $\mathbf{y}_t$ is a stationary VAR$_k(p)$ process with coefficient matrices $\mathbf{\Phi}_1, \ldots, \mathbf{\Phi}_p$ and covariance Σ, its spectral density matrix is given by

$$\begin{aligned} \mathbf{f}(\omega) &= [\mathbf{I} - \mathbf{\Phi}_1 e^{-i\omega} - \cdots - \mathbf{\Phi}_p e^{-i\omega p}]^{-1} \Sigma \times \\ & \times \{[\mathbf{I} - \mathbf{\Phi}_1 e^{-i\omega} - \cdots \mathbf{\Phi}_p e^{-i\omega p})]^{-1}\}^*. \end{aligned} \quad (9.19)$$

Similarly, for a VARMA$_k(p,q)$ stationary process with VAR coefficient matrices $\mathbf{\Phi}_1, \ldots, \mathbf{\Phi}_p$, MA coefficient matrices $\mathbf{\Theta}_1, \cdots, \mathbf{\Theta}_q$ and covariance Σ, the spectral density matrix is given by,

$$\mathbf{f}(\omega) = [\mathbf{I} - \mathbf{\Phi}_1 e^{-i\omega} - \cdots - \mathbf{\Phi}_p e^{-i\omega p}]^{-1} [\mathbf{I} - \mathbf{\Theta}_1 e^{-i\omega} - \cdots \mathbf{\Theta}_q e^{-i\omega q}]. \quad (9.20)$$

For TV-VAR and TV-VARMA process the instantaneous spectral density matrix $\mathbf{f}_t(\omega)$, can be computed via (9.19) and (9.20) using the time-varying VAR and VARMA coefficients, respectively, at each time t. If samples from the full posterior distribution of the TV-VAR or TV-VARMA are available in the time domain, the equations above provide a simple way of summarizing the posterior inference for these classes of models in the time-frequency domain.

9.7 Problems

1. Show the decomposition results for VAR models summarized in Section 9.1.3.
2. Simulate three-dimensional data $\mathbf{y}_t$ from the following model:

$$\mathbf{y}_t = \begin{pmatrix} 1 & 1 & 0 \\ 1 & 0 & 1 \\ 1 & 1 & 0 \end{pmatrix} \mathbf{x}_t + \boldsymbol{\epsilon}_t,$$

with $\boldsymbol{\epsilon}_t \sim N(\mathbf{0}, 4\mathbf{I}_3)$ and $\mathbf{x}_t = (x_{1,t}, x_{2,t}, x_{3,t})$ such that

$$\begin{aligned} x_{1,t} &= 0.95 x_{1,t-1} + \eta_{1,t}, \\ x_{2,t} &= 2 \times 0.95 \times \cos(2\pi/18) x_{2,t-1} - 0.95^2 x_{2,t-2} + \eta_{2,t}, \\ x_{3,t} &= 2 \times 0.95 \times \cos(2\pi/7) x_{3,t-1} - 0.95^2 x_{3,t-2} + \eta_{3,t}, \end{aligned}$$

where $\boldsymbol{\eta}_t = (\eta_{1,t}, \eta_{2,t}, \eta_{3,t})'$ and $\boldsymbol{\eta}_t \sim N(\mathbf{0}, \mathbf{I}_3)$. In other words, $x_{1,t}$ is an AR(1) process with coefficient 0.95, $x_{t,2}$ is a quasiperiodic AR(2) process with modulus 0.95 and wavelength 18, and $x_{t,3}$ is a quasiperiodic AR(2) process with modulus 0.95 and wavelength 7.

(a) Fit a VAR(p) model with the `ar` function in `R`. Choose the model order according to Akaike's information criterion (AIC) or the Bayesian information criterion (BIC).

(b) Compute and plot the time series decomposition of $\mathbf{y}_t$ based on your estimates of the VAR matrices of coefficients.

3. Consider a simple VAR$_2$(1) model for $\mathbf{x}_t = (x_{t,1}, x_{t,2})'$ given by $\mathbf{x}_t = \mathbf{G}\mathbf{x}_{t-1} + \boldsymbol{\omega}_t$ with $\boldsymbol{\omega}_t = (\omega_{t,1}, \omega_{t,2})' \sim N(\mathbf{0}, \mathbf{W})$. Denote the (i,j) element of $\mathbf{G}$ by g_{ij}, so the evolution equation element-by-element is

$$\begin{aligned} x_{t,1} &= g_{11}x_{t-1,1} + g_{12}x_{t-1,2} + \omega_{t,1}, \\ x_{t,2} &= g_{21}x_{t-1,1} + g_{22}x_{t-1,2} + \omega_{t,2}. \end{aligned}$$

(a) Write these equations in terms of backshift operators and manipulate them to eliminate $x_{t,2}$ in the equation for $x_{t,1}$; this will yield an equation for $x_{t,1}$ in terms of its lagged values and innovation terms-the implied *marginal model* for $x_{t,1}$ alone.

(b) Show that this leads to $\phi(B)x_{t,1} = \eta_t$ where $\phi(B) = 1 - \phi_1 B - \phi_2 B^2$ and η_t is a zero-mean, normal process with a lag-1 dependency.

(c) Is $x_{t,1}$ a Markov process?

(d) Give expressions for ϕ_1, ϕ_2, and η_t in terms of the VAR$_2$(1) model parameters and innovations.

(e) Verify the simple identities $\phi_1 = \text{trace}(\mathbf{G})$ and $\phi_2 = -|\mathbf{G}|$.

4. Two stationary, univariate AR(1) processes are driven by correlated innovation sequences. That is, we observe two processes y_t and z_t where

$$\begin{aligned} y_t &= \phi y_{t-1} + \epsilon_t \quad \epsilon_t \sim N(0, v), \\ z_t &= \gamma z_{t-1} + \eta_t, \quad \eta_t \sim N(0, w), \end{aligned}$$

where, as usual, ϵ_t and η_t are independent over time, i.e., ϵ_t and η_t are independent of ϵ_{t-k} and of η_{t-k} for any $k > 0$. However, the innovations of the two processes are contemporaneously cross-correlated, i.e., the two series are subject to a common influence. In particular, the vector $(\epsilon_t, \eta_t)'$ has a bivariate normal distribution with variance matrix $\mathbf{V} = \mathbf{A}\mathbf{A}'$ where $\mathbf{A}'$ is the upper triangular Cholesky component of $\mathbf{V}$. Define the 2$-$vector $\mathbf{x}_t$ time series by

$$\mathbf{x}_t = \mathbf{A}^{-1} \begin{pmatrix} y_t \\ z_t \end{pmatrix}.$$

Show that $\mathbf{x}_t$ is a VAR$_2$(1) process and identify the resulting 2×2 AR parameter matrix and innovations variance matrix. Discuss the implications, and comment on the assertion of a modeler who argues that we can always use a diagonal innovations variance matrix in VAR$_q$(1) models: do you agree?

5. Suppose the $q-$vector time series $\mathbf{y}_t$ follows a $\text{VAR}_q(p)$ model with

$$\mathbf{y}_t = \sum_{r=1}^{p} \mathbf{\Phi}_r \mathbf{y}_{t-r} + \boldsymbol{\nu}_t, \qquad \boldsymbol{\nu}_t \sim N(\mathbf{0}, \mathbf{V}),$$

with independent innovations $\boldsymbol{\nu}_t$ over time.

(a) Show that the model can be written as $\mathbf{y}_t = \mathbf{\Theta}'\mathbf{F}_t + \boldsymbol{\nu}_t$ for some regression vector $\mathbf{F}_t$ and *matrix* parameter $\mathbf{\Theta}$. Give the definitions of $\mathbf{F}_t$ and $\mathbf{\Theta}$ and comment on their dimensions. This result shows that the VAR model can be cast as a multivariate DLM with very specific structure. The following Chapter 10, particularly Section 10.2, discusses the important broader class of exchangeable time series (common components) models; this exercise shows that VAR models are a special subset of this broader class.

(b) How can you extend this model to include a nonzero level for the series? What changes to the definitions of $\mathbf{F}_t, \mathbf{\Theta}$ does this entail?

(c) Suppose $q = 400$ with the $y_{t,j}$ being 400 daily stock prices from the S&P500, and a modeler suggests a $\text{VAR}_q(3)$ model-up to 3 days lags might capture at least some aspects of price "momentum." Assuming the basic idea of such a model is attractive, what kinds of statistical issues do you see arising?

6. Suppose the $q-$vector $\mathbf{y}_t = (y_{t,1}, \ldots, y_{t,q})'$ time series is $\text{VAR}_q(1)$ with $\mathbf{y}_t = \mathbf{\Phi}\mathbf{y}_{t-1} + \boldsymbol{\nu}_t$ and $\boldsymbol{\nu}_t \sim N(\mathbf{0}, \mathbf{V})$ independently over time.

(a) Take any non singular $q \times q$ matrix $\mathbf{A}$ and consider the transformation to $\mathbf{x}_t = \mathbf{A}\mathbf{y}_t$. Show that this yields a $\text{VAR}_q(1)$ process $\mathbf{x}_t$ and give the expressions for the VAR coefficient matrix and innovations variance matrix.

(b) Show that you can find an upper triangular, unit diagonal $q \times q$ matrix $\mathbf{A}$, and a positive diagonal $q \times q$ matrix $\mathbf{U}$, i.e.,

$$\mathbf{A} = \begin{pmatrix} 1 & \alpha_{1,2} & \alpha_{1,3} & \alpha_{1,4} & \cdots & \alpha_{1,q} \\ 0 & 1 & \alpha_{2,3} & \alpha_{2,4} & \cdots & \alpha_{2,q} \\ 0 & 0 & 1 & \alpha_{3,4} & \cdots & \alpha_{3,q} \\ \vdots & \vdots & & \ddots & & \vdots \\ 0 & 0 & \cdots & \cdots & 1 & \alpha_{q-1,q} \\ 0 & 0 & \cdots & \cdots & \cdots & 1 \end{pmatrix}$$

and $\mathbf{U} = \text{diag}(u_1, u_2, \ldots, u_q)$ with each $u_i > 0$, such that the implied $\text{VAR}_q(1)$ process $\mathbf{x}_t = \mathbf{A}\mathbf{y}_t$ has innovations variance matrix $\mathbf{U}$.

(c) Discuss interpretations of this matrix $\mathbf{A}$.

(d) Show that this results in a set of q equations for the individual $y_{t,j}$ series of the form

$$y_{t,j} = \mathbf{F}_t'\boldsymbol{\theta}_j + \gamma_{t,j} + \epsilon_{t,j}, \qquad \epsilon_{t,j} \sim N(0, u_j)$$

over time t, where:

- $\mathbf{F}_t = \mathbf{y}_{t-1}$;
- $\boldsymbol{\theta}'_j$ is row j of $\mathbf{A\Phi}$;
- $\gamma_{t,q} = 0$ and, for $j = 1, \ldots, q-1$,

$$\gamma_{t,j} = \boldsymbol{\alpha}'_j \mathbf{y}_{t,pa(j)}$$

where the index set $pa(j) = \{j+1, \ldots, q\}$ so that the vector $\mathbf{y}_{t,pa(j)} = (y_{t,j+1}, \ldots, y_{t,q})'$.

Identify the vectors $\boldsymbol{\alpha}_j$ here.

(e) Discuss the above result, and consider how it might enable alternative analyses of VAR$_q$(1) models that allow alternative, and possible more flexible/richer classes of priors on model parameters.

(f) Suppose you perform such an analysis. A colleague points out that you could reorder the elements of $\mathbf{y}_t$ and redo the analysis, but would get different results without some serious thought about how to make the priors assumed compatible across the analyses. Comment on this point, particularly imagining problems that involve large values of q that are increasingly common.

(g) Suppose now you realize that the applied context of interest requires a time-varying model, i.e., the starting VAR$_q$(1) is to be replaced with a TV-VAR$_q$(1) model $\mathbf{y}_t = \boldsymbol{\Phi}_t \mathbf{y}_{t-1} + \boldsymbol{\nu}_t$ and $\boldsymbol{\nu}_t \sim N(\mathbf{0}, \mathbf{V}_t)$. How would you modify the above discussion?

Chapter 10

General classes of multivariate dynamic models

This chapter discusses a main class of multi-, and matrix-variate dynamic linear models, referred to as exchangeable time series models and/or as common component models. The full class includes traditional VARs (of chapter 9) as special cases, while allowing for broader ranges of structures for both univariate and cross-series dynamic relationships. Examples in financial forecasting and portfolio analysis highlight the use of such models. More recent extensions include statistical graphical model structuring of time-varying precision matrices that address questions of parsimony and scalability in the number of time series jointly modeled. The chapter concludes by contacting more recent developments such as structured simultaneous graphical dynamic models that extend traditional time-varying VAR models and intersect with dynamic simultaneous equations models. Further areas include multivariate and multi-scale dynamic models generally, with some specific examples of models for discrete/count time series, with forecasting applications in commercial applications as well as in dynamic network studies for monitoring and anomaly detection.

10.1 Theory of Multivariate and Matrix Normal DLMs

The general principles and theoretical framework of the univariate dynamic linear model (DLM) theory summarized in Chapter 4 extends to a broad class of multivariate DLMs with known observational and evolution variance-covariance matrices.

10.1.1 Multivariate Normal DLMs

For an $r \times 1$ vector time series with observations $\mathbf{y}_t$, a multivariate normal DLM is defined via $\{\mathbf{F}_t, \mathbf{G}_t, \mathbf{V}_t, \mathbf{W}_t\}$, where $\mathbf{F}_t$ is the $(p \times r)$ matrix, $\mathbf{G}_t$ is a $p \times p$ matrix, $\mathbf{V}_t$ is the $(r \times r)$ observational variance matrix, and $\mathbf{W}_t$ is the $(p \times p)$ system variance matrix. This implies that

$$\begin{aligned} \mathbf{y}_t &= \mathbf{F}_t' \boldsymbol{\theta}_t + \boldsymbol{\nu}_t, \quad \boldsymbol{\nu}_t \sim N(\mathbf{0}, \mathbf{V}_t), \\ \boldsymbol{\theta}_t &= \mathbf{G}_t \boldsymbol{\theta}_{t-1} + \boldsymbol{\omega}_t, \quad \boldsymbol{\omega}_t \sim N(\mathbf{0}, \mathbf{W}_t). \end{aligned} \tag{10.1}$$

Under the assumption that $\mathbf{V}_t$ and $\mathbf{W}_t$ are known and setting $(\boldsymbol{\theta}_0|\mathcal{D}_0) \sim N(\mathbf{m}_0, \mathbf{C}_0)$ for some (known) $\mathbf{m}_0$ and $\mathbf{C}_0$, it follows that $(\boldsymbol{\theta}_t|\mathcal{D}_{t-1}) \sim N(\mathbf{a}_t, \mathbf{R}_t)$, with $\mathbf{a}_t$ and $\mathbf{R}_t$ given by (4.6); $(\mathbf{y}_t|\mathcal{D}_{t-1}) \sim N(\mathbf{f}_t, \mathbf{Q}_t)$, with $\mathbf{f}_t = \mathbf{F}_t'\mathbf{a}_t$, and $\mathbf{Q}_t = \mathbf{F}_t'\mathbf{R}_t\mathbf{F}_t + \mathbf{V}_t$; and finally, $(\boldsymbol{\theta}_t|\mathcal{D}_t) \sim N(\mathbf{m}_t, \mathbf{C}_t)$, with $\mathbf{m}_t = \mathbf{a}_t + \mathbf{A}_t\mathbf{e}_t$, $\mathbf{C}_t = \mathbf{R}_t - \mathbf{A}_t\mathbf{Q}_t\mathbf{A}_t'$, $\mathbf{A}_t = \mathbf{R}_t\mathbf{F}_t\mathbf{Q}_t^{-1}$, and $\mathbf{e}_t = \mathbf{y}_t - \mathbf{f}_t$.

Extension of model fitting analysis to include inference about uncertain covariance elements can be done using Markov chain Monte Carlo (MCMC) methods or via multiprocess mixtures in some cases, while a conjugate analysis extension is available for a rich class of models with unknown covariance structures that include time-varying cases, or multivariate volatility models. The latter class of models are discussed and explored in Section 10.4 below.

Multivariate Gaussian and non-Gaussian DLMs have also been extended to model spatio-temporal data in which multiple time series are recorded at various geographical locations (see for example Section 8.4 of Banerjee, Carlin, and Gelfand 2004 and Lemos and Sansó 2009).

10.1.2 Matrix Normal DLMs and Exchangeable Time Series

Stemming from foundational developments in Quintana and West (1987) and Chapter 16 of West and Harrison (1997) is a class of models for vector and matrix-valued time series that are of increasing interest in applications in areas such as finance, econometrics, and environmental studies. Multivariate versions for vector series are developed more extensively in the following section, and represent a mature and widely used class of models. Matrix models have been less widely explored and used, but are of increasing interest as time series of matrix data become increasingly common, arising in areas such as panel studies and macroeconometrics (e.g., monthly data on several economic indicators across several states, countries, or economic sectors), and in increasingly data-rich areas of environmental science (e.g., time series of several related environmental measurements made at multiple geographic locations over time).

The model class involves multiple time series that follow individual DLMs having the same regression and state evolution structure over time, which explains why the terms *exchangeable time series* and *exchangeable component time series* are relevant. Building on earlier work and general theory in Quintana and West (1987) and West and Harrison (1997), the general matrix model framework is summarized in Wang and West (2009), with examples in analyses of macroeconomic time series; key theoretical aspects are detailed here. The material now discussed involves matrix normal distribution theory; see the supporting appendix material in Section 10.7.1.

Consider a time series of $r \times q$ matrix observations $\mathbf{Y}_t$ following the *matrix normal DLM* given by

$$\begin{aligned} \mathbf{Y}_t &= (\mathbf{I}_r \otimes \mathbf{F}_t')\boldsymbol{\Theta}_t + \mathbf{N}_t, \quad \mathbf{N}_t \sim N(\mathbf{0}, v_t\mathbf{U}, \mathbf{V}), \\ \boldsymbol{\Theta}_t &= (\mathbf{I}_r \otimes \mathbf{G}_t)\boldsymbol{\Theta}_{t-1} + \boldsymbol{\Omega}_t, \quad \boldsymbol{\Omega}_t \sim N(\mathbf{0}, \mathbf{U} \otimes \mathbf{W}_t, \mathbf{V}) \end{aligned} \tag{10.2}$$

for $t = 1, 2, \ldots,$ where *(a)* $\mathbf{Y}_t = (y_{t,i,j})$, the $r \times q$ matrix of observations on the rq univariate time series at time t; *(b)* $\boldsymbol{\Theta}_t$ is the $rp \times q$ state matrix comprised of $r \times q$ state vectors $\boldsymbol{\theta}_{t,i,j}$, each of dimension $p \times 1$; *(c)* $\boldsymbol{\Omega}_t$ is the $rp \times q$ matrix comprised of $r \times q$ state evolution vectors $\boldsymbol{\omega}_{t,i,j}$, each of dimension $p \times 1$; *(d)* $\mathbf{N}_t = (\nu_{t,i,j})$, the $r \times q$ matrix of observational errors; *(e)* v_t is a known, positive scale factor at time t; *(f)* $\mathbf{W}_t$ is the $p \times p$ state evolution covariance matrix at time t, common to all rq univariate time series; *(g)* for all t, the p-vector $\mathbf{F}_t$ and $p \times p$ state evolution matrix $\mathbf{G}_t$ are known. Also, $\boldsymbol{\Omega}_t$ follows a matrix-variate normal distribution with mean $\mathbf{0}$, left covariance matrix $\mathbf{U} \otimes \mathbf{W}_t$, and right covariance matrix $\mathbf{V}$.

The $r \times r$ matrix $\mathbf{U}$ and $q \times q$ matrix $\mathbf{V}$ induce correlation patterns among the observations, as well as among the state vectors, across all component series. Note that matrix normal models of any kind require a one-dimensional identifying constraint on parameters of $\mathbf{U}$ or $\mathbf{V}$. The simplest constraint, and that used and recommended by Wang and West (2009), for example, is to specify $u_{1,1} = 1$. The variance scale factors v_t provide flexibility in, for example, modeling outliers or other known scale variations, though often $v_t = 1$ for all t will be relevant.

The $r \times q$ univariate time series $y_{t,i,j}$ follow individual univariate DLMs

$$\begin{aligned} y_{t,i,j} &= \mathbf{F}_t'\boldsymbol{\theta}_{t,i,j} + \nu_{t,i,j}, \quad \nu_{t,i,j} \sim N(0, v_t u_{i,i} v_{j,j}), \\ \boldsymbol{\theta}_{t,i,j} &= \mathbf{G}_t\boldsymbol{\theta}_{t-1,i,j} + \boldsymbol{\omega}_{t,i,j}, \quad \boldsymbol{\omega}_{t,i,j} \sim N(\mathbf{0}, u_{i,i}v_{j,j}\mathbf{W}_t) \end{aligned} \tag{10.3}$$

for each i, j, and t. That is, $y_{t,i,j}$ follows the DLM defined by the quadruple $\{\mathbf{F}_t, \mathbf{G}_t, v_t u_{i,i} v_{j,j}, u_{i,i}v_{j,j}\mathbf{W}_t\}$. The vector $\mathbf{F}_t$ and matrix $\mathbf{G}_t$ are common across series. The correlation structures across the rq time series are induced by $\mathbf{U}$ and $\mathbf{V}$ and affect both the observation and evolution equations. Consider, for example, any two rows of $\mathbf{Y}_t$, namely $\mathbf{y}_{t,i,\star}$ and $\mathbf{y}_{t,j,\star}$; a large, positive value of $u_{i,j}$ implies that these two vector series will show concordant behavior in movements of both their state vectors and their observational variations. Similarly, $v_{h,k}$ defines correlations between the time series structures of the pair of columns $\mathbf{y}_{t,\star,h}$ and $\mathbf{y}_{t,\star,k}$ of $\mathbf{Y}_t$. In practice, the evolution variance matrix sequence $\mathbf{W}_t$-also common across the univariate series-will often be specified using discount factors.

Assuming $\mathbf{U}, \mathbf{V}$ and, for each t, $v_t, \mathbf{W}_t$ are specified, then the common component structure of this matrix DLM enables a closed form, forward filtering and retrospective analysis that neatly extends that for univariate and vector models. Assume an initial matrix normal prior $(\mathbf{\Theta}_0|\mathcal{D}_0) \sim N(\mathbf{M}_0, \mathbf{U} \otimes \mathbf{C}_0, \mathbf{V})$ for specified $rp \times q$ mean matrix $\mathbf{M}_0$ and $p \times p$ initial covariance matrix $\mathbf{C}_0$ for each of the individual state vectors $\boldsymbol{\theta}_{0,i,j}$. Then, for all $t = 1, 2, \dots,$ we can deduce the following updating equations.

At each time t, we have the following distributional results:

- Posterior at $t-1$: $(\mathbf{\Theta}_{t-1}|\mathcal{D}_{t-1}) \sim N(\mathbf{M}_{t-1}, \mathbf{U} \otimes \mathbf{C}_{t-1}, \mathbf{V})$.
- Prior at t: $(\mathbf{\Theta}_t|\mathcal{D}_{t-1}) \sim N(\mathbf{a}_t, \mathbf{U} \otimes \mathbf{R}_t, \mathbf{V})$ where $\mathbf{a}_t$ is the $rp \times q$ mean matrix $\mathbf{a}_t = (\mathbf{I}_r \otimes \mathbf{G}_t)\mathbf{M}_{t-1}$ and $\mathbf{R}_t$ is the $p \times p$ covariance matrix common across series and defined by $\mathbf{R}_t = \mathbf{G}_t \mathbf{C}_{t-1} \mathbf{G}_t' + \mathbf{W}_t$.
- One-step forecast at $t-1$: $(\mathbf{Y}_t|\mathcal{D}_{t-1}) \sim N(\mathbf{f}_t, q_t\mathbf{U}, \mathbf{V})$ with forecast mean $r \times q$ matrix $\mathbf{f}_t = (\mathbf{I}_r \otimes \mathbf{F}_t'\mathbf{G}_t)\mathbf{M}_{t-1}$ and scalar variance term $q_t = \mathbf{F}_t'\mathbf{R}_t\mathbf{F}_t + v_t$.
- Posterior at t: $(\mathbf{\Theta}_t|\mathcal{D}_t) \sim N(\mathbf{M}_t, \mathbf{U} \otimes \mathbf{C}_t, \mathbf{V})$ with $\mathbf{M}_t = \mathbf{a}_t + (\mathbf{I}_r \otimes \mathbf{A}_t)\mathbf{e}_t$ and $\mathbf{C}_t = \mathbf{R}_t - \mathbf{A}_t\mathbf{A}_t' q_t$ where $\mathbf{A}_t = \mathbf{R}_t\mathbf{F}_t/q_t$ and $\mathbf{e}_t = \mathbf{Y}_t - \mathbf{f}_t$ is the $r \times q$ forecast error matrix.

This theory stems from the application of the results for multivariate models applied to the vectorized observations $\text{vec}(\mathbf{Y}_t)$. The structure involves separability of covariances; for example, state posteriors have separable covariance structures in that $Cov(\text{vec}(\mathbf{\Theta}_t)|\mathcal{D}_t) = \mathbf{V} \otimes \mathbf{U} \otimes \mathbf{C}_t$. Note that the evolution/update equations for $\mathbf{R}_t$, $\mathbf{C}_t$, and q_t are precisely those of a single univariate DLM $\{\mathbf{F}_t, \mathbf{G}_t, v_t, \mathbf{W}_t\}$ whose structure is shared by each of the component series of the matrix model.

The sequential updating theory easily leads to the evaluation of implied forecast normal distributions $p(\mathbf{Y}_{t+h}|\mathcal{D}_t)$ for $h = 1, 2, \dots,$ and retrospective filtering and smoothing distributions for past states, namely $p(\mathbf{\Theta}_t|\mathcal{D}_T)$ for $t \le T$ made at time T. Details are omitted here and left to the reader as an exercise; full details in special, practicable, and widely used special cases of multivariate DLMs and exchangeable time series are developed in the next section.

The above theory is all conditional on known values of the two covariance matrices $\mathbf{U}, \mathbf{V}$. Wang and West (2009) develop MCMC analysis to embed the above model in an overall analysis that includes posterior simulation for $\mathbf{U}, \mathbf{V}$ and inference from the implied full posterior for a sequence of states jointly with $\mathbf{U}, \mathbf{V}$. Additional developments in their work extends the analysis to include graphical model structuring of one or both of the covariance matrices; see further discussion below in Section 10.5.

10.2 Multivariate DLMs and Exchangeable Time Series

A widely used special case is that of $r = 1$, when the above matrix DLM becomes a multivariate model for just the single first row of $\mathbf{Y}_t$. To simplify notation, and in concordance with earlier notation in Quintana and West (1987), West and Harrison (1997), and Carvalho and West (2007), for example, we drop the row index; the observation at time t is now $\mathbf{Y}_t \equiv \mathbf{y}_t'$ where $\mathbf{y}_t$ is a q-dimensional column vector, and we use similar simplifications to other aspects of notation that are clear in the model specification below. With this in mind, setting $r = 1$ in the model of Equation (10.2) and using $u_{1,1} = 1$ for identification, we have

$$\begin{aligned} \mathbf{y}_t' &= \mathbf{F}_t'\boldsymbol{\Theta}_t + \boldsymbol{\nu}_t', \quad & \boldsymbol{\nu}_t &\sim N(\mathbf{0}, v_t\mathbf{V}), \\ \boldsymbol{\Theta}_t &= \mathbf{G}_t\boldsymbol{\Theta}_{t-1} + \boldsymbol{\Omega}_t, \quad & \boldsymbol{\Omega}_t &\sim N(\mathbf{0}, \mathbf{W}_t, \mathbf{V}) \end{aligned} \tag{10.4}$$

where $\boldsymbol{\Theta}_t$ is now the $p \times q$ state matrix whose columns are the q state vectors $\boldsymbol{\theta}_{t,j}$ each of dimension $p \times 1$, and $\boldsymbol{\Omega}_t$ is the $p \times q$ matrix whose columns are the corresponding state evolution vectors $\boldsymbol{\omega}_{t,j}$, $j = 1 : q$.

In terms of scalar time series elements $\mathbf{y}_t' = (y_{t,1}, \dots, y_{t,q})'$, we have q component univariate DLMs,

$$\begin{aligned} y_{t,j} &= \mathbf{F}_t'\boldsymbol{\theta}_{t,j} + \nu_{t,j}, \quad & \nu_{t,j} &\sim N(0, v_t v_{j,j}), \\ \boldsymbol{\theta}_{t,j} &= \mathbf{G}_t\boldsymbol{\theta}_{t-1,j} + \boldsymbol{\omega}_{t,j}, \quad & \boldsymbol{\omega}_{t,j} &\sim N(\mathbf{0}, v_{j,j}\mathbf{W}_t) \end{aligned} \tag{10.5}$$

for each t. That is, $y_{t,j}$ follows the DLM $\{\mathbf{F}_t, \mathbf{G}_t, v_t v_{j,j}, v_{j,j}\mathbf{W}_t\}$. Relationships across series are induced by nonzero covariances in $\mathbf{V}$, namely $Cov(\nu_{t,i}, \nu_{t,j}) = v_t v_{i,j}$ and $Cov(\boldsymbol{\omega}_{t,i}, \boldsymbol{\omega}_{t,j}) = v_{i,j}\mathbf{W}_t$ for $i \neq j$.

10.2.1 Sequential Updating

The resulting forward filtering and forecasting theory is then the direct special case of that summarized in the previous section. At each time t, the results are the same but for the dimensional simplifications. That is, at each time t:

- Posterior at $t-1$: $(\boldsymbol{\Theta}_{t-1}|\mathcal{D}_{t-1}) \sim N(\mathbf{M}_{t-1}, \mathbf{C}_{t-1}, \mathbf{V})$ with $p \times q$ mean matrix $\mathbf{M}_{t-1}$ and $p \times p$ column covariance matrix $\mathbf{C}_{t-1}$.

- Prior at t: $(\boldsymbol{\Theta}_t|\mathcal{D}_{t-1}) \sim N(\mathbf{a}_t, \mathbf{R}_t, \mathbf{V})$ where $\mathbf{a}_t$ is the $p \times q$ mean matrix $\mathbf{a}_t = \mathbf{G}_t\mathbf{M}_{t-1}$ and $\mathbf{R}_t$ is the $p \times p$ column covariance matrix $\mathbf{R}_t = \mathbf{G}_t\mathbf{C}_{t-1}\mathbf{G}_t' + \mathbf{W}_t$.
- One-step forecast at $t-1$: $(\mathbf{y}_t|\mathcal{D}_{t-1}) \sim N(\mathbf{f}_t, q_t\mathbf{V})$ with forecast mean column q-vector $\mathbf{f}_t = \mathbf{a}_t'\mathbf{F}_t$ and scalar variance term $q_t = \mathbf{F}_t'\mathbf{R}_t\mathbf{F}_t + v_t$.
- Posterior at t: $(\boldsymbol{\Theta}_t|\mathcal{D}_t) \sim N(\mathbf{M}_t, \mathbf{C}_t, \mathbf{V})$ with $\mathbf{M}_t = \mathbf{a}_t + \mathbf{A}_t\mathbf{e}_t'$ and $\mathbf{C}_t = \mathbf{R}_t - \mathbf{A}_t\mathbf{A}_t'q_t$ with column p-vector of adaptive coefficients $\mathbf{A}_t = \mathbf{R}_t\mathbf{F}_t/q_t$ and where $\mathbf{e}_t = \mathbf{y}_t - \mathbf{f}_t$ is the column q-vector of forecast errors.

10.2.2 Forecasting and Retrospective Smoothing

Predictive distributions for step-ahead forecasting arise easily from the multivariate normal theory. Extrapolating from time t to time $t+h$, we have the following:

- $(\boldsymbol{\Theta}_{t+h}|\mathcal{D}_t) \sim N(\mathbf{a}_t(h), \mathbf{R}_t(h), \mathbf{V})$ where $\mathbf{a}_t(h)$ and $\mathbf{R}_t(h)$ are computed as follows: starting at $\mathbf{a}_t(0) = \mathbf{m}_t$ and $\mathbf{R}_t(0) = \mathbf{C}_t$, recursively compute $\mathbf{a}_t(k) = \mathbf{G}_{t+k}\mathbf{a}_t(k-1)$ and $\mathbf{R}_t(k) = \mathbf{G}_{t+k}\mathbf{R}_t(k-1)\mathbf{G}_{t+k}' + \mathbf{W}_{t+k}$ for $k = 1:h$.
- For $h = 1, 2, \ldots,$ $(\mathbf{y}_{t+h}|\mathcal{D}_t) \sim N(\mathbf{f}_t(h), q_t(h)\mathbf{V})$ with forecast mean vector $\mathbf{f}_t(h) = \mathbf{a}_t(h)'\mathbf{F}_{t+h}$ and scale factor $q_t(h) = \mathbf{F}_{t+h}'\mathbf{R}_t(h)\mathbf{F}_{t+h} + v_{t+h}$.

Retrospective filtering and smoothing for inference on past states uses distributions as follows. At time T looking back to times $t < T$:

$$(\boldsymbol{\Theta}_t|\mathcal{D}_T) \sim N(\mathbf{a}_T(t-T), \mathbf{R}_T(t-T), \mathbf{V}),$$

where $\mathbf{a}_t(t-T)$ and $\mathbf{R}_T(t-T)$ are computed via direct extensions of Equations (4.10) and (4.11) to the matrix states in the current models. Beginning with $\mathbf{a}_T(0) = \mathbf{m}_T$ and $\mathbf{R}_T(0) = \mathbf{C}_T$, these move backward in time over $t = T-1, T-2, \ldots,$ recursively computing

$$\begin{aligned} \mathbf{a}_T(t-T) &= \mathbf{M}_t - \mathbf{B}_t(\mathbf{a}_{t+1} - \mathbf{a}_T(t-T+1)), \\ \mathbf{R}_T(t-T) &= \mathbf{C}_t - \mathbf{B}_t(\mathbf{R}_{t+1} - \mathbf{R}_T(t-T+1))\mathbf{B}_t', \end{aligned} \tag{10.6}$$

where $\mathbf{B}_t = \mathbf{C}_t\mathbf{G}_{t+1}'\mathbf{R}_{t+1}^{-1}$ for each t.

Forecasting and retrospective analyses often uses posterior simulation to explore uncertainty about future time series paths and about historical state parameter trajectories. For retrospective posterior sampling, conditional distributions underlying computation of the smoothing Equations (10.6) are used. Beginning with a simulated draw $\boldsymbol{\Theta}_T \sim N(\mathbf{M}_T, \mathbf{C}_T, \mathbf{V})$ at time T, move backward in time over $t = T-1, T-2, \ldots,$ recursively sampling from

$$p(\boldsymbol{\Theta}_t|\boldsymbol{\Theta}_{t+1}, \mathcal{D}_T) \equiv p(\boldsymbol{\Theta}_t|\boldsymbol{\Theta}_{t+1}, \mathcal{D}_t),$$

at each step substituting the recent sampled matrix $\mathbf{\Theta}_{t+1}$ in the conditioning. This distribution is $\mathbf{\Theta}_t \sim N(\mathbf{M}_t^*, \mathbf{C}_t^*, \mathbf{V})$ where

$$\begin{aligned} \mathbf{M}_t^* &= \mathbf{M}_t + \mathbf{B}_t(\mathbf{\Theta}_{t+1} - \mathbf{a}_{t+1}), \\ \mathbf{C}_t^* &= \mathbf{C}_t - \mathbf{B}_t\mathbf{R}_{t+1}\mathbf{B}_t', \end{aligned} \tag{10.7}$$

and so is easily sampled.

Beyond the direct use in contexts where linear, normal models are appropriate, the above general theory may be embedded within richer model classes with mixture and other nonnormal structure, enabling broader use in problems having conditional normal structure. Using data transformations and/or MCMC methods to address posterior analysis under nonnormal structure that is not amenable to such approaches, creative applications such as to problems of compositional time have been developed (e.g., Quintana and West 1988; Cargnoni, Müller, and West 1997).

10.3 Learning Cross-Series Covariances

In the general matrix model of Section 10.1.2, admitting uncertainty about either $\mathbf{U}$ or $\mathbf{V}$ leads to the opportunity to extend the analysis via conjugate normal, inverse Wishart distribution theory. Much of the interest in applications has been in the special case of exchangeable time series models of Section 10.2, which involves just the uncertain cross-sectional covariance structure represented by the $q \times q$ covariance matrix $\mathbf{V}$. We summarize the extensions of analysis to learning this cross-series covariance structure here; the extension to the general case of matrix models in Section 10.1.2 (assuming known $\mathbf{U}$) is similar and details are left to the reader as an exercise.

We make a change of notation to explicitly reflect the fact that the cross-series covariance patterns are now uncertain and to be estimated, and also to agree with notation of earlier work. That is, we set $\mathbf{V} \equiv \mathbf{\Sigma}$, the now uncertain covariance matrix

$$\mathbf{\Sigma} = \begin{pmatrix} \sigma_1^2 & \sigma_{1,2} & \cdots & \sigma_{1,q} \\ \sigma_{2,1} & \sigma_2^2 & \cdots & \sigma_{2,q} \\ \vdots & \vdots & \ddots & \vdots \\ \sigma_{q,1} & \sigma_{q,2} & \cdots & \sigma_q^2 \end{pmatrix}.$$

We therefore have the model of Equation (10.4) in this new notation, viz.

$$\begin{aligned} \mathbf{y}_t' &= \mathbf{F}_t'\mathbf{\Theta}_t + \boldsymbol{\nu}_t', \quad & \boldsymbol{\nu}_t &\sim N(\mathbf{0}, v_t\mathbf{\Sigma}), \\ \mathbf{\Theta}_t &= \mathbf{G}_t\mathbf{\Theta}_{t-1} + \mathbf{\Omega}_t, \quad & \mathbf{\Omega}_t &\sim N(\mathbf{0}, \mathbf{W}_t, \mathbf{\Sigma}) \end{aligned} \tag{10.8}$$

for each $t \geq 1$.

The full theoretical development of Section 10.2 now applies with the notational change of $\mathbf{\Sigma}$ replacing $\mathbf{V}$; that section gives the relevant collections of prior, posterior, and forecast distributions all conditional on $\mathbf{\Sigma}$. The extensions here now add priors for $\mathbf{\Sigma}$ and define a complete forward filtering, forecasting, and retrospective analysis for time-varying state matrices and $\mathbf{\Sigma}$ jointly.

Adopt the conjugate matrix normal, inverse Wishart (matrix NIW, see Section 10.6.5) prior

$$(\mathbf{\Theta}_0, \mathbf{\Sigma}|\mathcal{D}_0) \sim NIW(\mathbf{M}_0, \mathbf{C}_0, n_0, \mathbf{D}_0)$$

where $n_0 > 0$ is the initial degrees of freedom of the marginal inverse Wishart (IW) prior $\mathbf{\Sigma} \sim IW(n_0, \mathbf{D}_0)$, and the $q \times q$ sum-of-squares matrix $\mathbf{D}_0$ defines the prior location. The harmonic mean of the prior is $\mathbf{E}(\mathbf{\Sigma}^{-1}|\mathcal{D}_0)^{-1} = \mathbf{D}_0/(n_0+q-1)$, while the mean is $\mathbf{E}(\mathbf{\Sigma}|\mathcal{D}_0) = \mathbf{D}_0/(n_0-2)$ in cases where $n_0 > 2$; also write $\mathbf{S}_0 = \mathbf{D}_0/n_0$.

10.3.1 Sequential Updating

The sequential model analysis results now extend to include covariance matrix learning, building on the conjugacy of the NIW prior. Full details were originally derived in Quintana (1985, 1987); see also extensions and more recent developments in Quintana and West (1987), West and Harrison (1997), Carvalho and West (2007), and Wang and West (2009).

At each time t, the results are as follows:

- Posterior at $t-1$: $(\mathbf{\Theta}_{t-1}, \mathbf{\Sigma}|\mathcal{D}_{t-1}) \sim NIW(\mathbf{M}_{t-1}, \mathbf{C}_{t-1}, n_{t-1}, \mathbf{D}_{t-1})$ with $p \times q$ mean matrix $\mathbf{M}_{t-1}$ and $p \times p$ column covariance matrix $\mathbf{C}_{t-1}$ for $\mathbf{\Theta}_{t-1}$, and $q \times q$ sum-of-squares matrix $\mathbf{D}_{t-1}$ for $\mathbf{\Sigma}$. Write $\mathbf{S}_{t-1} = \mathbf{D}_{t-1}/n_{t-1}$.
- Prior at t: $(\mathbf{\Theta}_t, \mathbf{\Sigma}|\mathcal{D}_{t-1}) \sim NIW(\mathbf{a}_t, \mathbf{R}_t, n_{t-1}, \mathbf{D}_{t-1})$ where $\mathbf{a}_t$ is the $p \times q$ mean matrix $\mathbf{a}_t = \mathbf{G}_t\mathbf{M}_{t-1}$ and $\mathbf{R}_t$ is the $p \times p$ column covariance matrix $\mathbf{R}_t = \mathbf{G}_t\mathbf{C}_{t-1}\mathbf{G}_t' + \mathbf{W}_t$.
- One-step forecast at $t-1$: $(\mathbf{y}_t|\mathcal{D}_{t-1}) \sim T_{n_{t-1}}(\mathbf{f}_t, q_t\mathbf{S}_{t-1})$ with forecast location given by the column q-vector $\mathbf{f}_t = \mathbf{a}_t'\mathbf{F}_t$ (the forecast mode, and mean if $n_{t-1} > 1$) and the scalar variance multiplier term $q_t = \mathbf{F}_t'\mathbf{R}_t\mathbf{F}_t + v_t$. Moving from the conditional normal $(\mathbf{y}_t|\mathbf{\Sigma}, \mathcal{D}_{t-1}) \sim N(\mathbf{f}_t, q_t\mathbf{\Sigma})$ to the marginal forecast distribution simply substitutes the estimate $\mathbf{S}_{t-1}$ for $\mathbf{\Sigma}$ and flattens the tails of the forecast distribution.
- Posterior at t: $(\mathbf{\Theta}_t, \mathbf{\Sigma}|\mathcal{D}_t) \sim NIW(\mathbf{M}_t, \mathbf{C}_t, n_t, \mathbf{D}_t)$ with $\mathbf{M}_t = \mathbf{a}_t + \mathbf{A}_t\mathbf{e}_t'$ and $\mathbf{C}_t = \mathbf{R}_t - \mathbf{A}_t\mathbf{A}_t'q_t$ based on the column p-vector of adaptive coefficients $\mathbf{A}_t = \mathbf{R}_t\mathbf{F}_t/q_t$, $\mathbf{e}_t = \mathbf{y}_t - \mathbf{f}_t$ the column q-vector of forecast

errors, $n_t = n_{t-1} + 1$ the updated degrees of freedom, and $\mathbf{D}_t = \mathbf{D}_{t-1} + \mathbf{e}_t\mathbf{e}_t'/q_t$ with corresponding updated estimate of $\boldsymbol{\Sigma}$ as $\mathbf{S}_t = \mathbf{D}_t/n_t$.

10.3.2 Forecasting and Retrospective Smoothing

Predictive distributions for step-ahead forecasting also follow from the general distributional theory of NIW models overlaid on the results of Section 10.2. Extrapolating from time t to time $t + h$, we have:

- $(\boldsymbol{\Theta}_{t+h}, \boldsymbol{\Sigma}|\mathcal{D}_t) \sim NIW(\mathbf{a}_t(h), \mathbf{R}_t(h), n_t, \mathbf{D}_t)$ with forecast distribution $(\mathbf{y}_{t+h}|\boldsymbol{\Sigma}, \mathcal{D}_t) \sim N(\mathbf{f}_t(h), q_t(h)\boldsymbol{\Sigma})$ where $\mathbf{a}_t(h), \mathbf{R}_t(h), \mathbf{f}_t(h)$, and $q_t(h)$ are computed exactly as in Section 10.2.2.
- Marginalization over $\boldsymbol{\Sigma}$ now leads to the one-step-ahead multivariate T forecast distribution $(\mathbf{y}_{t+h}|\mathcal{D}_t) \sim T_{n_t}(\mathbf{f}_t(h), q_t(h)\mathbf{S}_t)$.

Retrospective inference is based on the smoothed joint distribution of past states. At any time T looking back to times $t < T$, we have

$$(\boldsymbol{\Theta}_t, \boldsymbol{\Sigma}|\mathcal{D}_T) \sim NIW(\mathbf{a}_T(t-T), \mathbf{R}_T(t-T), n_T, \mathbf{D}_T),$$

where $\mathbf{a}_t(t-T)$ and $\mathbf{R}_T(t-T)$ are computed via Equations (10.6).

In simulation of predictive distributions, computations sequentially step through and sample each of the following distributions with conditioning variates at their previously sampled values:

- At time t, simulate from the current posterior via $\boldsymbol{\Sigma} \sim IW(n_t, \mathbf{D}_t)$ then $\boldsymbol{\Theta}_t \sim N(\mathbf{M}_t, \mathbf{C}_t, \boldsymbol{\Sigma})$;
- For $k = 1 : h$, simulate from the state evolution

  $$\boldsymbol{\Theta}_{t+k} \sim N(\mathbf{G}_{t+k}\boldsymbol{\Theta}_{t+k-1}, \mathbf{W}_{t+k}, \boldsymbol{\Sigma})$$

  and then from the observation model

  $$\mathbf{y}_{t+k} \sim N(\boldsymbol{\Theta}_{t+k}'\mathbf{F}_{t+k}, v_{t+k}\boldsymbol{\Sigma}).$$

For simulation of retrospective posteriors for states, analysis begins by simulating the matrix state and covariance matrix at time T, via a draw from $p(\boldsymbol{\Theta}_T, \boldsymbol{\Sigma}|\mathcal{D}_T)$ at time T. We then move backward in time over $t = T-1, T-2, \ldots,$ recursively sampling from

$$(\boldsymbol{\Theta}_t|\boldsymbol{\Theta}_{t+1}, \boldsymbol{\Sigma}, \mathcal{D}_t) \sim N(\mathbf{M}_t^*, \mathbf{C}_t^*, \boldsymbol{\Sigma}),$$

at each step substituting the recent sampled matrix $\boldsymbol{\Theta}_{t+1}$ in the conditioning, with elements $\mathbf{M}_t^*$ and $\mathbf{C}_t^*$ computed via Equations (10.7).

10.4 Time-Varying Covariance Matrices

10.4.1 Introductory Discussion

In areas such as financial time series modeling, matrix DLMs with stochastically time-varying variances and covariances among multiple time series are of broad practical importance and utility. A variety of approaches to modeling stochastic volatility are central to components of more elaborate forecasting and portfolio management models in financial modeling research and front-line applications; key aspects of various approaches are developed in Quintana and West (1987), Bollerslev, Chou, and Kroner (1992), Quintana (1992), Jacquier, Polson, and Rossi (1994), Putnam and Quintana (1994), Quintana, Chopra, and Putnam (1995), Quintana and Putnam (1996), Kim, Shephard, and Chib (1998), Aguilar, Huerta, Prado, and West (1999), Pitt and Shephard (1999b), Aguilar and West (2000), Polson and Tew (2000), Quintana, Lourdes, Aguilar, and Liu (2003), Carvalho and West (2007), and Wang and West (2009), for example. Some selections of recent review material include Shephard (2005) and Chib, Omori, and Asai (2009).

This section concerns approaches based on discounting methods and models that were pioneered by Quintana and West (1987, 1988) as the first published approach to multivariate stochastic volatility defined in the Ph.D. thesis of Quintana (1987) and an earlier unpublished technical report, Quintana (1985). The basic idea of discount factor models for evolution of covariance matrices builds on the established stochastic discounting methods that were introduced by Ameen and Harrison (1985a, 1985b) and developed further in the 1989 first edition of West and Harrison (1997) and, as discussed in Section 4.3.7, developed broadly in applications in the Bayesian time series and dynamic modeling literature. Following the early development and integration into practical time series forecasting and decision/portfolio applications, later theoretical developments generated a number of "random walk" evolution models that underlie discount methods. The following material defines the currently widely used discount model and method. We then describe and detail variants and more recent developments, and discuss and clarify theoretical issues relevant to future research and extensions of these approaches.

10.4.2 Wishart Matrix Discounting Models

The foundational concept in "locally smooth" stochastic volatility modeling is that variances and covariances will change in time, typically slowly and unpredictably in the short-term, suggestive of random walk-like behavior.

Discount factor-based methods build on the sequential updating in learning about uncertain parameters, introducing time-to-time point discounting of cumulated information as a mechanism to reflect information decay over time linked to parameters changing.

Refer back to the inverse Wishart updating of prior to posterior information summaries in the above multivariate models of Section 10.3. There the time $t-1$ prior is $(\mathbf{\Sigma}|\mathcal{D}_{t-1}) \sim IW(n_{t-1}, \mathbf{D}_{t-1})$; equivalently, $(\mathbf{\Phi}|\mathcal{D}_{t-1}) \sim W(h_{t-1}, \mathbf{D}_{t-1}^{-1})$ where $\mathbf{\Phi} = \mathbf{\Sigma}^{-1}$ is the precision matrix corresponding to covariance matrix $\mathbf{\Sigma}$ and $h_{t-1} = n_{t-1} + q - 1$. Here the degrees of freedom (either n_{t-1} or h_{t-1}) represent cumulated information, $\mathbf{D}_{t-1}$ represents the cumulated sum-of-squares sufficient statistic, with $E(\mathbf{\Phi}|\mathcal{D}_{t-1}) = h_{t-1}\mathbf{D}_{t-1}^{-1}$ giving the harmonic mean estimate of covariance $\mathbf{D}_{t-1}/h_{t-1}$.

Consider now a time-varying covariance matrix $\mathbf{\Sigma}_t$ at time t with implied time-varying precision matrix $\mathbf{\Phi}_t = \mathbf{\Sigma}_t^{-1}$. Discounted covariance estimation develops a stochastic evolution model by applying the discount factor to cumulated information. Specifically, between times $t-1$ and t information "decays" in evolving from $t-1$ to t based on a discount factor β, $(0 \le \beta \le 1)$, that defines the map of distributions

$$(\mathbf{\Phi}_{t-1}|\mathcal{D}_{t-1}) \sim W(h_{t-1}, \mathbf{D}_{t-1}^{-1}) \;\rightarrow\; (\mathbf{\Phi}_t|\mathcal{D}_{t-1}) \sim W(\beta h_{t-1}, (\beta \mathbf{D}_{t-1})^{-1}) \tag{10.9}$$

at each time t. Note that this evolution increases the uncertainty in the Wishart distribution for precision matrices by discounting the degrees of freedom h_{t-1}, "losing" $100(1-\beta)\%$ of the information and so increasing the spread of the distribution accordingly. As usual in discounting, β will tend to be high, close to one to reflect slow, steady change of covariance matrices over time, while a smaller value allows for and represents more radical change. Note further that the two distributions have similar general locations; for example, $h_{t-1}\mathbf{D}_{t-1}^{-1}$ is the mean of $p(\mathbf{\Phi}_{t-1}|\mathcal{D}_{t-1})$ and $p(\mathbf{\Phi}_t|\mathcal{D}_{t-1})$ above. In these senses, the evolution has an inherent random walk nature.

This discounting construction arises naturally (and historically) from the univariate model as described in Section 4.1.3, originally developed in Ameen and Harrison (1985a, 1985b) and extended in the 1989 first edition of West and Harrison (1997). When $q = 1$, the model simplifies with $h_t \equiv n_t$, so there is no ambiguity about definition of degrees of freedom, and the Wishart distribution reduces to a gamma distribution. Equations (10.9) then reduce to those of the discount model based on a beta-gamma evolution detailed in Section 4.3.7; see Equation (4.17) and related discussion. The following theory is precisely the multivariate extension of this univariate model, as the reader can verify is now obtained as the special case of $q = 1$.

10.4.3 Matrix Beta Evolution Model

The discounting method arises from a number of candidate stochastic evolution models for covariance matrices. Uhlig (1994) defined the first such model, showing that discounting as applied to Wishart distributions for precision matrices is implied under a class of models involving *singular* matrix-variate beta distributions as the random evolution noise terms. This was explored and developed in Quintana, Chopra, and Putnam (1995), who discussed variants of the model and opened up questions about flexibility and constraints. The matrix beta model in restricted form was later elaborated in Uhlig (1997), and variants of it are now widely used as discount factor-based components of time series and forecasting models. We summarize and exemplify the model here, and in the next section introduce some previously unpublished theory of key practical interest in retrospective time series analysis.

At time $t-1$, suppose $(\mathbf{\Phi}_{t-1}|\mathcal{D}_{t-1}) \sim W(h_{t-1}, \mathbf{D}_{t-1}^{-1})$ as in Equation (10.9), and that an evolution distribution

$$p(\mathbf{\Phi}_t|\mathbf{\Phi}_{t-1}, \mathcal{D}_{t-1})$$

is defined as follows. Given $\mathbf{\Phi}_{t-1}$, set $\mathbf{\Phi}_t = \mathbf{U}_{t-1}'\mathbf{\Gamma}_t\mathbf{U}_{t-1}/\beta$ where:

- $\mathbf{U}_{t-1}$ is any square root of $\mathbf{\Phi}_{t-1}$ so that $\mathbf{\Phi}_{t-1} = \mathbf{U}_{t-1}'\mathbf{U}_{t-1}$; the most convenient, and usual, is to take $\mathbf{U}_{t-1}$ as the upper triangular Cholesky component of $\mathbf{\Phi}_{t-1}$.
- $\mathbf{\Gamma}_t$ is a $q \times q$ matrix random quantity having a matrix beta distribution, denoted by

  $$(\mathbf{\Gamma}_t|\mathcal{D}_{t-1}) \sim Be(\beta h_{t-1}/2, (1-\beta)h_{t-1}/2).$$

  The notation for matrix beta distributions here follows, for example, Tan (1969) and Mitra (1970), specializing to the usual beta distribution when $q = 1$; note that Dawid (1981), who provides key theory for these and other models, uses a slightly different notation.

These results are valid when βh_{t-1} and $(1-\beta)h_{t-1}$ each exceed $q-1$, or when they are integral.

Using distribution theory from Dawid (1981)(see exercises in Section 10.8 below) it easily follows that the implied marginal distribution of $\mathbf{\Phi}_t$ is as defined in Equation (10.9), i.e., the matrix beta evolution model underlies the discount evolution method. This is important in defining the theory for retrospective analysis, in particular, as detailed below. First, we detail the forward filtering implications.

10.4.4 DLM Extension and Sequential Updating

With time-varying $\mathbf{\Sigma}_t$, the model of Equation (10.8) now becomes

$$\begin{aligned} \mathbf{y}'_t &= \mathbf{F}'_t\mathbf{\Theta}_t + \boldsymbol{\nu}'_t, \quad \boldsymbol{\nu}_t \sim N(\mathbf{0}, v_t\mathbf{\Sigma}_t), \\ \mathbf{\Theta}_t &= \mathbf{G}_t\mathbf{\Theta}_{t-1} + \mathbf{\Omega}_t, \quad \mathbf{\Omega}_t \sim N(\mathbf{0}, \mathbf{W}_t, \mathbf{\Sigma}_t) \end{aligned} \tag{10.10}$$

for each $t \geq 1$. Model completion now including the stochastic evolution of $\mathbf{\Sigma}_t$ involves one additional component: the covariance matrix undergoes its matrix beta evolution prior to the state matrix evolving through the state equation, so that the prior for time t becomes $(\mathbf{\Theta}_{t-1}, \mathbf{\Sigma}_t|\mathcal{D}_{t-1}) \sim NIW(\mathbf{M}_{t-1}, \mathbf{C}_{t-1}, \beta h_{t-1} - q + 1, \beta\mathbf{D}_{t-1})$.

The sequential analysis theory of Section 10.3.1 is now simply modified to incorporate the sequence of time-varying covariance matrices $\mathbf{\Sigma}_t$. The structure and notation for parameters of all state distributions remains the same, with $\mathbf{\Sigma}$ replaced by $\mathbf{\Sigma}_{t-1}$ or $\mathbf{\Sigma}_t$ as appropriate in conditional matrix normal priors and posteriors. These are now coupled with the corresponding inverse Wishart components based on the learning equations for $\mathbf{\Sigma}_t$ simply modified by the discount factor. That is,

- Posterior at $t-1$: $(\mathbf{\Phi}_{t-1}|\mathcal{D}_{t-1}) \sim W(h_{t-1}, \mathbf{D}_{t-1}^{-1})$, and with the corresponding $(\mathbf{\Sigma}_{t-1}|\mathcal{D}_{t-1}) \sim IW(n_{t-1}, \mathbf{D}_{t-1})$ where $n_{t-1} = h_{t-1} - q + 1$.
- Prior at t: $(\mathbf{\Phi}_t|\mathcal{D}_{t-1}) \sim W(\beta h_{t-1}, (\beta\mathbf{D}_{t-1})^{-1})$ and with the corresponding $(\mathbf{\Sigma}_t|\mathcal{D}_{t-1}) \sim IW(\beta h_{t-1} - q + 1, \beta\mathbf{D}_{t-1})$.
- Posterior at t: $(\mathbf{\Phi}_t|\mathcal{D}_t) \sim W(h_t, \mathbf{D}_t^{-1})$ where

  $$h_t = \beta h_{t-1} + 1 \quad \text{and} \quad \mathbf{D}_t = \beta\mathbf{D}_{t-1} + \mathbf{e}_t\mathbf{e}'_t/q_t,$$

  with corresponding $(\mathbf{\Sigma}_t|\mathcal{D}_t) \sim IW(n_t, \mathbf{D}_t)$ and where $n_t = h_t - q + 1$.

As an estimate of $\mathbf{\Sigma}_t$ at time t, the harmonic mean $\mathbf{D}_t/h_t$ is a weighted average of past scaled forecast errors that discounts historical square errors by a factor of β at each consecutive time point to reflect, at a practical level, the effects of variation over time. As t increases, h_t quickly converges to the limiting value $1/(1-\beta)$. Note that this implies additional constraints on β to maintain a valid model, since we require either $h_t > q-1$ or h_t be integral. The former constraint implies that β cannot be too small, $\beta > (q-2)/(q-1)$ defined by the limiting value.

10.4.5 Retrospective Analysis

Again the components of distribution theory for conditional normals of matrix states uses the retrospective theory laid out in Section 10.3.2, with conditional normals now modified to depend on the time-varying $\mathbf{\Sigma}_t$ replacing the constant matrix. However, though the sequential, forward filtering analysis maintains the conjugacy of NIW distributions, there is now no complete conjugate theory for retrospective analysis. Fortunately, simulation can be used based on the result below. This is practically critical in defining simulation of retrospective distributions for past states to underlie smoothing analyses based on posterior Monte Carlo samples. Previously unpublished, this is a cornerstone applied aspect of the analysis for evaluation of changes over time in historical trajectories of time-varying covariance matrices, as well as the implications on inference for past states.

At any time T looking back to times $t < T$, the full retrospective simulation analysis proceeds as follows:

- Simulate the matrix state and covariance matrix at time T, via a draw from $\mathbf{\Sigma}_T \sim IW(n_T, \mathbf{D}_T)$ followed by $\mathbf{\Theta}_T \sim N(\mathbf{M}_T, \mathbf{C}_T, \mathbf{\Sigma}_T)$ conditioning on the sampled $\mathbf{\Sigma}_T$.
- Step back in time over $t = T-1, T-2, \ldots,$ recursively sampling from

$$p(\mathbf{\Sigma}_t|\mathbf{\Sigma}_{t+1}, \mathcal{D}_T) \equiv p(\mathbf{\Sigma}_t|\mathbf{\Sigma}_{t+1}, \mathcal{D}_t) \tag{10.11}$$

and

$$(\mathbf{\Theta}_t|\mathbf{\Theta}_{t+1}, \mathbf{\Sigma}_{t+1}, \mathcal{D}_t) \sim N(\mathbf{M}_t^*, \mathbf{C}_t^*, \mathbf{\Sigma}_{t+1}),$$

with elements $\mathbf{M}_t^*$ and $\mathbf{C}_t^*$ as earlier defined (Equations 10.7) and based on the recently simulated matrices $\mathbf{\Theta}_{t+1}$ and $\mathbf{\Sigma}_{t+1}$.

Completing a sweep through this sampler generates a full posterior draw from the posterior of all historical states and covariance matrices together, conditional on the full data set $\mathcal{D}_T$.

To implement this requires simulation of the distributions in Equation (10.11). In fact, these are readily available from the matrix beta construction of the forward evolution model (see exercises in Section 10.8 below): we draw from the conditional distribution $p(\mathbf{\Sigma}_t|\mathbf{\Sigma}_{t+1}, \mathcal{D}_t)$ by setting $\mathbf{\Sigma}_t = \mathbf{\Phi}_t^{-1}$ with

$$\mathbf{\Phi}_t = \beta\mathbf{\Phi}_{t+1} + \mathbf{\Upsilon}_t \quad \text{where} \quad (\mathbf{\Upsilon}_t|\mathcal{D}_t) \sim W((1-\beta)h_t, \mathbf{D}_t^{-1}) \tag{10.12}$$

and where $\mathbf{\Upsilon}_t$ is independent of $\mathbf{\Phi}_{t+1}$.

10.4.6 Financial Time Series Volatility Example

10.4.6.1 Data and model

Figures 10.1–10.15 inclusive display some aspects of posterior inferences from analysis of a time series of daily international exchange rates over a

10 year period. For currency $i = 1 : q$ on day $t = 1 : T$ the exchange rate is $y_{t,i} = 100(P_{t,i}/P_{t-1,i} - 1)$ where $P_{t,i}$ is the daily closing spot price of the currency in \$USA. The $q = 12$ currencies are listed in Table 10.1 and the data run over a time period of $T = 2{,}566$ business days beginning in October 1986 and ending in August 1996. This time period is before the emergence of the euro, and several of the series are for what were then distinct European currencies. Plots of time series for several of the rate series appear in the figures.

Table 10.1 $q = 12$ *international currencies relative to the US dollar in multivariate stochastic volatility example.*

AUD	Australian Dollar
BEF	Belgian Franc
CAD	Canadian Dollar
FRF	French Franc
DEM	German Mark
JPY	Japanese Yen
NLG	Dutch Guilder
NZD	New Zealand Dollar
ESP	Spanish Peseta
SEK	Swedish Krone
CHF	Swiss Franc
GBP	British Pound Sterling

Analysis uses the simplest, multivariate steady model for the means of the series, simply tracking the very low levels of stochastic changes in mean level over time in the context of substantial, dominant variation. Denote by $\mu_{t,i}$ the mean level of series i at time t. The model is the special case of Equation (10.10) with $p = 1$, $\mathbf{F}_t' = 1$, $\boldsymbol{\Theta}_t = \boldsymbol{\theta}_t' = (\mu_{t,1}, \ldots, \mu_{t,q})'$, $v_t = 1$, and $\mathbf{G}_t = \mathbf{I}_q$. The matrix normal distributions reduce to multivariate normals, with elements $\mathbf{W}_t = w_t$ now scalars, and the model is

$$\begin{aligned} \mathbf{y}_t &= \boldsymbol{\theta}_t + \boldsymbol{\nu}_t, \quad \boldsymbol{\nu}_t \sim N(\mathbf{0}, \boldsymbol{\Sigma}_t), \\ \boldsymbol{\theta}_t &= \boldsymbol{\theta}_{t-1} + \boldsymbol{\omega}_t, \quad \boldsymbol{\omega}_t \sim N(\mathbf{0}, w_t\boldsymbol{\Sigma}_t), \end{aligned} \tag{10.13}$$

where w_t is defined via a single discount factor δ. All distributions for states and observational covariance matrices in the sequential updating and retrospective smoothing analyses are now multivariate normal, inverse Wishart. For example, the posterior at time t becomes $NIW(\boldsymbol{\theta}_t, \boldsymbol{\Sigma}_t | \mathbf{m}_t, c_t, n_t, \mathbf{D}_t)$ with scalar c_t, and so forth. The summary sequential updating equations at times $t-1$ to t are

$$\mathbf{m}_t = \mathbf{m}_{t-1} + A_t\mathbf{e}_t \;\text{ and }\; c_t = r_t - A_t^2 q_t,$$

with

$$\mathbf{D}_t = \beta \mathbf{D}_{t-1} + \mathbf{e}_t \mathbf{e}_t'/q_t, \quad h_t = \beta h_{t-1} + 1, \quad \text{and} \quad n_t = h_t - q + 1,$$

where $\mathbf{e}_t = \mathbf{y}_t - \mathbf{m}_{t-1}$, $r_t = c_{t-1} + w_t \equiv c_{t-1}/\delta$, $q_t = r_t + 1$, and $A_t = r_t/q_t$.

Analysis uses $\delta = 0.99$ for the state vector of local means and $\beta = 0.95$ for multivariate volatility discounting. Initial priors are vague, with $\mathbf{m}_0 = \mathbf{0}$, $c_0 = 100$, $n_0 = 20$, and $\mathbf{D}_0 = 20\mathbf{I}_q$.

Following forward filtering and updating over $t = 1 : T$, retrospective analysis updates the posterior summaries for the historical trajectory of states $\boldsymbol{\theta}_t$ and simulates from the posterior over the historical trajectory of covariance matrices $\boldsymbol{\Sigma}_t$. Under this simpler model, the summary equations for these retrospective computations are, over $t = (T-1) : 1$, given as follows. For the states, Equations (10.6) simplify to

$$\mathbf{a}_T(t-T) = (1-\delta)\mathbf{m}_t + \delta \mathbf{a}_T(t-T+1)$$

and

$$r_T(t-T) = (1-\delta)c_t + \delta^2 r_T(t-T+1).$$

Retrospective simulation of the $\boldsymbol{\Sigma}_t = \boldsymbol{\Phi}_t^{-1}$ sequence proceeds precisely as defined in Equation (10.12).

10.4.6.2 Trajectories of multivariate stochastic volatility

Figure 10.1 shows the time series of returns on £UK:\$USA (GBP) and ¥Japan:\$USA (JPY) exchange rates together with estimated volatilities. For the latter, the plots show a random selection of 50 historical trajectories of the individual time-varying standard deviations, $\sigma_{t,i}$ for i indexing GBP and JPY, respectively, from the retrospectively simulated posterior $p(\boldsymbol{\Sigma}_1, \ldots, \boldsymbol{\Sigma}_T | \mathcal{D}_T)$. Also shown is the approximate posterior mean of the volatility from the Monte Carlo samples.

This analysis may be compared with the alternative univariate stochastic volatility model analysis of Example 7.17 in Section 7.5.4; that example used only the latter 1,000 days of this full data set, from late 1992 onward. Estimated on different models and with different data, the estimated trajectories over the period in common are concordant, exhibiting increased volatility common to each currency at various times as well as currency-specific changes. Volatility is substantially higher than normal for GBP around the time of withdrawal of the United Kingdom from the European Monetary System (EMS) in late 1992, and thereafter, while later spurts of increased volatility in JPY associated with imposed target currency bands in EU are evident in late 1993. Further discussion of background economic circumstances appear in Quintana and Putnam (1996) and Aguilar, Huerta, Prado, and West (1999).

Similar plots for several other currencies appear in Figures 10.2 and 10.3. There are marked patterns of similarity in some of the changes in volatility across the core EU currencies DEM and FRF, some shared by GBP though with lesser concordance, as is to be expected. This points to the role and relevance of the exchangeable time series model in capturing aspects of common covariation across series while permitting individual patterns to overlay this. CAD is, in contrast, apparently quite distinct in having low levels of volatility that change far less than the other currencies, due to the tight linkages of the Canadian and USA economies and the resulting strong coupling of the CAD with the USA. AUD shares some patterns of volatility changes with GBP, though again is less strongly concordant with most of the other series illustrated.

Figure 10.4 shows a selection of similar plots for pairwise correlations in $\mathbf{\Sigma}_1, \ldots, \mathbf{\Sigma}_T$ among a few of the currencies. While GBP and DEM have high correlation, there is clear variation that is most marked and abrupt around times of major economic change, including the 1990s recessionary years and following the EMS events in late 1992. The patterns of change in positive correlation between GBP and JPY are more marked. The other displays evidence the lack of any real correlation at all between fluctuations of CAD and AUD, and the very tight and persistent correlation between core EU currencies DEM and FRF.

Though time variation in $\mathbf{\Sigma}_t$ is defined by a single discount factor, the model clearly has the ability to adapt to and isolate substantially different degrees of volatility changes-in covariation patterns as well as variation levels-evidenced across multiple time series.

10.4.6.3 Time-varying principal components analysis

Principal component (PC) decompositions of $\mathbf{\Sigma}_t$ shed more light on underlying commonalities and differences in patterns over time across the series. We compute principal components of each of the $\mathbf{\Sigma}_t$ matrices sampled from the full retrospective posterior distribution over $t = 1:T$, so that the computed values represent samples from the posterior of all elements of the PC decomposition. Plots show some aspects of posterior uncertainty by again just graphing values over time from 50 randomly selected posterior samples, with approximate posterior means superimposed.

The PC decomposition of each $\mathbf{\Sigma}_t$ is

$$\mathbf{\Sigma}_t = \mathbf{E}_t \mathbf{\Delta}_t \mathbf{E}_t' = \sum_{j=1}^{q} \delta_{t,j}^2 \mathbf{e}_{t,j} \mathbf{e}_{t,j}'$$

where $\mathbf{\Delta} = \text{diag}(\delta_{t,1}^2, \ldots, \delta_{t,q}^2)$ with $\delta_{t,j} \geq 0$ and in decreasing order, and

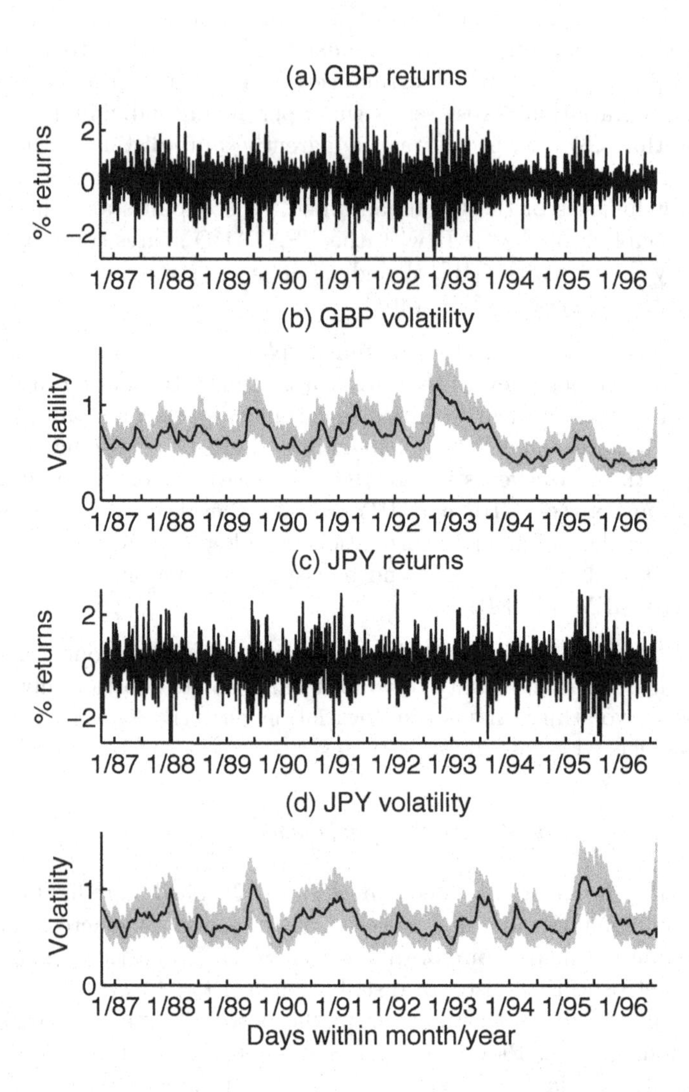

Figure 10.1 *(a) Returns on daily £UK:\$USA (GB pounds, or GBP) exchange rate over approximately 10 years, and (b) its estimated volatility; the full line indicates the posterior mean plotted over days $t = 1 : T$, and the grey shading is 50 similar time plots representing 50 randomly selected trajectories from the posterior. (c), (d) are the corresponding plots for returns and volatility for the ¥Japan:\$USA (Japanese Yen) exchange rates.*

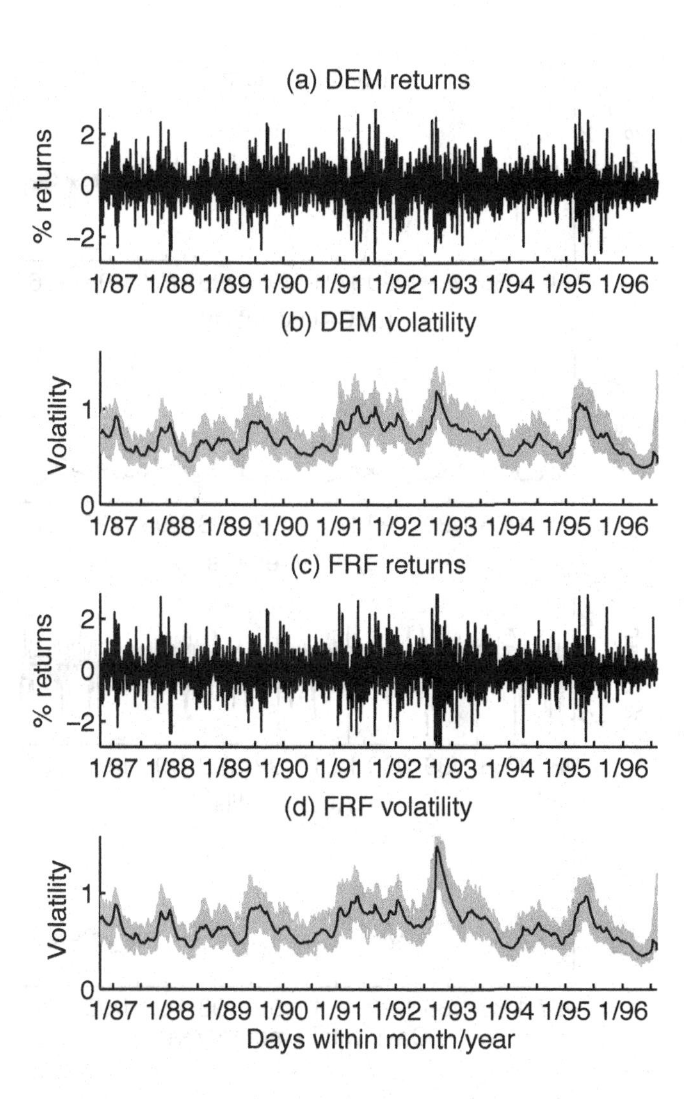

Figure 10.2 *(a) Returns on daily DM Germany:$USA (German mark, or DEM) exchange rate over approximately 10 years, and (b) its estimated volatility; the full line indicates the posterior mean plotted over days* $t = 1 : T$*, and the grey shading is 50 similar time plots representing 50 randomly selected trajectories from the posterior. (c), (d) are the corresponding plots for returns and volatility for the Fr France:$USA (French franc, or FRF) exchange rates.*

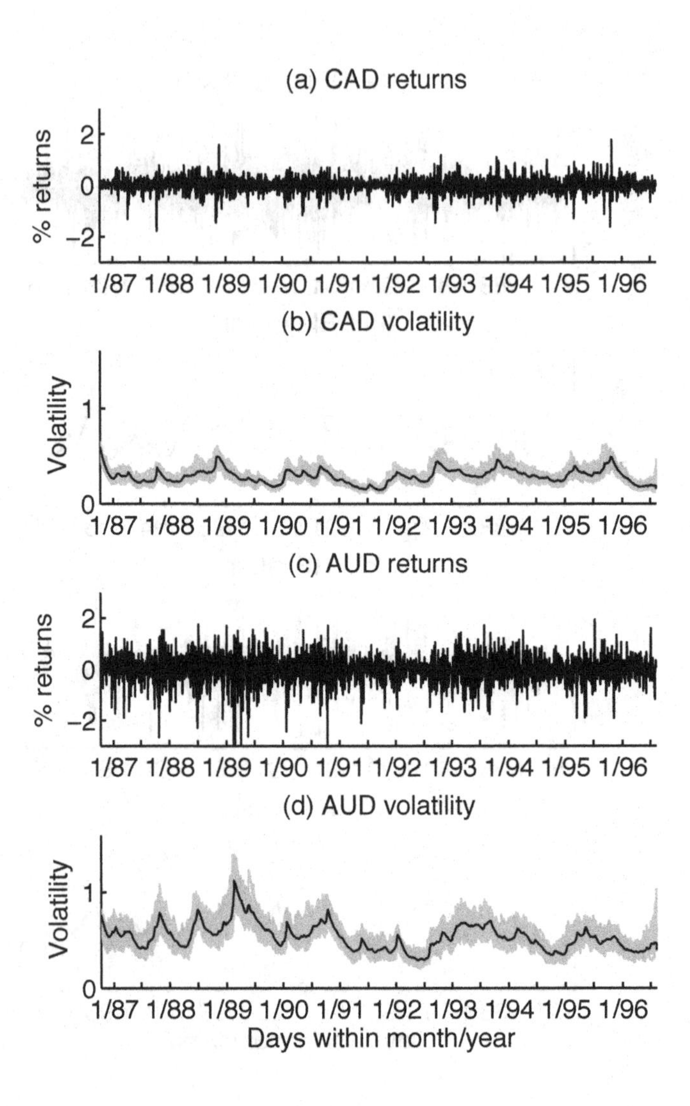

Figure 10.3 *(a) Returns on daily $Canada:$USA (Canadian dollar, or CAD) exchange rate over approximately 10 years, and (b) its estimated volatility; the full line indicates the posterior mean plotted over days $t = 1 : T$, and the grey shading is 50 similar time plots representing 50 randomly selected trajectories from the posterior. (c), (d) are the corresponding plots for returns and volatility for the $Australia:$USA (Australian dollar, or AUD) exchange rates.*

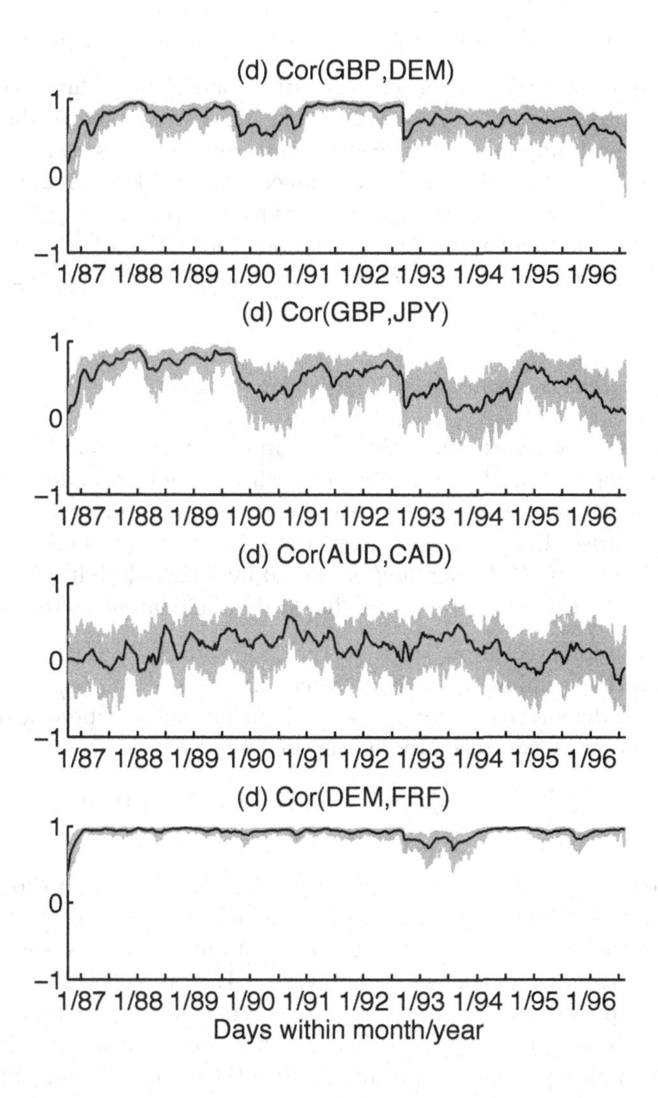

Figure 10.4 *Approximate posterior means and 50 posterior sample trajectories of estimated time-varying correlations between some of the 12 international exchange rate time series.*

where $\mathbf{E}_t$ is orthogonal with q-vector columns $\mathbf{e}_{t,1}, \ldots, \mathbf{e}_{t,q}$. The $\delta_{t,j}$ represent underlying volatility processes that impact on the overall patterns of multivariate volatility, while additional time variation in the $\mathbf{E}_t$ matrices modulates and modifies the role of these underlying latent processes.

Figure 10.5 plots the posterior summaries for the first three component volatility processes, $\delta_{t,j},\ j = 1, 2, 3$. The final frame shows the relative levels of overall impact of these components, and how this varies over time, in terms of the usual "% variation explained" values $100\delta_{t,j}^2/\mathbf{1}'\boldsymbol{\Delta}\mathbf{1}$ over time t for each $j = 1, 2, 3$. In this sense, these three components together explain two-thirds or more of the overall variation among the 12 time series, and so clearly reflect underlying economic and financial factors of regional and global import.

10.4.6.4 Latent components in multivariate volatility

The final sets of figures take this decomposition into underlying components further, linking the volatility of each currency to underlying latent volatilities, and extending to identify the corresponding latent components in return series. Based on the multivariate volatility tracking model, this is a flexible approach to defining empirical factors underlying common comovements in the series, i.e., a PC-based dynamic latent factor analysis.

For any random vector $\mathbf{x}_t = (x_{t,1}, \ldots, x_{t,q})'$ with $\mathbf{x}_t \sim N(\mathbf{0}, \boldsymbol{\Sigma}_t)$, the PC decomposition implies $\mathbf{x}_t = \mathbf{E}_t\mathbf{z}_t$ where $\mathbf{z}_t = (z_{t,1}, \ldots, z_{t,q})'$ and $z_{t,j} \sim N(0, \delta_{t,j}^2)$ independently over $j = 1 : q$. That is, each component of $\mathbf{x}_t$ and its corresponding variance are decomposed as

$$\begin{aligned} x_{t,i} &= \textstyle\sum_{j=1}^q x_{t,i,j} \quad \text{with} \quad x_{t,i,j} = e_{t,i,j} z_{t,j}, \\ V(x_{t,i}) &= \textstyle\sum_{j=1}^q u_{t,i,j}^2 \quad \text{with} \quad u_{t,i,j}^2 = e_{t,i,j}^2 \delta_{t,j}^2, \end{aligned}$$

and where $e_{t,i,j}$ is the (i, j)-th element of $\mathbf{E}_t$. The $z_{t,j}$ are orthogonal, latent components of the "output" series $\mathbf{x}_t$, with time-varying variances and also time variation in the "loadings" $e_{t,i,j}$ of output series on latent components. Given $\boldsymbol{\Sigma}_t$ and $\mathbf{x}_t$ we compute $\mathbf{z}_t = \mathbf{E}_t'\mathbf{x}_t$ directly. The evaluation of the components $x_{t,i,j}$ of the $x_{t,i}$ time series, and of the corresponding components $u_{t,i,j}^2$ of the $V(x_{t,i})$, can often represent a multiscale decomposition, with components varying substantially on different time scales as well as in amplitudes. This can be applied to evaluate empirical latent components in fitted residuals of the time series $\mathbf{y}_t$ at each time point t and using multiple $\boldsymbol{\Sigma}_t$ matrices from the posterior simulation sample. This generates a Monte Carlo approximation to the full joint posterior over the common components of multiple time series as well as their volatilities.

Figures 10.6 and 10.7 show aspects of such a decomposition for the GBP returns series using the Monte Carlo approximations to posterior means of the PC components from the multivariate analysis. Indexing GBP as series i, Figure 10.6 shows the data $x_{t,i} \equiv y_{t,i}$, the first two components $x_{t,i,j}, j = 1,2$, and then the sum of the remaining components $x_{t,i,j}, j = 3:12$. Plotted on the same scales, this clearly shows that the first component is a dominant factor of the GBP series, the second component impacts substantially with two or three "bursts" of activity around times of major economic change relevant to the UK, while the remaining 10 components contribute substantially. Figure 10.7 shows the corresponding components of the volatility, in terms of standard deviations $\sigma_{t,i}$ of the overall GBP series, then $u_{t,i,1}, u_{t,i,2}$, and, finally, the sum of the remaining $u_{t,i,j}$ $(j = 3:12)$. Also shown with the first two components are the corresponding $\delta_{t,j}$; since $u_{t,i,j} = |e_{t,i,j}|\delta_{t,j}$, this helps to understand the impact of changes in $|e_{t,i,j}|$ on the GBP volatility. The overall dominance of the first component is clear, and the nature of the bursty impact of component $j = 2$ is clearly highlighted. The lower frame further shows that the remaining components contribute most substantially also at the key time of major GBP volatility in late 1992/early 1993.

Figures 10.8 and 10.9 present similar graphs for the DEM series, showing that DEM basically defines, and is defined by, the first latent component $z_{t,1}$. The figures for FRF are similar, as they are for the other EU currencies; thus component $j = 1$ is an effective EU factor that plays a major, driving role in comovements of volatilities and returns among the EU countries. DEM shows very little relationship to the other components. Across several other EU currencies, however, small but meaningful associations exist with several high order components at various time periods, reflecting country and, perhaps, smaller trading block-specific effects. Figures 10.10 and 10.11 show the decompositions for the JPY series, indicating a strong but not dominant role for the EU component, and now a clear and marked role for the second component. The AUD decomposition in Figures 10.12 and 10.13 shows that the third and higher order components are most relevant, while component 2 plays a role; the relationship with the major EU component is very limited. For CAD, tightly linked to the USA, the two dominant components are almost negligible, there being limited volatility overall and much of it explained by higher order components and that is almost wholly idiosyncratic to CAD (see Figures 10.14 and 10.15).

10.4.7 Short-term Forecasting for Portfolio Decisions

The use of dynamic Bayesian forecasting models and Bayesian decision analysis in asset allocation problems has represented a major area of application of the models of this chapter for many years. This is exemplified in recent works of Quintana, Lourdes, Aguilar, and Liu (2003), Carvalho and West (2007), Quintana, Carvalho, Scott, and Costigliola (2010), for

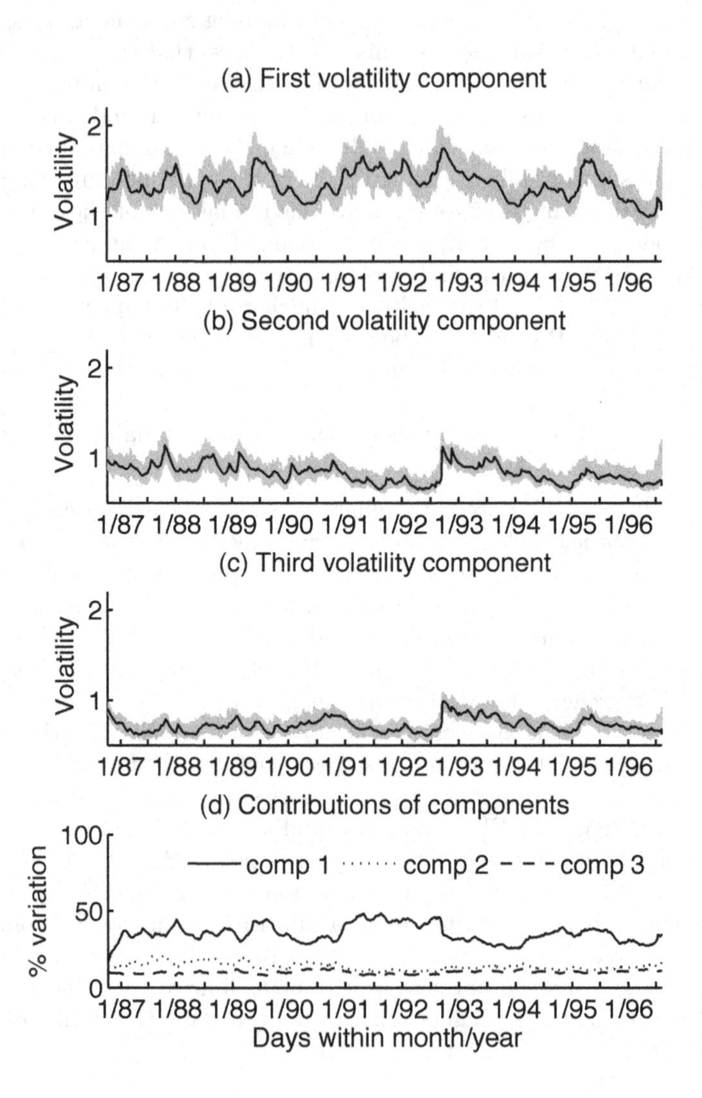

Figure 10.5 *Approximate posterior means and 50 posterior sample trajectories of estimated time-varying volatilities of latent components underlying the 12 international exchange rate time series, computed from PCA (principal components analysis) of the posterior sampled variance matrices* $\mathbf{\Sigma}_t$ *at each time point. (a), (b), and (c) plot the first three dominant components in volatility, while (d) shows approximate percentages of the total variation in* $\mathbf{\Sigma}_t$ *over time* $t = 1 : T$ *contributed by each of these three components.*

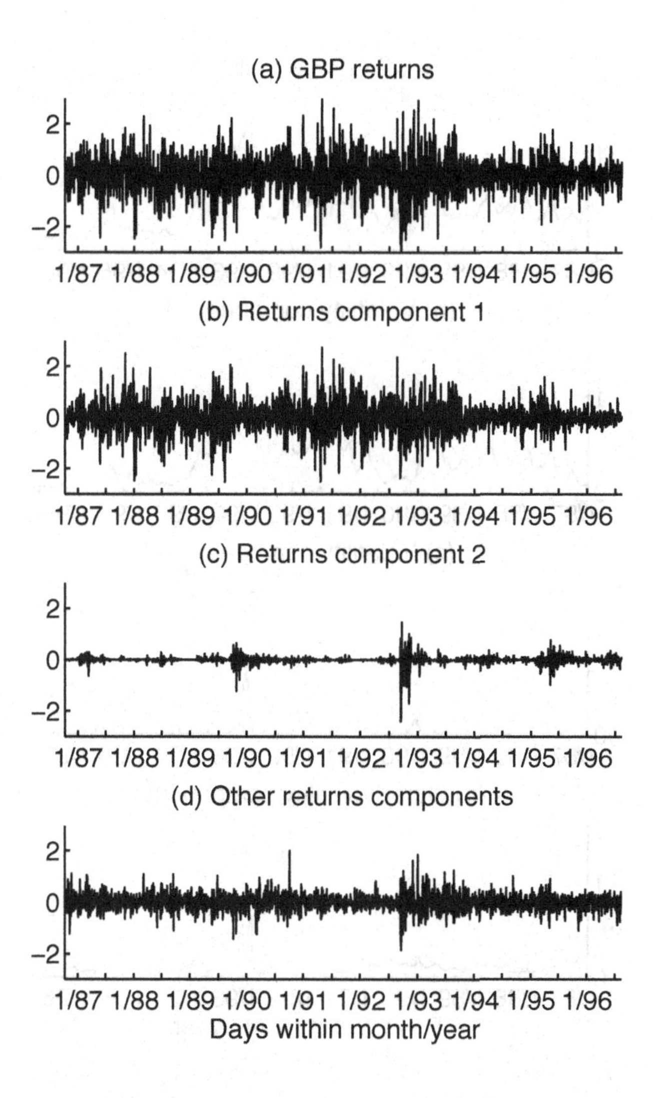

Figure 10.6 *(a) £UK:$USA returns time series and estimated values of latent components defined by volatility process decomposition over time. (b) and (c) show estimates of the two dominant components, while (d) shows the contributions of remaining components. Returns (a) are the direct sums of components (b), (c), and (d).*

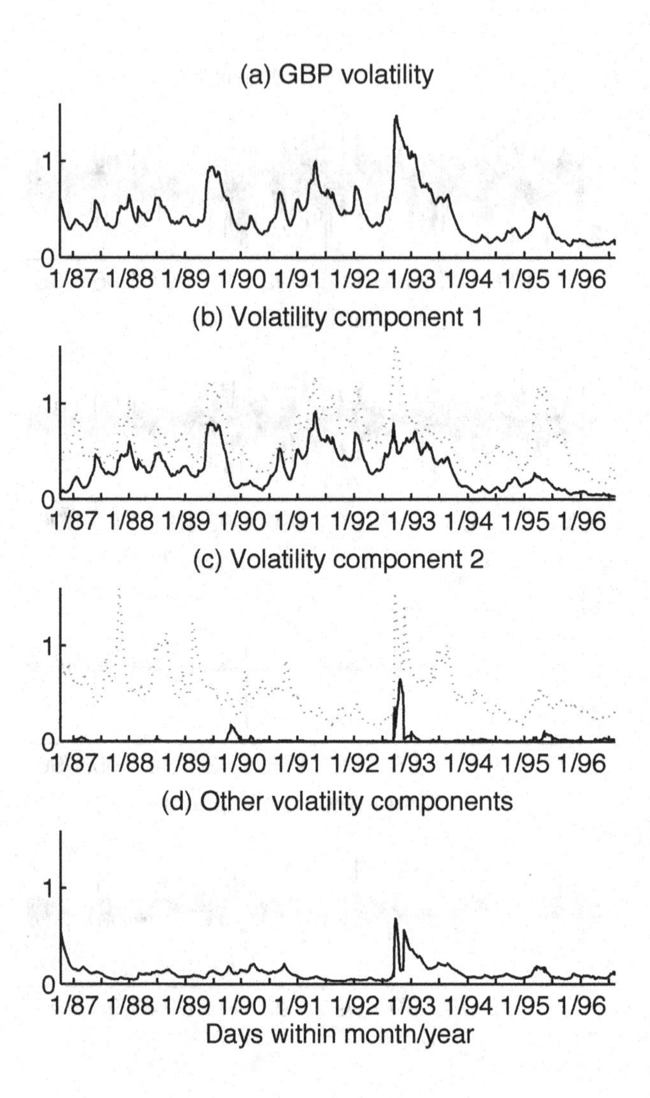

Figure 10.7 *Components of volatility in £UK:\$USA returns corresponding to return decompositions in Figure 10.6. The overall volatility* $\sigma_{t,i}$ *in this currency is made up of contributions from those graphed in (b), (c), and (d) as full lines; (b) and (c) are the* $u_{t,i,j}$ *arising from the two dominant latent returns processes* $(j = 1, 2)$*, while (d) comes from the remaining components. In (b) and (c), the grey dashed lines represent the actual component volatilities* $\delta_{t,j}$ $(j = 1, 2)$ *that are also plotted in Figure 10.5(b),(c); the full lines incorporate the modifying effects of the time-varying loadings* $|e_{t,i,j}|$.

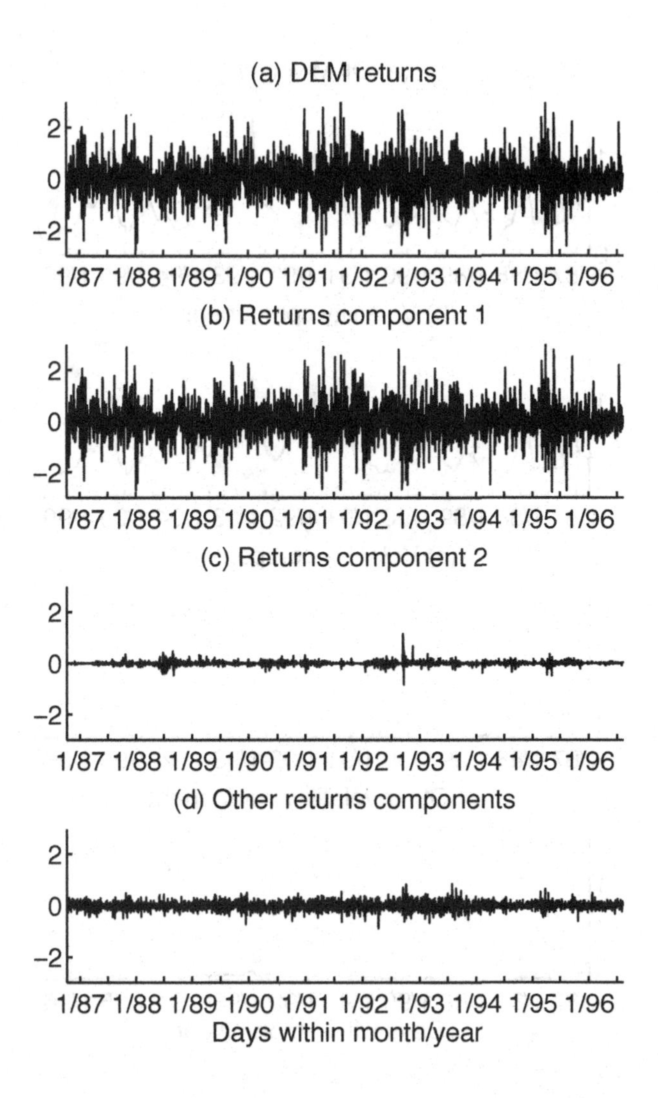

Figure 10.8 *DM Germany:$USA returns time series and estimated values of latent components, in a format similar to that of Figure 10.6.*

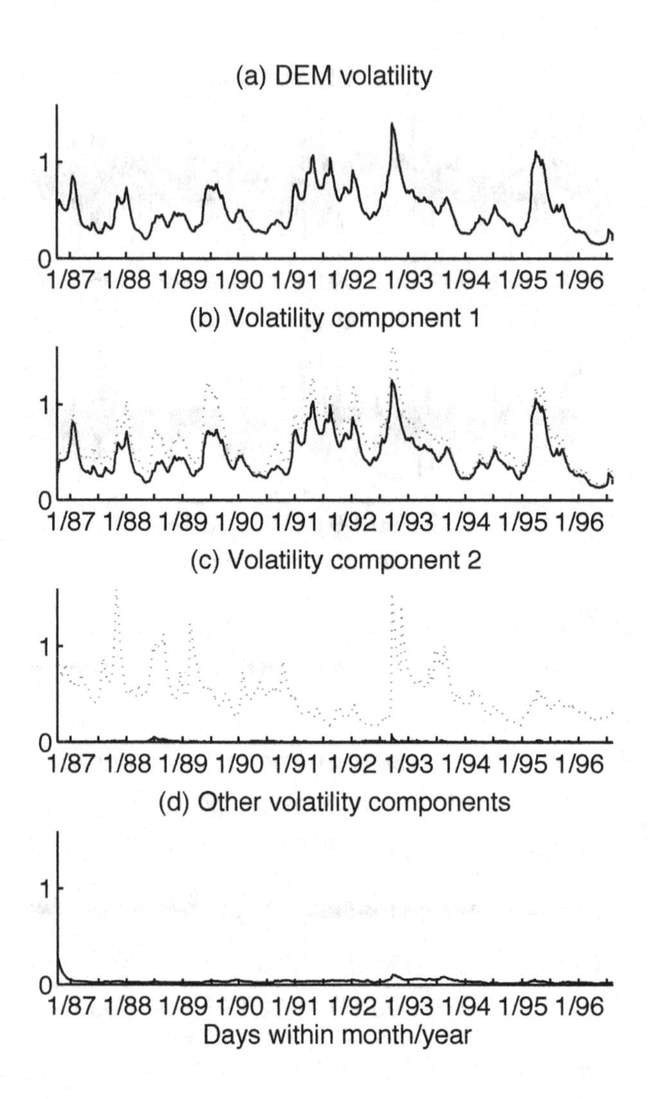

Figure 10.9 *Components of volatility in DM Germany:$USA returns corresponding to return decompositions in Figure 10.8, in a format similar to that of Figure 10.7.*

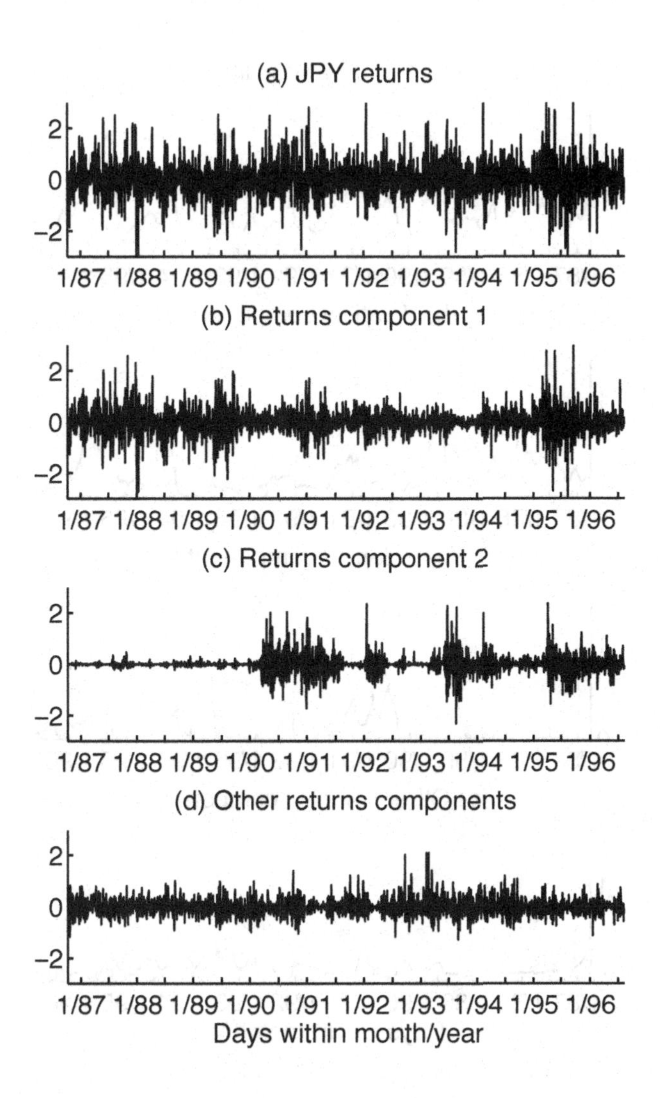

Figure 10.10 *¥Japan:$USA returns time series and estimated values of latent components, in a format similar to that of Figure 10.6.*

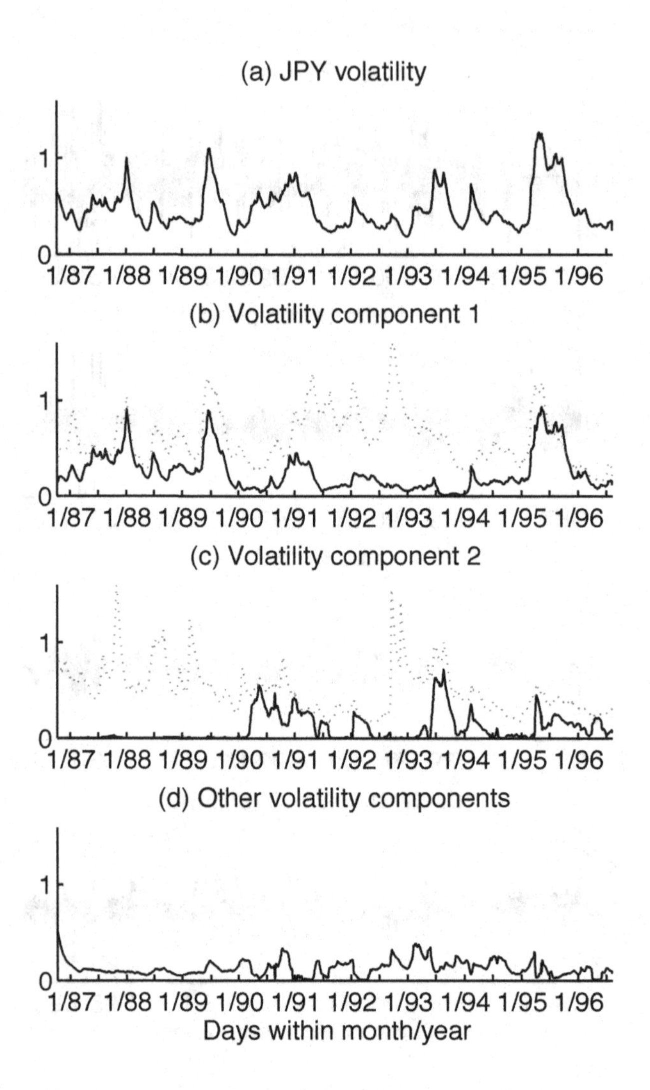

Figure 10.11 *Components of volatility in ¥Japan:\$USA returns corresponding to return decompositions in Figure 10.10, in a format similar to that of Figure 10.7.*

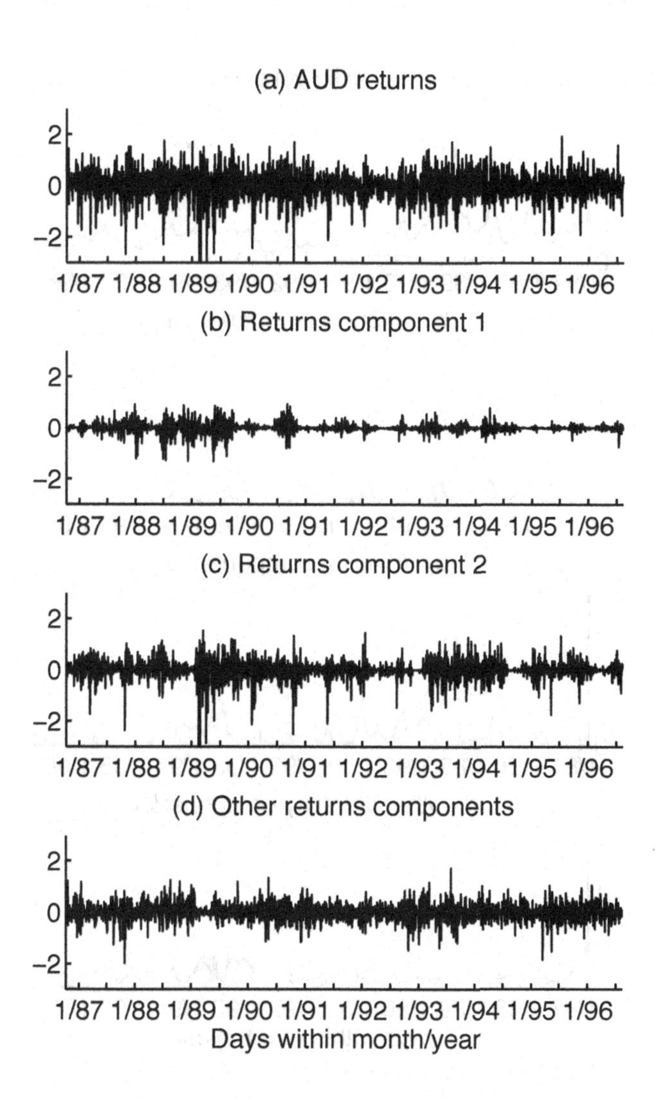

Figure 10.12 *\$Australia:\$USA returns time series and estimated values of latent components, in a format similar to that of Figure 10.6.*

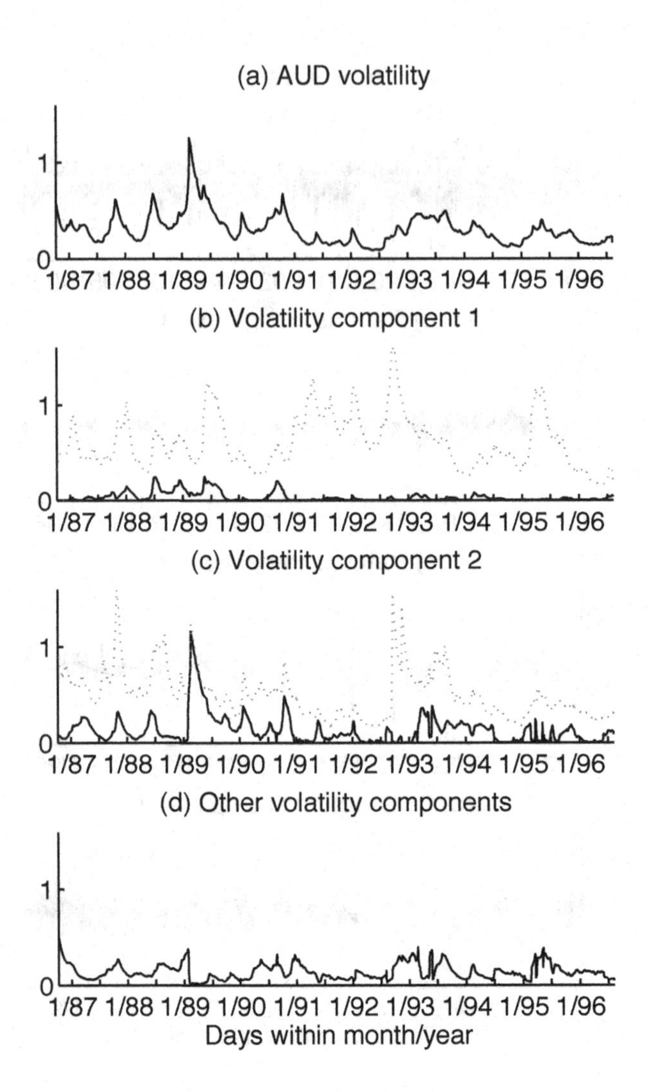

Figure 10.13 *Components of volatility in \$Australia:\$USA returns corresponding to return decompositions in Figure 10.12, in a format similar to that of Figure 10.6.*

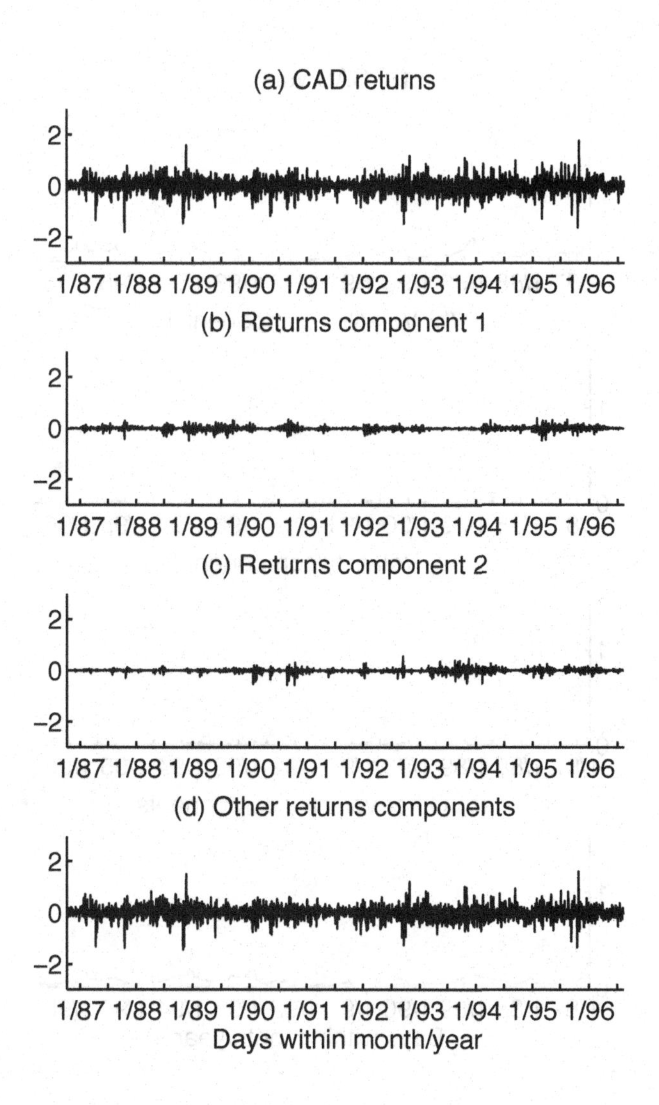

Figure 10.14 *\$Canada:\$USA returns time series and estimated values of latent components, in a format similar to that of Figure 10.7.*

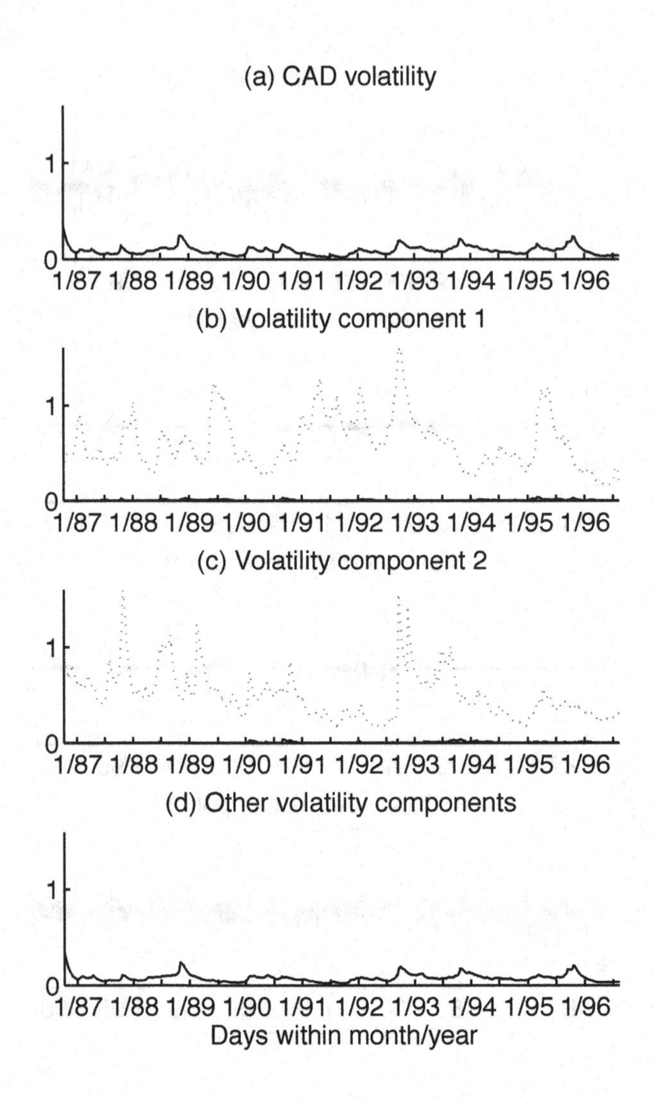

Figure 10.15 *Components of volatility in $Canada:$USA returns corresponding to return decompositions in Figure 10.14, in a format similar to that of Figure 10.6.*

example, as well as many references therein. Practicable models will include financial or econometric predictor variables in a dynamic regression component $\mathbf{F}_t'\mathbf{\Theta}_t$, rather than just the local level $\boldsymbol{\theta}_t$ here. Nevertheless, the example still serves to illustrate sequential one-step-ahead forecasting that feeds into ongoing portfolio revision decisions.

We apply optimal portfolio theory, as defined by Markowitz (1959), at each time t. This requires the one-step-ahead predictive moments $\mathbf{f}_t = E(\mathbf{y}_t|\mathcal{D}_{t-1})$ and $\mathbf{V}_t = V(\mathbf{y}_t|\mathcal{D}_{t-1})$. In this example model, the one-step-ahead predictive T distribution has moments

$$\mathbf{f}_t = \mathbf{m}_{t-1} \quad \text{and} \quad \mathbf{V}_t = q_t\mathbf{S}_{t-1}(\beta h_{t-1}/(\beta h_{t-1} - 2))$$

assuming the divisor is positive. Given a target return level r_t, the investor decision problem is to choose the q-vector of portfolio weights $\mathbf{w}_t$ to minimize the one-step-ahead portfolio variance $\mathbf{w}_t'\mathbf{V}_t\mathbf{w}_t$ subject to constraints $\mathbf{w}_t'\mathbf{f}_t = r_t$ and $\mathbf{w}_t'\mathbf{1} = 1$, i.e., find the least risky portfolio among those with the target mean return and subject to a fixed allocation of capital, where "risk" is measured in terms of the portfolio variance. We then observe the time t return and can track realized returns, as well as the corresponding risk measures in terms of volatilities $(\mathbf{w}_t'\mathbf{V}_t\mathbf{w}_t)^{1/2}$ at the chosen portfolio vector $\mathbf{w}_t$ sequentially over time.

The one-step-ahead optimization is achieved as follows. Write $\mathbf{K}_t = \mathbf{V}_t^{-1}$ for the forecast precision. Then

$$\mathbf{w}_t = \mathbf{K}_t(u_t\mathbf{f}_t + z_t\mathbf{1}) \qquad (10.14)$$

where $u_t = \mathbf{1}'\mathbf{K}_t\mathbf{g}_t$ and $z_t = -\mathbf{f}_t'\mathbf{K}_t\mathbf{g}_t$ with $\mathbf{g}_t = (\mathbf{1}r_t - \mathbf{f}_t)/d_t$ where $d_t = (\mathbf{1}'\mathbf{K}_t\mathbf{1})(\mathbf{f}_t'\mathbf{K}_t\mathbf{f}_t) - (\mathbf{1}'\mathbf{K}_t\mathbf{f}_t)^2$.

Figures 10.16 and 10.17 provide some summaries of two analyses using the DLM of the previous section and based on a target mean return $r_t = 0.02$ for each t. This serves to illustrate and compare the portfolio adaptation under a dynamic covariance model, using $\beta = 0.95$ as above, with a static model having $\mathbf{\Sigma}_t = \mathbf{\Sigma}$ for all t, i.e., $\beta = 1$. This illustrates the greater degree of adaptation over time in portfolio weight on a selection of three currencies-GBP, JPY, and DEM-when using the dynamic model. This is particularly notable for currency DEM. The effects of major economic events discussed above are also apparent in influencing the weights on GBP and JPY differently. Over time, cumulative returns under the dynamic model dominate those of the static model, though there are periods where this reverses. The overall dominance across the 10-year period is also consistent with statistical dominance in terms of Bayes's factors based on computing marginal model likelihoods under different values of β. The higher risk levels incurred in portfolio reallocations based on the dynamic model are naturally more volatile and for much of the time period higher than with the static model, as illustrated in Figure 10.17(b).

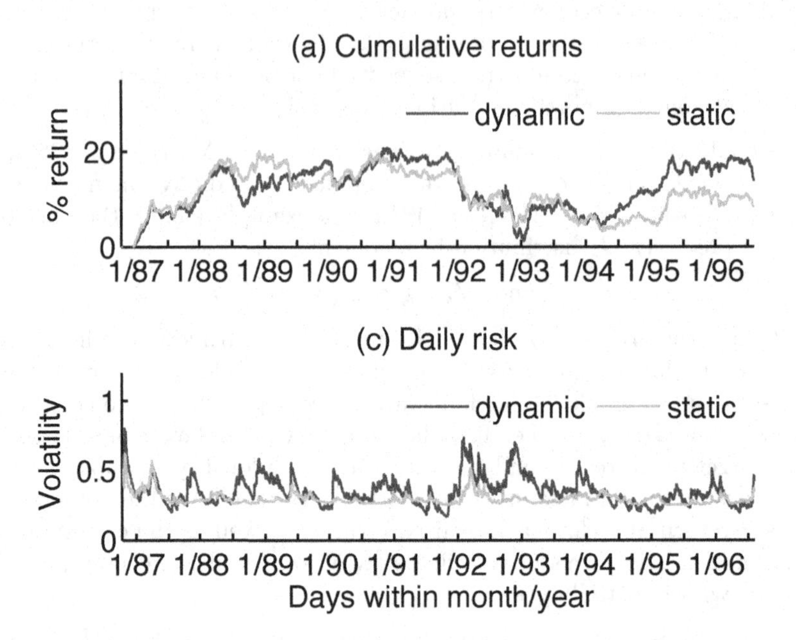

Figure 10.16 *Summaries of optimal portfolios in the forward filtering, one-step prediction and portfolio optimization example with* $q = 12$ *currency exchange rates. Analysis used the target mean return* $r_t = 0.02$ *for all* t *and the same prior and model settings as in the analyses above. (a) Cumulative returns from each of the two analyses: dynamic, with covariance matrix discount factor* $\beta = 0.95$, *and static, with* $\beta = 1$. *(b) On-line, daily risk measured in terms of the portfolio volatility* $(\mathbf{w}_t'\mathbf{V}_t\mathbf{w}_t)^{1/2}$ *with the optimal weight vector and one-step prediction covariance matrix* $\mathbf{V}_t$ *at each time* t, *again from each of the dynamic and static analyses.*

10.4.7.1 Additional comments and extensions

The discount method is widely used in forward filtering analyses without regard to constraints on values of β. This reflects its natural appeal as a model for tracking short-term random fluctuations in variances and covariances of multiple series, and in adapting to volatility changes that would otherwise potentially bias estimation of the underlying states and degrade short-term forecasts. Its roles and utility in financial portfolio studies earlier referenced strongly bear this out, as does its ease of implementation. In considering retrospective analysis, however, it becomes clear that the constraints can bind since the distribution theory relies strongly on the requirement of valid matrix beta distributions underlying the forward dis-

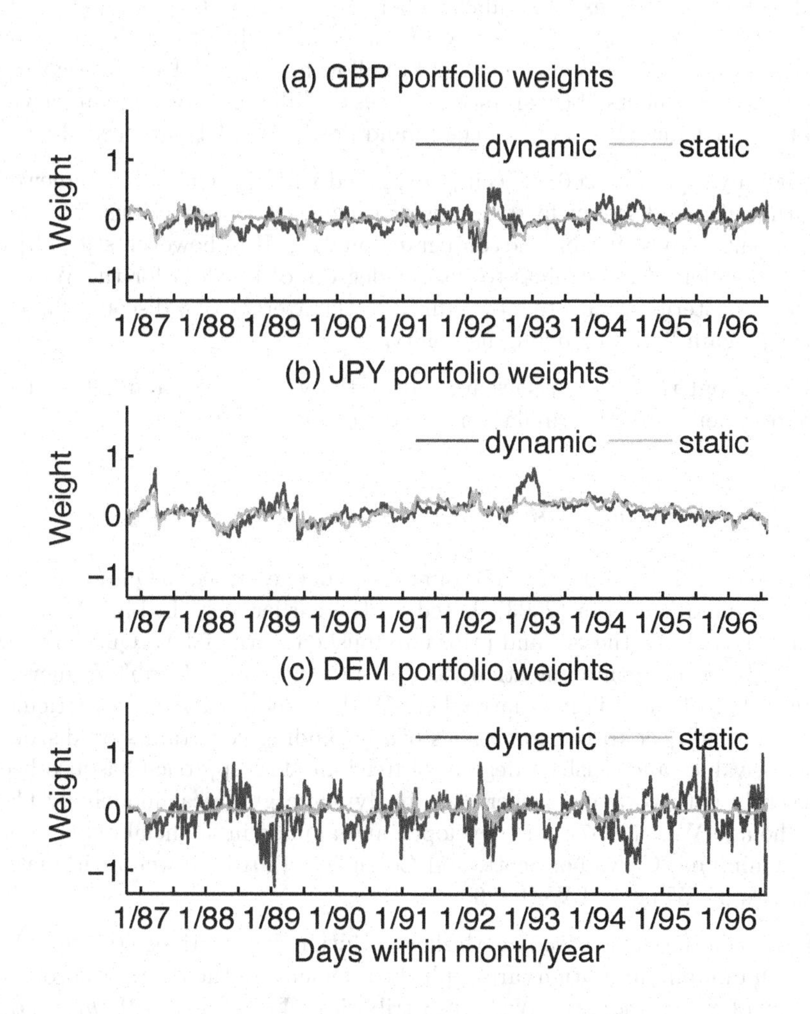

Figure 10.17 *Optimal portfolio weights for three of the $q = 12$ currencies in the forward filtering, one-step prediction and portfolio optimization example.*

count method. In models with increasingly large q, the constraints on the degrees of freedom parameter in relation to the discount factor β become harsh. With β typically near one, the constraint becomes $h_t > (q-1)/(1-\beta)$ or h_t integral for all t. This requirement is clear, for example, in Equation (10.12) to ensure valid Wishart distributions involved in retrospection. The former constraint quickly becomes impossible to satisfy in realistic models, requiring higher degrees of freedom that may be unreasonable, especially in defining initial priors. Hence the integer values constraint is more typically adopted, as in our example above where $q = 12$, $\beta = 0.95$, and $h_0 = 20$ so that $h_t = 20$ for all t. Further, we require the ability to modify discount factors over time, and also to extend to allow for potential stochastic variation in components of covariance matrices at different rates, and at varying rates over time. Hence more general and flexible models are needed.

Triantafyllopoulos (2008) recently extended this approach to a framework with multiple discount factors, one for each of the q dimensions. This nice generalization maintains the conjugate updating. It is, however, still subject to fundamental constraints to integer degrees of freedom for the Wishart priors/posteriors that are, now, linked to the average of q discount factors, which again limits modeling flexibility.

It turns out that these issues are substantially resolved via an alternative, more general model formulation, as now noted.

10.4.8 Beta-Bartlett Wishart Models for Stochastic Volatility

The alternative multivariate discounting model described in Quintana, Lourdes, Aguilar, and Liu (2003), based on the earlier work in Liu (2000, unpublished Ph.D. thesis), and prior unpublished work of J.M. Quintana and F. Li (personal communication), is a more general and flexible framework than that of the original matrix beta-Wishart model above. In particular, it is not subject to the sometimes overly binding constraints on discount factor values and Wishart degrees of freedom. The approach has also been recently extended to a richer class of dynamic graphical models by Carvalho and West (2007), and developed and applied in financial applications by Quintana, Carvalho, Scott, and Costigliola (2010) as well as in matrix models by Wang and West (2009).

This richer model is based on Markov evolution distributions $p(\mathbf{\Phi}_t|\mathbf{\Phi}_{t-1})$ in which random innovations are applied to elements of the Bartlett decomposition of $\mathbf{\Phi}_{t-1}$. Under the Wishart distribution $(\mathbf{\Phi}_{t-1}|\mathcal{D}_{t-1}) \sim W(h_{t-1}, \mathbf{D}_{t-1}^{-1})$ at time $t-1$, write $\mathbf{P}_{t-1}$ for the upper triangular Cholesky component of $\mathbf{D}_{t-1}^{-1}$. The Bartlett decomposition is

$$\mathbf{\Phi}_{t-1} = \mathbf{P}'_{t-1}\mathbf{U}'_{t-1}\mathbf{U}_{t-1}\mathbf{P}_{t-1}$$

where $\mathbf{U}_{t-1}$ is a $q \times q$ upper triangular matrix of random elements $u_{t-1,i,j}$ such that:

- $u_{t-1,i,i}^2 \sim \chi^2_{h_{t-1}-i+1} = G((h_{t-1} - i + 1)/2, 1/2)$, for $i = 1 : q$;
- above the diagonal, $u_{t-1,i,j} \sim N(0, 1)$, for $j = (i + 1) : q$ and $i = 1 : q$;
- the $u_{t-1,i,j}$ are mutually independent.

The Markovian evolution developed in Quintana, Lourdes, Aguilar, and Liu (2003) is

$$\mathbf{\Phi}_t = \beta^{-1}\mathbf{P}'_{t-1}\tilde{\mathbf{U}}'_t\tilde{\mathbf{U}}_t\mathbf{P}_{t-1} \tag{10.15}$$

for a discount factor $\beta \in (0,1)$ and where the new random matrix $\tilde{\mathbf{U}}_t$ is simply constructed from $\mathbf{U}_{t-1}$ using univariate beta-gamma stochastic volatility models analogous to the model of Section 4.3.7. A univariate evolution model is applied to each of the diagonal entries of $\mathbf{U}_{t-1}$ independently, leaving the off-diagonal terms untouched. Specifically, $\tilde{\mathbf{U}}_t$ is also upper triangular, has precisely the same upper off-diagonal elements $u_{t-1,i,j}$ as $\mathbf{U}_{t-1}$, but has diagonal elements

$$\tilde{u}_{t,i,i} = u_{t-1,i,i}\gamma_{t,i}^{1/2}$$

where the $\gamma_{t,i}$ are independent beta random quantities, all independent of the off-diagonal $u_{t-1,i,j}$. For $i = 1:q$, take

$$(\gamma_{t,i}|\mathcal{D}_{t-1}) \sim Be(\beta_{t,i}(h_{t-1}-i+1)/2, (1-\beta_{t,i})(h_{t-1}-i+1)/2)$$

where, for each i, $\beta_{t,i} = (\beta h_{t-1} - i + 1)/(h_{t-1} - i + 1)$. It follows that

$$(\tilde{u}^2_{t,i,i}|\mathcal{D}_{t-1}) \sim \chi^2_{\beta h_{t-1}-i+1} = G((\beta h_{t-1} - i + 1)/2, 1/2).$$

The result of this construction is that the evolution of Equation (10.15) implies

$$(\mathbf{\Phi}_t|\mathcal{D}_{t-1}) \sim W(\beta h_{t-1}, (\beta \mathbf{D}_{t-1})^{-1}).$$

Thus, the new beta-Bartlett Wishart model, based on the specification of a single discount factor β, induces precisely the same discounting structure as the original matrix beta model: reducing the Wishart degrees of freedom by a factor of β to increase uncertainty, while maintaining the location in terms of the mean of the precision matrices $h_t\mathbf{D}_t^{-1}$. However, the beta-Bartlett suffers from no constraints on the range of the discount factor, since the independent beta random shocks at each time t are defined for any value of $\beta \in (0,1)$.

10.4.8.1 Discount model variants

The construction can be modified in a number of ways, since the use of a set of independent univariate beta "shocks" provides a valid model for other choices of the discount strategy. That is, *any* degrees of freedom can be generated for the evolved precision/covariance matrix from the beta-Bartlett stochastic model. The general framework and supporting theory are left to the reader in the exercises in Section 10.8.

As one other key example, and variant on the above, rather than discounting the Wishart degrees of freedom h_{t-1} we may prefer to discount

$n_{t-1} = h_{t-1} - q + 1$, the "sample size equivalent" degrees of freedom parameter. The same construction applies, with Equation (10.15) modified to

$$\mathbf{\Phi}_t = b_t^{-1}\mathbf{P}'_{t-1}\tilde{\mathbf{U}}'_t\tilde{\mathbf{U}}_t\mathbf{P}_{t-1}$$

for a constant b_t to be chosen, and with the beta distributions of the shocks now having modified $\beta_{t,i}$ parameters, viz. $\beta_{t,i} = (\beta n_{t-1}+q-i)/(n_{t-1}+q-i)$ for $i = 1:q$. With this model, the $W(n_{t-1}+q-1, \mathbf{D}_{t-1}^{-1})$ distribution for $\mathbf{\Phi}_{t-1}$ evolves to

$$(\mathbf{\Phi}_t|\mathcal{D}_{t-1}) \sim W(\beta n_{t-1}+q-1, (b_t\mathbf{D}_{t-1})^{-1}).$$

Equivalently in terms of covariance matrices, the inverse Wishart posterior at $t-1$,

$$(\mathbf{\Sigma}_{t-1}|\mathcal{D}_{t-1}) \sim IW(n_{t-1}, \mathbf{D}_{t-1}),$$

evolves to

$$(\mathbf{\Sigma}_t|\mathcal{D}_{t-1}) \sim IW(\beta n_{t-1}, b_t\mathbf{D}_{t-1}).$$

Hence, in this case, taking $b_t = (\beta n_{t-1}+q-1)/(n_{t-1}+q-1)$ ensures that the mean of the precision matrix is unchanged through the evolution, being fixed at $(n_{t-1}+q-1)\mathbf{D}_{t-1}^{-1} = h_{t-1}\mathbf{D}_{t-1}^{-1}$. As pointed out in Quintana, Lourdes, Aguilar, and Liu (2003), various choices of b_t constrain different choices of location measures of the Wishart or inverse Wishart, and lead to variants of the general model. With relatively large degrees of freedom, the differences are small as $b_t \sim \beta$ as n_{t-1} increases, though the differences are more meaningful with small n_{t-1}.

10.4.8.2 Additional comments and current research areas

Among other things, the above development provides a basis and rationale for the existing broad, applied use of covariance matrix discounting for sequential, forward filtering with arbitrary values of discount factors, while opening up opportunities for more general models. Extensions may allow different discount factors to be used on different elements of the Cholesky diagonal elements at each time, so engendering a richer class of stochastic evolution models under which different covariance structures evolve at different rates. Further, the model is unconstrained in terms of allowing specification of time-dependent discount factors, such as to allow for more abrupt changes in aspects volatility to be incorporated into the model at some times than at others. This model class and approach are open to further development in these and other respects, that include the theoretical study of the distributions implied for retrospective learning, and the implementation of corresponding algorithms; that are, as yet, undeveloped.

Beyond the models and approaches above, recent developments have included interesting extensions of discount-based models, as earlier noted (e.g., Triantafyllopoulos 2008). In addition, there is growing interest in new models for dynamic covariance structures including modifications of discount-based models, and alternative forms inspired by these models. One interesting class of models introduced by Philipov and Glickman (2006a,b) involves variants in which $\mathbf{\Sigma}_t$ is constructed from $\mathbf{\Sigma}_{t-1}$ in a Markovian discount manner, but that generates models with stationary structure. Developments of this approach by these authors integrate these new *Wishart process* ideas into latent factor volatility models, building on some of the developments we now discuss in the following section.

10.5 Multivariate Dynamic Graphical Models

10.5.1 Gaussian Graphical Models

Modeling and inference with higher dimensional time series raises challenges to the ability to scale up analyses, with questions about statistical and computational efficiency as well as parsimony and the relevance of imposing increasing constraints on model parameters as their numbers grow. At a conceptual level, structured Gaussian graphical models constrain the parameters of covariance matrices; in the models of the preceding sections, the covariance matrices $\mathbf{\Sigma}_t$ are unconstrained, having $q(q+1)/2$ parameters at time t, though they are of course related over time. For larger q, as in portfolios with hundreds or thousands of time series (e.g., Polson and Tew 2000; and Carvalho and West 2007), or in systems biology models with hundreds or thousands of genes or proteins measured over time, the statistical and computational imperatives are to constrain to reduce dimension, while maintaining faithfulness to the patterns of dependencies evident in data.

Graphical models characterize conditional independencies via graphs; see foundational material in Whittaker (1990) and Lauritzen (1996), and supporting theory and ideas in the appendix material of Section 10.7.6 below. As exemplified in Jones, Dobra, Carvalho, Hans, Carter, and West (2005) and Jones and West (2005) in Gaussian models, complicated patterns of dependencies among increasing numbers of variables can often be explained by simpler, structured sets of *conditional* dependencies among smaller subsets of variables in the context of substantial patterns of *conditional independence.* In the Gaussian case, for a q-vector $\mathbf{x} = (x_1, \ldots, x_q)'$ with $\mathbf{x} \sim N(\boldsymbol{\mu}, \mathbf{\Sigma})$, suppose the precision matrix $\mathbf{\Omega} = \mathbf{\Sigma}^{-1}$ has some off-diagonal zeros, $\omega_{i,j} = 0$ for some pairs (i, j), $i \neq j$. The undirected, conditional independence graph representing this has q nodes $V = \{1, \ldots, q\}$ indexing and representing the variables x_i, and edges between pairs of nodes only for which $\omega_{i,j} \neq 0$. With E denoting the set of pairs (i, j) for which

$\omega_{i,j} \neq 0$, known as the edge set, the undirected graph is $G = (V, E)$. For each $i = 1, \ldots, q$, the set of variables indexed by $ne(i) = \{j, j = 1, \ldots, q \ : \ (i,j) \in E)\}$ is the set of neighbors of variable x_i; conditional on $x_{ne(i)} \equiv \{x_j,\ j \in ne(i)\}$, the variable x_i is conditionally independent of the other variables $\{x_k,\ k \notin ne(i)\}$.

The multivariate normal density $p(\mathbf{x}|\mathbf{\Sigma})$ factorizes over the graph G as a ratio of products of lower dimensional normal marginal distributions, viz.,

$$p(\mathbf{x}|\mathbf{\Sigma}, G) = \frac{\prod_{P \in \mathcal{P}} p(\mathbf{x}_P|\mathbf{\Sigma}_P)}{\prod_{S \in \mathcal{S}} p(\mathbf{x}_S|\mathbf{\Sigma}_S)}, \tag{10.16}$$

where $\mathbf{x}_P$ and $\mathbf{x}_S$ indicate subsets of variables in each of the set of prime components (P) and separators (S) of G, respectively. This engenders the ability to develop statistical computations on (intersecting) sets of lower dimensional distributions. Each separator S is a complete subgraph, with all variables conditionally *dependent* within S; that is, the marginal precision matrix $\mathbf{\Omega}_S = \mathbf{\Sigma}_S^{-1}$ has *no* zero off-diagonal elements. A decomposable graph is one in which the same is true for all prime components P. Much of the tractable theory and computation in graphical models applies primarily to decomposable cases, which we focus on here.

For the given graph G, Bayesian inference in multivariate and matrix-variate normal models is enabled via the conjugate class of normal, hyper-inverse Wishart distributions as detailed in the appendix of Section 10.7.6. Prior to posterior updating is a direct extension of the usual normal, inverse Wishart analysis now structured on G, and predictive distributions may be computed in closed form to provide input to graphical model uncertainty analysis; by exploring multiple candidate graphs G, the latter lead to marginal likelihood functions over graphs. Details are laid out algorithmically and with several examples in, for example, Giudici (1996), Giudici and Green (1999), Wong, Carter, and Kohn (2003), and Jones, Dobra, Carvalho, Hans, Carter, and West (2005). The latter references develop and detail MCMC-based computation for exploring uncertainty about the structure G, with-in Jones *et al.*-effective methods of stochastic search for higher dimensional problems that include efficient software.

10.5.2 Dynamic Graphical Models

Carvalho and West (2007) introduced a synthesis of matrix-variate DLMs with Gaussian graphical models to address issues of scalability with time series dimension q, and adaptability of the earlier multivariate DLMs to inherently structured and often sparse precision matrices $\mathbf{\Sigma}_t^{-1}$. The entire model framework, theory, and machinery of Section 10.3-sequential learning via forward filtering, forecasting, and retrospective analysis-extends to models in which the time-varying observation covariance matrices $\mathbf{\Sigma}_t$ are constrained by a graph G, now featuring normal, hyper-inverse Wishart (NHIW) distributions as priors and posteriors in an extension of the normal, inverse Wishart theory. This involves extension of the beta-Bartlett Wishart stochastic evolution model of Section 10.4.8 for forward filtering and updating.

We note the details for the vector time series case of Equation (10.10) and following discussion. Essentially, the entire discussion is the same but for the embedding of a graphical model based on a graph G, and the generalization of inverse Wishart to hyper-inverse Wishart (HIW) distributions. One point of difference with the summary updating distribution theory of Section 10.4.4 is that the HIW discounting is now based on the beta-Bartlett Wishart stochastic evolution model of Section 10.4.8, hence the inherent inverse Wishart degrees of freedom n_t is discounted each time, rather than $h_t = n_t - q + 1$. The model is

$$\begin{aligned} \mathbf{y}_t' &= \mathbf{F}_t'\mathbf{\Theta}_t + \boldsymbol{\nu}_t', \quad \boldsymbol{\nu}_t \sim N(\mathbf{0}, v_t\mathbf{\Sigma}_t), \\ \mathbf{\Theta}_t &= \mathbf{G}_t\mathbf{\Theta}_{t-1} + \mathbf{\Omega}_t, \quad \mathbf{\Omega}_t \sim N(\mathbf{0}, \mathbf{W}_t, \mathbf{\Sigma}_t) \end{aligned}$$

at each time t, with $\mathbf{\Sigma}_t$ now constrained by G. The sequential analysis theory is as follows:

- Posterior at $t-1$:

$$(\mathbf{\Theta}_{t-1}, \mathbf{\Sigma}_{t-1}|\mathcal{D}_{t-1}) \sim NHIW(\mathbf{M}_{t-1}, \mathbf{C}_{t-1}, n_{t-1}, \mathbf{D}_{t-1}).$$

- Prior at t :
  - $\mathbf{\Sigma}_{t-1}$ first evolves to

$$(\mathbf{\Sigma}_t|\mathcal{D}_{t-1}) \sim HIW_G(\beta n_{t-1}, \beta\mathbf{D}_{t-1})$$

    based on a discount factor $\beta \in (0,1)$, so that

$$(\mathbf{\Theta}_{t-1}, \mathbf{\Sigma}_t|\mathcal{D}_{t-1}) \sim NHIW_G(\mathbf{M}_{t-1}, \mathbf{C}_{t-1}, \beta n_{t-1}, \beta\mathbf{D}_{t-1}).$$

  - $\mathbf{\Theta}_t$ then evolves via the state equation, giving time t prior

$$(\mathbf{\Theta}_t, \mathbf{\Sigma}_t|\mathcal{D}_{t-1}) \sim NHIW_G(\mathbf{a}_t, \mathbf{R}_t, \beta n_{t-1}, \beta\mathbf{D}_{t-1})$$

    where $\mathbf{a}_t = \mathbf{G}_t\mathbf{M}_{t-1}$ and $\mathbf{R}_t = \mathbf{G}_t\mathbf{C}_{t-1}\mathbf{G}_t' + \mathbf{W}_t$ as before.
- Posterior at t : $(\mathbf{\Theta}_t, \mathbf{\Sigma}_t|\mathcal{D}_t) \sim NHIW(\mathbf{M}_t, \mathbf{C}_t, n_t, \mathbf{D}_t)$ where, as before:
  - For $\mathbf{\Theta}_t$: $\mathbf{M}_t = \mathbf{a}_t + \mathbf{A}_t\mathbf{e}_t'$ and $\mathbf{C}_t = \mathbf{R}_t - \mathbf{A}_t\mathbf{A}_t'q_t$ with adaptive coefficient vector $\mathbf{A}_t = \mathbf{R}_t\mathbf{F}_t/q_t$, one-step forecast variance multiplier $q_t = \mathbf{F}_t'\mathbf{R}_t\mathbf{F}_t + v_t$, and one-step forecast error $\mathbf{e}_t = \mathbf{y}_t - \mathbf{f}_t$ based on the point forecast vector $\mathbf{f}_t = \mathbf{a}_t'\mathbf{F}_t$.
  - For $\mathbf{\Sigma}_t$: $n_t = \beta n_{t-1} + 1$ and $\mathbf{D}_t = \beta\mathbf{D}_{t-1} + \mathbf{e}_t\mathbf{e}_t'/q_t$.

For model uncertainty analysis, the graphical model marginal likelihood is delivered from the sequential updating as the product of components at each time t. That is,

$$p(\mathbf{y}_1, \ldots, \mathbf{y}_T | G, \mathcal{D}_0) = \prod_{t=1}^{T} p(\mathbf{y}_t | G, \mathcal{D}_{t-1})$$

where each component is evaluated using a closed form expression that results from the conjugacy of the NHIW structure. See Carvalho and West (2007) for full details that simply graft over to dynamic models the core methodology of Giudici and Green (1999). In the example below, shotgun stochastic search over graphs as developed in Jones, Dobra, Carvalho, Hans, Carter, and West (2005) was used to explore and generate a profile of the posterior over G, evaluating unnormalized posterior probabilities on each candidate graph via these marginal likelihood values combined with a specified prior.

Example 10.1 *Dynamic graphical structure in mutual fund time series.* Figure 10.18 shows time series of monthly returns on a selection of $q = 14$ Vanguard mutual funds over several years in the early 2000s, for a total of $T = 87$ months. The above analysis was applied using the steady model of Equation (10.13) to track level and volatility changes under a given graphical model G, and to explore graphical model space as noted above using the shotgun stochastic search method and computational tools. Analysis initialized with relatively vague priors based on $\mathbf{M}_0 = \mathbf{0}$, $c_0 = 100$, $n_0 = 1$, and $\mathbf{D}_0 = \mathbf{I}_{14}$, and used discount factors $\delta = 0.99$ for the local level and $\beta = 0.95$ for $\mathbf{\Sigma}_t$. The prior over graphs used the standard independent edge inclusion probability $Pr(i \sim j) = 1/q$.

Stochastic search over graphs G, initialized at the empty graph with no edges, proceeds as described in Jones, Dobra, Carvalho, Hans, Carter, and West (2005). At each iterate standing at the "current" graph G, this looks at all one-edge neighboring graphs-each graph that is a single one-edge in/out different to G-and evaluates the unnormalized posterior probability of all such graphs. The next step in the iterative search then samples from that set of graphs with respect to these conditional posterior probabilities. Having run through many iterations, at each step adding to the growing list of graphs visited together with their unnormalized posterior probabilities, this generates a list of many graphs together with probabilities that are used to define a conditional (on graphs visited) posterior for inferences and model averaging.

From these computations, the most probable graph identified-a verifiable posterior mode-is illustrated by the 0/1 graph showing edge included, in Figure 10.19. The Vanguard funds are labeled with names that identify some aspects of the investment context and strategy for the funds (growth, US domestic, international, etc.) as well as whether the funds are actively administered by select fund managers (the first nine funds in the order listed) compared to relatively simply administered "index tracking" funds (the last five funds listed, marked "Index"). Figure 10.20 shows a similar image but now indicating levels of the Monte Carlo estimates of posterior probabilities of edge inclusion for all edges. The posterior mode agrees well with these posterior probabilities; in fact, the modal graph identified is also precisely the median posterior graph in that the approximate posterior edge inclusion probability exceeds 0.5 for each edge in the modal graph, and is below 0.5 for each edge missing. Importantly, the concordance of structure with the fund types is evident. There is strong conditional dependency among small subsets of funds that are expected to be more closely related: the group of index funds, the clique of three somewhat entrepreneurially managed international funds (Global Equity, International Value, International Explorer), and the more conservatively managed and US domestic growth funds. There are also dependencies strongly indicated between some of the managed growth and index growth funds, and between managed funds with strong international and especially Asian focuses and the Pacific stock index fund. There are also clear indications of strong patterns of conditional independence, with about 50% of the edges having negligible or very low probability, indicative of the relevance of the graphical model structure. Further studies of this nature have confirmed the dominance of graphical dynamic models with increasing sparsity as q increases in terms of portfolio decisions in this context of mutual funds, with models of both monthly and daily returns (Reeson, Carvalho, and West 2009).

Additional examples of a similar nature appear in Carvalho and West (2007). Sequential portfolio applications illustrate the impact of graphical model structuring in a real practical sense. By appropriately reducing dimension to sparser precision matrices, even in the context of then having to expand the model space to consider multiple graphs and address graphical model uncertainty, the analysis is able to more efficiently and adequately capture aspects of variation in patterns of covariation among series over time, and this is demonstrably important in modifying the resulting short-term portfolio decisions. In examples including one with $q = 500$ variables, the S&P 500 index, Carvalho and West (2007) show that selecting and posterior averaging over even a relatively small number of "data consistent" graphs can generate improved short-term adaptability that results in increased performance of portfolios. Further, the reduced dimension of parameters implied by sparser graphical models leads to generally lower realized portfolio risk and lower volatility of time trajectories of portfolio weights, which are both very positive attributes in a real-world sequential portfolio revision context. The examples also demonstrate that the methodology can be implemented with reasonably high-dimensional time series based on currently available approaches to searching over graphs G

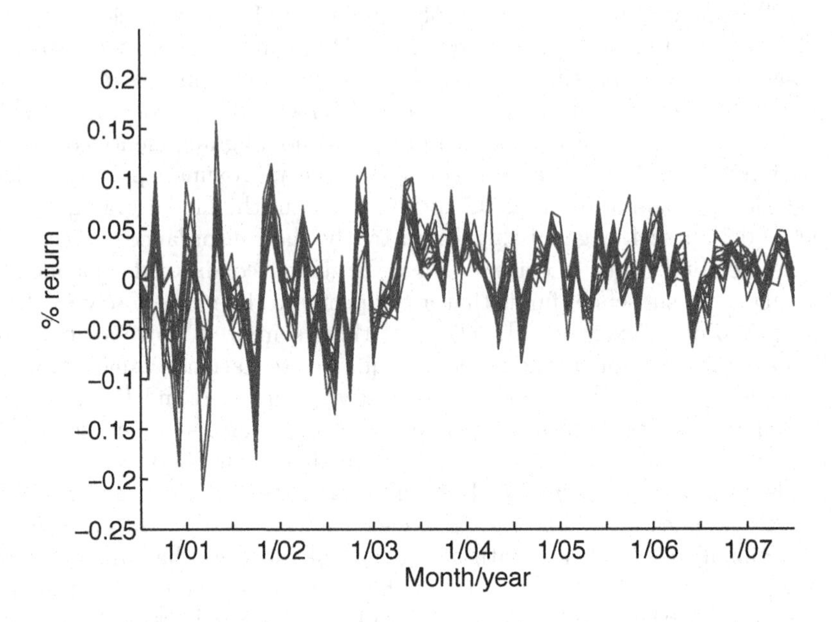

Figure 10.18 *Time series data of monthly returns on 14 Vanguard mutual funds.*

using stochastic search methods. Additional examples in analyses of macroeconomic matrix time series appear in Wang and West (2009), who extend this modeling approach to matrix time series. A core example there involves a matrix DLM with locally varying trends coupled with seasonal components, representing a novel application of the broader class of dynamic graphical models.

10.6 Selected recent developments

Advances in dynamic graphical modeling have been partly motivated by challenges of scaling analysis to higher-dimensional time series. Selected recent developments with this core goal have also been based on the idea of coupling together sets of univariate models, already noted in the context of dynamic dependence network models (DDNMs-Zhao, Xie, and West 2016) in the previous chapter, Section 9.4 and references there. Some links to recent work that exploits the concept of decouple/recouple in new ways, and in contexts of nonnormal as well as increasingly high-dimensional time series, are noted here for readers to explore in the recent literature. A main point of reference to the overall decouple/recouple concept; a starting point toward more development of the models simply noted here is West (2020).

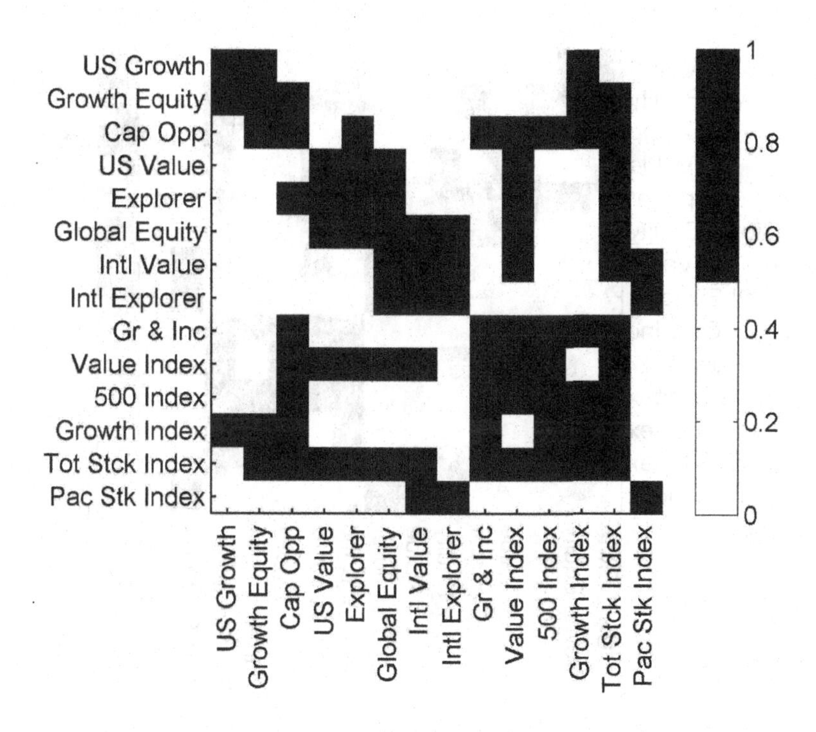

Figure 10.19 *Image of posterior modal graph (highest probability graph identified in stochastic search over graphical models) in the dynamic graphical model analysis of the 14 Vanguard mutual fund time series. In this analysis, this modal graph happens to coincide with the "approximate median posterior probability graph," i.e., the graph including only edges having posterior inclusion probability exceeding 0.5.*

10.6.1 Simultaneous Graphical Dynamic Models

The classes of *simultaneous graphical dynamic linear models (SGDLMs)* generalize DDNMs by allowing each univariate time series to be a contemporaneous predictor of any other. DDNMs are defined on a specific ordering of the series with the contemporaneous relationships based on an acyclic, directed graphical model structure underlying the precision matrix, related to the recursive or ordered specification of structured TV-VAR models often used in econometric applications. SGDLMs relax this and so are not

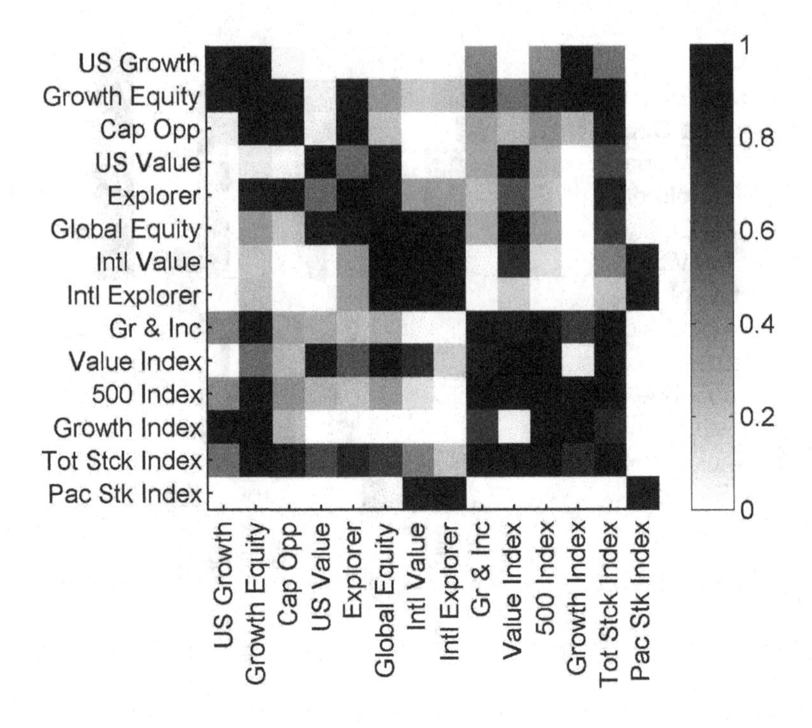

Figure 10.20 *Image of approximate posterior edge inclusion probabilities in the dynamic graphical model analysis of the 14 Vanguard mutual fund time series.*

dependent at all on a series ordering; the structures allow for cycles in the resulting graphical model of contemporaneous dependencies, akin to the general approach to structural or simultaneous equations models in econometrics, but now with time-varying parameters and volatilities. The map from structural to reduced form of the model ensures a uniquely specified model in spite of potential cyclical dependencies. The resulting Bayesian filtering and forecasting analysis is then, however, more complex. While DDNMs can be analyzed as a set of decoupled univariate DLMs and then recoupled for forecasting, a non recursive SGDLM involves additional technical and computational questions. As defined and developed in Gruber and West (2016) and Gruber and West (2017), this can be effectively and efficiently done using simple importance sampling and variational Bayes's steps at each filtering time point. Some of the major benefits and opportunities of this framework can be seen in financial time series forecasting and portfolio decision examples in these papers.

10.6.2 Models for Multivariate Time Series of Counts

Problems of monitoring and forecasting discrete time series, and notably many related time series of counts, arise more and more commonly in many applied fields. Some areas of increased interest include consumer behavior in a range of socio-economic contexts, various natural and biological systems, and commercial and economic problems of analysis and forecasting of discrete outcomes (e.g., Cargnoni, Müller, and West 1997, Yelland 2009, Alves, Gamerman, and Ferreira 2010, Terui and Ban 2014, Chen and Lee 2017, Aktekin, Polson, and Soyer 2018, and Glynn, Tokdar, Banks, and Howard 2019). Some current challenges include addressing questions of modeling simultaneously at different scales as well as of integrating information across series and scales, and the central question of scaling multivariate models computationally while maintaining modeling flexibility.

The recent, general state-space models of Berry and West (2020) and Berry, Helman, and West (2020) are motivated by challenges in one broadly interesting area: that of forecasting consumer demand and sales in supermarkets on a daily basis. Here the scale of time series is high, with many thousands of items on sale across many supermarkets, while the resulting time series are inherently dependent. The need in such settings is for flexible, customizable models to be applied to sales and/or demand for individual items, while recognizing the complexities in cross-item dependencies over time. This is addressed with classes of time series models for multivariate non negative counts that exploit and extend dynamic generalized linear models (DGLMs: see Section 4.4, as well as West, Harrison, and Migon 1985, and chapter 14 of West and Harrison 1997). Cross-series linkages are modeled using a multiscale factor approach. The *dynamic count mixture models (DCMMs)* of Berry and West (2020) combine (i) dynamic models for binary time series of sale versus no sale, with (ii) shifted Poisson DGLMs for the numbers of units of the item sold conditional on a sale. The *dynamic binary cascade models (DBCMs)* of Berry, Helman, and West (2020) extend this by first forecasting transactions for a specific item and then combining with dynamic models for numbers of items sold per transaction; the latter can improve forecasting when items are bought in various numbers per customer.

For sales and demand for each specific item over time, these decoupled univariate models are easy to fit and use in forecasting. The state vectors can involve local trends, seasonality, and regression terms as usual. The multivariate structure is represented by one or more common factors used as dynamic regression variables in each of the univariate models; these factors are estimated and predicted via an external or aggregate model, and projected onto each of the univariate models. Conditional on the factors, the univariate models are decoupled; the coupling is via the common effects the factors induce. The analysis scales effectively linearly in the number of univariate time series, while nevertheless representing full, coherent multivariate forecast models and predictive distributions.

While motivated by commercial applications, the theory and methodology underlying these specific models is applicable in many fields. The detailed forecasting examples in these papers show some of the potential to exploit the decouple/recouple concept in addressing large-scale, complex, and dynamic discrete data generating systems.

10.6.3 Models for Flows on Dynamic Networks

Time series of integer counts derived from flows of traffic between and among nodes in a network arise in fields such as internet traffic, computer networks generally, physical traffic flows on transportation networks, and multiple areas involving social network in social sciences and biostatistical studies. As with the above context of multivariate non negative count series in commercial forecasting, network flows involve large-scale count time series. While Bayesian models have been developed for network tomography and physical traffic flow forecasting (e.g., Tebaldi and West 1998, Congdon 2000, Tebaldi, West, and Karr 2002, Anacleto, Queen, and Albers 2013, Hazelton 2015), increasingly large dynamic network flow problems motivate new approaches.

Dynamic models in Chen, Irie, Banks, Haslinger, Thomas, and West (2018) and Chen, Banks, and West (2019) introduced systems of Poisson DGLMs for flows between pairs of nodes in a defined network, as well as for flows into/out of the network, over time. Again, efficient filtering for network monitoring and, in some applications, short-term forecasting is a requirement, so that decoupled sets of DGLMs are attractive. In these papers, sets of DGLMs are used to characterize the inherent variability and stochastic structure in flows, and to provide efficient and effective monitoring for changes or anomalies. The applied context generating these models is internet traffic in e-commerce, where interventions are called for based on triggering automatic Bayesian monitoring signals. Recoupling across the network is based on the concept of Bayesian model emulation. Here the sets of decoupled univariate model states as they vary over time are deterministically mapped to a structured model containing time-varying node effects for traffic intensity in/out as well as node-node interaction terms.

In dynamic network studies elsewhere, forecasting may be of interest but primary goals often include characterizing normal patterns of stochastic variation in flows, monitoring and adapting models to changes over time, and informing decisions based on patterns of changes. Networks are increasingly large; internet and social networks can involve hundreds or thousands of nodes, and are effectively unbounded in any practical sense from the viewpoint of statistical modeling. The development in Chen, Irie, Banks, Haslinger, Thomas, and West (2018) and Chen, Banks, and West (2019) define flexible multivariate models exploiting the decouple/recouple concept in new ways to enable scalability. As with the DCMM/DBCM framework, the resulting dynamic network flow model analysis scales as efficiently as possible, in this case with computational load proportional to the number of pairs of nodes.

10.6.4 Dynamic Multiscale Models

Section 10.6.2 noted the use of multiscale factor modeling concepts in Berry and West (2020) and Berry, Helman, and West (2020). The broader area of dynamic multiscale modeling includes the computationally scalable approach using multiscale factorizations (Ferreira, Bertolde, and Holan 2010, Ferreira, Holan, and Bertolde 2011, Hoegh, Ferreira, and Leman 2016, Fonseca and Ferreira 2017, Ferreira 2020, Elkhouly and Ferreira 2021). Models based on this approach can be effective when the time series data are collected on a graph or network structure that may be represented (exactly or approximately) by a multiscale tree structure (Ferreira 2002, Ferreira and Lee 2007). The tree structure enables the organization of a multivariate time series into multiple levels of resolution. A key practical benefit is that the computational cost of analysis of such a dynamic multiscale model grows only linearly with the number of nodes in the graph or network underlying the time series.

Consider, for example, financial data such as quarterly revenue from companies listed in the S&P 500 stock market index. We may organize these companies by economic sector in a multiscale tree structure. For example, one branch of the tree might have oil companies, another branch electricity companies, and both branches would connect through a higher level branch of energy companies. In this way, the S&P 500 companies may be organized into multiple levels or scales of resolution: the finest level is revenues of the individual companies, the first aggregate level has revenues by sector, and the most aggregated level has just total revenue of all S&P 500 companies.

There are three key steps involved in the development of these dynamic multiscale models: (i) to decompose the multivariate time series into many empirical multiscale coefficient vectors; (ii) to represent the time series of each empirical coefficient vector with a dynamic model; (iii) to combine the results of the analyses for the many coefficients to apply to the original multivariate time series at multiple resolution scales, including the original finest resolution level. This approach is inherently related to multiscale concepts discussed in Section 10.6 and following West (2020). The development of dynamic multiscale models discussed here can thus be regarded as building on the broader decouple/recouple concept in multivariate modeling, as noted in Ferreira (2020).

Dynamic multiscale models are widely applicable. One main area of application is spatiotemporal modeling where multivariate time series data are spatially indexed. Multiscale analysis is useful when spatial grids on which the data are observed are partitioned into nested grids at multiple spatial resolution scales (Ferreira 2002, Ferreira and Lee 2007). These classes of models have been developed for spatiotemporal Gaussian data (Ferreira, Bertolde, and Holan 2010, Ferreira, Holan, and Bertolde 2011) and Poisson data (Hoegh, Ferreira, and Leman 2016, Fonseca and Ferreira 2017) when each spatial gridblock (or node in a graph) has one variable at a time. More recently, Elkhouly and Ferreira (2021) developed dynamic multiscale spatiotemporal Gaussian models for the case when each spatial gridblock (or node in a graph) has multiple variables of interest at a time. In the context of the S&P 500 example above, the models of Elkhouly and Ferreira (2021) would be useful when considering multiple variables for each company such as revenue, debt, and expenses.

There are other multiscale models and methods useful for time series analysis that do not use multiscale decompositions. For example, Ferreira, West, Lee, and Higdon (2006) developed multiscale and hidden resolution time series models useful for problems with observations at several different levels of temporal resolution. In addition to time series analysis, these models have been used to define priors for permeability fields in the context of fluid flow through porous media (Ferreira, Bi, West, Lee, and Higdon 2003). As another example, Holan, Toth, Ferreira, and Karr (2010) developed a Bayesian multiscale multiple imputation method for time series observed at different resolution levels with data declared as missing, and hence removed from analysis, for confidentiality. Detailed discussions of Bayesian multiscale modeling, computations, and concepts can be found in Ferreira and Lee (2007), Ferreira (2020), and references therein.

10.7 Appendix

On distributional notation, we use here modern, standard notation for Wishart and inverse Wishart distributions, and for related joint normal, inverse Wisharts, building on recent and now popular usage. This replaces earlier notations in Bayesian multivariate analysis and time series, such as in Box and Tiao (1973), Press (1982), and West and Harrison (1997). The latter reference, for example, uses the notation $W_{\cdot}^{-1}(\cdot)$ to denote an inverse Wishart distribution, whereas we use here the standard notation $IW(\cdot,\cdot)$, as detailed below. Similarly, the earlier notation for a matrix normal, inverse Wishart distribution such as $NW_{\cdot}^{-1}(\cdot,\cdot,\cdot)$ is now deprecated in favor of the more standard $NIW(\cdot,\cdot,\cdot,\cdot)$, again as detailed below.

Throughout we use the notation $\text{etr}\{\mathbf{A}\}$ to denote $\exp\{\text{trace}(\mathbf{A})\}$ for any square matrix $\mathbf{A}$.

10.7.1 The Matrix Normal Distribution

The $r \times q$ random matrix $\mathbf{\Theta}$ has a matrix normal distribution, denoted by $\mathbf{\Theta} \sim N(\mathbf{M}, \mathbf{U}, \mathbf{V})$, when its density function is given by

$$p(\mathbf{\Theta}) = (2\pi)^{-rq/2}|\mathbf{U}|^{-q/2}|\mathbf{V}|^{-r/2} \times \text{etr}\{-(\mathbf{\Theta}-\mathbf{M})'\mathbf{U}^{-1}(\mathbf{\Theta}-\mathbf{M})\mathbf{V}^{-1}/2\}$$

with mean matrix $\mathbf{M}$ $(r \times q)$, column (or left) variance matrix $\mathbf{U}$, $(r \times r)$, and row (or right) variance matrix $\mathbf{V}$, $(q \times q)$. The distribution is defined when either or both of the variance matrices are nonnegative definite, and it is nonsingular if and only if each variance matrix is positive definite.

Matrices $\mathbf{\Theta} = (\theta_{i,j})$ and $\mathbf{M} = (m_{i,j})$ have rows $\boldsymbol{\theta}_{i,\star}$ and $\mathbf{m}_{i,\star}$, and columns $\boldsymbol{\theta}_{\star,j}$ and $\mathbf{m}_{\star,j}$ while the variance matrices have elements $\mathbf{U} = (u_{i,j})$ and $\mathbf{V} = (v_{i,j})$, for $i = 1,\ldots,r$, $j = 1,\ldots,q$. All marginal and conditional distributions of elements of $\mathbf{\Theta}$ are normal: $p(\mathbf{\Theta})$ has multivariate normal margins for rows, $\boldsymbol{\theta}_{i,\star}' \sim N(\mathbf{m}_{i,\star}', u_{i,i}\mathbf{V})$, and for columns, $\boldsymbol{\theta}_{\star,j} \sim N(\mathbf{m}_{\star,j}, v_{j,j}\mathbf{U})$, for $i = 1 : r$ and $j = 1 : q$. For any two rows (i,s), $Cov(\boldsymbol{\theta}_{i,\star}', \boldsymbol{\theta}_{s,\star}') = u_{i,s}\mathbf{V}$ and for any two columns (j,t), $Cov(\boldsymbol{\theta}_{\star j}, \boldsymbol{\theta}_{\star,t}) = v_{j,t}\mathbf{U}$. The marginal distribution of any pair of elements $\theta_{i,j}, \theta_{s,t}$ is bivariate normal with $Cov(\theta_{i,j}, \theta_{s,t}) = u_{i,s}v_{j,t}$. Stacking columns of each of $\mathbf{\Theta}$ and $\mathbf{M}$ into $rq \times 1$ vectors $\text{vec}(\mathbf{\Theta})$ and $\text{vec}(\mathbf{M})$ yields a multivariate normal $\text{vec}(\mathbf{\Theta}) \sim N(\text{vec}(\mathbf{M}), \mathbf{V} \otimes \mathbf{U})$ where $\otimes$ denotes Kronecker product.

The notation is sometimes modified to explicitly reflect the dimensions, viz. $\mathbf{\Theta} \sim N_{r,q}(\mathbf{M}, \mathbf{U}, \mathbf{V})$ and $\text{vec}(\mathbf{\Theta}) \sim N_{rq}(\text{vec}(\mathbf{M}), \mathbf{V} \otimes \mathbf{U})$.

10.7.2 The Wishart Distribution

The $q \times q$ positive definite and symmetric matrix $\mathbf{\Omega}$ has a Wishart distribution $\mathbf{\Omega} \sim W(h, A)$ when its density function is

$$p(\mathbf{\Omega}) = c|\mathbf{\Omega}|^{(h-q-1)/2}\text{etr}\{-\mathbf{\Omega}\mathbf{A}^{-1}/2\}$$

where c is a constant given by

$$c^{-1} = |\mathbf{A}|^{h/2} 2^{hq/2} \pi^{q(q-1)/4} \prod_{i=1}^{q} \Gamma((h+1-i)/2).$$

Here $h \geq q$ is the degrees of freedom and $\mathbf{A}$ is a $q \times q$ positive definite and symmetric matrix $\mathbf{A}$. The mean is $E(\mathbf{\Omega}) = h\mathbf{A}$ and, if $h > q+1$, $E(\mathbf{\Omega}^{-1}) = \mathbf{A}^{-1}/(h-q-1)$. The notation is sometimes modified to explicitly reflect the dimension, via $\mathbf{\Omega} \sim W_q(h, \mathbf{A})$.

The density above is valid for any real-valued degrees of freedom $h \geq q$, while the distribution exists for integer degrees of freedom $0 < h < q$. In the latter case, the distribution is singular with a modified density function defined and positive only on a reduced space of matrices $\mathbf{\Omega}$ of rank $h < q$.

10.7.3 The Inverse Wishart Distribution

The $q \times q$ positive definite and symmetric matrix $\mathbf{\Sigma}$ has an inverse Wishart distribution $\mathbf{\Sigma} \sim IW(n, \mathbf{D})$ when its density function is

$$p(\mathbf{\Sigma}) = c|\mathbf{\Sigma}|^{-(q+n/2)} \mathrm{etr}\{-\mathbf{\Sigma}^{-1}\mathbf{D}/2\}$$

where c is the constant given by

$$c^{-1} = |\mathbf{D}|^{-(n+q-1)/2} 2^{(n+q-1)q/2} \pi^{q(q-1)/4} \prod_{i=1}^{q} \Gamma((n+q-i)/2).$$

Here $n > 0$ and $\mathbf{D}$ is a $q \times q$ positive definite and symmetric matrix, referred to as the sum-of-squares parameter matrix. If $n > 2$ then the mean is defined and is $E(\mathbf{\Sigma}) = \mathbf{D}/(n-2)$. The notation is sometimes modified to explicitly reflect the dimension, via $\mathbf{\Sigma} \sim IW_q(n, \mathbf{D})$.

The inverse Wishart and Wishart are related by $\mathbf{\Omega} = \mathbf{\Sigma}^{-1}$ when $\mathbf{\Sigma} \sim IW(n, \mathbf{D})$ and $\mathbf{\Omega} \sim W(h, \mathbf{A})$ with $h = n+q-1$ and $\mathbf{A} = \mathbf{D}^{-1}$. The map from Wishart to inverse Wishart, and back, is derived by direct transformation, using the Jacobians

$$\left|\frac{\partial \mathbf{\Omega}}{\partial \mathbf{\Sigma}}\right| = |\mathbf{\Sigma}|^{-(q+1)} \text{ and } \left|\frac{\partial \mathbf{\Sigma}}{\partial \mathbf{\Omega}}\right| = |\mathbf{\Omega}|^{-(q+1)}.$$

The constant c in the densities of each is the same, expressed in terms of either parameterization $(h, \mathbf{A})$ or $(n, \mathbf{D})$. Note that both h and $n = h-q+1$ are often referred to as degrees of freedom, and it is important to avoid notational confusion.

10.7.3.1 Point estimates of variance matrices

Priors and posteriors for variance matrices are often derived as conditional inverse Wishart distributions. Inference will often use simulation as well as specific point estimates of variance matrices under such distributions. When $\mathbf{\Sigma} \sim IW(n, \mathbf{D})$, two standard point estimates of $\mathbf{\Sigma}$ are the mean $E(\mathbf{\Sigma}) = \mathbf{D}/(n-2)$ (when $n > 2$) and the harmonic mean $E(\mathbf{\Sigma}^{-1})^{-1} = \mathbf{D}/(n+q-1)$. These are two standard choices of point estimates of $\mathbf{\Sigma}$. When the dimension q increases, the divisors $n-2$ and $n+q-1$ become practically different; in practical work with empirical results being viewed purely subjectively, it is often the case that the mean is viewed as defining overestimates of variance levels in $\mathbf{\Sigma}$, while the harmonic mean generates underestimates, especially for larger q. Practical compromises take alternative divisors, such as $\mathbf{S} = \mathbf{D}/n$ that arise naturally in Bayesian analyses under conjugate normal, inverse Wishart priors and posteriors.

10.7.4 The Normal, Inverse Wishart Distribution

The $q \times 1$ vector $\boldsymbol{\theta}$ and $q \times q$ positive definite and symmetric matrix $\mathbf{\Sigma}$ have a normal, inverse Wishart distribution $(\boldsymbol{\theta}, \mathbf{\Sigma}) \sim NIW(\mathbf{m}, c, n, \mathbf{D})$ when $(\boldsymbol{\theta}|\mathbf{\Sigma}) \sim N(\mathbf{m}, c\mathbf{\Sigma})$ and $\mathbf{\Sigma} \sim IW(n, \mathbf{D})$, where $c > 0$.

Under this distribution, the implied marginal for $\boldsymbol{\theta}$ is a multivariate T distribution with n degrees of freedom, $\boldsymbol{\theta} \sim T_n(\mathbf{m}, c\mathbf{S})$ where $\mathbf{S} = \mathbf{D}/n$; the distribution has mode $\mathbf{m}$ that is also the mean if $n > 1$, and covariance matrix $c\mathbf{S}n/(n-2)$ if $n > 2$. The density function is

$$p(\boldsymbol{\theta}) = k\{1 + (\boldsymbol{\theta} - \mathbf{m})'(c\mathbf{S})^{-1}(\boldsymbol{\theta} - \mathbf{m})/n\}^{-(q+n)/2}$$

with normalizing constant defined by

$$k^{-1} = |c\mathbf{S}|^{1/2}(n\pi)^{q/2}\Gamma(n/2)/\Gamma((n+q)/2).$$

This q-variate T distribution parallels the conditional normal $p(\boldsymbol{\theta}|\mathbf{\Sigma})$; moving from conditional to the marginal simply substitutes the estimate $\mathbf{S}$ for $\mathbf{\Sigma}$ and flattens the tails by moving from normal to T.

10.7.5 The Matrix Normal, Inverse Wishart Distribution

The $r \times q$ matrix $\mathbf{\Theta}$ and $q \times q$ positive definite and symmetric matrix $\mathbf{\Sigma}$ have a matrix normal, inverse Wishart distribution $(\mathbf{\Theta}, \mathbf{\Sigma}) \sim NIW(\mathbf{M}, \mathbf{C}, n, \mathbf{D})$ when $(\mathbf{\Theta}|\mathbf{\Sigma}) \sim N(\mathbf{M}, \mathbf{C}, \mathbf{\Sigma})$ and $\mathbf{\Sigma} \sim IW(n, \mathbf{D})$, where $\mathbf{C}$ is an $r \times r$ variance matrix, the column variance matrix of the conditional matrix normal distribution of $(\mathbf{\Theta}|\mathbf{\Sigma})$.

For any row $i = 1 : r$, $(\boldsymbol{\theta}'_{i,\star}, \boldsymbol{\Sigma}) \sim NIW(\mathbf{m}_{i,\star}, c_{i,i}, n, \mathbf{D})$. Further, any pair of rows i, s has $Cov(\boldsymbol{\theta}'_{i,\star}, \boldsymbol{\theta}'_{s,\star}|\boldsymbol{\Sigma}) = c_{i,s}\boldsymbol{\Sigma}$.

The implied marginal for $\boldsymbol{\Theta}$ is a matrix T distribution-the extension to matrix variates of the multivariate T distribution. For example, for any row $i = 1 : r$, $\boldsymbol{\theta}'_{i,\star} \sim T_n(\mathbf{m}_{i,\star}, c_{i,i}\mathbf{S})$ with $\mathbf{S} = \mathbf{D}/n$, simply by reference to Section 10.7.4 above.

Extensive additional development of matrix normal, inverse Wishart and related distribution theory is given in Dawid (1981), with foundational material in Press (1982).

10.7.6 Hyper-Inverse Wishart Distributions

Central aspects of conjugate Bayesian analysis for inference on covariance matrices in Gaussian graphical models involve the class of hyper-inverse Wishart (HIW) distributions, the natural extension of Wisharts to graphical models (Dawid and Lauritzen 1993; Roverato 2002). We restrict attention to decomposable models that provide the main practicable methods due to tractability of computations of prior-posterior and predictive distributions, as illustrated, for example, in Giudici and Green (1999) and Jones, Dobra, Carvalho, Hans, Carter, and West (2005).

10.7.6.1 Decomposable graphical models

Suppose the q-vector $\mathbf{x} \sim N(\boldsymbol{\mu}, \boldsymbol{\Sigma})$ has conditional independencies encoded by a graph $G = (V, E)$ with node set $V = \{1, \ldots, q\}$ representing univariate elements of $\mathbf{x}$ and edge set $E = \{(i, j) : i \sim j\}$ where $i \sim j$ implies an edge between i and j in G, there being no such edge otherwise. The neighbor set of node i is $ne(i) = \{j : (i, j) \in E\}$. Then $\boldsymbol{\Omega} = \boldsymbol{\Sigma}^{-1}$ has zero off-diagonal elements $\omega_{i,j}$ for all $(i, j) \notin E$, and nonzero elements elsewhere. For each i, let $\mathbf{x}_{-i}$ represent all variables except x_i, and $\mathbf{x}_{ne(i)}$ be the subvector of elements x_j for $j \in ne(i)$. Then $p(x_i|\mathbf{x}_{-i}) \equiv p(x_i|\mathbf{x}_{ne(i)})$; the off-diagonal zeros in row i of $\boldsymbol{\Omega}$ define and are defined by this set of conditional independencies. The $q \times q$ covariance matrix $\boldsymbol{\Sigma}$ is constrained by the graph due to these zeros in the precision matrix.

Take G to be decomposable, so G can be decomposed into a set of intersecting prime components $\mathcal{P} = \{P_1, \ldots, P_g\}$ linked by separators $\mathcal{S} = \{S_2, \ldots, S_g\}$ where: (a) each P_i is a nonempty, complete subgraph of G, and (b) for $i = 2 : g$, $S_i = P_i \cap \{\cup_{j<i} P_j\}$, and so each S_i is also a complete subgraph of G. Collectively, P, S define the components of G.

10.7.6.2 The hyper-inverse Wishart distribution

The hyper-inverse Wishart distribution for $\mathbf{\Sigma}$, $\mathbf{\Sigma} \sim HIW(n, \mathbf{D})$, is based on $n > 0$ degrees of freedom and sum-of-squares parameter matrix $\mathbf{D}$, a positive definite and symmetric $q \times q$ matrix. Notation sometimes makes the graph explicit via $\mathbf{\Sigma} \sim HIW_G(n, \mathbf{D})$. This is the unique hyper-Markov distribution having margins on the components of G that are inverse Wishart, and that are consistent in the sense that the margin on each of P_{i-1} and P_i on their separating intersection S_i is the same.

In detail, for each component $P \in \mathcal{P}$, write $\mathbf{x}_P$ for the subvector of $\mathbf{x}$ of variables in P, $\mathbf{x}_P \sim N(\boldsymbol{\mu}_P, \Sigma_P)$ where $\mathbf{\Sigma}_P$ is the corresponding submatrix of $\mathbf{\Sigma}$. Under $\mathbf{\Sigma} \sim HIW(n, \mathbf{D})$, the marginal for $\mathbf{\Sigma}_P$ is inverse Wishart, $\mathbf{\Sigma}_P \sim IW(n, \mathbf{D}_P)$ where $\mathbf{D}_P$ is the corresponding submatrix of the sum-of-squares matrix $\mathbf{D}$. Similarly, $\mathbf{x}_S \sim IW(n, \mathbf{D}_S)$ for each $S \in \mathcal{S}$.

10.7.6.3 Prior and posterior HIW distributions

The density of $\mathbf{x}$ factorizes as a ratio of products of terms on the components of G, viz.

$$p(\mathbf{x}|\boldsymbol{\mu}, \mathbf{\Sigma}) = \frac{\prod_{P \in \mathcal{P}} N(\mathbf{x}_P|\boldsymbol{\mu}_P, \mathbf{\Sigma}_P)}{\prod_{S \in \mathcal{S}} N(\mathbf{x}_S|\boldsymbol{\mu}_S, \mathbf{\Sigma}_S)}.$$

Bayesian inference is enabled by the fact that the HIW distribution is the conjugate prior for likelihood functions comprised of products of this form. Specifically, with $\mathbf{\Sigma} \sim HIW(n, \mathbf{D})$, the prior density function is

$$p(\mathbf{\Sigma}) = \frac{\prod_{P \in \mathcal{P}} IW(\mathbf{\Sigma}_P|n, \mathbf{D}_P)}{\prod_{S \in \mathcal{S}} IW(\mathbf{\Sigma}_S|n, \mathbf{D}_S)}$$

where each term is inverse Wishart. As a result, the posterior is $(\mathbf{\Sigma}|\mathbf{x}) \sim HIW(n+1, \mathbf{D} + (\mathbf{x} - \boldsymbol{\mu})(\mathbf{x} - \boldsymbol{\mu})')$ with corresponding generalizations based on samples of size more than one.

10.7.6.4 Normal, hyper-inverse Wishart distributions

The direct and simple extension of the normal, inverse Wishart distribution theory to the graphical models is immediate. The $r \times q$ matrix $\mathbf{\Theta}$ and $q \times q$ positive definite and symmetric matrix $\mathbf{\Sigma}$ have a matrix normal, hyper-inverse Wishart (NHIW) distribution $(\mathbf{\Theta}, \mathbf{\Sigma}) \sim NIW(\mathbf{M}, \mathbf{C}, n, \mathbf{D})$ (NHIW) when $(\mathbf{\Theta}|\mathbf{\Sigma}) \sim N(\mathbf{M}, \mathbf{C}, \mathbf{\Sigma})$ for some parameters $\mathbf{M}, \mathbf{C}$, and $\mathbf{\Sigma} \sim HIW_G(n, \mathbf{D})$ on the graph G.

In the special case of the multivariate NHIW distribution when $\mathbf{\Theta} \equiv \boldsymbol{\theta}$, a vector, we have $(\boldsymbol{\theta}, \mathbf{\Sigma}) \sim NHIW(\mathbf{m}, c, n, \mathbf{D})$ under which $(\boldsymbol{\theta}|\mathbf{\Sigma}) \sim N(\mathbf{m}, c\mathbf{\Sigma})$ with c now a scalar.

10.8 Problems

1. In the exchangeable time series model of Section 10.3, suppose that $\mathbf{G}_t = \mathbf{I}_p$ for all t, and that the $\mathbf{W}_t$ sequence is defined via a single discount factor δ so $\mathbf{R}_t = \mathbf{C}_{t-1}/\delta$ for all t. This models evolution of all state parameters as steady models, or random walks, evolving at a constant discount rate. Show that the retrospective smoothing equations simplify to

$$\begin{aligned} \mathbf{a}_T(t-T) &= (1-\delta)\mathbf{M}_t + \delta\mathbf{a}_T(t-T+1), \qquad (10.17)\\ \mathbf{R}_T(t-T) &= (1-\delta)\mathbf{C}_t + \delta^2\mathbf{R}_T(t-T+1). \qquad (10.18)\end{aligned}$$

2. Consider two independent $q \times q$ Wishart matrices $\mathbf{S}_1 \sim W(\nu_1, \mathbf{A})$ and $\mathbf{S}_2 \sim W(\nu_2, \mathbf{A})$ where $\nu_1 = \beta h$ and $\nu_2 = (1-\beta)h$ for some $h > q-1$ and $\beta \in (0,1)$.
   - What is the distribution of $\mathbf{S} = \mathbf{S}_1 + \mathbf{S}_2$?
   - Use Corollary 2 of Dawid (1981) to show that $\mathbf{S}_1 = \mathbf{U}'\mathbf{\Gamma}\mathbf{U}$ where $\mathbf{U}$ satisfies $\mathbf{S} = \mathbf{U}'\mathbf{U}$ and $\mathbf{\Gamma}$ has the matrix beta distribution $\mathbf{\Gamma} \sim Be(\beta h/2, (1-\beta)h/2)$.
   - Use the above result to verify the matrix beta evolution of precision matrices of Section 10.4.3.
   - Further, use the direct identity $\mathbf{S} = \mathbf{S}_1 + \mathbf{S}_2$ and the implied construction of $p(\mathbf{S}|\mathbf{S}_1)$ to simply verify the retrospective filtering theory of Equation (10.12).
3. Simplify the retrospective filtering of Equation (10.12) for the case of a univariate time series, $q = 1$. Describe how the retrospectively simulated values of past time-varying precisions now depend on a sequence of random "retrospective shocks" that have $\chi^2_{h_t}$ distributions ($t = (T-1):1$).
4. Use the retrospective filtering analysis of Equation (10.12) to derive a retrospective recursive equation for computing the sequence of estimates $E(\mathbf{\Phi}_t|\mathcal{D}_T)$ over $t = (T-1):1$.
5. Develop software to implement the forward filtering/sequential updating analysis, and the retrospective smoothing computations, for the exchangeable time series DLM with time-varying observational covariance matrix under the matrix beta evolution model. Explore the exchange rate time series data in your code development, aiming to reproduce aspects of the example detailed in Section 10.4.6.
6. Verify the optimal portfolio construction theory for one-step-ahead portfolio decisions summarized in Section 10.4.7. That is, suppose that the predictive moments $\mathbf{f}_t = E(\mathbf{y}_t|\mathcal{D}_{t-1})$ and $\mathbf{V}_t = V(\mathbf{y}_t|\mathcal{D}_{t-1})$ are available, with forecast precision matrix $\mathbf{K}_t = \mathbf{V}_t^{-1}$. The constrained optimization problem is to find $\mathbf{w}_t$ such that

$$\mathbf{w}_t = \arg\min_{\mathbf{w}}(\mathbf{w}'\mathbf{V}_t\mathbf{w}) \quad \text{subject to: } \mathbf{w}'\mathbf{f}_t = r_t \text{ and } \mathbf{w}'\mathbf{1} = 1.$$

Using Lagrange multipliers, solve this optimization and verify that the results are as summarized in Equation (10.14).

7. Verify the Wishart evolution distribution theory of Section 10.4.8 in the following general setting.
   The $q \times q$ precision matrix $\mathbf{\Phi}$ has the Wishart distribution $\mathbf{\Phi} \sim W(h, \mathbf{A})$ for some degrees of freedom $h = n + q - 1$ where $n > 0$ so that $h > q - 1$, and where $\mathbf{A}$ is the inverse sum-of-squares matrix. The Bartlett decomposition, often used for simulation of Wishart matrices as well as theoretical developments (e.g., Odell and Feiveson 1966), is that $\mathbf{\Phi} = \mathbf{P}'\mathbf{U}'\mathbf{U}\mathbf{P}$ where $\mathbf{P}$ is the upper triangular Cholesky component of $\mathbf{A}$, so that $\mathbf{A} = \mathbf{P}'\mathbf{P}$, and

$$\mathbf{U} = \begin{pmatrix} u_{1,1} & u_{1,2} & u_{1,3} & \cdots & u_{1,q} \\ 0 & u_{2,2} & u_{2,3} & \cdots & u_{2,q} \\ 0 & 0 & u_{3,3} & \cdots & u_{3,q} \\ \vdots & \vdots & \vdots & \ddots & \vdots \\ 0 & 0 & \cdots & 0 & u_{q,q} \end{pmatrix}$$

   where the nonzero entries are independent random quantities with $u_{i,j} \sim N(0,1)$ for $1 \leq i < j \leq q$ and $u_{i,i} = \sqrt{\kappa_i}$ where $\kappa_i \sim \chi^2_{h-i+1}$ for $i = 1, \ldots, q$.
   Define $\tilde{\mathbf{\Phi}} = b^{-1}\mathbf{P}'\tilde{\mathbf{U}}'\tilde{\mathbf{U}}\mathbf{P}$ where $b > 0$ and $\tilde{\mathbf{U}}$ is an upper triangular matrix with the same off-diagonal elements as $\mathbf{U}$, and diagonal elements $\tilde{u}_{i,i} = \sqrt{\tilde{\kappa}_i}$ where $\tilde{\kappa}_i = \kappa_i \eta_i$ with $\eta_i \sim Be(\beta_i(h-i+1)/2, (1-\beta_i)(h-i+1)/2)$ for some constants $\beta_i \in (0,1)$, for $i = 1 : q$. The η_i are mutually independent, and independent of the normal $u_{i,j}$.
   - Show that, for $i = 1 : q$, $\tilde{\kappa}_i \sim \chi^2_{\beta_i(h-i+1)}$ independently over i and independently of the upper, off-diagonal elements of $\tilde{\mathbf{U}}$.
   - Deduce that $\tilde{\mathbf{\Phi}} \sim W(k, b^{-1}\mathbf{A})$ when $\beta_i = (k+i-1)/(h+i-1)$ for each $i = 1, \ldots, q$, and where $0 < k < h$.
   - Confirm that the results of Section 10.4.8 for stochastic models of evolution of precision matrices based on discounting either degrees of freedom parameter, h_{t-1} or n_{t-1}, are delivered as special cases. Identify the values of (k, b) and the β_i in each of these special cases.

Chapter 11

Latent factor models

Latent factor models for multivariate time series assume an underlying dependence structure induced by a common latent term written as a product of an unknown matrix of factor loadings and a vector of common factors. Latent factor models are widely used in diverse fields such as psychology, finance, and environmental science. This chapter introduces foundational concepts related to Bayesian latent factor models, and presents dynamic latent factor models, factor stochastic volatility models, and spatiotemporal dynamic factor models. Markov Chain Monte Carlo methods for these models are developed and illustrated with applications in finance and environmental science.

11.1 Introduction

The general area of latent factor modeling is vibrant and growing in multivariate time series analysis as it is in other areas of applied statistics, partly based on advances in the ability to fit and explore increasingly structured models via simulation and optimization methods enabled by computational advances (e.g., West 2003, Lopes and West 2004). Section 8.1.2 introduced and exemplified simple latent factor models in time series; the core concept is that of dynamic regression of multiple time series on predictors that are themselves unobserved, i.e., latent, aiming to isolate common underlying features in multiple time series—a model-based extension of the ideas underlying dynamic principal components as illustrated in Section 10.4.6.3. Model-based approaches open the path to introducing more structure into the factor components, with opportunity for predictive extensions, as well as defining reduced dimensional explanations of what may be a high dimensional time series. The literature is large and extensive in applications;

some key historical references include factor modeling for dimension reduction using autoregressive moving average (ARMA) models in Peña and Box (1987) and Tiao and Tsay (1989), and in the major growth area of multivariate factor modeling in financial time series the foundational works of Harvey, Ruiz, and Shephard (1994), Jacquier, Polson, and Rossi (1994, 1995) and, later, Kim, Shephard, and Chib (1998). The financial multivariate volatility area continues to motivate factor modeling developments, based on the extensions of these early ideas to structured volatility models for short-term volatility forecasting and portfolio analysis; see Aguilar, Huerta, Prado, and West (1999), Pitt and Shephard (1999b), and Aguilar and West (2000), for key developments and examples.

This chapter reviews Bayesian latent factor models. Introduced by Spearman (1904), factor analysis has been widely used in diverse fields such as psychology, medicine, economics, finance, and environmental science. In its simplest form, factor analysis models an observed vector of variables of interest as a sum of a latent systematic term and an error vector. The latent systematic term is written as the product of an unknown matrix of factor loadings and a vector of latent factors. The error vector is assumed to have a diagonal covariance matrix. Thus, the dependence structure among the observed variables is induced by the latent factors. Typically, the number of observed variables of interest is much larger than the number of latent factors, and thus the factor analysis provides a way to reduce the dimension of the problem being studied. In addition, as illustrated by the applications in Sections 11.2.7, 11.3.1, 11.4.2, and 11.5.1, the latent factors are usually associated with underlying common processes and thus factor analysis sheds light on and facilitates meaningful interpretation of the dependence structure among observed variables.

Here we review model-based approaches that allow the introduction of meaningful probabilistic structure in the factor loadings and in the common factors. Such probabilistic structure on unknown quantities is more naturally interpreted within a Bayesian paradigm. In addition, statistical inference for these highly structured latent factor models is greatly facilitated by Bayesian simulation-based approaches implemented with Markov chain Monte Carlo algorithms. Therefore, in this chapter we focus on Bayesian factor models.

We distinguish between static factor models and dynamic factor models. In static factor models, the matrix of factor loadings is fixed through time and the common factors associated with the vectors of observations are assumed to be independent. In contrast, dynamic factor models may allow for temporal evolution of the matrix of factor loadings and/or temporal dependence among the common factors. Thus, dynamic factor models are more adequate for time series analysis, providing more precise estimates of factor loadings and common factors, and allowing for temporal prediction of future observations. Section 11.2 presents static factor models to introduce notation and general issues related to factor models such as identifiability, Bayesian computation, and selection of number of factors. Section 11.3 introduces dynamic latent factor models.

In its simplest form, factor analysis is related to principal component analysis (PCA). Principal components are orthogonal linear combinations of the variables of interest that have special properties. For example, the first principal component is the normalized linear combination with largest variance. The second principal component is, among the normalized linear combinations orthogonal to the first principal component, the one with largest variance. And so on. A standard result is that the population principal components are the eigenvectors of the population covariance matrix (Anderson 2003). In addition, the variances of the population principal components are the respective eigenvalues of the population covariance matrix. Further, the MLEs of the population principal components and their variances are respectively the eigenvectors and eigenvalues of the sample covariance matrix multiplied by $(n-1)/n$. Thus, usual PCA analysis estimates the population principal components and their variances with the eigenvectors and eigenvalues of the sample covariance matrix. Finally, the statistician usually focuses on the few principal components that explain most of the variability of the variables of interest.

Hence, PCA analysis and factor analysis have different objectives – PCA analysis aims at estimating linear combinations that have high variance whereas factor analysis aims at estimating latent factors that explain the dependence structure. However, in practice the values of the estimated principal components and of the factor loadings MLE (under an orthogonality identifiability constraint) may be fairly similar. Under a factor model, the smaller the variances of the errors the closer these estimates will be, with the estimated principal components being noisier than the estimated factor loadings. Thus, it is common practice for statisticians to perform an exploratory data analysis using PCA before doing a factor analysis. Such PCA-based exploratory data analysis may be used not only in simpler contexts but also to gain insight into more complex problems.

For example, within the context of multivariate stochastic volatility, Section 10.4.6.3 computes principal components of matrices that are draws from the posterior distribution of time-varying covariance matrices modeled with a matrix beta evolution model. As discussed in Sections 10.4.3 and 10.4.4, this discount-factor-based evolution model for covariance matrices has a relatively simple random walk nature. As such, the matrix beta evolution model does not account for possible existence of a small number of common latent components in multivariate stochastic volatility. When supported by the observed data, multivariate stochastic volatility models based on a small number of common latent components may yield more precise estimates of the stochastic volatility, lead to key insights on the multivariate stochastic volatility process, and provide more precise predictions. In that regard, for the exchange rates application presented in Section 10.4.6.3, the exploratory PCA analysis of the posterior draws provides empirical evidence of the existence of a small number of latent common factors that may explain the time-varying dependence structure in the multivariate time series. Section 11.4 presents factor stochastic volatility models that use dynamic latent common factors to model time-varying covariance matrices.

Section 11.5 presents dynamic factor models for spatiotemporal data, that is, data collected in multiple spatial locations over time. This type of data is common in a wide range of areas such as business, economics, agriculture, epidemiology, and environmental science. These data may be classified as areal data or point-referenced data (Banerjee, Carlin, and Gelfand 2014). Areal data arise when the region of interest is partitioned into subregions and each data point corresponds to one of the subregions. An example of areal data is the per capita income in the state of Virginia per county. Point-referenced data arise when the region of interest is sampled at specific spatial locations. An example of point-referenced data is the data on ozone levels measured by pollution monitoring stations in the United States. Typically, Markov random field models (also known as conditional autoregressive models) are used for areal data and continuous Gaussian process models are used for point-referenced data. Specifically, Section 11.5 presents dynamic factor models for point-referenced data by assigning a continuous Gaussian process prior to the elements of the matrix of factor loadings. This has many benefits such as the ability to produce estimated maps of the latent process of interest at each time point, as well as predictions for any future time point at any spatial location (sampled or not sampled) within the region of interest. Finally, we note that extensions of these ideas to spatiotemporal areal data are straightforward and assign a Markov random field prior to the elements of the matrix of factor loadings.

Finally, Section 11.6 presents other extensions of dynamic latent factor models and recent developments.

11.2 Static Factor Models

11.2.1 1-Factor Case

The developments in this section assume that the variables of interest have been centered. Thus, the static factor model below describes the variance-covariance structure of the variables of interest. Assume that a sample of size n is obtained on r variables of interest. Let $\mathbf{y}_1, \ldots, \mathbf{y}_n$ be the sampled

r-dimensional vectors. To introduce notation and provide intuition, let us consider a static factor model with one factor

$$\mathbf{y}_i = \mathbf{B}x_i + \mathbf{v}_i, \tag{11.1}$$

where $\mathbf{B} = (b_1, \ldots, b_r)'$ is the r-dimensional vector of factor loadings, and x_i is the i-th realization of a one-dimensional latent factor process. In addition, $\mathbf{v}_i$ is the r-dimensional error vector for the i-th observation. Further, the error vectors $\mathbf{v}_1, \ldots, \mathbf{v}_n$ are assumed to be i.i.d. $N(\mathbf{0}, \mathbf{V})$, where $\mathbf{V} = \text{diag}(\sigma_1^2, \ldots, \sigma_r^2)$. Typically, $\sigma_1^2, \ldots, \sigma_r^2$ are called idiosyncratic variances. Finally, the latent factor realizations $x_1, \ldots, x_n$ are assumed to be i.i.d. $N(0, H)$. Therefore, $\mathbf{B}x_i$ is a latent systematic term that induces covariation among the r variables of interest, and the remaining variability not captured by $\mathbf{B}x_i$ specific to each variable of interest is represented by the idiosyncratic variances $\sigma_1^2, \ldots, \sigma_r^2$.

If $x_1, \ldots, x_n$ were known, then (11.1) would be a multivariate regression model. In that case, conditional on x_i, the second-order moments would be $V(y_{ij}|x_i) = \sigma_j^2$ and $Cov(y_{ij}, y_{ik}|x_i) = 0$. With known $x_1, \ldots, x_n$, the factor loadings $b_1, \ldots, b_r$ would be the regression coefficients that measure the impact of the univariate factor x_i on each element of the vector $\mathbf{y}_i$. However, for statistical factor models, $x_1, \ldots, x_n$ are not observable.

Thus, to obtain the first- and second-order moments of the observation vector $\mathbf{y}_i$ we have to integrate out x_i. By doing that, we find that the expected value of $\mathbf{y}_i$ is $E(\mathbf{y}_i) = \mathbf{0}$ and that is why we need to center the variables of interest before using model (11.1). In addition, the covariance matrix of $\mathbf{y}_i$ is $Cov(\mathbf{y}_i) = H\mathbf{B}\mathbf{B}' + \mathbf{V}$. Hence, this factor model with one factor induces a very specific covariance structure with $V(y_{ij}) = Hb_j^2 + \sigma_j^2$ and $Cov(y_{ij}, y_{ik}) = Hb_jb_k$. Therefore, the factor loadings vector $\mathbf{B} = (b_1, \ldots, b_r)'$ plays a critical role in the covariance structure of $\mathbf{y}_i$, with larger b_j implying larger variances and covariances.

Alternatively, we can write model (11.1) in matrix notation

$$\mathbf{Y} = \mathbf{X}\mathbf{B}' + \mathbf{v}, \tag{11.2}$$

where $\mathbf{Y} = (\mathbf{y}_1, ..., \mathbf{y}_n)'$ is the $n \times r$ matrix of observations, $\mathbf{X} = (x_1, \ldots, x_n)'$ is the n dimensional vector of latent factors, and $\mathbf{v} = (\mathbf{v}_1', ..., \mathbf{v}_n')'$ is the $n \times r$ matrix of errors.

Identifiability is an important issue in factor models. Note that in the model given in Equation (11.1), if we define $\mathbf{B}^* = a\mathbf{B}$ and $x_i^* = x_i/a$ for a scalar $a \neq 0$, then $\mathbf{y}_i = \mathbf{B}^*x_i^* + \mathbf{v}_i$ is exactly the same model for the observations as the model given in Equation (11.1). There are many ways to resolve this identifiability problem. One simple and intuitive way that we use here is to set the first element of $\mathbf{B}$ to be equal to 1, that is, $b_1 = 1$. This choice is a particular case of the hierarchical structural constraint introduced in Section 11.2.4. Another way to resolve the identifiability problem is to assume that the length of the vector $\mathbf{B}$ is one, that is, $\sum_{i=1}^{r} b_i^2 = 1$. Although this is relatively straightforward to implement computationally for the case of one factor, its extension to more than one factor in a Bayesian framework is far from trivial. We return to this issue in Section 11.2.4.

Further, we assume that the other factor loadings in the vector $\mathbf{B}$ are unconstrained. In particular, in a Bayesian framework we have to assume a probability model for the factor loadings. Here we consider for the factor loadings $b_2, \ldots, b_r$ independent priors $N(0, \tau^2)$, where τ^2 is the factor loadings variance. The factor loadings variance controls the magnitude of the factor loadings $b_2, \ldots, b_r$.

There are many possible choices for the priors for the unknown parameters of factor model (11.1). Here we favor conditionally conjugate priors because they facilitate the design of MCMC algorithms to explore the posterior distribution. For the latent factor variance H we assume $IG(n_H/2, n_H s_H^2/2)$. For the idiosyncratic variances $\sigma_1^2, \ldots, \sigma_j^2$ we assume independent priors $IG(n_\sigma/2, n_\sigma s_\sigma^2/2)$. Finally, for the factor loadings variance τ^2 we assume the prior $IG(n_\tau/2, n_\tau s_\tau^2/2)$.

11.2.2 MCMC for Factor Models with One Factor

Posterior exploration for the one-factor model can be performed in a straightforward manner with the Gibbs sampler, that is, by iteratively sampling from the full conditional distributions of the unknown model quantities.

The full conditional distribution of the latent factor x_i, $i = 1, \ldots, n$, is the following Gaussian distribution

$$N\left(\frac{\sum_{j=1}^{r} \sigma_j^{-2} b_j y_{ij}}{H^{-1} + \sum_{j=1}^{r} \sigma_j^{-2} b_j^2}, \frac{1}{H^{-1} + \sum_{j=1}^{r} \sigma_j^{-2} b_j^2}\right). \tag{11.3}$$

The full conditional distribution of the factor loading b_j, $j = 2, \ldots, r$, is

$$N\left(\frac{\sum_{i=1}^{n} x_i y_{ij}}{\sigma_j^2 \tau^{-2} + \sum_{i=1}^{n} x_i^2}, \frac{\sigma_j^2}{\sigma_j^2 \tau^{-2} + \sum_{i=1}^{n} x_i^2}\right). \tag{11.4}$$

The full conditional distribution of the idiosyncratic variance σ_j^2, $j = 1, \ldots, r$, is

$$IG\left(\frac{n_\sigma + n}{2}, \frac{n_\sigma s_\sigma^2 + \sum_{i=1}^{n}(y_{ij} - b_j x_i)^2}{2}\right). \tag{11.5}$$

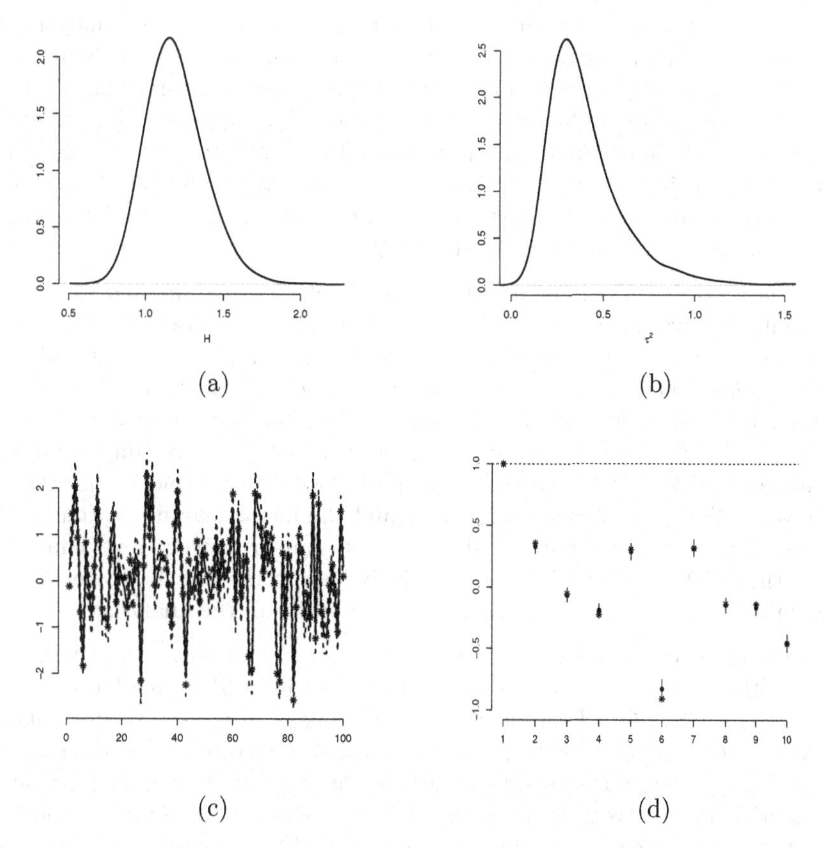

Figure 11.1 *Simulated example. (a) Posterior density of latent factor variance H. (b) Posterior density of factor loadings variance τ^2. (c) Plot of posterior mean (solid line), 95% credible intervals (dashed lines), and true values (stars) common factors x_i. (d) Posterior means (circles), 95% credible intervals (vertical lines), and true values (stars) of factor loadings b_j.*

The full conditional distribution of the factor loadings variance τ^2 is

$$IG\left(\frac{n_\tau + r - 1}{2}, \frac{n_\tau s_\tau^2 + \sum_{j=2}^r b_j^2}{2}\right). \tag{11.6}$$

Finally, the latent factor variance H has full conditional distribution

$$IG\left(\frac{n_H + n}{2}, \frac{n_H s_H^2 + \sum_{i=1}^n x_i^2}{2}\right). \tag{11.7}$$

We recommend that factor analysis with the above Gibbs sampler and conjugate prior specifications be applied to standardized variables. When applied to standardized variables, factor analysis becomes an analysis of the correlation matrix (as opposed to the covariance matrix). This in turn facilitates the assignment of priors for the parameters. In particular, for the factor analysis of standardized variables, the following prior hyperparameters impart little information and work well in practice: $n_\tau = 1$, $n_\tau s_\tau^2 = 1$, $n_H = 1$, $n_H s_H^2 = 1$, $n_\sigma = 2.2$, and $n_\sigma s_\sigma^2 = 0.1$. We note that this choice of values for the hyperparameters n_σ and $n_\sigma s_\sigma^2$ in the prior of σ_i^2 are those recommended by Lopes and West (2004).

We have implemented the above Gibbs sampler for a simulated example. Specifically, we have simulated a sample of $n = 100$ observation vectors of dimension $r = 10$ from the factor model (11.1). For the magnitude of the observations to be similar to those of standardized variables, we have simulated the data using the following values for the parameters of the model: $\tau^2 = 0.25$, $\sigma_1^2 = \ldots = \sigma_r^2 = 0.1$, and $H = 1$. We have run the Gibbs sampler for 5,000 iterations. Visual inspection of the trace plots of posterior draws of the latent factor variance H and the factor loadings variance τ^2 (not shown) indicate that the above Gibbs sampler converges very fast in less than 100 iterations. We conservatively discard the first 1,000 iterations as burn-in and use the remaining 4,000 iterations in the analysis.

Panels (a) and (b) of Figure 11.1 present posterior densities of H and τ^2, respectively. The posterior mean of H is 1.19 and a 95% credible interval for H is (0.88, 1.60). Further, the posterior mean of τ^2 is 0.41 and a 95% credible interval for τ^2 is (0.16, 0.98). Hence, the credible intervals contain the true values of H and τ^2. Panels (c) and (d) of Figure 11.1 present posterior means, credible intervals, and true values of the latent factors x_i and the factor loadings b_j, which are clearly well estimated. Therefore, the Gibbs sampler algorithm combined with the vague prior specification do a good job at recovering the generating mechanism of the simulated data.

11.2.3 Example: A 1-Factor Model for Temperature

To illustrate the use of the factor model with one factor, here we apply that model to the analysis of troposphere daily average air temperature at 9 different altitudes measured in terms of pressure levels from 1000 millibars to 250 millibars at longitude -100 and latitude 55 over Canada from June 1 to August 31, 2015. These data come from the NCEP Reanalysis dataset (Kalnay et al., 1996) provided by NOAA/OAR/ESRL PSD, Boulder, Colorado, USA, that is available from the NOAA web site at http://www.esrl.noaa.gov/psd/.

Figure 11.2 depicts the lowest 50 kilometers of Earth's atmosphere that are divided into two layers: the troposphere and the stratosphere. In the

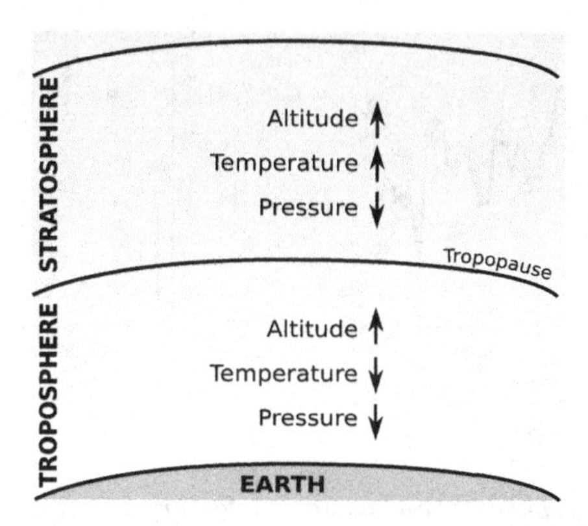

Figure 11.2 *Representation of the lowest 50 kilometers of Earth's atmosphere divided into the troposphere and the stratosphere.*

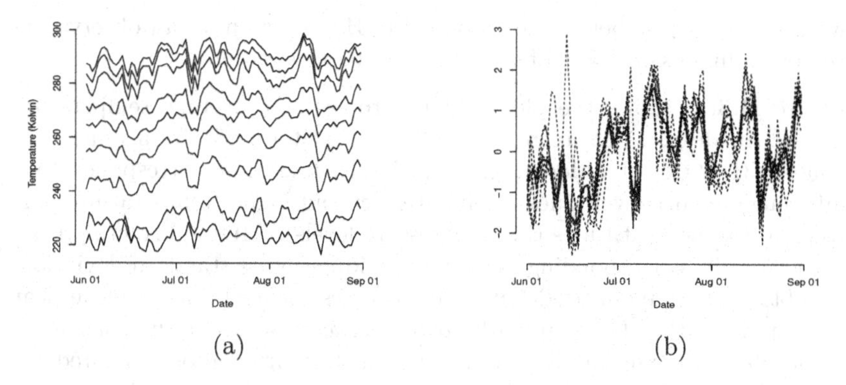

(a) (b)

Figure 11.3 *Troposphere temperature. (a) Time series plot of daily average troposphere temperature at longitude -100° and latitude 55° from June 1 to August 31, 2015, at different pressure levels. Each line corresponds to a different pressure level; (b) Time series plot of the several standardized temperature series (dashed lines) and the estimated common factor (bold solid line).*

troposphere temperatures usually drop with increase in altitude, while in the stratosphere temperatures increase with increase in altitude. The interface between the troposphere and the stratosphere is called the tropopause. The tropopause is located at an altitude corresponding to a pressure level of about 200 millibars. Here we consider pressure levels of 1000, 925, 850, 700, 600, 500, 400, 300, and 250 millibars, thus the lowest pressure level

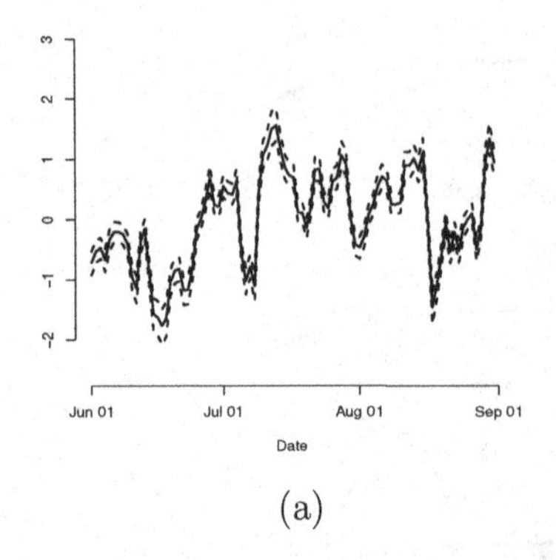

(a)

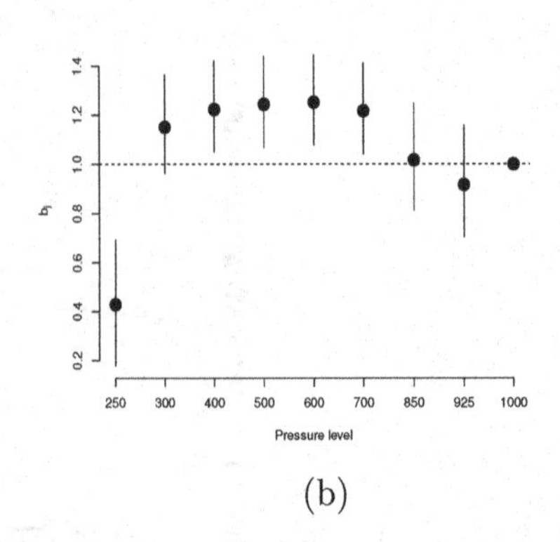

(b)

Figure 11.4 *Troposphere temperature – 1-factor model. (a) Time series plot of common factor posterior mean (solid line) and 95% credible intervals (dashed lines); (b) Posterior means (circles) and 95% credible intervals (vertical lines) of factor loadings b_j for each pressure level.*

we consider is just below the tropopause. Hence, we have an observation vector of dimension $r = 9$ observed at $n = 92$ time points.

Figure 11.3(a) shows the plot of daily troposphere average temperature (in Kelvin degrees) at longitude -100° and latitude 55° from June 1 to August 31, 2015, at different pressure levels. Each line corresponds to a different pressure level; lines rarely intersect and higher temperatures correspond to lower altitudes (higher pressure levels). After subtracting from each series its corresponding mean and dividing by its standard deviation, we obtain the standardized temperature series (dashed lines) that appear in Figure 11.3(b). The standardized temperature series at the considered troposphere altitudes move in tandem and, thus, are highly correlated.

Here we apply the 1-factor model to the standardized daily temperatures. As mentioned in Section 11.2.2, one advantage of analyzing standardized variables is that it is more straightforward to assign priors. We use here the same values for the prior hyperparameters used in the simulated example in Section 11.2.2: $n_\tau = 1$, $n_\tau s_\tau^2 = 1$, $n_H = 1$, $n_H s_H^2 = 1$, $n_\sigma = 2.2$, and $n_\sigma s_\sigma^2 = 0.1$. These priors impart little information in the analysis of standardized variables and allow the data to speak by themselves. We implemented a Gibbs sampler with 5,000 iterations and 1,000 iterations as burnin.

Figure 11.3(b) presents the estimated common factor (bold solid line). The estimated common factor captures the overall common temporal behavior of the several temperature series. To assess the uncertainty in the estimation of the common factor, Figure 11.4(a) plots the common factor posterior mean (solid line) as well as 95% credible intervals (dashed lines). The credible intervals are fairly narrow compared to the overall variation of the common factor, indicating that the common factor is estimated with relatively high precision.

Finally, Figure 11.4(b) presents posterior mean (circle) and 95% credible interval (vertical line) of factor loadings b_j for each pressure level. Note that $b_1 = 1$ corresponds to the lowest altitude considered in this analysis with pressure level of 1000 millibars. Thus, by model construction there is no uncertainty related to b_1 and all the other factor loadings should be analyzed relative to $b_1 = 1$. The plot clearly shows that the factor loadings for pressure levels of 925 and 850 millibars are not significantly different from 1, indicating that altitudes corresponding to these pressure levels are well represented by the common factor. In contrast, the factor loading for pressure level of 250 millibars is significantly lower than 1, indicating that, as we approach the tropopause, the air temperature becomes less correlated with the air temperature at lower troposphere altitudes.

11.2.4 Factor Models with Multiple Factors

A factor model with k factors extends naturally from a one-factor model. In this case, the latent factor is a k-dimensional vector. Specifically, the k-factor model can be written as

$$\mathbf{y}_i = \mathbf{B}\mathbf{x}_i + \mathbf{v}_i, \tag{11.8}$$

where again $\mathbf{y}_i$ is the i-th observation of the r-dimensional vector of interest. Now, $\mathbf{B}$ is the $r \times k$ matrix of factor loadings and $\mathbf{x}_i$ is the i-th realization of the k-dimensional latent factor process. In addition, $\mathbf{v}_i$ is the r-dimensional error vector for the i-th observation. Similarly to the one-factor model, the error vectors $\mathbf{v}_1, \ldots, \mathbf{v}_n$ are assumed to be i.i.d. $N(\mathbf{0}, \mathbf{V})$, where $\mathbf{V} = \text{diag}(\sigma_1^2, \ldots, \sigma_r^2)$ and $\sigma_1^2, \ldots, \sigma_r^2$ are the idiosyncratic variances. Finally, for the k-factor model the latent factor realizations $\mathbf{x}_1, \ldots, \mathbf{x}_n$ are assumed to be i.i.d. $N(\mathbf{0}, \mathbf{H})$ with $\mathbf{H} = \text{diag}(h_1, \ldots, h_k)$. Typically $k < r$, that is, the number of latent factors is smaller (often much smaller) than the number of observed variables. Hence, the latent structure has a lower dimension than the vector of observed variables. Therefore, the k-factor model provides a dimension reduction that often facilitates understanding and interpretation of the underlying process under study.

The likelihood function for the k-factor model is

$$f(\mathbf{Y}|\mathbf{B}, \mathbf{x}, \mathbf{H}, \mathbf{V}) \propto |\mathbf{V}|^{-n/2} \exp\left\{ -\frac{1}{2} \sum_{i=1}^{n} (\mathbf{y}_i - \mathbf{B}\mathbf{x}_i)' \mathbf{V}^{-1} (\mathbf{y}_i - \mathbf{B}\mathbf{x}_i) \right\}. \tag{11.9}$$

A key issue in the implementation of the k-factor model is identifiability. To ensure identifiability, one needs to impose constraints on the factor loadings matrix $\mathbf{B}$. There are many ways to impose constraints. For example, in frequentist factor analysis a common way to make the model identifiable is to assume three conditions: the columns of $\mathbf{B}$ are orthogonal; each column of $\mathbf{B}$ is a unit vector; and the factors are ordered in decreasing order of factor variance. Implementing these three conditions at the same time in a Bayesian framework would be much more difficult than implementing the constraint that we now describe. Specifically, here we assume that $\mathbf{B}$ is a full rank block lower triangular matrix with diagonal elements equal to 1. Such a block lower triangular constraint is known as a hierarchical structural constraint and has been used in Bayesian factor analysis by, among others, Geweke and Zhou (1996), Aguilar and West (2000), and Lopes and West (2004). Specifically, we assume that the factor loadings matrix is given by

$$\mathbf{B} = \begin{pmatrix} 1 & 0 & 0 & \cdots & 0 \\ b_{21} & 1 & 0 & \cdots & 0 \\ b_{31} & b_{32} & 1 & \cdots & 0 \\ \vdots & \vdots & \vdots & \ddots & \vdots \\ b_{k1} & b_{k2} & b_{k3} & \cdots & 1 \\ b_{k+1,1} & b_{k+1,2} & b_{k+1,3} & \cdots & b_{k+1,k} \\ \vdots & \vdots & \vdots & & \vdots \\ b_{r1} & b_{r2} & b_{r3} & \cdots & b_{rk} \end{pmatrix} \tag{11.10}$$

The hierarchical structural constraint on the factor loadings matrix $\mathbf{B}$ has important implications for the interpretation of the factors. Specifically, the order of the variables in the observation vector $\mathbf{y}_i$ defines the factors. For example, the first variable is equal to the first factor plus a noise term, the second variable is equal to b_{21} times the first factor plus the second factor plus a noise term, etc. Thus, the order of the variables may influence the interpretation of the factors. In addition, the order of the variables may influence the choice of the number of factors k (e.g., see Aguilar and West 2000, and Lopes and West 2004). However, for a fixed number of factors k, the ordering has no influence either on the covariance matrix of $\mathbf{y}_i$ or on forecasts.

The order of the variables in the observation vector $\mathbf{y}_i$ is a modeling decision. Because this order influences the interpretation of the factors, we recommend if possible to order the variables based on prior knowledge of their relative importance and possible relationships. For example, consider a hypothetical three-factor model for the time series of gross domestic product (GDP) for several countries in the Americas, Asia, and Europe. Because the United States has by far the largest GDP in the world, we would choose as the first variable the GDP of the United States. In addition, because China has the second largest GDP we would use the Chinese GDP as the second variable. The choice for the third variable is not as straightforward. We note that even though Japan has the third largest GDP, its economy is highly connected to the economies of China and the United States. The third variable should represent an economic block not yet represented by China and the United States, which in this example would be Europe. Thus, for the third variable we would choose one of the European countries with the largest GDP such as for example Germany or Great Britain. Note that the ordering is not set in stone; an analyst at the Bank of Japan may find it useful to use the order United States, Japan, and Great Britain. We provide more discussion on the choice of ordering in Section 11.2.7 where we analyze the troposphere temperature data with a k-factor model.

Similarly to the one-factor model, for the k-factor model we use conditionally conjugate priors. These priors allow the probabilistic representation of many degrees of prior information. In addition, the use of conditionally conjugate priors allows for straightforward and efficient MCMC algorithms. As in the one-factor model, for the idiosyncratic variances $\sigma_1^2, \ldots, \sigma_j^2$ we assume independent priors $IG(n_\sigma/2, n_\sigma s_\sigma^2/2)$; and for the factor loadings variance τ^2 we assume the prior $IG(n_\tau/2, n_\tau s_\tau^2/2)$. In addition, for the k-factor model, the factor loadings b_{jl}, $l = 1, \ldots, k$, $j = l+1, \ldots, r$, are assumed *a priori* to be i.i.d. $N(0, \tau^2)$. Finally, the latent factor variances $h_1, \ldots, h_k$ are assumed to be i.i.d. $IG(n_H/2, n_H s_H^2/2)$.

11.2.5 MCMC for the k-Factor Model

We explore the posterior distribution for the k-factor model with a straightforward Gibbs sampler.

The latent factor vector $\mathbf{x}_i$, $i = 1, \ldots, n$, is simulated from its full conditional distribution

$$N\left(\left(\mathbf{H}^{-1} + \mathbf{B}'\mathbf{V}^{-1}\mathbf{B}\right)^{-1}\mathbf{B}'\mathbf{V}^{-1}\mathbf{y}_i, \left(\mathbf{H}^{-1} + \mathbf{B}'\mathbf{V}^{-1}\mathbf{B}\right)^{-1}\right). \tag{11.11}$$

For the factor loadings matrix $\mathbf{B}$, we consider one row at a time and simulate its unrestricted elements from their joint full conditional distribution. Let $\mathbf{b}_{j.} = (\mathbf{b}_{j1}, \ldots, \mathbf{b}_{jk})'$ be the vector that contains the j-th row of $\mathbf{B}$, $j = 1, \ldots, r$. Note the restrictions $b_{jj} = 1$ and $b_{jl} = 0$, $j = 1, \ldots, r$, $l > j$. Let $\mathbf{b}_{j.}^*$ be the vector of unrestricted elements of $\mathbf{b}_{j.}$. Thus, for $2 \leq j \leq k$, $\mathbf{b}_{j.}^* = (\mathbf{b}_{j1}, \ldots, \mathbf{b}_{j,j-1})'$. In addition, for $j = k+1, \ldots, r$, $\mathbf{b}_{j.}^* = \mathbf{b}_{j.}$. Hence, when simulating the unrestricted elements of each row of the factor loadings matrix $\mathbf{B}$, we have two types of full conditional distributions depending on whether or not the row index j is larger than the number of factors k.

Specifically, the full conditional distribution of $\mathbf{b}_{j.}^*$ is $N(\mathbf{m}_j, \mathbf{C}_j)$, where for $j = 2, \ldots, k$

$$\mathbf{C}_j = \left(\tau^{-2}\mathbf{I}_{j-1} + \sigma_j^{-2}\sum_{i=1}^{n}\mathbf{x}_{i,1:(j-1)}\mathbf{x}'_{i,1:(j-1)}\right)^{-1}, \quad (11.12)$$

$$\mathbf{m}_j = \sigma_j^{-2}\mathbf{C}_j\sum_{i=1}^{n}(y_{ij} - x_{ij})\mathbf{x}_{i,1:(j-1)}, \quad (11.13)$$

and for $j = k+1, \ldots, r$

$$\mathbf{C}_j = \left(\tau^{-2}\mathbf{I}_k + \sigma_j^{-2}\sum_{i=1}^{n}\mathbf{x}_i\mathbf{x}'_i\right)^{-1}, \quad (11.14)$$

$$\mathbf{m}_j = \sigma_j^{-2}\mathbf{C}_j\sum_{i=1}^{n}y_{ij}\mathbf{x}_i. \quad (11.15)$$

For the idiosyncratic variance σ_j^2, $j = 1, \ldots, r$, the full conditional distribution is

$$IG\left(\frac{n_\sigma + n}{2}, \frac{n_\sigma s_\sigma^2 + \sum_{i=1}^{n}(y_{ij} - \mathbf{b}'_{j.}\mathbf{x}_i)^2}{2}\right). \quad (11.16)$$

For the factor loadings variance τ^2, the full conditional distribution is

$$IG\left(\frac{2n_\tau + k(2r - k - 1)}{4}, \frac{n_\tau s_\tau^2 + \sum_{l=1}^{k}\sum_{j=l+1}^{r} b_{jl}^2}{2}\right). \quad (11.17)$$

For the latent factor variance h_l, $l = 1, \ldots, k$, the full conditional distribution is

$$IG\left(\frac{n_H + n}{2}, \frac{n_H s_H^2 + \sum_{i=1}^{n} x_{il}^2}{2}\right). \quad (11.18)$$

Finally, we note that for the case when the number of factors is equal to one, all the above full conditional distributions specialize to the full conditional distributions presented in Section 11.2.2.

11.2.6 Selection of Number of Factors

A crucial step in factor analysis is the selection of the number of factors.

When using the k-factor model described in Section 11.2.4, there is an upper bound for the number of factors. This upper bound results from the fact that a covariance matrix for a q-dimensional random vector has at most $q(1 + q)/2$ free parameters. Moreover, note that the covariance matrix implied by the k-factor model in Equation (11.8) is equal to $\mathbf{BHB}'+$

$\mathbf{V}$. With the hierarchical structural constraint on $\mathbf{B}$, the latter covariance matrix has number of free parameters equal to $qk + k + q - k(1+k)/2$, which should be less or equal to $q(1+q)/2$. As a consequence, the number of factors k has to satisfy the inequality $k^2 - (2q+1)k + q(q-1) \geq 0$, which implies $k \leq q + 0.5 - 0.5\sqrt{1+8q}$ (Aguilar and West 2000).

Lopes and West (2004) performed an extensive simulation study for factor models to assess the performance of several model selection criteria to select the number of factors. These criteria included the AIC and the BIC, as well as several approximations to the marginal data density. While the AIC tends to choose more factors than necessary, Lopes (2003) and Lopes and West (2004) have indicated that a Laplace-Metropolis approximation to the marginal data density is reliable and easy to implement.

The Laplace-Metropolis approximation can be computed in a straightforward way from the MCMC output. Specifically for the k-factor model, the algorithm presented in Section 11.2.5 provides an MCMC sample from the posterior distribution of the vector $\boldsymbol{\theta}_k$ that collects all the unknown parameters in $(\mathbf{B}, \mathbf{H}, \mathbf{V}, \tau^2)$. Let d be dimension of $\boldsymbol{\theta}_k$, $\hat{\boldsymbol{\theta}}_k$ be the MCMC draw that maximizes $p(y|k, \boldsymbol{\theta}_k)p(\boldsymbol{\theta}_k|k)$, and $\boldsymbol{\Psi}$ be the posterior covariance matrix of $\boldsymbol{\theta}_k$ computed from the MCMC output. Then, the Metropolis-Laplace estimator of the marginal data density is (Lewis and Raftery 1997)

$$(2\pi)^{d/2}|\boldsymbol{\Psi}|^{1/2}p(y|k, \hat{\boldsymbol{\theta}}_k)p(\hat{\boldsymbol{\theta}}_k|k),$$

where $p(\mathbf{y}|k, \hat{\boldsymbol{\theta}}_k)$ is the integrated likelihood function obtained by integrating out the latent factor vectors $\mathbf{x}_1, \ldots, \mathbf{x}_n$. This integrated likelihood function is given by (Lopes and West 2004)

$$p(\mathbf{y}|k, \hat{\boldsymbol{\theta}}_k) = (2\pi)^{-nr/2}|\mathbf{BHB}' + \mathbf{V}|^{-n/2} \exp\left\{-\frac{1}{2}\sum_{i=1}^{n} \mathbf{y}_i'(\mathbf{BHB}' + \mathbf{V})^{-1}\mathbf{y}_i\right\}.$$

For implementation, we note that the Metropolis-Laplace approximation is based on a Gaussian approximation. Thus, for the approximation to work more properly, we transform the parameters that correspond to variances using a logarithm transformation before applying the Metropolis-Laplace approximation.

11.2.7 Example: A k-Factor Model for Temperature

Here we consider the troposphere air temperature data previously analyzed with a one-factor model in Section 11.2.3 and reanalyze it with k-factor models.

First, we choose the order of the variables. From Figure 11.4(b), we note that in the one-factor model the variable with the largest estimated factor loading is the standardized temperature at altitude pressure level of 600 millibars. In addition, the temperature at that altitude seems to behave similarly to temperatures at other troposphere mid-altitudes. Thus, we assign the first position in the observation vector $\mathbf{y}_i$ to the standardized temperature at altitude pressure level of 600 millibars. We assign the second position to the standardized temperature at altitude pressure level of 1000 millibars that seems to behave similarly to temperatures at other low troposphere altitudes. And we assign the third position to the standardized temperature at altitude pressure level of 250 millibars, which is close to the tropopause and seems to behave differently from the temperatures at lower altitudes. The order for the remaining variables is the original order from lower to upper altitudes.

Table 11.1 *Troposphere temperature. Logarithm of marginal data density for factor models with distinct number of factors.*

	Number of factors				
	1	2	3	4	5
Log marginal density	89.7	319.2	326.3	310.0	300.6

A fundamental question is how many factors to include in the model. Note that for $q = 9$ variables, the upper bound on the number of factors is $q+0.5-0.5\sqrt{1+8q} = 5.23$. Thus, in this application the number of factors is less or equal to 5. Table 11.1 presents the logarithm of the marginal data density computed using the Metropolis-Laplace estimator described in Section 11.2.6 for models with number of factors 1 to 5. Note that in the logarithm scale a difference of 7 is extremely large. Therefore, the 3-factor model is by far the best.

Further information on the impact of the number of factors in the explanatory ability of the k-factor model is provided by Figure 11.5. This figure presents posterior means and 95% credible intervals for the idiosyncratic variances for the temperatures at each altitude pressure level based on factor models with 1, 2, and 3 factors. A model with one factor captures mostly the variation in the temperature at the mid-troposphere. Adding a second factor helps to capture the temperature variation in the lower troposphere. Finally, adding the third factor captures much of the variation in the upper troposphere. Henceforth, we present an analysis for the factor model with 3 factors.

Figure 11.6 presents the factor loadings for the 3 factors. The first factor in the 3-factor model represents a weighted average of the troposphere temperatures and resembles the only factor in the 1-factor model discussed in

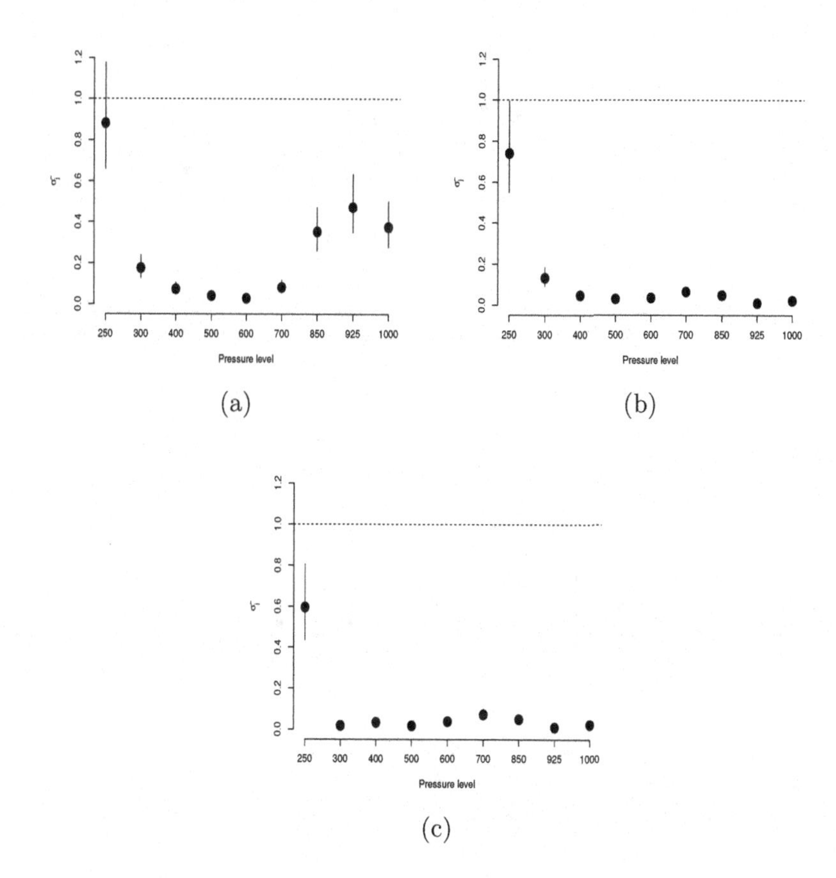

Figure 11.5 *Troposphere temperature – k-factor models. Posterior means (circles) and 95% credible intervals (vertical lines) for idiosyncratic variances based on factor models with (a) 1, (b) 2, and (c) 3 factors.*

Section 11.2.3. The main difference is that the factor loading for temperature at altitude pressure level 600 now is equal to 1, and accordingly the other factor loadings have been shifted downward. The loadings of the first factor are the largest for altitude pressure levels 400 to 700, being estimated at about 1, confirming the importance of the first factor for temperatures at mid-troposphere. The first-factor loadings for altitude pressure levels 850 to 1000 are somehow lower, at about 0.8. And the first-factor loading for pressure level 250 is much lower, being estimated at about 0.375, making it clear that the first latent factor is less important for temperatures in the high troposphere. The second factor represents a contrast between the

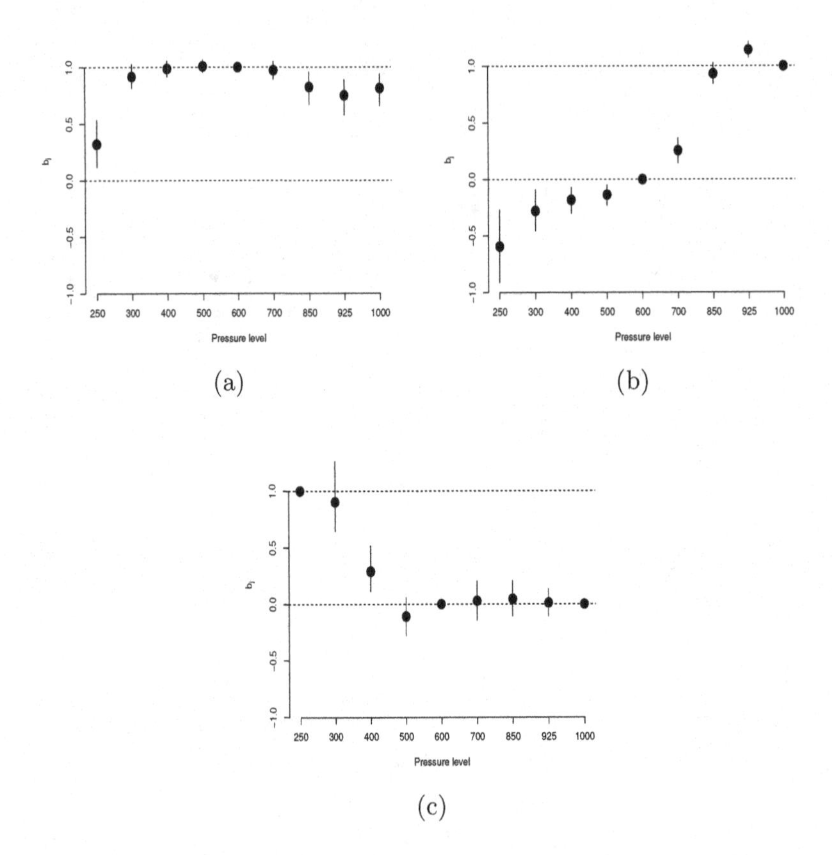

Figure 11.6 *Troposphere temperature – 3-factor model. Posterior means (circles) and 95% credible intervals (vertical lines) of factor loadings for (a) first factor, (b) second factor, and (c) third factor.*

temperatures in the low troposphere and those in the upper troposphere. Finally, the third factor represents temperatures in the upper troposphere.

Figure 11.7 presents time series plots of common factors posterior means (solid line) and 95% credible intervals. Three features emerge from these plots. First, the first factor has the largest variability, followed by the second and the third factors. Second, the 95% credible intervals are fairly tight for the first and second factors suggesting high precision in their estimation. Further, the third factor credible intervals are wider, indicating less precision in the estimation of the third factor. The third feature is that all common factors seem to have temporal dependence. Thus, by including a dynamic evolution for the common factors using for example AR or TVAR

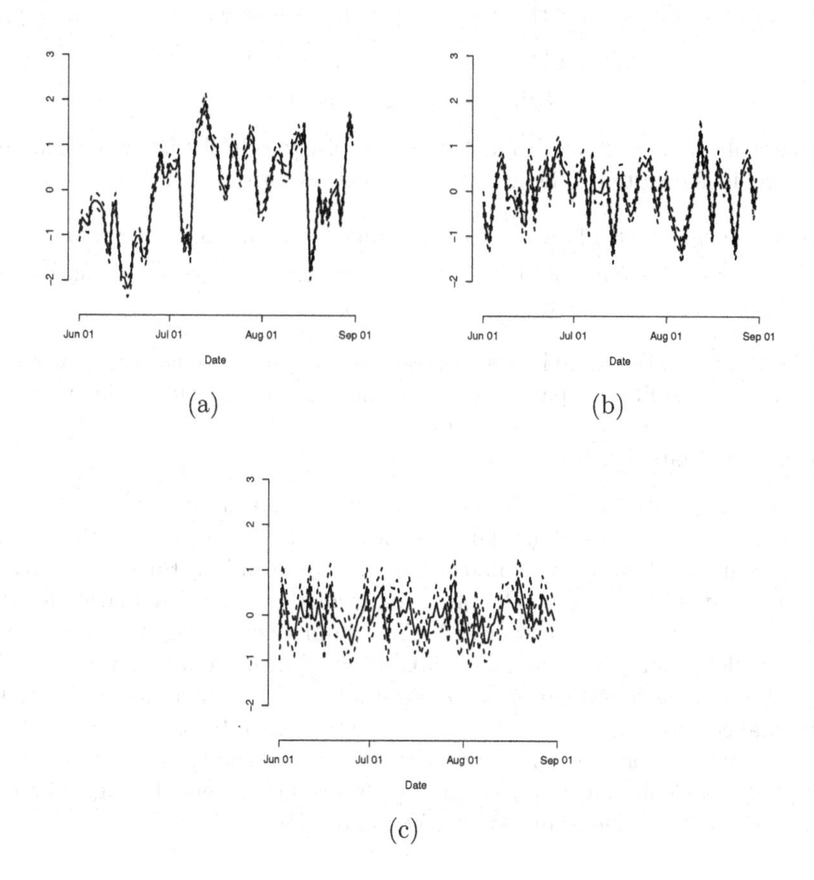

Figure 11.7 *Troposphere temperature – 3-factor model. Time series plots of common factors posterior means (solid line) and 95% credible intervals for (a) first factor, (b) second factor, and (c) third factor.*

models, we can gain estimation efficiency by borrowing strength across time, as well as we can increase the ability to predict future observations. This motivates the subject of multivariate dynamic latent factor models that we present in Section 11.3.

11.3 Multivariate Dynamic Latent Factor Models

In a multivariate DLM context, one class of dynamic factor models adds latent structure to the basic multivariate DLM of Equation (10.1) in Section 10.1.1. That is, for the $r \times 1$ vector time series with observations $\mathbf{y}_t$,

$$\begin{aligned}\mathbf{y}_t &= \mathbf{F}_t'\boldsymbol{\theta}_t + \mathbf{B}_t\mathbf{x}_t + \boldsymbol{\nu}_t, \quad \boldsymbol{\nu}_t \sim N(\mathbf{0}, \mathbf{V}_t),\\ \boldsymbol{\theta}_t &= \mathbf{G}_t\boldsymbol{\theta}_{t-1} + \boldsymbol{\omega}_t, \quad \boldsymbol{\omega}_t \sim N(\mathbf{0}, \mathbf{W}_t),\end{aligned} \tag{11.19}$$

where all elements apart from $\mathbf{B}_t\mathbf{x}_t$ are precisely as in the DLMs considered in previous chapters. The additional terms comprise:

- $\mathbf{x}_t = (x_{t,1} \dots, x_{t,k})'$, a k-vector dynamic latent factor process, and
- the $r \times k$ dynamic factor loadings matrix $\mathbf{B}_t$, providing regression coefficients that map factors to observations.

The central notion here is that, beyond the potential to predict $\mathbf{y}_t$ as modeled in $\boldsymbol{\theta}_t$, additional patterns of variation and correlation among the elements of $\mathbf{y}_t$ may be explained by a small number ($k < r$ and, typically, $k << r$) of latent factor processes.

Under assumptions of a normal distribution, or a normal Markov evolution model — a subsidiary DLM — for $\mathbf{x}_t$, the model is a conditionally Gaussian DLM when $\mathbf{B}_t$ is known, and so the standard Bayesian analysis methods can be exploited. Coupling conditional posterior simulation for the $\mathbf{B}_t$ matrices with this in overall MCMC approaches yields the framework for model fitting. Now standard MCMC approaches exist and are widely used in models for which $\mathbf{B}_t = \mathbf{B}$, constant for all t, including applications in finance and other areas; this builds on advances in Bayesian computation for standard, nontime series factor models and structured prior modeling to impose identification constraints in latent factor models (e.g., Geweke and Zhou 1996, Lopes and West 2004, and references therein).

11.3.1 Example: A Dynamic 3-Factor Model for Temperature

Here we consider again the working example of troposphere temperature. An analysis of the sample autocorrelation functions of the estimated common factors depicted in Figure 11.7 for the 3-factor model indicates that each of the three common factors may be better modeled with autoregressive processes. Specifically, the exploratory analysis suggests autoregressive processes of order 2 for the first two common factors and an autoregressive process of order 1 for the third common factor. Thus, we consider the following special case of the dynamic factor model in Equation (11.19) for troposphere temperature

$$\begin{aligned}\mathbf{y}_t &= \mathbf{B}\mathbf{x}_t + \boldsymbol{\nu}_t, \quad \boldsymbol{\nu}_t \sim N(\mathbf{0}, \mathbf{V}),\\ \mathbf{x}_t &= \mathbf{F}'\boldsymbol{\theta}_t,\\ \boldsymbol{\theta}_t &= \mathbf{G}\boldsymbol{\theta}_{t-1} + \boldsymbol{\omega}_t, \quad \boldsymbol{\omega}_t \sim N(\mathbf{0}, \mathbf{W}),\end{aligned} \tag{11.20}$$

where $\mathbf{x}_t$ is a k-dimensional vector of common factors with l-th element x_{tl} following an AR(p_l) process. As in Section 11.2.4 we assume that $\mathbf{V} = \text{diag}(\sigma_1^2, \ldots, \sigma_r^2)$, where $\sigma_1^2, \ldots, \sigma_r^2$ are idiosyncratic variances.

The matrices $\mathbf{F}$, $\mathbf{G}$, and $\mathbf{W}$ in Equation (11.20) are block-diagonal matrices with the l-th blocks representing in state-space form the AR(p_l) process for x_{tl}. Specifically, $\mathbf{F}' = \text{blockdiag}(\mathbf{F}_1', \ldots, \mathbf{F}_k')$ is a k by $(\sum_{l=1}^k p_l)$ blockdiagonal matrix where $\mathbf{F}_l' = (1, \mathbf{0}_{p_l-1}')$ is a p_l-dimensional row vector, $l = 1, \ldots, k$. In addition, the evolution matrix is of the form $\mathbf{G} = \text{blockdiag}(\mathbf{G}_1, \ldots, \mathbf{G}_k)$ where $\mathbf{G}_l$ is a p_l by p_l matrix such that

$$\mathbf{G}_l = \begin{pmatrix} \phi_{l1} & \phi_{l2} & \ldots & \phi_{l,p_l-1} & \phi_{lp_l} \\ 1 & 0 & \ldots & 0 & 0 \\ 0 & 1 & \ldots & 0 & 0 \\ \vdots & & \ddots & & \vdots \\ 0 & 0 & \ldots & 1 & 0 \end{pmatrix}.$$

Finally, the evolution covariance matrix $\mathbf{W}$ is a $(\sum_{l=1}^k p_l)$ by $(\sum_{l=1}^k p_l)$ blockdiagonal covariance matrix of the form $\mathbf{W} = \text{blockdiag}(\mathbf{W}_1, \ldots, \mathbf{W}_k)$ where $\mathbf{W}_l$ is a p_l by p_l matrix having entries all zero except for the first element $(\mathbf{W})_{1,1} = w_l > 0$, the variance of the AR innovations for the l-th common factor.

We explore the posterior distribution for the dynamic k-factor model using a Gibbs sampler. Specifically, simulations of b_{jl}, τ^2, and σ_j^2 are performed analogously to the static k-factor model. Simulation of $\boldsymbol{\theta}_{1:T}$ is performed using the efficient forward filter backward sampler (Carter and Kohn 1994; Frühwirth-Schnatter 1994). Finally, simulations of ϕ_l and w_l are performed as in the AR(p_l) model.

Figure 11.8 presents posterior means and 95% credible intervals for the factor loadings for the dynamic 3-factor model. The factor loadings of the first two factors are indistinguishable from those of the static 3-factor model (Figure 11.6). Specifically, the first factor represents a weighted average of the troposphere temperatures, and the second factor represents a contrast between the temperatures in the low troposphere and those in the upper troposphere. However, the factor loadings for the third factor in the dynamic model are fairly different from those in the static model. While in both models the third factor represents a weighted average of upper troposphere temperatures, the weights differ quite a bit. Note that, by construction, in both models the loading for temperature at altitude pressure level 250 millibars is equal to one. In addition, in the static model the third factor has a posterior mean for the loading of the temperature at 300 millibars that is very close to one (the credible interval contains one), and a smaller loading for temperature at 400 millibars. In contrast, in the

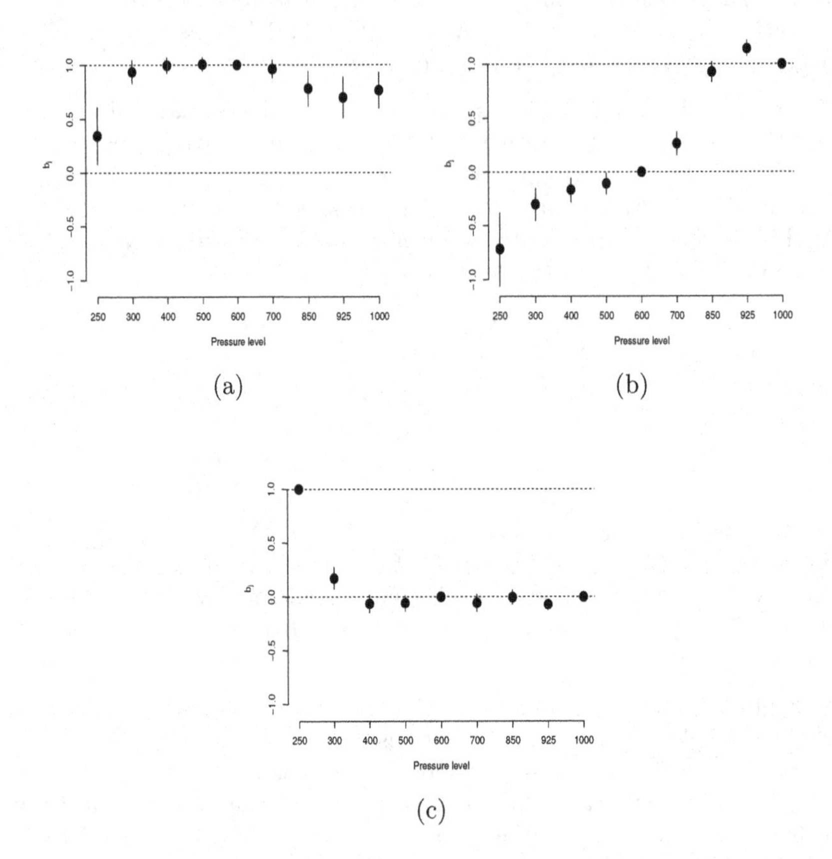

Figure 11.8 *Troposphere temperature – dynamic 3-factor model. Posterior means (circles) and 95% credible intervals (vertical lines) of factor loadings for (a) first factor, (b) second factor, and (c) third factor.*

dynamic model the loading for temperature at 300 millibars is significantly smaller than one, and the loading for temperature at 400 millibars is not significantly different than zero. Therefore, in the dynamic 3-factor model the third factor represents to a great extent the temperature at altitude pressure level 250 millibars and to a lesser extent the temperature at 300 millibars.

Figure 11.9 presents time series plots of the three common factors' posterior means (solid line) and 95% credible intervals (dashed lines). The first two common factors are indistinguishable from those of the static 3-factor model (Figure 11.7). However, the third factor in the dynamic model is very different from that in the static model. One of the main differences is that,

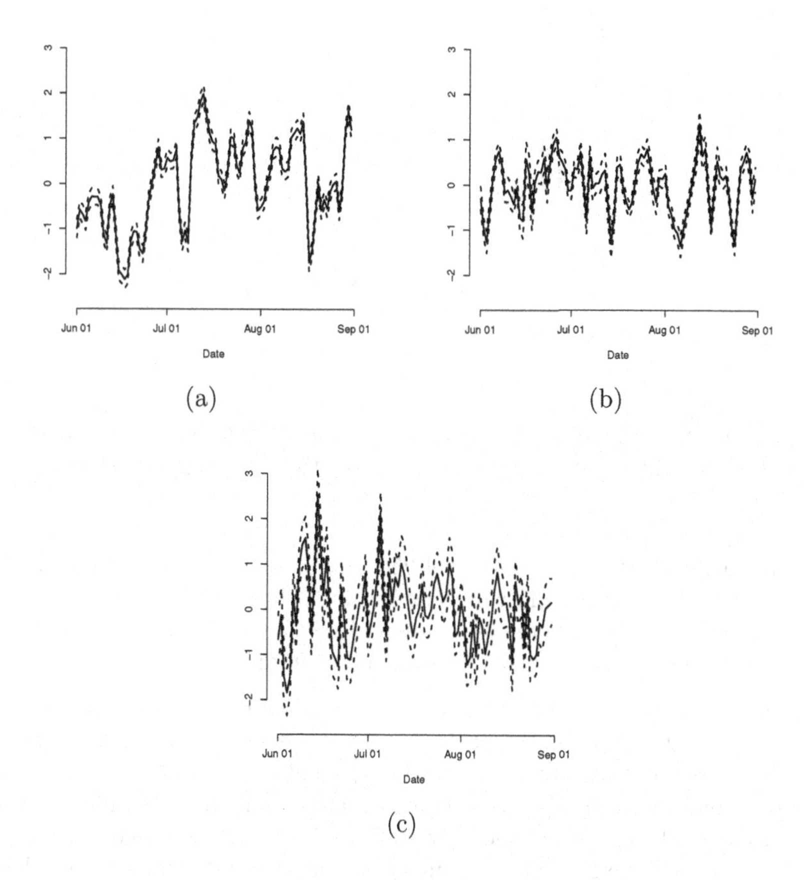

Figure 11.9 *Troposphere temperature – dynamic 3-factor model. Time series plots of common factors posterior means (solid line) and 95% credible intervals for (a) first factor, (b) second factor, and (c) third factor.*

when compared to the static model, the third factor in the dynamic model has a much stronger temporal dependence. Specifically, fitting a first-order autoregressive model to the estimated static third factor yields an estimated autoregressive coefficient of 0.28. In contrast, the posterior mean of the autoregressive coefficient for the third factor in the dynamic model is equal to 0.49. Finally, the other main difference is that the third factor in the dynamic model has a much larger marginal variance.

Figure 11.10 presents the idiosyncratic variances based on the dynamic 3-factor model for the temperatures at different altitude pressure levels. When compared to the idiosyncratic variances presented in Figure 11.5, there is an incredibly large decrease in the idiosyncratic variance corre-

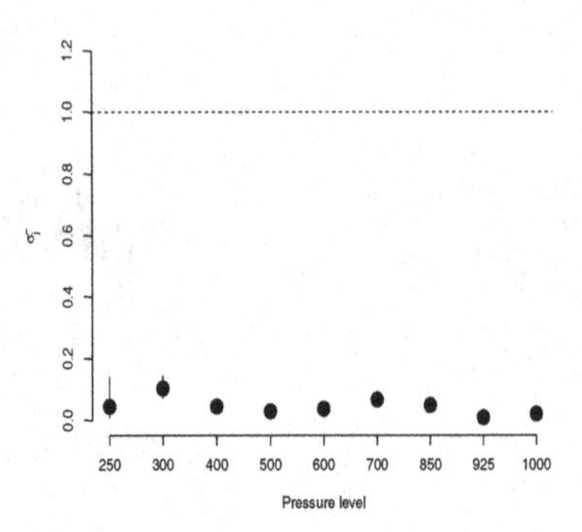

Figure 11.10 *Troposphere temperature – dynamic 3-factor model. Posterior means (circles) and 95% credible intervals (vertical lines) of idiosyncratic variances.*

sponding to altitude pressure level 250 millibars. Also worth notice is the fact that the idiosyncratic variances for the other altitude pressure levels remain fairly small. Therefore, compared to the static factor models, the dynamic 3-factor model provides substantial improvement.

Figure 11.11 presents the posterior densities of the autoregressive coefficients for the dynamic common factors. There are several facts worth notice. First, the five autoregressive coefficients are significantly far from zero, reinforcing the need for common factors that are dynamic. Second, the roots of the characteristic polynomial for the first and second common factors are complex conjugates, implying that these common factors have stochastic cycles. Specifically, the posterior means for the autoregressive coefficients are $\widehat{\phi}_{11} = 1.13$, $\widehat{\phi}_{12} = -0.36$, $\widehat{\phi}_{21} = 0.75$, $\widehat{\phi}_{22} = -0.26$, and $\widehat{\phi}_{31} = 0.49$. Thus, for the first common factor, the estimated period is 19.3 days and the estimated modulus of reciprocal roots is 0.59. For the second common factor, the estimated period is 8.4 days with estimated modulus of reciprocal roots equal to 0.51. Hence, the first common factor has a period more than twice as long as that of the second factor. However, the estimated moduli of the reciprocal roots for the three dynamic common factors are all less than 0.60. Therefore, the forecast functions for each of these three common factors decay to zero fairly fast.

An important advantage of the dynamic factor model when compared to the static factor model is the ability to yield meaningful temporal predictions. These temporal predictions may be either for the common factors or for the several original time series. For example, panels (a), (b) and (c) of Figure 11.12 present, in addition to point and interval estimates for the

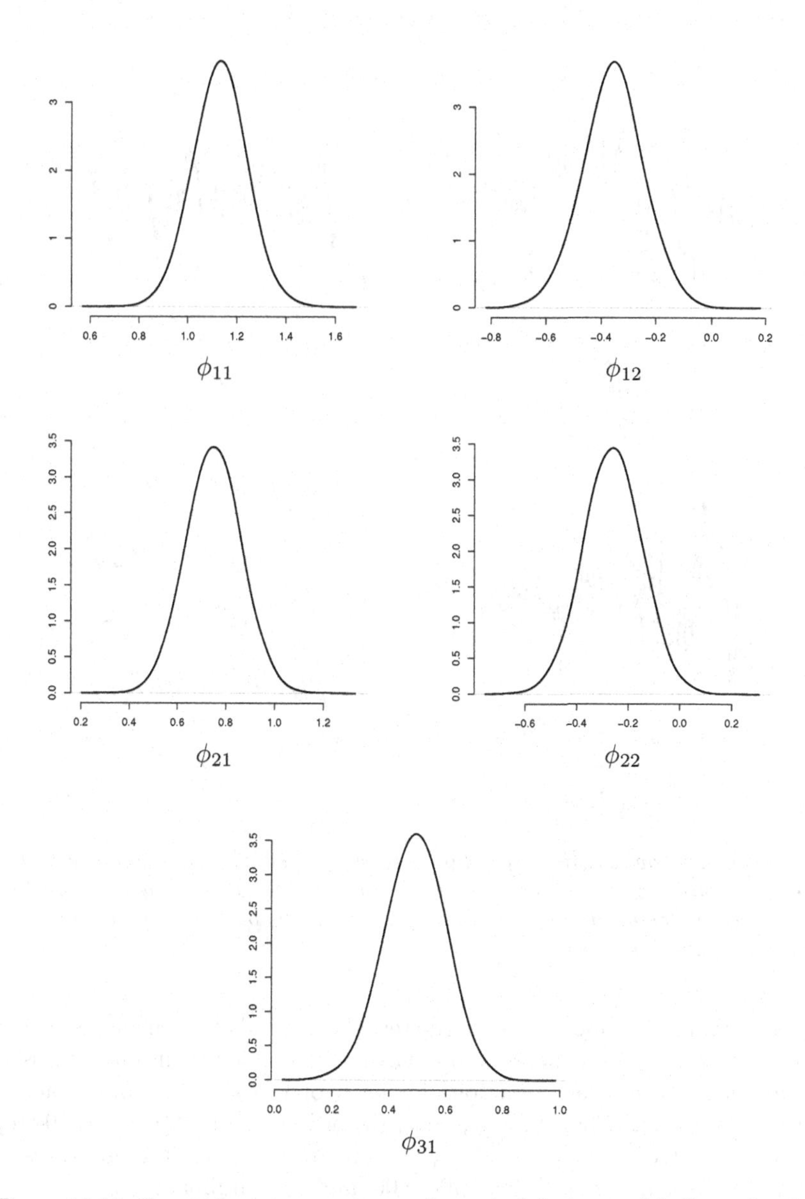

Figure 11.11 *Troposphere temperature – dynamic 3-factor model. Posterior densities of autoregressive coefficients for common factors.*

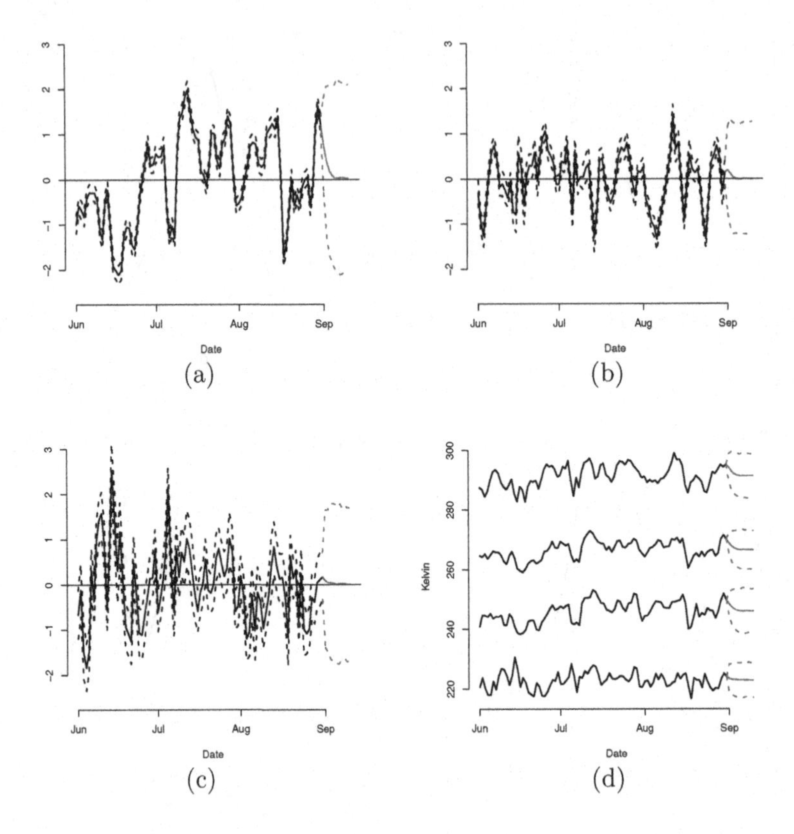

Figure 11.12 *Troposphere temperature – 10-step ahead predictive means and 95% predictive intervals for common factors 1 (panel a), 2 (panel b), and 3 (panel c), as well as for temperatures at altitudes pressure levels of 250, 400, 600 and 1000 millibars (panel d).*

three dynamic common factors, 10-step ahead predictive means and 95% predictive intervals. As noted above when we discussed their autoregressive coefficents, the forecast functions for each of the three common factors decay to zero fairly fast. Finally, panel (d) of Figure 11.12 presents 10-step ahead predictive means and 95% predictive intervals for temperatures at altitudes pressure levels of 250, 400, 600, and 1000 milibars.

An important quantity to consider when comparing models is the proportion of unexplained variance. Because the variables are standardized, the proportion of unexplained variance is defined as $\sum_{i=1}^{r} \sigma_i^2/r$. As this is a function of the idiosyncratic variances, we can easily obtain a posterior

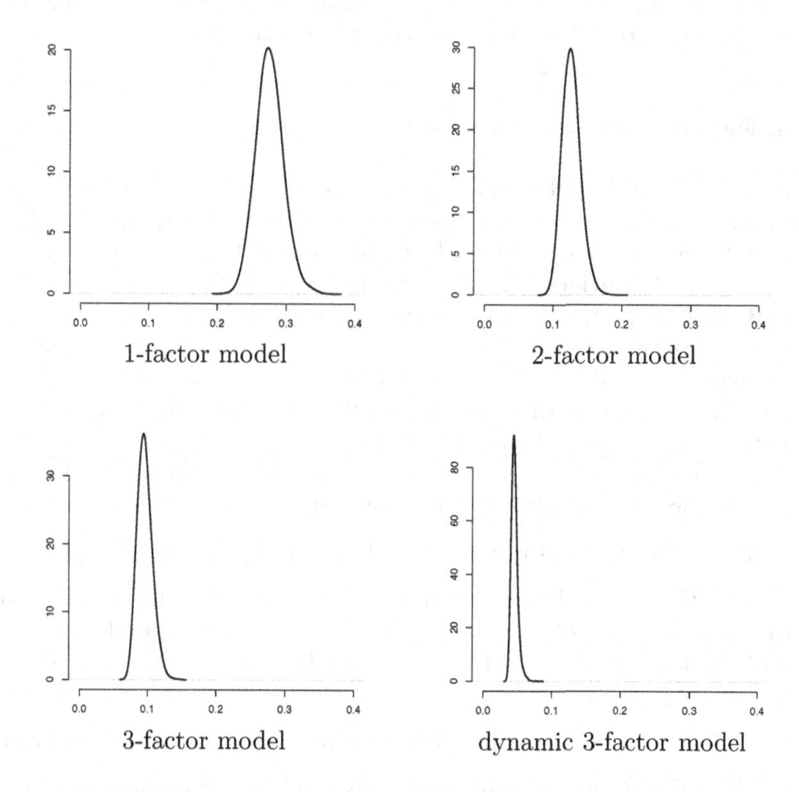

Figure 11.13 *Troposphere temperature – posterior densities for the proportion of unexplained variance under different factor models.*

sample for the proportion of unexplained variance. Posterior means for the proportion of unexplained variance for the static factor models with 1, 2, and 3 factors are equal to 0.275, 0.126, and 0.096, respectively; for the dynamic 3-factor model, the posterior mean is 0.045. In addition, Figure 11.13 presents posterior densities of the proportion of unexplained variance by static 1-, 2-, and 3-factor models, as well as by the dynamic 3-factor model. It is clear that as the number of factors increases from 1 to 3 in the static factor models, the proportion of unexplained variance decreases substantially. Posterior histograms for static factor models with 4 or more factors (not shown) do not exhibit substantial reduction in the proportion of unexplained variance, which is in agreement with the posterior probabilities of the several static factor models. However, including dynamics in the common factors in a dynamic 3-factor model leads to substantial reduction in the proportion of unexplained variance. Therefore, for the troposphere temperature data, an analysis of the proportion of unexplained variance clearly indicates the need for the dynamic factor model.

11.4 Factor Stochastic Volatility

A key area in which multivariate dynamic latent factor models have flourished is in volatility modeling, as noted above. The developments in Aguilar, Huerta, Prado, and West (1999), Pitt and Shephard (1999b), and Aguilar and West (2000) provided MCMC-based model fitting approaches in models with $\mathbf{B}_t = \mathbf{B}$, fixed for all t, and in which the $\mathbf{x}_t$ represent zero-mean factors whose covariance matrices are time-varying and induce overall volatility patterns in $\mathbf{y}_t$. If volatility is heavily determined by a few underlying factor processes, and if this underlying volatility can be modeled, there is potential for improved adaptability in a sequential analysis, and therefore potential to improve decisions such as arise in portfolio allocation. Take $\mathbf{x}_t$ to be conditionally independent over time with

$$\mathbf{x}_t \sim N(\mathbf{0}, \mathbf{H}_t) \quad \text{and where} \quad \mathbf{H}_t = \text{diag}\{\exp(\lambda_{t1}), \ldots, \exp(\lambda_{tk})\};$$

here λ_{tl} is the log of the instantaneous variance of the l-th latent factor at time t. Write $\boldsymbol{\lambda}_t = (\lambda_{t1}, \ldots, \lambda_{tk})'$. The above references developed analyses of what are now standard models in which $\boldsymbol{\lambda}_t$ is a VAR(1) (vector autoregressive) process,

$$\boldsymbol{\lambda}_t = \boldsymbol{\mu} + \boldsymbol{\Phi}(\boldsymbol{\lambda}_{t-1} - \boldsymbol{\mu}) + \boldsymbol{\xi}_t \qquad (11.21)$$

driven by independent innovations $\boldsymbol{\xi}_t \sim N(\mathbf{0}, \mathbf{U})$ for some covariance matrix $\mathbf{U}$.

Usually, the AR coefficient matrix $\boldsymbol{\Phi}$ is assumed to be diagonal, that is, $\boldsymbol{\Phi} = \text{diag}(\boldsymbol{\phi})$ where $\boldsymbol{\phi} = (\phi_1, \ldots, \phi_k)'$ is the vector of autoregressive coefficients for the log of the instantaneous variances for the latent factors. Further, we note that it is usually assumed that $0 < \phi_l < 1$, $l = 1, \ldots, k$, which implies stationarity of the log of the instantaneous variance of each latent factor. Denote by $\mathbf{W}$ the marginal covariance matrix of $\boldsymbol{\lambda}_t$ implied by Equation (11.21). Thus, because of stationarity, $\mathbf{W}$ is the solution to $\mathbf{W} = \boldsymbol{\Phi}\mathbf{W}\boldsymbol{\Phi} + \mathbf{U}$. Straightforward matrix algebra shows that $\mathbf{W} = \mathbf{U} + \sum_{i=1}^{\infty} \boldsymbol{\Phi}^i \mathbf{U} \boldsymbol{\Phi}^i$. Further, denote by $\oslash$ element-wise division. Then, when $\boldsymbol{\Phi}$ is a diagonal matrix, we can apply standard results for geometric series to obtain $\mathbf{W} = \mathbf{U} \oslash (\mathbf{1}_k \mathbf{1}_k' - \boldsymbol{\phi}\boldsymbol{\phi}')$.

In addition, for the observational errors ν_t, assume a diagonal covariance matrix $\mathbf{V}_t = \text{diag}(v_{t1}, \ldots, v_{tr})$ where the idiosyncratic variances $v_{t1}, \ldots, v_{tr}$ follow independent first-order stochastic volatility models. That is, let $\eta_{ti} = \log(v_{ti})$ follow the first-order autoregressive model

$$\eta_{ti} = \alpha_i + \rho_i(\eta_{t-1,i} - \alpha_i) + \zeta_{ti}, \quad \zeta_{ti} \sim N(0, s_i), \qquad (11.22)$$

where $0 < \rho_i < 1$ and $\eta_{t1}, \ldots, \eta_{tr}$ are mutually independent.

This is one class of extensions of the standard univariate stochastic volatility model of Section 7.5, and many variants are clearly possible. Aguilar, Huerta, Prado, and West (1999) and Aguilar and West (2000) explicitly admitted contemporaneous dependencies between the innovations impacting volatilities by using and estimating a nondiagonal matrix $\mathbf{U}$, finding that to be strongly supported in analyses of financial time series data, including exchange rate series, and to lead to models with improved performance in short term portfolio optimization analyses. The application we consider in Section 11.4.2 also provides evidence of the need of a nondiagonal covariance matrix $\mathbf{U}$. These analyses also allow for volatility modeling in the observation error terms, and can be extended to integrate time-varying covariances using discount models. Evidently, fitting and use of these more highly structured models of multivariate volatility require highly customized computational methods as well as care in prior specification; the above references discuss both of these aspects in detail, and lay out explicit, algorithmic development of custom MCMC methods. We discuss these computational methods in Section 11.4.1 and illustrate their application with an analysis of exchange rate series in Section 11.4.2

Excellent and more recent discussion, and review of later developments, can be found in select chapters of Shephard (2005), in Chib, Nadari, and Shephard (2005), and in the review paper of Chib, Omori, and Asai (2009). Further extensions of the overall approach to model and incorporate time variation in the factor loadings matrices $\mathbf{B}_t$, a challenging but practically most important next stage of development of these models, have been introduced by Lopes (2007) and Lopes and Carvalho (2007), the latter incorporating stochastic jumps via Markov switching regime models.

11.4.1 Computations

Exploration of the posterior distribution for the factor stochastic volatility model may be accomplished with Markov chain Monte Carlo algorithms.

Simulation of parameters associated with latent factors are analogous to the simulation of the parameters of the k-factor model discussed in Section 11.2.4. These include the simulation of the matrix of factor loadings $\mathbf{B}$, the latent common factors $\mathbf{x}_t$, and the variance of factor loadings τ^2.

Simulation of parameters associated with the idiosyncratic log-volatilities are analogous to the simulation of the parameters of univariate stochastic volatility models discussed in Section 7.5. These include the simulation of the idiosyncratic log-volatilities η_{tj}, the autoregressive coefficients $\rho_1, \ldots, \rho_r$, the means $\alpha_1, \ldots, \alpha_r$ of the idiosyncratic log-volatilities, and the variances $s_1, \ldots, s_r$ of the innovations. In addition, computations usually rely on the approximation described in Section 7.5 using a mixture of normals with corresponding indicator variables γ_{tj}^{η}.

Simulation of parameters associated with the log-volatilities for the common factors are analogous to the simulation of parameters of multivariate stochastic volatility models as proposed in Aguilar and West (2000). These include the simulation of the log-volatilities of the common latent factors λ_{tl}, the autoregressive coefficients $\phi_1, \ldots, \phi_k$, the mean vector of log-volatilities $\boldsymbol{\mu}$, the covariance matrix of innovations $\mathbf{U}$, and the indicator variables for the mixture of normals γ_{tl}^{λ}. Because these computations do not appear in previous chapters, here we describe the simulations of parameters associated with the log-volatilities for the common factors.

Computations for the parameters associated with the log-volatilities for the common factors are based on a normal mixture approximation. Specifically, let $x_{tl}^* = \log(x_{tl}^2)$. Then $x_{tl}^* = \lambda_{tl} + \nu_{tl}^*$ with $\nu_{tl}^* = \log \kappa_{tl}$ and $\kappa_{tl} \sim \chi_1^2$. The distribution of ν_{tl}^* can be well approximated with a mixture of normals as proposed by Kim, Shephard, and Chib (1998) and described in Section 7.5. Specifically, $\nu_{tl} \overset{a}{\sim} \sum_{j=1}^{J} q_j N(b_j^*, w_j^*)$ where the values of b_j^* and w_j^* are given in Section 7.5. We then introduce indicator variables $\gamma_{tl}^{\lambda} \in \{1, \ldots, J\}$ such that $\nu_{tl} | \gamma_{tl}^{\lambda} = j \sim N(b_j^*, w_j^*)$ and $P(\gamma_{tl}^{\lambda} = j) = q_j$ a priori.

Within the MCMC algorithm, the indicator variables γ_{tl}^{λ} are simulated from their discrete full conditional distribution with

$$P(\gamma_{tl}^{\lambda} = j | x_{tl}, \lambda_{tl}) \propto q_j w_j^{*-0.5} \exp\left\{ -\frac{1}{2w_j^*}(x_{tl}^* - b_j^* - \lambda_{tl})^2 \right\}.$$

The vector of autoregressive coefficients $\boldsymbol{\phi} = (\phi_1, \ldots, \phi_k)'$ has a full conditional density proportional to $c(\boldsymbol{\phi}) N(\boldsymbol{\phi} | \mathbf{b}, \mathbf{B})$, with $0 < \phi_1 < 1, \ldots, 0 < \phi_k < 1$, where $c(\boldsymbol{\phi}) = |\mathbf{W}|^{-0.5} \exp\{-0.5 tr(\mathbf{W}^{-1} \boldsymbol{\lambda}_1^* \boldsymbol{\lambda}_1^{*\prime})\}$, $\mathbf{B}^{-1} = \mathbf{C}_{\phi}^{-1} + \sum_{t=2}^{T} \mathbf{E}_{t-1}' \mathbf{U}^{-1} \mathbf{E}_{t-1}$ and $\mathbf{b} = \mathbf{B}\left(\mathbf{C}_{\phi}^{-1} \mathbf{m}_{\phi} + \sum_{t=2}^{T} \mathbf{E}_{t-1}' \mathbf{U}^{-1} \boldsymbol{\lambda}_t^*\right)$, $\boldsymbol{\lambda}_t^* = \boldsymbol{\lambda}_t - \boldsymbol{\mu}$, and $\mathbf{E}_t = \mathrm{diag}(\boldsymbol{\lambda}_t^*)$. An effective way to simulate $\boldsymbol{\phi}$ is with a Metropolis-Hastings step that simulates a proposal $\boldsymbol{\phi}^*$ from the normal distribution $N(\mathbf{b}, \mathbf{B})$ with the constraints $0 < \phi_1 < 1, \ldots, 0 < \phi_k < 1$ and accepts the proposal with probability equal to $\min(1, c(\boldsymbol{\phi}^*)/c(\boldsymbol{\phi}^{(current)}))$.

The mean vector of log-volatilities $\boldsymbol{\mu}$ is simulated with a Gibbs step from its full conditional distribution $N(\mathbf{m}_{\mu}^*, \mathbf{C}_{\mu}^*)$, where the covariance matrix is $\mathbf{C}_{\mu}^* = \left\{\mathbf{C}_{\mu} + \mathbf{W}^{-1} + \ (T-1)(\mathbf{I} - \boldsymbol{\Phi})' \mathbf{U}^{-1} (\mathbf{I} - \boldsymbol{\Phi})\right\}^{-1}$ and the mean vector is $\mathbf{m}_{\mu}^* = \mathbf{C}_{\mu}^* \left\{\mathbf{C}_{\mu} \mathbf{m}_{\mu} + \mathbf{W}^{-1} \boldsymbol{\lambda}_1 + (\mathbf{I} - \boldsymbol{\Phi})' \mathbf{U}^{-1} \sum_{t=2}^{T} (\boldsymbol{\lambda}_t - \boldsymbol{\Phi} \boldsymbol{\lambda}_{t-1})\right\}$.

The covariance matrix of innovations $\mathbf{U}$ has a full conditional density proportional to $a(\mathbf{U}) IW(\mathbf{U} | r_U^*, r_U^* R_U^*)$, where the number of degrees of freedom is $r_U^* = r_U + T - 1$, the scale matrix is $r_U^* R_U^* = r_U R_U + \sum_{t=2}^{T} [\boldsymbol{\lambda}_t - \{\mathbf{m} + \boldsymbol{\Phi}(\boldsymbol{\lambda}_{t-1} - \mathbf{m})\}][\boldsymbol{\lambda}_t - \{\mathbf{m} + \boldsymbol{\Phi}(\boldsymbol{\lambda}_{t-1} - \mathbf{m})\}]'$ and $a(\mathbf{U}) = |\mathbf{W}|^{-0.5} \exp\{-0.5(\boldsymbol{\lambda}_1 - \mathbf{m})' \mathbf{W}^{-1} (\boldsymbol{\lambda}_1 - \mathbf{m})\}$, where recall that $\mathbf{W} = \mathbf{U} \oslash (\mathbf{1}_k \mathbf{1}_k' - \boldsymbol{\phi}\boldsymbol{\phi}')$. Hence, $\mathbf{U}$ may be simulated with a Metropolis-Hastings step that proposes $\mathbf{U}^*$ with distribution $IW(r_U^*, r_U^* R_U^*)$ and accepts the proposal with probability $\min(1, a(\mathbf{U}*)/a(\mathbf{U}^{(current)}))$.

Finally, $\boldsymbol{\lambda}_{1:T}$ may be efficiently simulated with an FFBS algorithm applied to the dynamic linear model

$$\mathbf{x}_t^* = \boldsymbol{\lambda}_t + \begin{pmatrix} b^*_{\gamma_{t1}^{\lambda}} \\ \vdots \\ b^*_{\gamma_{tk}^{\lambda}} \end{pmatrix} + \boldsymbol{\nu}_t^*,$$

$$\boldsymbol{\lambda}_t = \mathbf{m} + \boldsymbol{\Phi}(\boldsymbol{\lambda}_{t-1} - \mathbf{m}) + \boldsymbol{\xi}_t,$$

where $\boldsymbol{\nu}_t^* \sim N(\mathbf{0}, \text{diag}(w^*_{\gamma_{t1}^{\lambda}}, \ldots, w^*_{\gamma_{tk}^{\lambda}}))$ and $\boldsymbol{\xi}_t \sim N(\mathbf{0}, \mathbf{U})$.

11.4.2 Factor Stochastic Volatility Model for Exchange Rates

Figure 11.14 presents the daily returns of exchange rates from August 1, 2000, to December 30, 2011 of the following ten currencies to the United States Dollar: Euro, British Pound, Japanese Yen, Australian Dollar, Swiss Franc, New Zealand Dollar, Canadian Dollar, Norwegian Krone, Singapore Dollar, and South African Rand. A notable feature for these ten exchange rates returns is the substantial increase in volatility during the financial crisis of 2008–2009. To estimate the common volatility features among these ten time series as well as to estimate the remaining idiosyncratic volatilities, we have fit the factor stochastic volatility model with three latent common factors.

Following the suggestions of Aguilar and West (2000) for a similar application to exchange rates data, we have used the following priors for the model parameters. For the mean vector of log-volatilities $\boldsymbol{\mu}$, we assume the prior $\boldsymbol{\mu} \sim N(\mathbf{m}_\mu, \mathbf{C}_\mu)$ with $\mathbf{C}_\mu = 1.5^2 \mathbf{I}$ and $\mathbf{m}_\mu = -15 \mathbf{1}_k$. For the covariance matrix of innovations $\mathbf{U}$, we assume an inverse Wishart prior $\mathbf{U} \sim IW(r_U, r_U \mathbf{R}_U)$ with $r_U = 5$ and $\mathbf{R}_U = 0.0015 * (\mathbf{I}_k + 0.5(\mathbf{1}_k \mathbf{1}_k' - \mathbf{I}_k))$. For the autoregressive coefficients $\phi_1, \ldots, \phi_k$ and $\rho_1, \ldots, \rho_r$, we assume independent Gaussian priors $N(m_\phi, C_\phi)$ truncated to the interval $(0, 1)$, with $C_\phi = 0.0025$ and $m_\phi = 0.95$. For the means $\alpha_1, \ldots, \alpha_r$ of the idiosyncratic log-volatilities, we assume independent Gaussian priors $N(m_\alpha, C_\alpha)$ with $m_\alpha = -15$ and $C_\alpha = 1.5^2$. For the variances $s_1, \ldots, s_r$ of the innovations of the idiosyncratic log-volatilities, we assume independent inverse gamma priors $IG(0.5 n_s, 0.5 n_s s_s^2)$ with $n_s = 5$ and $n_s s_s^2 = 0.003 \times 4$, which imply prior mean equal to 0.003. Finally, for the variance of factor loadings τ^2

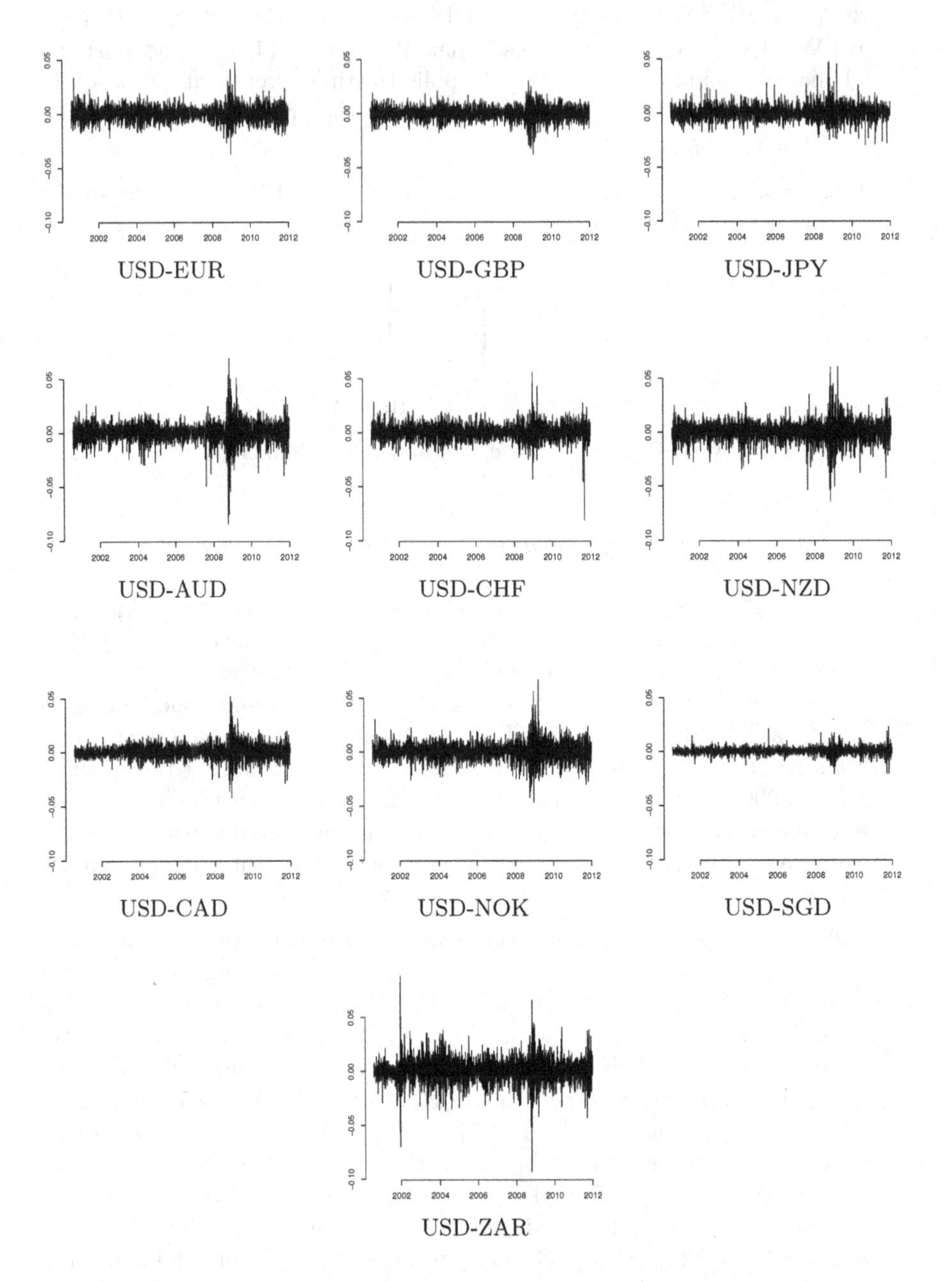

Figure 11.14 *Time series of returns daily returns of exchange rates from August 1, 2000, to December 30, 2011 of the following ten currencies to the United States Dollar: Euro, British Pound, Japanese Yen, Australian Dollar, Swiss Franc, New Zealand Dollar, Canadian Dollar, Norwegian Krone, Singapore Dollar, and South African Rand.*

we assume an inverse Gamma prior $IG(0.5n_\tau, 0.5n_\tau s_\tau^2)$ with $n_\tau = 1$ and $n_\tau s_\tau^2 = 1$. We have run the MCMC algorithm for 20,000 iterations and have discarded 10,000 for burnin, keeping 10,000 to perform posterior inference.

Figure 11.15 presents the posterior means and 95% credible intervals for the factor loadings, as well as time series of the three latent factors. The first factor may be interpreted as representing the overall common volatility in the exchange rates market, with positive factor loadings for all currencies. The largest factor loadings are for the Euro (by construction equal to one), for the Swiss Franc, and for the Norwegian Krone, clearly showing the importance of the exchange rates between the United States and European countries in the global exchange rates market. Australian Dollar, New Zealand Dollar and South African Rand also have substantial factor loadings, showing that these currencies are strongly related to the European currencies. The British Pound, Canadian Dollar and the Japanese Yen have somehow lower factor loadings, and the Singapore Dollar has much lower factor loading. Finally, the plot of the first latent factor over time clearly shows the large increase in global exchange rates market volatility during the financial crisis of 2008–2009. The second factor is largely dominated by the Australian Dollar and the New Zealand Dollar, with somehow moderate factor loadings for the Canadian Dollar and the South African Rand, and not practically significant factor loadings for the other currencies. The time series plot of the second latent factor shows increased volatility toward the end of 2007 and substantially higher volatility during 2008 and 2009. The third factor is a contrast between the Japanese Yen and the Swiss Franc with positive factor loadings versus the South African Rand with a negative factor loading. The time series plot of the third latent factor shows not much volatility up to mid 2007, then progressively increasing volatility culminating with very large volatility in 2008 and 2009, and the volatility somehow decreasing after 2009 but not returning to its low volatility levels from before the financial crisis.

Figure 11.16 shows the posterior densities for the elements of the mean vector of log-volatilities $\boldsymbol{\mu}$, as well as the time series of posterior means and 95% credible intervals for the log-volatilities $\lambda_{1:T,l}$ of the common latent factors. The means of the log-volatility of the first and second factors are about the same, whereas the mean of the log-volatility of the third factor is much smaller. Specifically, the posterior means of the mean log-volatility μ_l are equal to -10.6, -10.8, and -13.0 for the first, second, and third factors, respectively. The posterior probabilities that the means of the log-volatility of the first and second factors are larger than that of the third factor are $p(\mu_1 > \mu_3|\mathbf{y}_{1:T}) = 0.9983$ and $p(\mu_2 > \mu_3|\mathbf{y}_{1:T}) = 0.9999$, respectively. Therefore, the first and second factors explain much more of the common volatility of these ten return time series than the third factor. Further, the increase in the total common volatility obtained by adding a

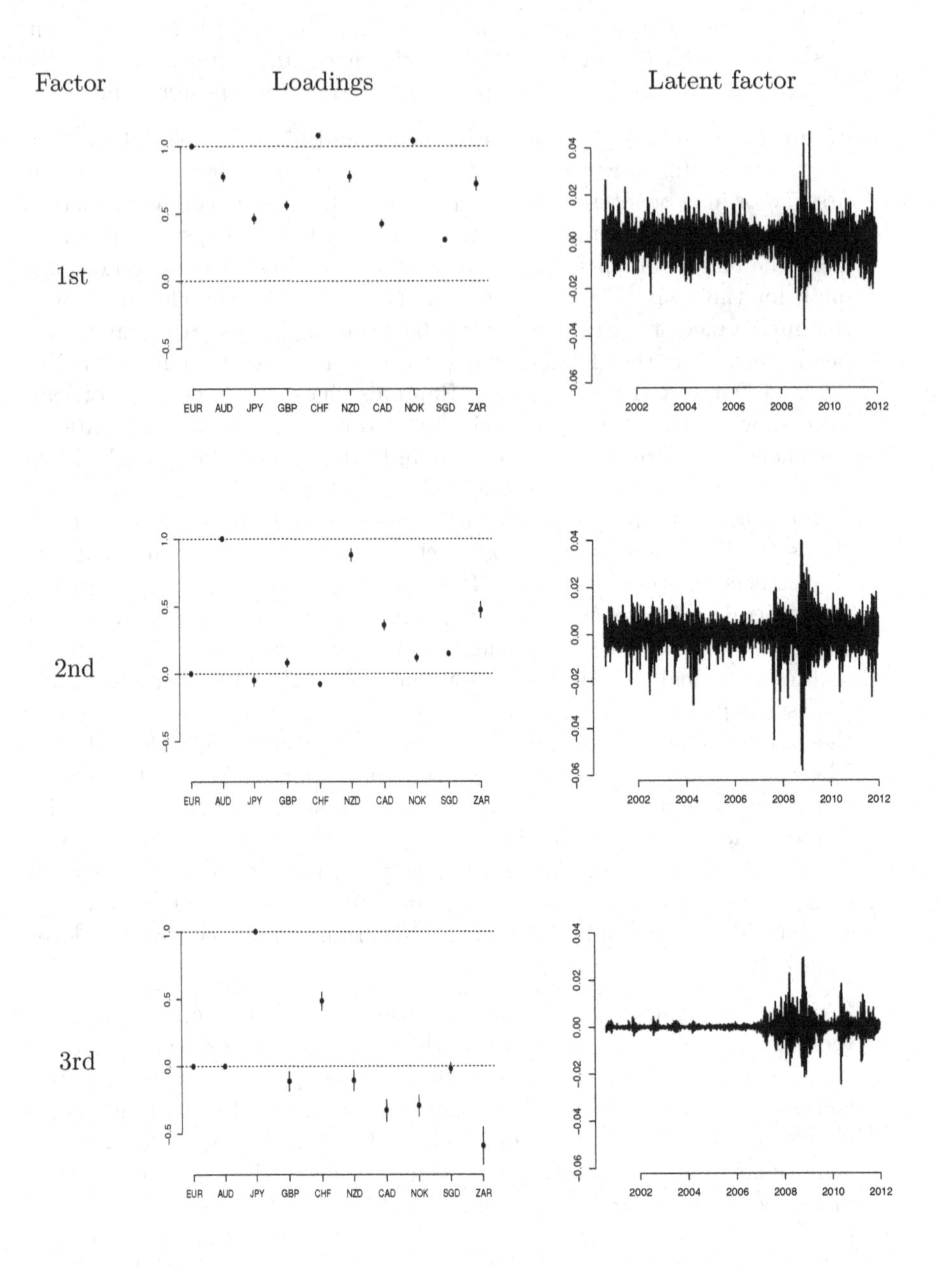

Figure 11.15 *FX factor stochastic volatility: Posterior means (circles) and 95% credible intervals (vertical lines) for the factor loadings, and time series of latent factors.*

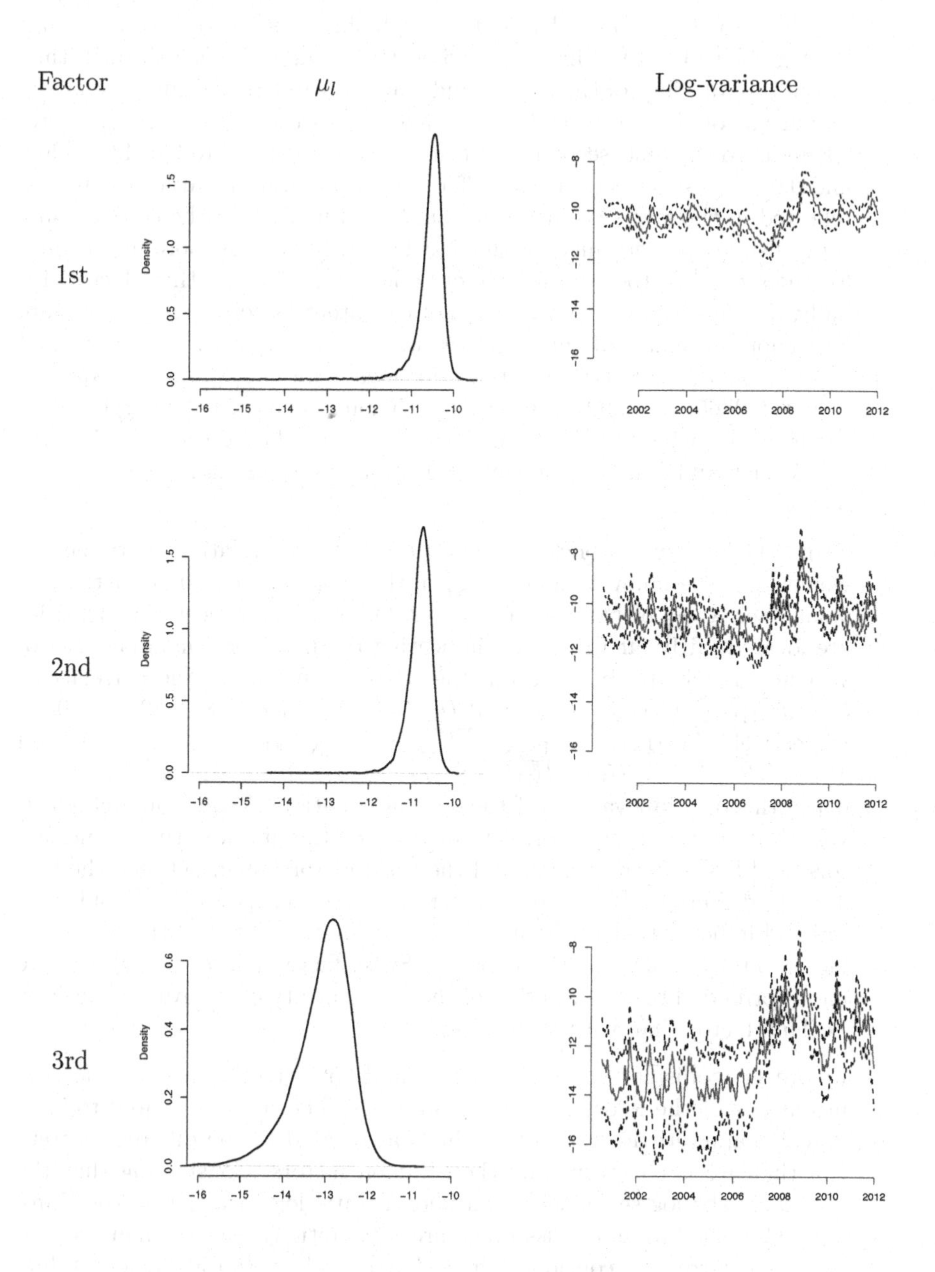

Figure 11.16 *FX factor stochastic volatility: Posterior density for the elements of the mean vector of log-volatilities $\boldsymbol{\mu}$, and posterior means and 95% credible intervals for the log-volatilities $\lambda_{1:T,l}$ of the common latent factors.*

fourth factor to the model would be negligible. Also worth notice is that the log-volatility of the first factor does not change as much through time as the log-volatility of the second and third factors. Specifically, the posterior means of the variances U_{11}, U_{22}, and U_{33} of the innovations $\boldsymbol{\xi}_{lt}$ corresponding to the first, second, and third factors are equal to 0.0049, 0.0269, and 0.0489, respectively. These differences in innovation variances are also reflected in the levels of uncertainty in the estimation of the common factors, with less uncertainty for the first factor, somewhat more uncertainty for the second factor, and a lot more uncertainty for the third factor. Finally, the autoregressive coefficients of the latent factors' log-variances are larger for the first and third factors, with $p(\phi_1 > \phi_2|\mathbf{y}_{1:T}) = 0.9886$ and $p(\phi_3 > \phi_2|\mathbf{y}_{1:T}) = 0.9412$ and posterior means for ϕ_1, ϕ_2, and ϕ_3 equal to 0.9938, 0.9865, and 0.9919, respectively. Therefore, as visually suggested by Figure 11.16, when compared to the second latent factor the log-volatilites of the first and third latent factors exhibit more persistent temporal behavior.

Figure 11.17 displays posterior densities for the correlations between the elements of the innovation vector $\boldsymbol{\xi}_t$ of the evolution equation for the log-volatilities $\lambda_{1:T,l}$ of the common latent factors. These posterior densities are located fairly far from zero, indicating the need for a nondiagonal covariance matrix $\mathbf{U}$. In addition, the posterior means of the correlations $U_{12}/\sqrt{U_{11}U_{22}}$, $U_{13}/\sqrt{U_{11}U_{33}}$, and $U_{23}/\sqrt{U_{22}U_{33}}$ are 0.78, 0.62, and 0.79, respectively. Further, $P(U_{12}/\sqrt{U_{11}U_{22}} > U_{13}/\sqrt{U_{11}U_{33}}|\mathbf{y}_{1:T}) = 0.96$ and $P(U_{23}/\sqrt{U_{22}U_{33}} > U_{13}/\sqrt{U_{11}U_{33}}|\mathbf{y}_{1:T}) = 0.87$. Hence, there is some evidence that the innovations of the system equation of common factor log-volatilities have stronger correlations between the first and the second factors, and between the second and the third factors, than between the first and third factors. Therefore, these three correlations are fairly substantial and indicate that the evolutions of the log-volatilities of the European factor (first factor) and the Japanese-Swiss factor (third factor) may be more correlated to the evolution of the log-volatility of the Australian-New Zealand factor than with themselves.

Figure 11.18 displays posterior means and 95% credible intervals for the means α_js of the idiosyncratic log-variances. The α_js are related to individual characteristics of the volatilities associated to the different currencies that are not captured by the common factors. Considering that the α_js are in the log scale, the mean idiosyncratic log-variances of the Euro, Australian Dollar, and Swiss Franc are substantially smaller than those of the other currencies considered here. Further, the mean idiosyncratic log-variance of Singapore is also small and partially reflects the fact, which is visually clear from Figure 11.14, that among all the considered currencies, the returns of USD-SGD have the smallest volatility. Finally, the impact of the different levels of the α_js can be seen in Figure 11.19, which presents time series posterior means and 95% credible intervals for the idiosyncratic log-variances η_i.

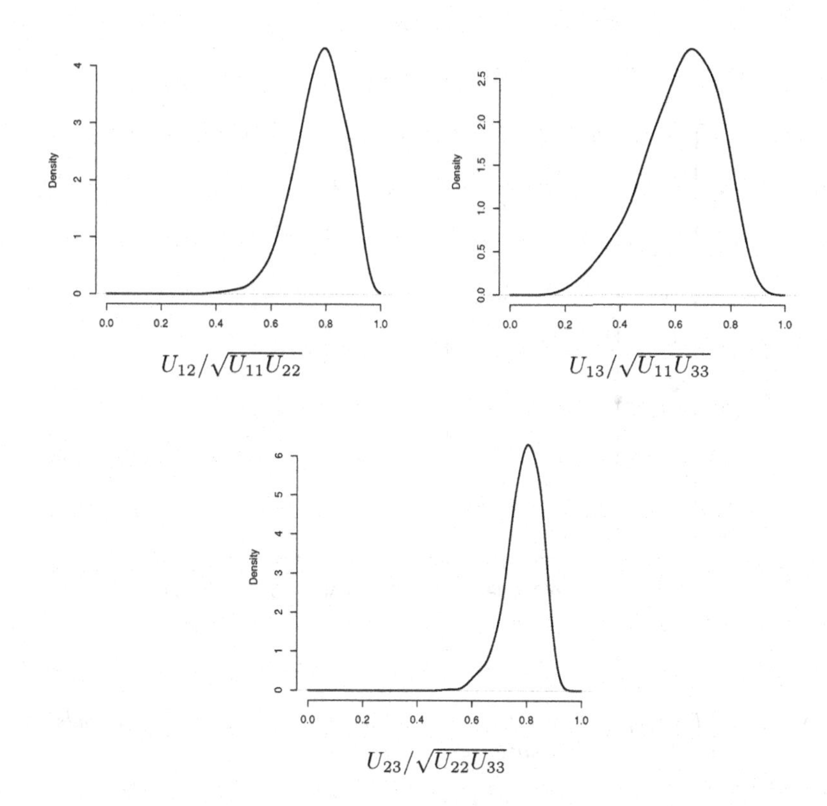

Figure 11.17 *FX application: Posterior densities for the correlations between the elements of the innovation vector* $\boldsymbol{\xi}_t$ *of the evolution equation for the log-volatilities* $\lambda_{1:T,l}$ *of the common latent factors.*

11.5 Spatiotemporal Dynamic Factor Models

Here we consider a class of spatiotemporal dynamic factor models proposed by Lopes, Salazar, and Gamerman (2008). Specifically, consider a spatiotemporal process observed at T time points at r locations $s_1, \ldots, s_r$. Let $y_{ti} = y_t(s_i)$ be the observation at time t and location s_i and $\mathbf{y}_t = (y_{t1}, \ldots, y_{tr})'$ be the vector of observations at time t. The framework is general enough for the number and spatial locations of the observations to vary with time, but to keep notation simple we assume that the number and spatial locations of the observations remain the same for all time points.

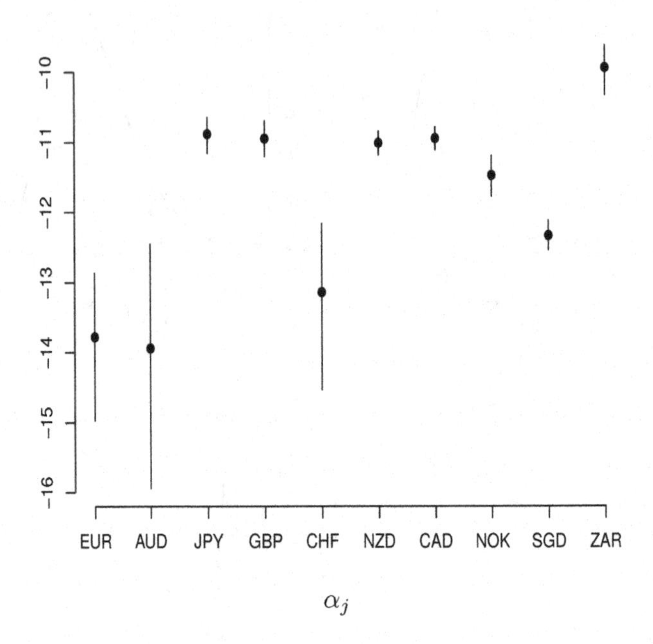

Figure 11.18 *FX application: Posterior means and 95% credible intervals for mean α_j of idiosyncratic log-variances.*

A general spatiotemporal dynamic factor model is

$$\mathbf{y}_t = \mathbf{F}_t\boldsymbol{\theta}_t + \mathbf{B}\mathbf{x}_t + \boldsymbol{\nu}_t, \quad \boldsymbol{\nu}_t \sim N(\mathbf{0}, \mathbf{V}_t), \tag{11.23}$$

$$\boldsymbol{\theta}_t = \mathbf{G}\boldsymbol{\theta}_{t-1} + \boldsymbol{\omega}_t, \quad \boldsymbol{\omega}_t \sim N(\mathbf{0}, \mathbf{W}), \tag{11.24}$$

$$\mathbf{x}_t = \mathbf{H}_t\mathbf{x}_{t-1} + \boldsymbol{\xi}_t, \quad \boldsymbol{\xi}_t \sim N(\mathbf{0}, \mathbf{Z}_t), \tag{11.25}$$

where $\mathbf{F}_t\boldsymbol{\theta}_t$ is the mean vector at time t, $\mathbf{x}_t$ is the realization of the k-dimensional latent factor process at time t, $\mathbf{B}$ is the $r \times k$ matrix of factor loadings, and $\mathbf{H}_t$ is the evolution matrix of the latent factor at time t.

Let $\mathbf{b}_{.l} = (\mathbf{b}_{1l}, \ldots, \mathbf{b}_{rl})'$ be the vector that contains the l-th column of $\mathbf{B}$, $l = 1, \ldots, k$. Note that the i-th element $\mathbf{b}_{il}$ of $\mathbf{b}_{.l}$ corresponds to location s_i. Hence, $\mathbf{b}_{.l}$ may be interpreted as the vector of spatial factor loadings for the l-th factor. Lopes, Salazar, and Gamerman (2008) assume that the spatial factor loadings for the l-th factor follow a Gaussian process. Assume that such Gaussian process has possibly spatially varying mean function $\mu_l(.)$, sill τ_l^2, and spatial correlation function $\rho_l(d; \boldsymbol{\lambda}_l)$,

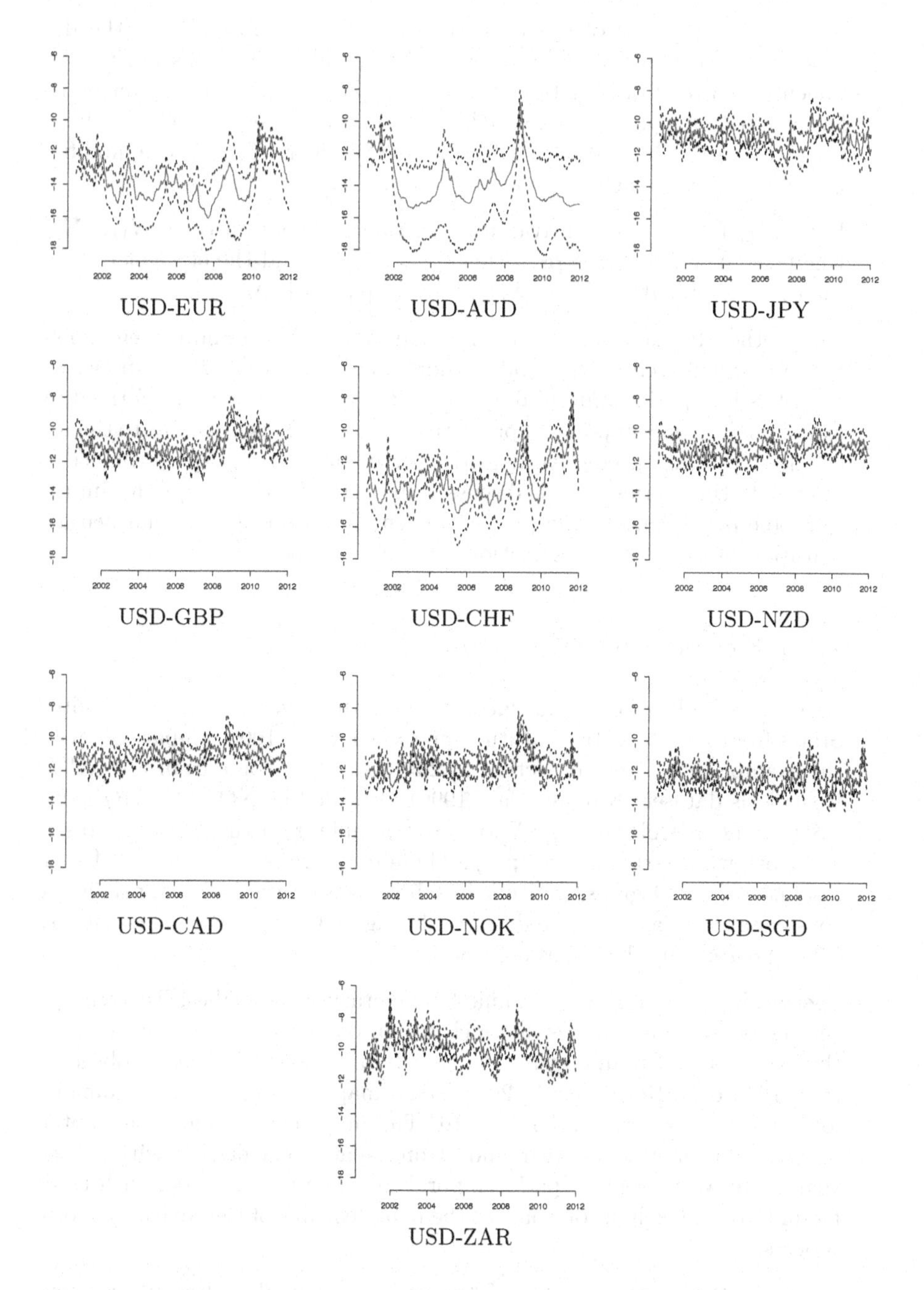

Figure 11.19 *FX application: Posterior means and 95% credible intervals for the idiosyncratic log-variances* η_i.

where $\boldsymbol{\lambda}_l$ is a vector of spatial correlation parameters and d is the distance between two points. Examples of correlation functions are the exponential correlation function $\rho(d;\lambda) = \exp(-d/\lambda)$, the Gaussian correlation function $\rho(d;\lambda) = \exp(-d^2/\lambda^2)$, and the Matern correlation function $\rho(d;\lambda,\kappa) = 2^{1-\kappa}\{\Gamma(\kappa)\}^{-1}(d/\lambda)^{\kappa}\mathcal{K}_{\kappa}(d/\lambda)$, where $\mathcal{K}_{\kappa}(.)$ is the modified Bessel function of the second kind and order κ.

Hence, $\mathbf{b}_{.l}$ follows the multivariate Gaussian distribution $\mathbf{b}_{.l} \sim N(\boldsymbol{\mu}_l, \boldsymbol{\Sigma}_l)$, where the mean vector is $\boldsymbol{\mu}_l = (\mu(s_1), \ldots, \mu(s_r))'$ and the element (i,j) of the covariance matrix $\boldsymbol{\Sigma}_l$ is $\{\boldsymbol{\Sigma}_l\}_{ij} = \tau_l^2 \rho_l(||s_i - s_j||; \boldsymbol{\lambda}_l)$.

As for other dynamic spatiotemporal models (e.g., Vivar and Ferreira 2009, Fonseca and Ferreira 2017, and Elkhouly and Ferreira 2021), model diagnostics for spatiotemporal dynamic factor models may be performed by examining the one-step-ahead predictive densities. This is similar to the use of one-step-ahead predictive densities for monitoring and performance of univariate time series models discussed in Section 4.3.8, except that in the spatiotemporal context the predictive densities are the joint multivariate densities for all spatial observations at each time point.

11.5.1 Example: Temperature Over the Eastern USA

Here we consider the annual mean temperature over the Eastern United States from 1948 to 2016 at altitude pressure level of 1000 milibars. As with the example presented in Section 11.2.3, these data come from the NCEP Reanalysis dataset (Kalnay et al., 1996) provided by NOAA/OAR/ESRL PSD, Boulder, Colorado, USA, that is available from the NOAA web site at http://www.esrl.noaa.gov/psd/. These data are on a regular grid with resolution of 2.5 degrees longitude by 2.5 degrees latitude. To be specific, we consider as the Eastern United States the region located between longitudes 270° and 290° and latitudes 30° and 45°.

Specifically, we analyze the annual temperature anomalies. To compute the temperature anomalies, for each region on the grid, we subtract from the annual mean temperature the long-term mean temperature observed from 1948 to 2016. Figure 11.20 presents maps of temperature anomalies for every 9 years from 1953 to 2016. The maps clearly show substantial variability from year to year and strong spatial correlation within each year. Here we present a spatiotemporal dynamic factor model with three factors that sheds light on some of the main features of this spatiotemporal process.

Let $\mathbf{y}_t$ be the vectorized field of temperature anomalies. Because we consider temperature anomalies, we do not include in the model the mean term

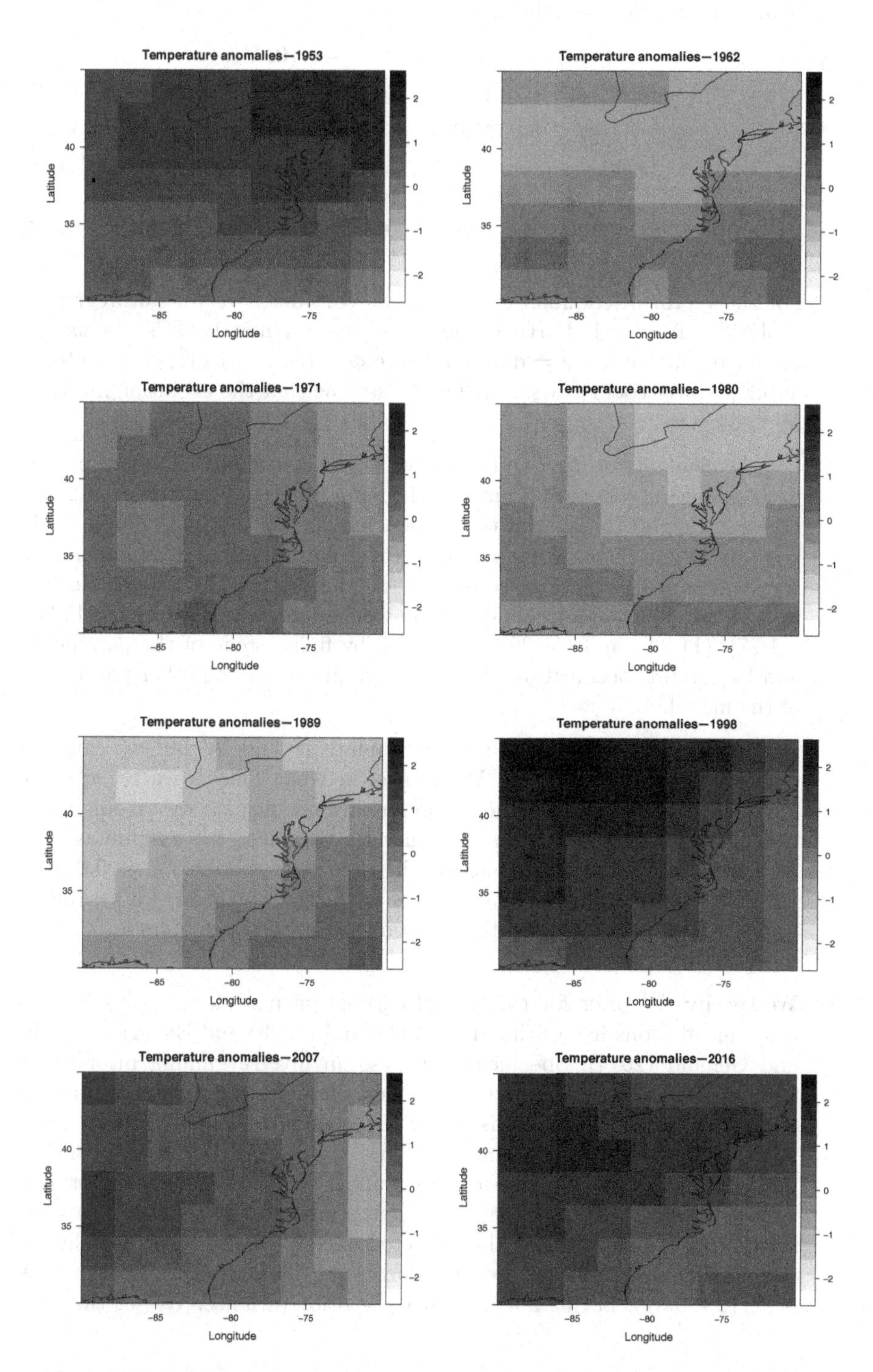

Figure 11.20 *Annual temperature anomalies in the Eastern United States.*

$\mathbf{F}_t\boldsymbol{\theta}_t$. We consider the following model:

$$\begin{aligned}\mathbf{y}_t &= \mathbf{B}\mathbf{x}_t + \boldsymbol{\nu}_t, \;\; \boldsymbol{\nu}_t \sim N(\mathbf{0}, \mathbf{V}), & (11.26)\\ \mathbf{x}_t &= \boldsymbol{\Phi}\mathbf{x}_{t-1} + \boldsymbol{\omega}_t, \;\; \boldsymbol{\omega}_t \sim N(\mathbf{0}, \mathbf{W}), & (11.27)\\ \mathbf{b}_{.l} &\sim N(\mathbf{0}, \boldsymbol{\Sigma}_l), & (11.28)\\ \{\boldsymbol{\Sigma}_l\}_{ij} &= \tau_l^2 \rho_l(||s_i - s_j||; \boldsymbol{\lambda}_l). & (11.29)\end{aligned}$$

As in previous sections, here again $\mathbf{V} = \text{diag}(\sigma_1^2, \ldots, \sigma_r^2)$, where $\sigma_1^2, \ldots, \sigma_r^2$ are idiosyncratic variances.

For the vector of common factors $\mathbf{x}_t$, we consider a vector autoregressive evolution of order 1. Further, the AR coefficient matrix $\boldsymbol{\Phi}$ is assumed to be diagonal, that is, $\boldsymbol{\Phi} = \text{diag}(\boldsymbol{\phi})$ where $\boldsymbol{\phi} = (\phi_1, \ldots, \phi_k)'$ is the vector of autoregressive coefficients. Further, we assume that the evolution matrix is $\mathbf{W} = \text{diag}(w_1, \ldots, w_k)$.

For the elements of the l-th column of $\mathbf{B}$, we consider a Gaussian process with a Matérn correlation matrix with smoothness parameter $\kappa = 2$. Let $\mathbf{R}_l = \boldsymbol{\Sigma}_l/\tau_l^2$ be the spatial correlation matrix for the l-th column of $\mathbf{B}$. Here we follow Lopes, Salazar, and Gamerman (2008) and do not impose the hierarchical structural constraint on the matrix $\mathbf{B}$ discussed in Section 11.2.4. Thus, identifiability of the model defined by Equations (11.26), (11.27), (11.28), and (11.29) is achieved by fixing some of the parameters and by careful specification of informative priors for the other parameters of the model.

For example, we cannot separately estimate τ_l^2 and $\mathbf{W}$ but we can only estimate their product $\tau_l^2 \times \mathbf{W}$. Hence, in what follows we set $\tau_l^2 = 1$. Further, for the priors for the autoregressive coefficients we assume $m_\phi = 0.95$ and $C_\phi = 0.04$. For the idiosyncratic variances, we assume as prior an $IG(n_\sigma/2, n_\sigma s_\sigma^2/2)$ distribution with hyperparameters $n_\sigma = 0.01$ and $n_\sigma s_\sigma^2 = 0.01$. For the variances of the evolution equation, we assume as prior an $IG(n_w/2, n_w s_w^2/2)$ distribution with hyperparameters $n_w = 6$ and $n_w s_w^2 = 0.1$.

We specify the prior for the spatial correlation parameter λ_l inspired by recommendations from Schmidt and Gelfand (2003) and Banerjee, Carlin, and Gelfand (2014). Specifically, we use an inverse gamma prior for λ_l with hyperparameters α and β specified based on the effective range of the correlation function. It is easier to think in term of the prior mean $\beta/(\alpha-1)$ and the prior variance $\beta^2/\{(\alpha-1)^2(\alpha-2)\}$. Let us first consider the prior mean. In the case of the Matérn correlation with smoothness parameter $\kappa = 2$, the effective range is approximately $5.37 \times \lambda_l$. Hence, we set the prior mean so that the expected effective range a priori is half of the maximum distance d_s between locations. Thus, the prior mean of λ_l is $d_s/(2 \times 5.37)$. Let us now consider the prior variance. We set the prior variance such that the effective range is with high prior probability, say 0.975, less than d_s. Hence, using a crude Gaussian approximation, we set the prior variance such that the sum of the prior mean and twice the prior standard deviation is equal to the effective range. With these considerations and after some algebra, we find the prior hyperparameters for λ_l to be $\alpha = 6$ and $\beta = 5 \times d_s/(2 \times 5.37)$.

Computations are for the most part similar to those of the dynamic factor model considered in Section 11.3.1, with a specialized Gibbs sampler to explore the posterior distribution of the model parameters. However, the simulation of the columns of the matrix $\mathbf{B}$ and the simulation of the related Gaussian processes hyperparameters are specific to the spatiotemporal dynamic factor model. We simulate each spatial correlation parameter λ_l, $l = 1, \ldots, k$, one at a time using a Metropolis step for the logarithm of λ_l with proposal distribution $N(\log(\lambda_l^{(curr)}), \delta_\lambda)$, where $\lambda_l^{(curr)}$ is the current value of λ_l. For the dataset considered here, we found that setting the tuning parameter $\delta_\lambda = 0.01$ works well.

Finally, we simulate the columns of $\mathbf{B}$ from their joint full conditional distribution. First, rewrite Equation (11.26) as $\mathbf{y}_t = \mathbf{B}^* \mathbf{x}_t^* + \boldsymbol{\nu}_t$ where $\mathbf{x}_t^* = \mathbf{x}_t' \otimes \mathbf{I}_r$ and $\mathbf{b}^* = (b_{.1}', b_{.2}', \ldots, b_{.k}')'$. The implied prior for $\mathbf{b}^*$ is $N(\mathbf{0}, \boldsymbol{\Sigma}_{b^*})$ where $\boldsymbol{\Sigma}_{b^*} = \text{blockdiag}(\boldsymbol{\Sigma}_1, \ldots, \boldsymbol{\Sigma}_k)$. Therefore, the full conditional distribution for $\mathbf{b}^*$ is $N(\mathbf{m}_{b^*}, \mathbf{C}_{b^*})$, where

$$
\begin{aligned}
\mathbf{C}_{b^*}^{-1} &= \sum_{t=1}^{n} (\mathbf{x}_t^*)' \mathbf{V}^{-1} \mathbf{x}_t^* + \boldsymbol{\Sigma}_{b^*}^{-1}, \\
\mathbf{m}_{b^*} &= \mathbf{C}_{b^*} \left\{ \sum_{t=1}^{n} (\mathbf{x}_t^*)' \mathbf{V}^{-1} \mathbf{y}_t \right\}.
\end{aligned}
$$

After simulation, the new values of $\mathbf{b}_{.1}, \mathbf{b}_{.2}, \ldots, \mathbf{b}_{.k}$ in $\mathbf{b}^*$ are rearranged to form the simulated matrix of factor loadings $\mathbf{B} = (\mathbf{b}_{.1}, \mathbf{b}_{.2}, \ldots, \mathbf{b}_{.k})$. We have run the Gibbs sampler for 5,000 iterations, discarding the first 3,000 iterations as burnin.

Figure 11.21 presents the maps of factor loadings for the three factors. Each of these maps is obtained by taking the corresponding column of the posterior mean of the matrix of factor loadings $\mathbf{B}$ and transforming it into a matrix. All the factor loadings of the first factor have the same negative sign. That indicates that the first factor represents an overall weighted mean of the temperature anomalies. The second factor is a contrast between the north and the south portions of the study region. An interesting feature of the second factor is that it takes into account the particular geography of the East Coast of the United States. Finally, the third factor is a contrast between the east and west portions of the study region.

Figure 11.22 shows time series plots of posterior means (solid line) and 95% credible intervals for the first, second, and third common factors. These plots should be interpreted in conjunction with the understanding gained

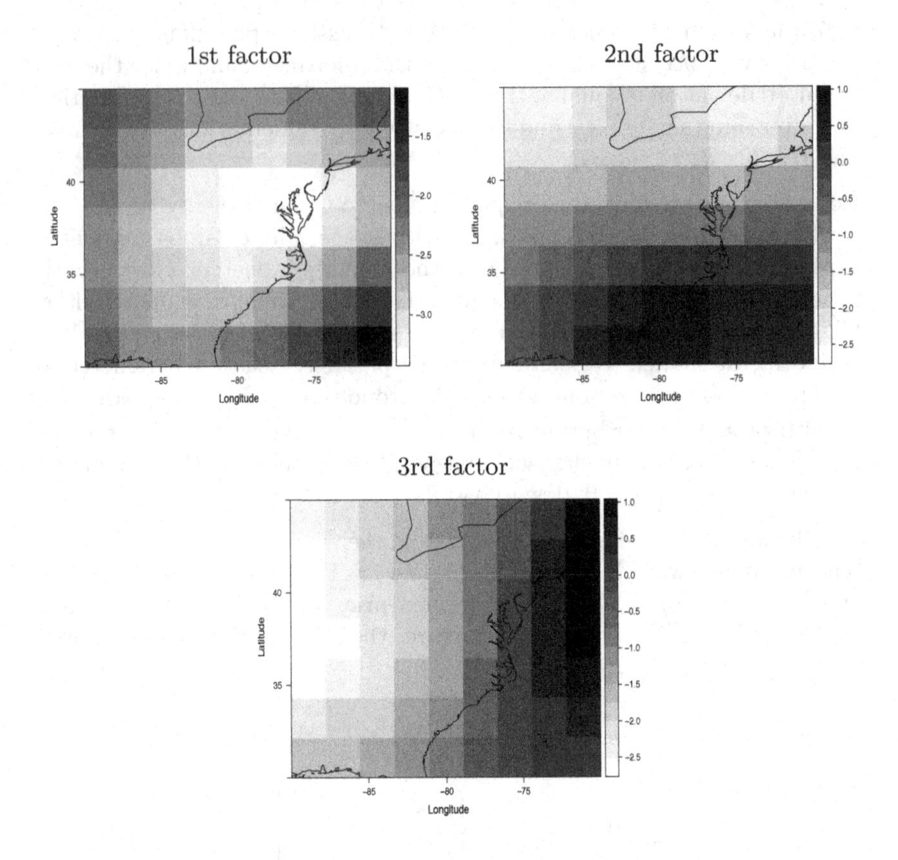

Figure 11.21 *Factor loadings for annual temperature anomalies in the Eastern United States.*

through the analysis of Figure 11.21 about the meaning of each factor. Because the first factor is the negative of an overall weighted mean, the first common factor being negative in a given year means that year was warmer than usual. Hence, there are three features worth notice in the time series of the first factor. First, the temperatures from 1948 to the mid 1950s were warmer than usual in the study region. Second, the behavior of the overall mean anomaly in the study region from the end of the 1960s to the end of the 1970s was visually different from the behavior in the rest of the study period. Finally, the overall mean temperature in the study region from 2010 to 2016 seems to have been warmer than the temperature from the 1980s to 2010. But we also note that the first common factor has substantial temporal variability.

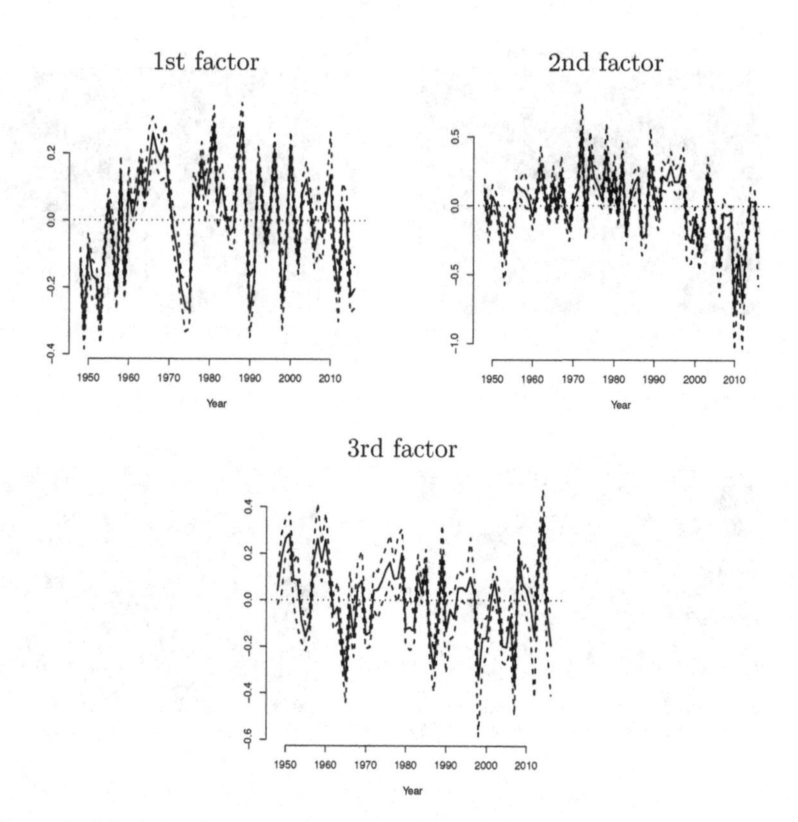

Figure 11.22 *Annual temperature anomalies – 3-factor model. Time series plots of common factors posterior means (solid line) and 95% credible intervals for (a) first factor, (b) second factor, and (c) third factor.*

The time series plot of the second common factor may present a more interesting story. There may have been a structural change sometime between the end of the 1990s and the beginning of the 2000s. Specifically, while the variability seems to be about the same for the periods before and after the year 2000, the mean level seems to have changed. The mean level of the second common factor was about zero before 2000, and seems to have shifted to be negative after 2000. What does that mean? We recall that the analysis of the factor loadings has indicated that the second common factor is a contrast between the north and south temperature anomalies. In addition, the northern factor loadings are negative whereas the southern loadings are positive. Hence, the change in mean level of the second common factor indicates that the temperatures in the north have increased more than the temperatures in the south.

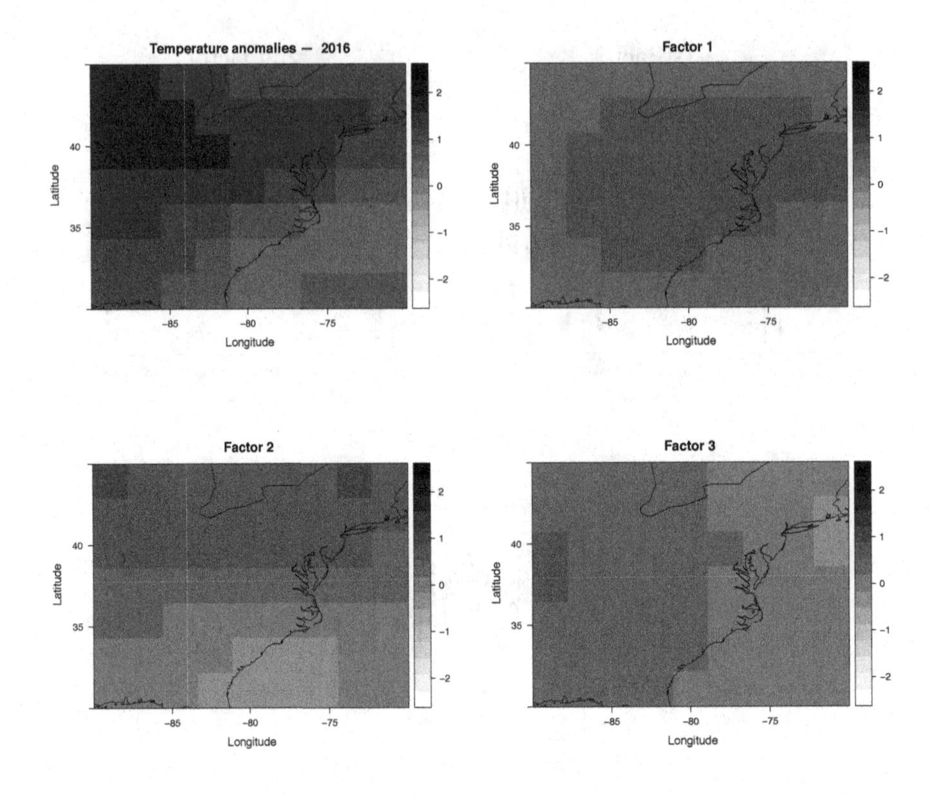

Figure 11.23 *Factorization of temperature anomalies for 2016. Top left panel: temperature anomalies in 2016. Other panels: contributions of each factor to the pattern observed in 2016.*

Finally, the time series plot of the third common factor does not seem to indicate any major change in the contrast between the east and west of the study region.

Another interesting plot to analyze comes about from expanding Equation (11.26) into the sum of the contributions of each of the three factors:

$$\mathbf{y}_t = \mathbf{x}_{t1}\mathbf{b}_{.1} + \mathbf{x}_{t2}\mathbf{b}_{.2} + \mathbf{x}_{t3}\mathbf{b}_{.3} + \boldsymbol{\nu}_t.$$

Specifically, the contribution of the l-th factor is $\mathbf{x}_{tl}\mathbf{b}_{.l}$ and can be plotted in a map for each of the years of our study period. Figure 11.23 presents such factorization of the temperature anomalies for 2016, which was a particularly warm year. The top left panel presents the temperature anomalies in 2016 and the other panels present the contributions of each of the three factors. From the figure, it is clear that not only the first factor contributed

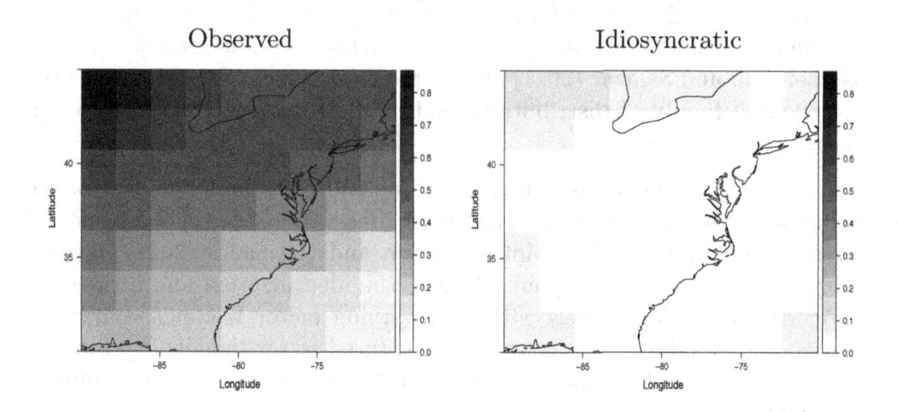

Figure 11.24 *Temperature anomalies. Observed marginal variances (left panel) and idiosyncratic variances (right panel).*

warmer temperatures for the whole region, but the second and third factors also contributed warmer temperatures respectively to the north and west of the study region.

Finally, Figure 11.24 presents the maps of observed marginal variances (left panel) and idiosyncratic variances (right panel). An interesting feature is that the marginal variances are larger in the north and smaller over the Atlantic Ocean. This is consistent with the fact that large bodies of water usually have a moderating effect on temperature. In addition, a comparison of the maps of observed marginal variances and idiosyncratic variances show a substantial decrease in the unexplained variability for this process. Further, comparing the totals of observed marginal variances and of idiosyncratic variances informs that the spatiotemporal dynamic factor model with three factors explains about 90% of the total variability.

11.6 Other Extensions and Recent Developments

There have been numerous extensions and recent developments in the area of dynamic factor models. Calder (2007) introduces dynamic factor models for multivariate spatiotemporal data motivated by environmental applications. The models proposed by Calder (2007) are similar to those considered in Section 11.5, the main distinctions being that she considers multivariate processes and the spatial dependence among factor loadings is modeled through process convolutions (Higdon, 1998). Also related to the models considered in Section 11.5, Liu and West (2009) integrate spatial structure into dynamic models driven by latent processes, using a somewhat different model formulation, motivated by multivariate time series applications in computer model emulation in engineering and applied science. Lopes, Gamerman, and Salazar (2011) extend the models from Section 11.5 to the exponential family of distributions and develop an application to rainfall data.

Motivated by applications in finance, Lopes and Carvalho (2007) introduce a factor stochastic volatility model with time-varying factor loadings and Markov switching regimes. Motta, Hafner, and von Sachs (2011) investigate, from an asymptotic point of view, the identification and estimation properties of factor models with time-varying factor loadings. Carvalho, Lopes, and Aguilar (2011) introduce structured factor models for stock returns that specify the matrix of factor loadings based on specific company information.

Topical issues at the current research frontiers include modeling and computational advances in factor analysis generally that represent an interest in developing priors over factor loadings matrices that are sparse, i.e., have many zeros relative to the number of nonzero entries, for reasons of parsimony in scaling to higher dimensional time series dimension r, and also non-Gaussian distributions for latent factors based on Bayesian nonparametric models (West 2003; Carvalho, Lucas, Wang, Chang, Nevins, and West 2008; Yoshida and West 2010; Rocková and George 2016). Much of this development is based on spike-and-slab priors which have also been used in time series models for global shrinkage of parameter subsets, that is, thresholding them to zero for all time points (Carvalho and West 2007; George, Sun, and Ni 2008; Wang 2010; Korobilis 2011).

A breakthrough in Bayesian sparse factor modeling for time series analysis has come with the development by Nakajima and West (2013a) of latent thresholding as a general approach to dynamic sparsity modeling, that is, time-varying thresholding of parameters to zero. Using this approach, Nakajima and West (2013b) develop latent thresholding dynamic factor models where the factor loadings vary through time and are dynamically thresholded. Nakajima and West (2013b) demonstrate the usefulness of these models for the analysis of multivariate foreign exchange time series and stock price time series. Zhou, Nakajima, and West (2014) extend latent thresholding dynamic factor models to allow dependencies among the latent factor processes. Nakajima and West (2017) adapt latent thresholding dynamic factor models to dynamic transfer response analysis of multivariate electroencephalographic data.

The currently very active status of research in the general area of dynamic factor models reflects the increasing growth of applied interest in highly structured dynamic models in multiple fields.

11.7 Problems

1. Assume the one-factor model given by Equation (11.1). Derive the full conditional distributions for the unknown model quantities given in Section 11.2.2.
2. Consider the k-factor model given by Equation (11.8). Show that if no constraints are imposed on $\mathbf{B}$ or $\mathbf{x}_i$ then the model is not identifiable.
3. Consider the k-factor model presented in Section 11.2.4. Derive the full conditional distributions for the unknown model quantities given in Section 11.2.5.
4. Consider the one-factor model given by Equation (11.1).
   (a) Simulate a sample of 200 observations of 12-dimensional vectors assuming $\sigma_j^2 = 0.1 + 0.01 \times j, j = 1, \ldots, 12$, $H = 1$, and $\tau^2 = 0.36$.
   (b) Implement the MCMC algorithm given in Section 11.2.2 to explore the posterior distribution.
   (c) Analyze the simulated data with the MCMC algorithm using the following prior hyperparameters: $n_\tau = 1$, $n_\tau s_\tau^2 = 1$, $n_H = 1$, $n_H s_H^2 = 1$, $n_\sigma = 2.2$, and $n_\sigma s_\sigma^2 = 0.1$.
5. In the context of selection of number of factors of Section 11.2.6, show that the integrated likelihood function for the parameters of the k-factor model obtained by integrating out the latent factor vectors $\mathbf{x}_1, \ldots, \mathbf{x}_n$ is given by (Lopes and West 2004)

$$p(\mathbf{y}|k, \hat{\theta}_k) = (2\pi)^{-nr/2}|\mathbf{BHB}'+\mathbf{V}|^{-n/2} \exp\left\{-\frac{1}{2}\sum_{i=1}^{n} \mathbf{y}_i'(\mathbf{BHB}' + \mathbf{V})^{-1}\mathbf{y}_i\right\}.$$

6. Consider the k-factor model given in Section 11.2.4.
   (a) Simulate a sample of 200 observations of 15-dimensional vectors with $k = 3$ factors assuming $\sigma_j^2 = 0.1 + 0.02 \times j, j = 1, \ldots, 12$, $h_1 = 1.5$, $h_2 = 0.5$, $h_3 = 0.25$, and $\tau^2 = 0.36$.
   (b) Implement the MCMC algorithm presented in Section 11.2.5 to explore the posterior distribution.
   (c) Implement the Metropolis-Laplace estimator of the marginal data density presented in Section 11.2.6.
   (d) Use the MCMC algorithm to fit factor models with number of factors equal to $1, 2, \ldots, 5$ to the simulated data. Consider the following prior hyperparameters: $n_\tau = 1$, $n_\tau s_\tau^2 = 1$, $n_H = 1$, $n_H s_H^2 = 1$, $n_\sigma = 2.2$, and $n_\sigma s_\sigma^2 = 0.1$.
   (e) Use the Metropolis-Laplace estimator of the marginal data density to select the number of factors.

Bibliography

Aguilar, O., G. Huerta, R. Prado, and M. West (1999). Bayesian inference on latent structure in time series (with discussion). In J. M. Bernardo, J. O. Berger, A. P. Dawid, and A. F. M. Smith (Eds.), *Bayesian Statistics 6*, pp. 3–26. Oxford: Oxford University Press.

Aguilar, O. and M. West (2000). Bayesian dynamic factor models and portfolio allocation. *Journal of Business and Economic Statistics 18*, 338–357.

Akaike, H. (1969). Fitting autoregressive models for prediction. *Annals of the Institute of Statistical Mathematical 21*, 243–247.

Akaike, H. (1974). A new look at statistical model identification. *IEEE Transactions on Automatic Control AC-19*, 716–723.

Aktekin, T., N. G. Polson, and R. Soyer (2018). Sequential Bayesian analysis of multivariate count data. *Bayesian Analysis 13*, 385–409.

Alves, M. B., D. Gamerman, and M. A. R. Ferreira (2010). Transfer functions in dynamic generalized linear models. *Statistical Modelling 10*, 3–40.

Ameen, J. R. M. and P. J. Harrison (1985a). Discount Bayesian multiprocess modelling with cusums. In O. D. Anderson (Ed.), *Time Series Analysis: Theory and Practice 5*, pp. 117–134. Amsterdam: North-Holland.

Ameen, J. R. M. and P. J. Harrison (1985b). Normal discount Bayesian models (with discussion). In J. M. Bernardo, M. H. DeGroot, D. V. Lindleys, and A. F. M. Smith (Eds.), *Bayesian Statistics 2*, pp. 271–298. Amsterdam: North-Holland and Valencia University Press.

Anacleto, O., C. M. Queen, and C. J. Albers (2013). Multivariate forecasting of road traffic flows in the presence of heteroscedasticity and measurement errors. *Journal of the Royal Statistical Society (Series C: Applied Statistics) 62*, 251–270.

Anderson, T. W. (2003). *An Introduction to Multivariate Statistical Analysis* (3rd ed.). New York, NY: Wiley.

Andrieu, C. and A. Doucet (2003). On-line expectation-maximization type algorithms for parameter estimation in general state-space models. In *Proceedings of the International Conference on Acoustics, Speech, and Signal Processing*, pp. 69–72. New York, NY: Institute of Electrical and Electronics Engineers.

Andrieu, C., A. Doucet, and V. Tadić (2005). Online simulation-based methods for parameter estimation in non-linear non-Gaussian state-space models. In *Proceedings of IEEE Conference on Decision and Control*, pp. 332–337. New York, NY: Institute of Electrical and Electronics Engineers.

Banerjee, S., B. P. Carlin, and A. E. Gelfand (2004). *Hierarchical Modeling and Analysis for Spatial Data.* New York, NY: Chapman & Hall.

Banerjee, S., B. P. Carlin, and A. E. Gelfand (2014). *Hierarchical Modeling and Analysis for Spatial Data* (2nd ed.). Boca Raton, FL: Chapman & Hall / CRC.

Barndorff-Nielsen, O. E. and G. Schou (1973). On the reparameterization of autoregressive models by partial autocorrelations. *Journal of Multivariate Analysis 3*, 408–419.

Barnett, G., R. Kohn, and S. Sheather (1996). Bayesian estimation of an autoregressive model using Markov chain Monte Carlo. *Journal of Econometrics 74*, 237–254.

Barnett, G., R. Kohn, and S. Sheather (1997). Robust Bayesian estimation of autoregressive-moving-average models. *Journal of Time Series Analysis 18*, 11–28.

Benati, L. (2008). The "Great Moderation" in the United Kingdom. *Journal of Money, Credit and Banking 40*, 121–147.

Benati, L. and P. Surico (2008). Evolving U.S. monetary policy and the decline of inflation predictability. *Journal of the European Economic Association 6*, 643–646.

Beran, J. (1994). *Statistics for Long Memory Processes.* New York, NY: Chapman & Hall.

Berry, L. R., P. Helman, and M. West (2020). Probabilistic forecasting of heterogeneous consumer transaction-sales time series. *International Journal of Forecasting 36*, 552–569.

Berry, L. R. and M. West (2020). Bayesian forecasting of many count-valued time series. *Journal of Business and Economic Statistics 38*, 872–887.

Berzuini, C., N. Best, W. R. Gilks, and C. Larizza (1997). Dynamic conditional independence models and Markov chain Monte Carlo methods. *Journal of the American Statistical Association 92*, 1403–1412.

Bloomfield, P. (2000). *Fourier Analysis of Time Series: An Introduction* (2nd ed.). New York, NY: John Wiley & Sons.

Bodkin, R. G., L. R. Klein, and K. Marwah (1991). *A History of Macroeconometric Model-Building.* Aldershot, UK: Edward Elgar.

Bollerslev, T., R. Chou, and K. Kroner (1992). ARCH modeling in finance. *Journal of Econometrics 52*, 5–59.

Box, G. E. P., G. M. Jenkins, G. C. Reinsel, and G. M. Ljung (2015). *Time Series Analysis: Forecasting and Control* (5th ed.). New Jersey: John Wiley & Sons.

Box, G. E. P. and G. C. Tiao (1973). *Bayesian Inference in Statistical Analysis.* Massachusetts: Addison-Wesley.

Bretthorst, L. G. (1988). *Bayesian Spectral Analysis and Parameter Estimation.* New York, NY: Springer-Verlag.

Bretthorst, L. G. (1990). Bayesian analysis III. Examples relevant to NMR. *Journal of Magnetic Resonance 88*, 571–595.

Brockwell, P. J. and R. A. Davis (1991). *Time Series: Theory and Methods* (2nd ed.). New York, NY: Springer-Verlag.

Brockwell, P. J. and R. A. Davis (2002). *Introduction to Time Series and Forecasting* (2nd ed.). New York, NY: Springer-Verlag.

Brooks, S. and A. Gelman (1998). General methods for monitoring convergence of iterative simulations. *Journal of Computational and Graphical Statistics 7*, 434–455.

Cadonna, A., A. Kottas, and R. Prado (2017). Bayesian mixture modeling for spectral density estimation. *Statistics and Probability Letters 125*, 189–195.

Cadonna, A., A. Kottas, and R. Prado (2019). Bayesian spectral modeling for multiple time series. *Journal of the American Statistical Association 114*, 1838–1853.

Calder, C. (2007). Dynamic factor process convolution models for multivariate space-time data with application to air quality assessment. *Environmental and Ecological Statistics 14*, 229–247.

Cappé, O., S. Godsill, and E. Moulines (2007). An overview of existing methods and recent advances in sequential Monte Carlo. *Proceedings of the IEEE 95*, 899–924.

Cappé, O., E. Moulines, and T. Rydén (2005). *Inference in Hidden Markov Models.* New York, NY: Springer-Verlag.

Cargnoni, C., P. Müller, and M. West (1997). Bayesian forecasting of multinomial time series through conditionally Gaussian dynamic models. *Journal of the American Statistical Association 92*, 640–647.

Carlin, B. P., N. G. Polson, and D. S. Stoffer (1992). A Monte Carlo approach to nonnormal nonlinear state-space modelling. *Journal of the American Statistical Association 87*, 493–500.

Carter, C. K. and R. Kohn (1994). Gibbs sampling for state space models. *Biometrika 81*, 541–553.

Carter, C. K. and R. Kohn (1997). Semiparametric Bayesian inference for time series with mixed spectra. *Journal of the Royal Statistical Society (Series B: Methodological) 59*, 255–268.

Carvalho, C. M., M. S. Johannes, H. F. Lopes, and N. G. Polson (2010). Particle learning and smoothing. *Statistical Science 25*, 88–106.

Carvalho, C. M. and H. F. Lopes (2007). Simulation-based sequential analysis of Markov switching stochastic volatility models. *Computational Statistics and Data Analysis 51*, 4526–4542.

Carvalho, C. M., H. F. Lopes, and O. Aguilar (2011). Dynamic stock selection strategies: A structured factor model framework. In J. M. Bernardo, M. J. Bayarri, J. O. Berger, A. P. Dawid, D. Heckerman, A. F. M. Smith, and M. West (Eds.), *Bayesian Statistics 9*, Oxford, pp. 69–90. Oxford University Press.

Carvalho, C. M., J. E. Lucas, Q. Wang, J. Chang, J. R. Nevins, and M. West (2008). High-dimensional sparse factor modelling – Applications in gene expression genomics. *Journal of the American Statistical Association 103*, 1438–1456.

Carvalho, C. M. and M. West (2007). Dynamic matrix-variate graphical models. *Bayesian Analysis 2*, 69–98.

Chan, N. H. and G. Petris (2000). Long memory stochastic volatility: A Bayesian approach. *Communications in Statistics: Theory and Methods 29*, 1367–1378.

Chatfield, C. (1996). *The Analysis of Time Series: An Introduction* (5th ed.). London: Chapman & Hall.

Chen, C. W. S. and S. Lee (2017). Bayesian causality test for integer-valued time series models with applications to climate and crime data. *Journal of the Royal of Statistical Society (Series C: Applied Statistics) 66*, 797–814.

Chen, X., D. Banks, and M. West (2019). Bayesian dynamic modeling and monitoring of network flows. *Network Science 7*, 292–318.

Chen, X., K. Irie, D. Banks, R. Haslinger, J. Thomas, and M. West (2018). Scalable Bayesian modeling, monitoring and analysis of dynamic network flow data. *Journal of the American Statistical Association 113*, 519–533.

Chen, Y. (1997). *Bayesian Time Series: Financial Models and Spectral Analysis.* PhD thesis: Institute of Statistics and Decision Sciences, Duke University, Durham, NC, USA.

Chib, S. and E. Greenberg (1994). Bayes inference in regression models with ARMA(p,q) errors. *Journal of Econometrics 64*, 183–206.

Chib, S., F. Nadari, and N. Shephard (2002). Markov chain Monte Carlo methods for stochastic volatility models. *Journal of Econometrics 108*, 281–316.

Chib, S., F. Nadari, and N. Shephard (2005). Analysis of high dimensional multivariate stochastic volatility models. *Journal of Econometrics 134*, 341–371.

Chib, S., Y. Omori, and M. Asai (2009). Multivariate stochastic volatility. In T. G. Andersen, R. A. Davis, J. P. Kreiss, and T. Mikosch (Eds.), *Handbook of Financial Time Series*, pp. 365–400. New York, NY: Springer-Verlag.

Choudhuri, N., S. Ghosal, and A. Roy (2004). Bayesian estimation of the spectral density of a time series. *Journal of the American Statistical Association 99*, 1050–1059.

Cleveland, W. (1979). Robust locally weighted regression and smoothing scatterplots. *Journal of the American Statistical Association 74*, 829–836.

Cleveland, W. S. and S. J. Devlin (1988). Locally weighted regression: An approach to regression analysis by local fitting. *Journal of the American Statistical Association 83*, 596–610.

Cogley, T. and T. J. Sargent (2005). Drifts and volatilities: Monetary policies and outcomes in the post WWII U.S. *Review of Economic Dynamics 8*, 262–302.

Congdon, P. (2000). A Bayesian approach to prediction using the gravity model, with an application to patient flow modeling. *Geographical Analysis 32*, 205–224.

Contreras-Cristán, A., E. Gutiérrez-Peña, and S. Walker (2006). A note on whittle's likelihood. *Communications in Statistics – Simulation and Computation 35*, 857–875.

Costa, L., J. Q. Smith, T. Nichols, J. Cussens, E. P. Duff, and T. R. Makin (2015). Searching multiregression dynamic models of resting-state fMRI networks using integer programming. *Bayesian Analysis 10*, 441–478.

Dawid, A. P. (1981). Some matrix-variate distribution theory: Notational considerations and Bayesian application. *Biometrika 68*, 265–274.

Dawid, A. P. and S. L. Lauritzen (1993). Hyper-Markov laws in the statistical analysis of decomposable graphical models. *The Annals of Statistics 3*, 1272–1317.

Del Moral, P., A. Doucet, and A. Jasra (2006). Sequential Monte Carlo samplers. *Journal of the Royal Statistical Society (Series B: Methodological) 68*, 411–436.

Del Moral, P., A. Jasra, and A. Doucet (2007). Sequential Monte Carlo for Bayesian computation (with discussion). In J. M. Bernardo, M. J. Bayarri, J. O. Berger, A. P. David, D. Heckerman, A. F. M. Smith, and M. West (Eds.), *Bayesian Statistics 8*. Oxford: Oxford University Press.

Dempster, A. P., N. M. Laird, and D. B. Rubin (1977). Maximum likelihood from incomplete data via EM algorithm. *Journal of the Royal Statistical Society (Series B: Methodological) 39*, 1–18.

Diggle, P. (1990). *Time Series: A Biostatistical Introduction.* Oxford: Oxford University Press.

Doucet, A., N. de Freitas, and N. J. Gordon (2001). *Sequential Monte Carlo Methods in Practice.* New York, NY: Springer-Verlag.

Doucet, A., S. J. Godsill, and C. Andrieu (2000). On sequential Monte Carlo sampling methods for Bayesian filtering. *Statistics and Computing 10*, 197–208.

Durbin, J. (1960). Estimation of parameters in time series regression models. *Journal of the Royal Statistical Society (Series B: Methodological) 22*, 139–153.

Durbin, J. and S. J. Koopman (2001). *Time Series Analysis by State Space Methods.* Oxford: Oxford University Press.

Elerian, O., S. Chib, and N. Shephard (2001). Likelihood inference for discretely observed non-linear diffusions. *Econometrica 69*, 959–993.

Elkhouly, M. and M. A. R. Ferreira (2021). Dynamic multiscale spatiotemporal models for multivariate Gaussian data. *Spatial Statistics 41*, 100475.

Engle, R. F. (1982). Autoregressive conditional heteroscedasticity with estimates of the variance of United Kingdom inflation. *Econometrica 50*, 987–1007.

Fan, J. and Q. Yao (2003). *Nonlinear Time Series Analysis.* New York, NY: Springer-Verlag.

Fearnhead, P. (2002). MCMC, sufficient statistics and particle filters. *Journal of Computational and Graphical Statistics 11*, 848–862.

Ferreira, M. A. R. (2002). *Bayesian Multi-scale Modeling.* PhD thesis, Institute of Statistics and Decision Sciences, Duke University, Durham, NC, USA.

Ferreira, M. A. R. (2020). Bayesian spatial and spatiotemporal models based on multiscale factorizations. *Wiley Interdisciplinary Reviews: Computational Statistics*, e1509.

Ferreira, M. A. R., A. I. Bertolde, and S. Holan (2010). Analysis of economic data with multi-scale spatio-temporal models. In A. O'Hagan and M. West (Eds.), *Handbook of Applied Bayesian Analysis*, pp. 295–318. Oxford: Oxford University Press.

Ferreira, M. A. R., Z. Bi, M. West, H. K. H. Lee, and D. Higdon (2003). Multi-scale modelling of 1-D permeability fields. In J. Bernardo, M. J. Bayarri, J. O. Berger, A. P. Dawid, D. Heckerman, A. F. M. Smith, and M. West (Eds.), *Bayesian Statistics 7*, pp. 519–527. Oxford: Oxford University Press.

Ferreira, M. A. R. and D. Gamerman (2000). Dynamic generalized linear models. In D. K. Dey, S. K. Ghosh, and B. K. Mallick (Eds.), *Generalized Linear Models: A Bayesian Perspective*, pp. 57–72. New York, NY: Marcel Dekker.

Ferreira, M. A. R., D. Gamerman, and H. S. Migon (1997). Bayesian dynamic hierarchical models: Covariance matrices estimation and nonnormality. *Brazilian Journal of Probability and Statistics 11*, 67–79.

Ferreira, M. A. R., S. H. Holan, and A. I. Bertolde (2011). Dynamic multiscale spatio-temporal models for Gaussian areal data. *Journal of the Royal Statistical Society (Series B: Methodological) 73*, 663–688.

Ferreira, M. A. R. and H. K. H. Lee (2007). *Multiscale Modeling: A Bayesian Perspective.* Springer Series in Statistics. New York, NY: Springer.

Ferreira, M. A. R., M. West, H. K. H. Lee, and D. Higdon (2006). Multi-scale and hidden resolution time series models. *Bayesian Analysis 1*, 947–968.

Fonseca, T. C. O. and M. A. R. Ferreira (2017). Dynamic multiscale spatiotemporal models for Poisson data. *Journal of the American Statistical Association 112*, 215–234.

Fonseca, T. C. O., M. A. R. Ferreira, and H. S. Migon (2008). Objective Bayesian analysis for the Student-t regression model. *Biometrika 95*, 325–333.

Fox, E. B., E. B. Sudderth, M. I. Jordan, and A. S. Willsky (2009). Nonparametric Bayesian identification of jump systems with sparse dependencies. In *Proceedings of the 15th IFAC Symposium on System Identification.*

Fox, E. B., E. B. Sudderth, M. I. Jordan, and A. S. Willsky (2011). Bayesian nonparametric inference of switching dynamic linear models. *IEEE Transactions on Signal Processing 59*, 1569–1585.

Frühwirth-Schnatter, S. (1994). Data augmentation and dynamic linear models. *Journal of Time Series Analysis 15*, 183–202.

Frühwirth-Schnatter, S. (2001). Markov chain Monte Carlo estimation of classical and dynamic switching and mixture models. *Journal of the American Statistical Association 96*, 194–209.

Frühwirth-Schnatter, S. (2006). *Finite Mixture and Markov Switching Models*. New York, NY: Springer-Verlag.

Gamerman, D. (1998). Markov chain Monte Carlo for dynamic generalised linear models. *Biometrika 85*, 215–227.

Gamerman, D. and H. F. Lopes (2006). *Markov Chain Monte Carlo: Stochastic Simulation for Bayesian Inference* (2nd ed.). London: Chapman & Hall.

Gao, B., H. Ombao, and M. Ho (2009). Cluster analysis for nonstationary time series. In S. Chow, E. Ferrer, and F. Hsieh (Eds.), *Statistical Methods for Modeling Human Dynamics: An Interdisciplinary Dialogue*, Chapter 4, pp. 85–122. New York, NY: Psychology Press, Taylor and Francis Group.

Gelman, A., J. B. Carlin, H. S. Stern, D. B. Dunson, A. Vehtari, and D. B. Rubin (2014). *Bayesian Data Analysis* (3rd ed.). New York, NY: Chapman & Hall/CRC Press.

Gelman, A. and D. B. Rubin (1992). Inference from iterative simulation using multiple sequences. *Statistical Science 7*, 457–472.

Geman, S. and D. Geman (1984). Stochastic relaxation, Gibbs distributions and the Bayesian restoration of images. *IEEE Transactions on Pattern Analysis and Machine Intelligence 6*, 721–741.

George, E. I., D. Sun, and S. Ni (2008). Bayesian stochastic search for VAR model restrictions. *Journal of Econometrics 142*, 553–580.

Geweke, J. F. (1992). Evaluating the accuracy of sampling-based approaches to calculating posterior moments. In J. M. Bernardo, J. O. Berger, A. P. Dawid, and A. F. M. Smith (Eds.), *Bayesian Statistics 4*, pp. 169–194. Oxford: Oxford University Press.

Geweke, J. F. and S. Porter-Hudak (1983). The estimation and application of long memory time series models. *Journal of Time Series Analysis 4*, 221–238.

Geweke, J. F. and K. J. Singleton (1981). Latent variable models for time series: A frequency domain approach to the Permanent Income Hypothesis. *Journal of Econometrics 17*, 287–304.

Geweke, J. F. and G. Zhou (1996). Measuring the pricing error of the arbitrage pricing theory. *Review of Financial Studies 9*, 557–587.

Giudici, P. (1996). Learning in graphical Gaussian models. In J. M. Bernado, J. O. Berger, A. P. Dawid, and A. F. M. Smith (Eds.), *Bayesian Statistics 5*, pp. 621–628. Oxford: Oxford University Press.

Giudici, P. and P. J. Green (1999). Decomposable graphical Gaussian model determination. *Biometrika 86*, 785–801.

Glynn, C., S. T. Tokdar, D. L. Banks, and B. Howard (2019). Bayesian analysis of dynamic linear topic models. *Bayesian Analysis 14*, 53–80.

Godsill, S. J. (1997). Bayesian enhancement of speech and audio signals which can be modelled as ARMA processes. *International Statistical Review 65*, 1–21.

Godsill, S. J., A. Doucet, and M. West (2004). Monte Carlo smoothing for non-linear time series. *Journal of the American Statistical Association 99*, 156–168.

Gordon, N. J., D. Salmond, and A. F. M. Smith (1993). Novel approach to nonlinear/non-Gaussian Bayesian state estimation. *IEEE Proceedings F, Radar and Signal Processing 140*, 107–113.

Granger, C. W. and R. Joyeux (1980). An introduction to long-memory time series models and fractional differencing. *Journal of Time Series Analysis 4*, 221–238.

Green, M. and P. J. Harrison (1973). Fashion forecasting for a mail order company. *Operational Research Quarterly 24*, 193–205.

Green, P. J. (1995). Reversible jump Markov chain Monte Carlo computation and Bayesian model determination. *Biometrika 82*, 711–732.

Greenberg, E. (2008). *Introduction to Bayesian Econometrics.* New York, NY: Cambridge University Press.

Gruber, L. F. and M. West (2016). GPU-accelerated Bayesian learning in simultaneous graphical dynamic linear models. *Bayesian Analysis 11*, 125–149.

Gruber, L. F. and M. West (2017). Bayesian forecasting and scalable multivariate volatility analysis using simultaneous graphical dynamic linear models. *Econometrics and Statistics 3*, 3–22.

Guhaniyogi, R. and D. Dunson (2015). Bayesian compressed regression. *Journal of the American Statistical Association 110*, 1500–1514.

Hahn, P. R. and C. M. Carvalho (2015). Decoupling shrinkage and selection in Bayesian linear models: A posterior summary perspective. *Journal of the American Statistical Association 110*, 435–448.

Hamilton, J. D. (1994). *Time Series Analysis.* New Jersey: Princeton University Press.

Harrison, P. J. and C. Stevens (1971). A Bayesian approach to short-term forecasting. *Operational Research Quarterly 22*, 342–362.

Harrison, P. J. and C. Stevens (1976). Bayesian forecasting (with discussion). *Journal of the Royal Statistical Society (Series B: Methodological) 38*, 205–247.

Harrison, P. J. and M. West (1987). Practical Bayesian forecasting. *The Statistician 36*, 115–125.

Harvey, A. C. (1981). *Time Series Models.* London: Philip Allan.

Harvey, A. C. (1991). *Forecasting, Structural Time Series Models and the Kalman Filter.* Cambridge: Cambridge University Press.

Harvey, A. C., E. Ruiz, and N. Shephard (1994). Multivariate stochastic variance models. *Review of Economic Studies 61*, 247–264.

Hastings, W. K. (1970). Monte Carlo sampling methods using Markov chains and its applications. *Biometrika 57*, 97–109.

Hazelton, M. L. (2015). Network tomography for integer-valued traffic. *Annals of Applied Statistics 9*, 474–506.

Heidelberger, P. and P. Welch (1983). Simulation run length control in the presence of an initial transient. *Operations Research 31*, 1109–1144.

Heyse, J. and W. W. S. Wei (1985). Inverse and partial lag autocorrelation for vector time series. In *ASA Proceedings of Business and Economic Statistics Section*, pp. 233–237.

Higdon, D. M. (1998). A process-convolution approach to modeling temperatures in the North Atlantic ocean. *Journal of Environmental and Ecological Statistics 5*, 173–190.

Hillmer, S. C. and G. C. Tiao (1979). Likelihood function of stationary multiple autoregressive moving average processes. *Journal of the American Statistical Association 74*, 652–660.

Hoegh, A., M. A. R. Ferreira, and S. Leman (2016). Spatiotemporal model fusion: Multiscale modelling of civil unrest. *Journal of the Royal Statistical Society (Series C: Applied Statistics) 65*, 529–545.

Holan, S., T. McElroy, and S. Chakraborty (2009). A Bayesian approach to estimating the long memory parameter. *Bayesian Analysis 4*, 159–190.

Holan, S. H., D. Toth, M. A. R. Ferreira, and A. F. Karr (2010). Bayesian multiscale multiple imputation with implications for data confidentiality. *Journal of the American Statistical Association 105*, 564–577.

Hosking, J. R. M. (1981). Fractional differencing. *Biometrika 68*, 165–176.

Huan, H., H. Ombao, and D. S. Stoffer (2004). Classification and discrimination of non-stationary time series using the SLEX model. *Journal of the American Statistical Association 99*, 763–774.

Huber, F., G. Koop, and L. Onorante (2020). Inducing sparsity and shrinkage in time-varying parameter models. *Journal of Business & Economic Statistics*.

Huerta, G. (1998). *Bayesian Analysis of Latent Structure in Time Series Models.* PhD thesis, Duke University, Durham, N.C., USA.

Huerta, G., W. Jiang, and M. A. Tanner (2001). A comment on the art of data augmentation. *Journal of Computational and Graphical Statistics 10*, 82–89.

Huerta, G., W. Jiang, and M. A. Tanner (2003). Time series modeling via hierarchical mixtures. *Statistica Sinica 23*, 1097–1118.

Huerta, G. and M. West (1999a). Bayesian inference on periodicities and component spectral structure in time series. *Journal of Time Series Analysis 20*, 401–416.

Huerta, G. and M. West (1999b). Priors and component structures in autoregressive time series models. *Journal of the Royal Statistical Society (Series B: Methodological) 61*, 881–899.

Hurst, H. (1951). Long term storage capacity of reservoirs. *Transactions of the American Society of Civil Engineers 116*, 778–808.

Hürzeler, M. and H. Künsch (2001). Approximating and maximizing the likelihood for general SSM. In A. Doucet, N. de Freitas, and N. J. Gordon (Eds.), *Sequential Monte Carlo Methods in Practice*, pp. 159–175. New York, NY: Springer-Verlag.

Irie, K. and M. West (2019). Bayesian emulation for multi-step optimization in decision problems. *Bayesian Analysis 14*, 137–160.

Ito, K. and K. Xiong (2000). Gaussian filters for nonlinear filtering problems. *IEEE Transactions on Automatic Control 45*, 910–927.

Jacquier, E., N. G. Polson, and P. E. Rossi (1994). Bayesian analysis of stochastic volatility models. *Journal of Business and Economic Statistics 12*, 371–389.

Jacquier, E., N. G. Polson, and P. E. Rossi (1995). Models and priors for multivariate stochastic volatility. Technical report, Graduate School of Business, University of Chicago.

Jazwinski, A. (1970). *Stochastic Processes and Filtering Theory.* New York, NY: Academic Press.

Jeffreys, H. S. (1961). *Theory of Probability* (3rd ed.). Oxford: Clarendon Press (1st ed., 1939).

Jochmann, M., G. Koop, and R. Strachan (2010). Bayesian forecasting using stochastic search variable selection in a VAR subject to breaks. *International Journal of Forecasting 26*, 326–347.

Johannes, M. and N. G. Polson (2008). Exact particle filtering and learning. Technical report, Graduate School of Business, University of Chicago.

Jones, B., A. Dobra, C. M. Carvalho, C. Hans, C. Carter, and M. West (2005). Experiments in stochastic computation for high-dimensional graphical models. *Statistical Science 20*, 388–400.

Jones, B. and M. West (2005). Covariance decomposition in undirected Gaussian graphical models. *Biometrika 92*, 779–786.

Jordan, M. I. and R. A. Jacobs (1994). Hierarchical mixtures of experts and the EM algorithm. *Neural Computation 6*, 181–214.

Juang, B. H. and L. R. Rabiner (1985). Mixture autoregressive hidden Markov models for speech signals. *IEEE Transactions on Acoustics, Speech, and Signal Processing 33*, 1404–1413.

Julier, S. J. and J. K. Uhlmann (1997). A new extension of the Kalman filter to nonlinear systems. In *AeroSense: The 11th International Symposium on Aerospace/Defense Sensing, Simulation and Control*, pp. 182–193. International Society for Optics and Photonics.

Kakizawa, Y., R. H. Shumway, and M. Taniguchi (1998). Discrimination and clustering for multivariate time series. *Journal of the American Statistical Association 93*, 328–340.

Kalman, R. E. (1960). A new approach to linear filtering and prediction problems. *Journal of Basic Engineering 82*, 35–45.

Kalnay, E., M. Kanamitsu, R. Kistler, W. Collins, D. Deaven, L. Gandin, M. Iredell, S. Saha, G. White, J. Woollen, et al. (1996). The NCEP/NCAR 40-year reanalysis project. *Bulletin of the American Meteorological Society 77*, 437-470.

Kastner, G. and F. Huber (2020). Sparse Bayesian vector autoregressions in huge dimensions. *Journal of Forecasting 39*, 1142–1165.

Kendall, M. G. and J. K. Ord (1990). *Time Series* (3rd ed.). Sevenoaks: Edward Arnold.

Kendall, M. G., A. Stuart, and J. K. Ord (1983). *The Advanced Theory of Statistics* (4th ed., Vol. 3). London: Griffin.

Kim, S., N. Shephard, and S. Chib (1998). Stochastic volatility: Likelihood inference and comparison with ARCH models. *Review of Economic Studies 65*, 361–393.

Kirch, C., M. Edwards, A. Meier, and R. Meyer (2017). Beyond Whittle: Non-parametric correction of a parametric likelihood with a focus on Bayesian time series analysis. *Bayesian Analysis 14*(4), 1–37.

Kitagawa, G. (1996). Monte Carlo filter and smoother for non-Gaussian nonlinear state space models. *Journal of Computational and Graphical Statistics 5*, 1–25.

Kitagawa, G. and W. Gersch (1985). A smoothness priors time varying AR coefficient modeling of non-stationary time series. *IEEE Transactions on Automatic Control 30*, 48–56.

Kitagawa, G. and W. Gersch (1996a). *Smoothness Priors Analysis of Time Series.* Lecture Notes in Statistics. New York, NY: Springer-Verlag.

Kitagawa, G. and W. Gersch (1996b). *Smoothness Priors Analysis of Time Series.* Lecture Notes in Statistics. New York: Springer-Verlag.

Knaus, P., A. Bitto-Nemling, A. Cadonna, and S. Frühwirth-Schnatter (2019). Shrinkage in the time-varying parameter model framework using the R package shrinkTVP. *arXiv: Econometrics*.

Kohn, R. and C. F. Ansley (1985). Efficient estimation and prediction in time series regression models. *Biometrika 72*, 694–697.

Kong, A., J. S. Liu, and W. H. Wong (1994). Sequential imputations and Bayesian missing data problems. *Journal of the American Statistical Association 89*, 278–288.

Koop, G. and D. Korobilis (2010). Bayesian multivariate time series methods for empirical macroeconomics. *Foundations and Trends in Econometrics 3*, 267–358.

Koop, G. and D. Korobilis (2013). Large time-varying parameter VARs. *Journal of Econometrics 177*, 185–198.

Koop, G., D. Korobilis, and D. Pettenuzzo (2019). Bayesian compressed vector autoregressions. *Journal of Econometrics 210*, 135–154.

Koop, G., R. Leon-Gonzalez, and R. W. Strachan (2009). On the evolution of the monetary policy transmission mechanism. *Journal of Economic Dynamics and Control 33*, 997–1017.

Korobilis, D. (2011). VAR forecasting using Bayesian variable selection. *Journal of Applied Econometrics 28*, 204–230.

Krolzig, H. M. (1997). *Markov-Switching Vector Autoregressions.* Lecture Notes in Econometrics and Mathematical Systems. New York, NY: Springer-Verlag.

Krueger, F. (2015). *bvarsv: Bayesian analysis of a vector autoregressive model with stochastic volatility and time-varying parameters.* R package version 1.1.

Krystal, A. D., R. Prado, and M. West (1999). New methods of time series analysis of non-stationary EEG data: Eigenstructure decompositions of time-varying autoregressions. *Clinical Neurophysiology 110*, 2197–2206.

Lauritzen, S. L. (1996). *Graphical Models.* Oxford: Clarendon Press.

Lemos, R. and B. Sansó (2009). A spatio-temporal model for mean, anomaly, and trend fields of North Atlantic sea surface temperature (with discussion). *Journal of the American Statistical Association 104*, 5–18.

Levinson, N. (1947). The Wiener (root mean square) error criterion in filter and design prediction. *Journal of Mathematical Physics 25*, 262–278.

Lewis, S. M. and A. E. Raftery (1997). Estimating Bayes factors via posterior simulation with the Laplace-metropolis estimator. *Journal of the American Statistical Association 92*, 648–655.

Li, Z. and R. T. Krafty (2018). Adaptive Bayesian power spectrum analysis of multivariate nonstationary time series. *Journal of the American Statistical Association 114*, 453–465.

Li, Z. and R. T. Krafty (2019). Adaptive Bayesian time-frequency analysis of multivariate time series. *Journal of the American Statistical Association 114*, 453–465.

Liseo, B., D. Marinucci, and L. Petrella (2001). Bayesian semiparametric inference on long-range dependence. *Biometrika 88*, 1089–1104.

Liu, F. and M. West (2009). A dynamic modelling strategy for Bayesian computer model emulation. *Bayesian Analysis 4*, 393–412.

Liu, J. (2000). *Bayesian Time Series: Analysis Methods Using Simulation-based Computation.* PhD thesis, Institute of Statistics and Decision Sciences, Duke University, Durham, NC, USA.

Liu, J. and M. West (2001). Combined parameter and state estimation in simulation-based filtering. In A. Doucet, N. de Freitas, and N. J. Gordon (Eds.), *Sequential Monte Carlo Methods in Practice*, pp. 197–217. New York, NY: Springer-Verlag.

Liu, J. S. (1996). Metropolized independent sampling with comparisons to rejection sampling and importance sampling. *Statistical Computing 6*, 113–119.

Liu, J. S. and R. Chen (1995). Blind deconvolution via sequential imputation. *Journal of the American Statistical Association 90*, 567–576.

Liu, J. S. and R. Chen (1998). Sequential Monte Carlo methods for dynamic systems. *Journal of the American Statistical Association 93*, 1032–1044.

Liu, J. S. and R. Chen (2000). Mixture Kalman filters. *Journal of the Royal Statistical Society (Series B: Methodological) 62*, 493–508.

Lopes, H. F. (2000). *Bayesian Analysis in Latent Factor and Longitudi-*

nal Models. PhD thesis: Institute of Statistics and Decision Sciences, Duke University, Durham, NC, USA.

Lopes, H. F. (2003). Expected posterior priors in factor analysis. *Brazilian Journal of Probability and Statistics 17*, 91–105.

Lopes, H. F. (2007). Factor stochastic volatility with time varying loadings. *Estadistica 57*, 75–91.

Lopes, H. F. and C. M. Carvalho (2007). Factor stochastic volatility with time varying loadings and Markov switching regimes. *Journal of Statistical Planning and Inference 137*, 3082–3091.

Lopes, H. F., D. Gamerman, and E. Salazar (2011). Generalized spatial dynamic factor analysis. *Computational Statistics and Data Analysis 55*, 1319–1330.

Lopes, H. F., R. E. McCulloch, and R. S. Tsay (2018). Parsimony inducing priors for large scale state-space models. Technical Report 2018-08, Booth School of Business, University of Chicago, Chicago, Illinois.

Lopes, H. F., E. Salazar, and D. Gamerman (2008). Spatial dynamic factor analysis. *Bayesian Analysis 3*, 1–34.

Lopes, H. F. and M. West (2004). Bayesian model assessment in factor analysis. *Statistica Sinica 14*, 41–67.

Lütkepohl, H. (2005). *New Introduction to Multiple Time Series Analysis*. Heidelberg: Springer-Verlag.

Macaro, C. and R. Prado (2014). Spectral decompositions of multiple time series: A Bayesian non-parametric approach. *Psychometrika 79*, 105–129.

Mandelbrot, B. B. and J. V. van Ness (1968). Fractional Brownian motions, fractional noises and applications. *SIAM Review 10*, 422–437.

Mandelbrot, B. B. and J. R. Wallis (1968). Noah, Joseph and operational hydrology. *Water Resources Research 4*, 909–918.

Markowitz, H. M. (1959). *Portfolio Selection: Efficient Diversification of Investments*. New York, NY: John Wiley & Sons.

Marriott, J. M., N. Ravishanker, A. E. Gelfand, and J. Pai (1996). Bayesian analysis of ARMA processes: Complete sampling-based inference under exact likelihoods. In D. A. Berry, K. M. Chaloner, and J. F. Geweke (Eds.), *Bayesian Analysis in Statistics and Econometrics: Essays in Honor of Arnold Zellner*, pp. 243–256. New York, NY: John Wiley & Sons.

Marriott, J. M. and A. F. M. Smith (1992). Reparametrization aspects of numerical Bayesian methodology for autoregressive moving-average models. *Journal of Time Series Analysis 13*, 327–343.

McCoy, E. J. and D. A. Stephens (2004). Bayesian time series analysis of periodic behaviour and spectral structure. *International Journal of Forecasting 20*, 713–730.

McCulloch, R. E. and R. S. Tsay (1994). Bayesian inference of trend- and difference-stationarity. *Economic Theory 10*, 596–608.

Meier, A., C. Kirch, M. C. Edwards, and R. Meyer (2018). *beyondWhittle: Bayesian spectral inference for stationary time series.* R package version 1.1.

Metropolis, N., A. W. Rosenbluth, M. N. Rosenbluth, A. H. Teller, and E. Teller (1953). Equations of state calculations by fast computing machines. *Journal of Chemical Physics 21*, 1087–1091.

Migon, H. S., D. Gamerman, H. F. Lopes, and M. A. R. Ferreira (2005). Dynamic models. In D. Dey and C. R. Rao (Eds.), *Bayesian Thinking, Modeling and Computation*, Handbook of Statistics, pp. 553–588. Amsterdam: North Holland.

Mitra, S. K. (1970). A density-free approach to the matrix variate beta distribution. *Sankhyā A 32*, 81–88.

Molenaar, P. C. M., J. G. de Gooijer, and B. Schmitz (1992). Dynamic factor analysis of nonstationary multivariate time series. *Psychometrika 57*, 333–349.

Monahan, J. F. (1983). Fully Bayesian analysis of ARMA time series models. *Journal of Econometrics 21*, 307–331.

Monahan, J. F. (1984). A note on enforcing stationarity in autoregressive moving average models. *Biometrika 71*, 403–404.

Motta, G., C. M. Hafner, and R. von Sachs (2011). Locally stationary factor models: Identification and nonparametric estimation. *Econometric Theory*, 1279–1319.

Nakajima, J., M. Kasuya, and T. Watanabe (2011). Bayesian analysis of time-varying parameter vector autoregressive model for the Japanese economy and monetary policy. *Journal of the Japanese and International Economies 25*, 225–245.

Nakajima, J. and M. West (2013a). Bayesian analysis of latent threshold dynamic models. *Journal of Business and Economic Statistics 31*, 151–164.

Nakajima, J. and M. West (2013b). Bayesian dynamic factor models: Latent threshold approach. *Journal of Financial Econometrics 11*, 116–153.

Nakajima, J. and M. West (2015). Dynamic network signal processing using latent threshold models. *Digital Signal Processing 47*, 6–15.

Nakajima, J. and M. West (2017). Dynamics and sparsity in latent threshold factor models: A study in multivariate EEG signal processing. *Brazilian Journal of Probability and Statistics 31*, 701–731.

Niemi, J. B. and M. West (2010). Adaptive mixture modelling Metropolis methods for Bayesian analysis of non-linear state-space models. *Journal of Computational and Graphical Statistics 19*, 260–280. PMC2887612.

Nieto-Barajas, L. and A. Contreras-Cristán (2014). A Bayesian nonparametric approach for time series clustering. *Bayesian Analysis 9*, 147–170.

Odell, P. and A. Feiveson (1966). A numerical procedure to generate a sample covariance matrix. *Journal of the American Statistical Association 61*, 199–203.

Ombao, H., J. Raz, R. von Sachs, and B. Malow (2001). Automatic statistical analysis of bivariate nonstationary time series. *Journal of the American Statistical Association 96*, 543–560.

Ombao, H., R. von Sachs, and W. Guo (2005). SLEX analysis of multivariate nonstationary time series. *Journal of the American Statistical Association 100*, 519–531.

Pamminger, C. and S. Frühwirth-Schnatter (2010). Bayesian clustering of categorical time series using finite mixtures of Markov chain models. *Bayesian Analysis 5*, 345–368.

Peña, D. and G. E. P. Box (1987). Identifying a simplifying structure in time series. *Journal of the American Statistical Association 82*, 836–843.

Percival, D. B. and A. T. Walden (1993). *Spectral Analysis for Physical Applications: Multitaper and Conventional Univariate Techniques.* Cambridge: Cambridge University Press.

Percival, D. B. and A. T. Walden (2006). *Wavelet Methods for Time Series Analysis.* New York, NY: Cambridge University Press.

Petris, G. (1997). *Bayesian Analysis of Long Memory Time Series.* PhD thesis: Institute of Statistics and Decision Sciences, Duke University, Durham, NC, USA.

Petris, G., S. Petrone, and P. Campagnoli (2009). *Dynamic Linear Models with R.* New York, NY: Springer-Verlag.

Petris, G. and M. West (1996). Bayesian spectral analysis of long memory time series. In *Proceedings of the 1996 Joint Statistical Meetings.* Arlington, VA: American Statistical Association.

Petris, G. and M. West (1998). Bayesian time series modelling and prediction with long-range dependence. Technical report, Institute of Statistics and Decision Science, Duke University.

Philipov, A. and M. E. Glickman (2006a). Factor multivariate stochastic volatility via Wishart processes. *Econometric Reviews 25*, 311–334.

Philipov, A. and M. E. Glickman (2006b). Multivariate stochastic volatility via Wishart processes. *Journal of Business and Economic Statistics 24*, 313–328.

Pinheiro, J. C. and D. M. Bates (1996). Unconstrained parametrizations for variance-covariance matrices. *Statistics and Computing 6*, 289–296.

Pitt, M. K. and N. Shephard (1999a). Filtering via simulation: Auxiliary variable particle filter. *Journal of the American Statistical Association 94*, 590–599.

Pitt, M. K. and N. Shephard (1999b). Time varying covariances: A factor stochastic volatility approach (with discussion). In J. M. Bernardo, J. O. Berger, A. P. Dawid, and A. F. M. Smith (Eds.), *Bayesian Statistics 4*, pp. 547–570. Oxford: Oxford University Press.

Plummer, M., N. Best, K. Cowles, and K. Vines (2006). CODA: Convergence diagnosis and output analysis for MCMC. *R News 6*(1), 7–11.

Pole, A., M. West, and P. J. Harrison (1994). *Applied Bayesian Forecasting and Time Series Analysis.* New York, NY: Chapman & Hall.

Polson, N. G., J. R. Stroud, and P. Müller (2008). Practical filtering with sequential parameter learning. *Journal of the Royal Statistical Society (Series B: Methodological) 39*, 413–428.

Polson, N. G. and B. Tew (2000). Bayesian portfolio selection: An empirical analysis of the S&P500 index 1970–1996. *Journal of Business and Economic Statistics 18*, 164–173.

Prado, R. (1998). *Latent Structure in Non-stationary Time Series.* PhD thesis, Institute of Statistics and Decision Sciences, Duke University, Durham, NC, USA.

Prado, R. (2009). Characterization of latent structure in brain signals. In S. Chow, E. Ferrer, and F. Hsieh (Eds.), *Statistical Methods for Modeling Human Dynamics: An Interdisciplinary Dialogue*, pp. 123–153. New York, NY: Psychology Press, Taylor and Francis Group.

Prado, R. (2010). Multi-state models for mental fatigue. In A. O'Hagan and M. West (Eds.), *The Handbook of Applied Bayesian Analysis*, pp. 845–874. Oxford: Oxford University Press.

Prado, R. and G. Huerta (2002). Time-varying autoregressions with model order uncertainty. *Journal of Time Series Analysis 23*, 599–618.

Prado, R., F. J. Molina, and G. Huerta (2006). Multivariate time series modeling and classification via hierarchical VAR mixtures. *Computational Statistics and Data Analysis 51*, 1445–1462.

Prado, R. and M. West (1997). Exploratory modelling of multiple non-stationary time series: Latent process structure and decompositions. In T. Gregoire (Ed.), *Modelling Longitudinal and Spatially Correlated Data*, pp. 349–362. New York, NY: Springer-Verlag.

Prado, R., M. West, and A. D. Krystal (2001). Multi-channel EEG analyses via dynamic regression models with time-varying lag/lead structure. *Journal of the Royal Statistical Society (Series C: Applied Statistics) 50*, 95–109.

Press, S. J. (1982). *Applied Multivariate Analysis: Using Bayesian and Frequentist Methods of Inference.* California: Krieger.

Priestley, M. B. (1994). *Spectral Analysis and Time Series* (8th ed.). New York, NY: Academic Press.

Priestley, M. B., T. Subba-Rao, and H. Tong (1974). Applications of principal components analysis and factor analysis in the identification of multivariable systems. *IEEE Transactions on Automatic Control 19*, 730–734.

Primiceri, G. E. (2005). Time varying structural vector autoregressions and monetary policy. *Review of Economic Studies 72*, 821–852.

Putnam, B. H. and J. M. Quintana (1994). New Bayesian statistical approaches to estimating and evaluating models of exchange rates determination. In *Proceedings of the ASA Section on Bayesian Statistical Science.* Arlington, VA: American Statistical Association.

Queen, C. M. (1994). Using the multiregression dynamic model to forecast brand sales in a competitive product market. *Journal of the Royal Statistical Society (Series D: The Statistician) 43*, 87–98.

Queen, C. M. and J. Q. Smith (1993). Multiregression dynamic models. *Journal of the Royal Statistical Society (Series B: Methodological) 55*, 849–870.

Queen, C. M., B. J. Wright, and C. J. Albers (2008). Forecast covariances in the linear multiregression dynamic model. *Journal of Forecasting 27*, 175–191.

Quintana, J. and B. H. Putnam (1996). Debating currency markets efficiency using multiple-factor models. In *Proceedings of the ASA Section on Bayesian Statistical Science, Joint Statistical Meetings*, pp. 55–80. Arlington, VA: American Statistical Association.

Quintana, J. M. (1985). A dynamic linear matrix-variate regression model. Technical report, University of Warwick, UK.

Quintana, J. M. (1987). *Multivariate Bayesian Forecasting Models.* PhD thesis: Department of Statistics, University of Warwick, UK.

Quintana, J. M. (1992). Optimal portfolios of forward currency contracts. In J. O. Berger, J. M. Bernardo, A. P. Dawid, and A. F. M. Smith (Eds.), *Bayesian Statistics 4*, pp. 753–762. Oxford: Oxford University Press.

Quintana, J. M., C. M. Carvalho, J. Scott, and T. Costigliola (2010). Futures markets, Bayesian forecasting and risk modeling. In A. O'Hagan and M. West (Eds.), *The Handbook of Applied Bayesian Analysis*, pp. 343–365. Oxford: Oxford University Press.

Quintana, J. M., V. K. Chopra, and B. H. Putnam (1995). Global asset allocation: Stretching returns by shrinking forecasts. In *Proceedings of the ASA Section on Bayesian Statistical Science.* Arlington, VA: American Statistical Association.

Quintana, J. M., V. Lourdes, O. Aguilar, and J. Liu (2003). Global gambling. In J. M. Bernardo, M. J. Bayarri, J. O. Berger, A. P. Dawid, D. Heckerman, A. F. M. Smith, and M. West (Eds.), *Bayesian Statistics 7*, pp. 349–368. Oxford: Oxford University Press.

Quintana, J. M. and M. West (1987). Multivariate time series analysis: New techniques applied to international exchange rate data. *The Statistician 36*, 275–281.

Quintana, J. M. and M. West (1988). Time series analysis of compositional data. In J. M. Bernardo, M. H. DeGroot, D. V. Lindley, and A. F. M. Smith (Eds.), *Bayesian Statistics 3*, pp. 747–756. Oxford: Oxford University Press.

R Core Team (2018). *R: A Language and Environment for Statistical Computing.* Vienna, Austria: R Foundation for Statistical Computing.

Raftery, A. E. and S. M. Lewis (1992). How many iterations in the Gibbs sampler? In J. M. Bernardo, J. O. Berger, A. P. Dawid, and A. F. M. Smith (Eds.), *Bayesian Statistics 4*, pp. 763–774. Oxford: Oxford University Press.

Rao, R. and W. Tirtotjondro (1996). Investigation of changes in characteristics of hydrological time series by Bayesian methods. *Stochastic Hydrology and Hydraulics 10*, 295–317.

Reeson, C., C. M. Carvalho, and M. West (2009). Dynamic graphical models and portfolio allocations for structured mutual funds. Discussion Paper 2009-27, Duke University, Duke University, Durham, North Carolina.

Reinsel, G. C. (1993). *Elements of Multivariate Time Series Analysis* (2nd ed.). New York, NY: Springer-Verlag.

Robert, C. P. and G. Casella (2005). *Monte Carlo Statistical Methods* (2nd ed.). New York, NY: Springer-Verlag.

Rocková, V. and E. I. George (2016). Fast Bayesian factor analysis via automatic rotations to sparsity. *Journal of the American Statistical Association 111*(516), 1608–1622.

Rosen, O. and D. S. Stoffer (2007). Automatic estimation of multivariate spectra via smoothing splines. *Biometrika 94*, 1–11.

Rosen, O., D. S. Stoffer, and S. Wood (2009). Local spectral analysis via a Bayesian mixture of smoothing splines. *Journal of the American Statistical Association 104*, 249–262.

Rosen, O., S. Wood, and D. Stoffer (2017). *BayesSpec: Bayesian Spectral Analysis Techniques.* R package version 0.5.3.

Rosen, O., S. Wood, and D. S. Stoffer (2012). AdaptSPEC: Adaptive spectral estimation of non-stationary time series. *Journal of the American Statistical Association 107*, 1575–1589.

Roverato, A. (2002). Hyper-inverse Wishart distribution for non-decomposable graphs and its application to Bayesian inference for Gaussian graphical models. *Scandinavian Journal of Statistics 29*, 391–411.

Rubin, D. B. (1988). Using the SIR algorithm to simulate posterior distributions. In J. M. Bernardo, M. DeGroot, D. Lindley, and A. F. M. Smith (Eds.), *Bayesian Statistics 3*, pp. 395–402. Oxford: Oxford University Press.

Schmidt, A. M. and A. E. Gelfand (2003). A Bayesian coregionalization approach for multivariate pollutant data. *Journal of Geophysical Research: Atmospheres 108*(D24).

Schwarz, G. (1978). Estimating the dimension of the model. *Annals of Statistics 6*, 461–464.

Shephard, N. (1994). Local scale models: State-space alternative to integrated GARCH models. *Journal of Econometrics 60*, 181–202.

Shephard, N. (Ed.) (2005). *Stochastic Volatility: Selected Readings.* Oxford: Oxford University Press.

Shumway, R. H. and D. S. Stoffer (2017). *Time Series Analysis and Its Applications: With R Examples* (4th ed.). New York, NY: Springer-Verlag.

Smith, A. F. M. and M. West (1983). Monitoring renal transplants: An application of the multi-process Kalman filter. *Biometrics 39*, 867–878.

Smith, B. J. (2007). boa: an R package for MCMC output convergence assessment and posterior inference. *Journal of Statistical Software 21*, 1–37.

Smith, M. and R. Kohn (2002). Parsimonious covariance matrix estimation for longitudinal data. *Journal of the American Statistical Association 97*, 1141–1153.

Spearman, C. (1904). "General Intelligence," objectively determined and measured. *The American Journal of Psychology 15*, 201–292.

Stoffer, D. S. (1999). Detecting common signals in multiple time series using the spectral envelope. *Journal of the American Statistical Association 94*, 1341–1356.

Storvik, G. (2002). Particle filters for state-space models with the presence of unknown static parameters. *IEEE Transactions on Signal Processing 50*, 281–289.

Tan, W. Y. (1969). Some results on multivariate regression analysis. *Nanta Mathematica 3*, 54–71.

Tebaldi, C. and M. West (1998). Bayesian inference on network traffic using link count data. *Journal of the American Statistical Association 93*, 557–573.

Tebaldi, C., M. West, and A. F. Karr (2002). Statistical analyses of freeway traffic flows. *Journal of Forecasting 21*, 39–68.

Terui, N. and M. Ban (2014). Multivariate time series model with hierarchical structure for over-dispersed discrete outcomes. *Journal of Forecasting 33*, 379–390.

Tiao, G. C. (2001a). Univariate autoregressive moving average models. In D. Peña, G. C. Tiao, and R. S. Tsay (Eds.), *A Course in Time Series Analysis*. New York, NY: John Wiley & Sons.

Tiao, G. C. (2001b). Vector ARMA models. In D. Peña, G. C. Tiao, and R. S. Tsay (Eds.), *A Course in Time Series Analysis*. New York, NY: John Wiley & Sons.

Tiao, G. C. and G. E. P. Box (1981). Modeling multiple time series with applications. *Journal of the American Statistical Association 76*, 802–816.

Tiao, G. C. and R. S. Tsay (1989). Model specification in multivariate time series with applications (with discussion). *Journal of the Royal Statistical Society (Series B: Methodological) 51*, 157–213.

Tong, H. (1983). *Threshold Models in Non-linear Time Series Analysis*. New York, NY: Springer-Verlag.

Tong, H. (1990). *Non-linear Time Series: A Dynamical Systems Approach*. Oxford Statistical Science Series. Oxford: Oxford University Press.

Trejo, L. J., K. Knuth, R. Prado, R. Rosipal, K. Kubitz, R. Kochavi, B. Matthews, and Y. Zhang (2007). EEG-based estimation of mental fatigue: Convergent evidence for a three-state model. In D. Schmorrow and L. Reeves (Eds.), *Augmented Cognition, HCII 2007, LNAI 4565*, pp. 201–211. New York: Springer LNCS.

Trejo, L. J., R. Kochavi, K. Kubitz, L. D. Montgomery, R. Rosipal, and B. Matthews (2006). EEG-based estimation of mental fatigue. Technical report, neurodia.com.

Triantafyllopoulos, K. (2007). Covariance estimation for multivariate conditionally Gaussian dynamic linear models. *Journal of Forecasting 26*, 551–569.

Triantafyllopoulos, K. (2008). Multivariate stochastic volatility with Bayesian dynamic linear models. *Journal of Statistical Planning and Inference 138*, 1021–1037.

Troutman, B. M. (1979). Some results in periodic autoregression. *Biometrika 66*, 219–228.

Uhlig, H. (1994). On singular Wishart and singular multivariate beta distributions. *Annals of Statistics 22*, 395–405.

Uhlig, H. (1997). Bayesian vector autoregressions with stochastic volatility. *Econometrica 1*, 59–73.

van der Merwe, R., A. Doucet, N. de Freitas, and E. Wan (2000). The unscented particle filter. In T. K. Leen, T. G. Dietterich, and V. Tresp (Eds.), *Advances in Neural Information Processing Systems.* Boston, MA: MIT Press.

Vidakovic, B. (1999). *Statistical Modeling by Wavelets.* New York, NY: John Wiley & Sons, Texts in Statistics.

Villagrán, A. and G. Huerta (2006). Bayesian inference on mixture-of-experts for estimation of stochastic volatility. In T. Fomby, D. Terrell, and R. Carter Hill (Eds.), *Econometric Analysis of Financial and Economic Time Series Part B. Advances in Econometrics*, Volume 20, pp. 277–296. Elsevier/JAI Press, Amsterdam.

Vivar, J. C. and M. A. R. Ferreira (2009). Spatio-temporal models for Gaussian areal data. *Journal of Computational and Graphical Statistics 18*, 658–674.

Wang, H. (2010). Sparse seemingly unrelated regression modelling: Applications in finance and econometrics. *Computational Statistics and Data Analysis 54*, 2866–2877.

Wang, H. and M. West (2009). Bayesian analysis of matrix normal graphical models. *Biometrika 96*, 821–834.

Wei, G. C. G. and M. A. Tanner (1990). A Monte Carlo implementation of the EM algorithm and the poor man's data augmentation algorithm. *Journal of the American Statistical Association 85*, 699–704.

Wei, W. S. (2019). *Multivariate Time Series Analysis and Applications.* Hoboken, NJ: Wiley Series in Probability and Statistics.

Weiner, R. D. and A. D. Krystal (1993). EEG monitoring of ECT seizures. In C. Coffey (Ed.), *The Clinical Science of Electroconvulsive Therapy*, pp. 93–109. Washington, DC: American Psychiatric Press.

West, M. (1993a). Approximating posterior distributions by mixtures. *Journal of the Royal Statistical Society (Series B: Methodological) 55*, 409–422.

West, M. (1993b). Mixture models, Monte Carlo, Bayesian updating and dynamic models. In J. Newton (Ed.), *Computing Science and Statistics: Proceedings of the 24th Symposium of the Interface*, pp. 325–333. Fairfax Station, VA: Interface Foundation of North America.

West, M. (1997a). Bayesian time series: Models and computations for the analysis of time series in the physical sciences. In K. Hanson and R. Silver (Eds.), *Maximum Entropy and Bayesian Methods 15*, pp. 23–34. Kluwer, New York.

West, M. (1997b). Modelling and robustness issues in Bayesian time series analysis (with discussion). In J. O. Berger, B. Betrò, E. Moreno, L. R. Pericchi, F. Ruggeri, G. Salinetti, and L. Wasserman (Eds.), *Bayesian Robustness*, IMS Monographs, Hayward, CA, pp. 231–252. Institute of Mathematical Statistics.

West, M. (1997c). Time series decomposition. *Biometrika 84*, 489–94.

West, M. (2003). Bayesian factor regression models in the "large p, small n" paradigm. In J. M. Bernardo, M. J. Bayarri, J. O. Berger, A. P. Dawid, D. Heckerman, A. F. M. Smith, and M. West (Eds.), *Bayesian Statistics 7*, pp. 723–732. Oxford: Oxford University Press.

West, M. (2020). Bayesian forecasting of multivariate time series: Scalability, structure uncertainty and decisions (with discussion). *Annals of the Institute of Statistical Mathematics 72*, 1–44.

West, M. and P. J. Harrison (1986). Monitoring and adaptation in Bayesian forecasting models. *Journal of the American Statistical Association 81*, 741–750.

West, M. and P. J. Harrison (1989). Subjective intervention in formal models. *Journal of Forecasting 8*, 33–53.

West, M. and P. J. Harrison (1997). *Bayesian Forecasting and Dynamic Models* (2nd ed.). New York, NY: Springer-Verlag.

West, M., P. J. Harrison, and H. S. Migon (1985). Dynamic generalised linear models and Bayesian forecasting (with discussion). *Journal of the American Statistical Association 80*, 73–97.

West, M., R. Prado, and A. D. Krystal (1999). Evaluation and comparison of EEG traces: Latent structure in nonstationary time series. *Journal of the American Statistical Association 94*, 1083–1095.

Whittaker, J. (1990). *Graphical Models in Applied Multivariate Statistics.* Chichester: John Wiley & Sons.

Wong, C. S. and W. K. Li (2000). On a mixture autoregressive model. *Journal of the Royal Statistical Society (Series B: Methodological) 62*, 95–115.

Wong, C. S. and W. K. Li (2001). On a mixture autoregressive conditional heteroscedastic model. *Journal of the American Statistical Association 96*, 982–995.

Wong, F., C. K. Carter, and R. Kohn (2003). Efficient estimation of covariance selection models. *Biometrika 90*, 809–830.

Yang, W., S. Holan, and C. Wikle (2016). Bayesian lattice filters for time-varying autoregression and time-frequency analysis. *Bayesian Analysis 11(4)*, 977–1003.

Yelland, P. M. (2009). Bayesian forecasting for low-count time series using state-space models: An empirical evaluation for inventory management. *International Journal of Production Economics 118*, 95–103.

Yoshida, R. and M. West (2010). Bayesian learning in sparse graphical factor models via annealed entropy. *Journal of Machine Learning Research 11*, 1771–1798. PMC2947451.

Zellner, A. (1996). *An Introduction to Bayesian Inference in Econometrics.* New York, NY: John Wiley & Sons.

Zhang, S. (2016). Adaptive spectral estimation for nonstationary multivariate time series. *Computational Statistics and Data Analysis 103*, 330–349.

Zhao, W. and R. Prado (2020). Efficient Bayesian PARCOR approaches for dynamic modeling of multivariate time series. *Journal of Time Series Analysis 41*, 759–784.

Zhao, Z. Y., M. Xie, and M. West (2016). Dynamic dependence networks: Financial time series forecasting and portfolio decisions (with discussion). *Applied Stochastic Models in Business and Industry 32*, 311–339.

Zhou, X., J. Nakajima, and M. West (2014). Bayesian forecasting and portfolio decisions using dynamic dependent sparse factor models. *International Journal of Forecasting 30*, 963–980.

Author Index

Subject Index